Biological Psychiatry

PRINCIPLES OF MEDICAL BIOLOGY

Editors: ***E. EDWARD BITTAR,*** *Department of Physiology*
University of Wisconsin, Madison
NEVILLE BITTAR, *Gemini Scientific*
Madison, Wisconsin

Biological Psychiatry

Edited by ***E. EDWARD BITTAR***
Department of Physiology
University of Wisconsin
Madison, Wisconsin

NEVILLE BITTAR
Gemini Scientific
Madison, Wisconsin

JAI PRESS INC.
Stamford, Connecticut

Library of Congress Cataloging-in-Publication Data

Biological Psychiatry / edited by E. Edward Bittar, Neville Bittar
p.; cm. — (Principles of medical biology ; v. 14)
Includes bibliographical references and index.
ISBN 1-55938-819-6
1. Biological psychiatry. 2. Brain—Pathophysiology. 3. Psychiatry. I. Bittar, E. Edward. II. Bittar, Neville. III. Series.
[DNLM: 1. Mental Disorders—physiopathology. 2. Brain—physiology. 3. Psychotropic Drugs—pharmacology. WM 140 P9751 2000]
RC341.P8926 2000
616.89—dc21 99-088148
CIP

100 Prospect Street
Stamford, Connecticut 06901

ISBN: 1-55938-819-6

Library of Congress Catalog Number: 99-088148

Manufactured in the United States of America

Transferred to digital printing 2005

Printed and bound by Antony Rowe Ltd, Eastbourne

CONTENTS

LIST OF CONTRIBUTORS

Paul F. Brain	School of Biological Sciences University of Wales Swansea Swansea, Wales, England
William Byne	Department of Psychiatry V. A. Medical Center Bronx, New York
Robert Cancro	Department of Psychiatry New York University Medical Center Tisch Hospital New York, New York
Sharon C. Cheetham	Knoll Pharmaceuticals Research and Development Nottingham, England
James C.-Y. Chou	Department of Psychiatry Bellevue Hospital New York, New York
Brenda K. Colasanti	Department of Pharmacology and Toxicology West Virginia University Health Sciences Morgantown, West Virginia
David L. Dunner	Department of Psychiatry University of Washington Seattle, Washington
Kathryn Evans	Academic Unit of Psychiatry St. Charles Hospital London, England
Christopher Gillberg	Section of Child and Adolescent Psychiatry University of Goteborg Goteborg, Sweden

Danya Glaser
Department of Psychological Medicine
Great Ormond St. Hospital for Children
London, England

David S. Goldbloom
Clark Institute of Psychiatry
University of Toronto
Toronto, Ontario, Canada

David J. Heal
Knoll Pharmaceuticals Research and Development
Nottingham, England

Alexey B. Kampov-Polevoy
Bowles Center for Alcoholism Studies
University of North Carolina
Chapel Hill, North Carolina

Carol E. Kaplan
Fleming Nuffield Unit
Newcastle City Health Trust
Newcastle upon Tyne, England

Eileen Kemmether
Department of Psychiatry
V. A. Medical Center
Bronx, New York

B. E. Leonard
Pharmacology Department
University College
Galway, Ireland

Michael J. Lyons
Department of Psychiatry
Massachusetts Mental Health Center
Harvard Medical School
Boston, Massachusetts

Kevin M. Malone
Department of Adult Psychiatry
Mater Hospital
Dublin, Ireland

Andrew H. Miller
Department of Psychiatry and Behavioral Sciences
Emory University
Atlanta, Georgia

Roger T. Mulder — Department of Psychological Medicine, Christchurch School of Medicine, Christchurch, New Zealand

John J. Neumaier — Department of Psychiatry, Harborview Medical Center, Seattle, Washington

David H. Overstreet — Bowles Center for Alcoholism Studies, University of North Carolina, Chapel Hill, North Carolina

Jaak Panksepp — Department of Psychology, Bowling Green State University, Bowling Green, Ohio

Carmine M. Pariante — Department of Psychiatry and Behavioral Sciences, Emory University School of Medicine, Atlanta, Georgia

Terence W. Picton — Rotman Research Institute, Baycrest Centre, North York, Ontario, Canada

Jan Scott — Division of Psychiatry, Gartnavel Royal Hospital, Glasgow, Scotland

Jan Stern — 15 Mulgrave Road, Sutton, Surrey 5M2 6LJ, England

Donald T. Stuss — Departments of Medicine and Psychology, University of Toronto, Toronto, Ontario, Canada

Anne Thompson — Fleming Nuffield Unit, Newcastle City Health Trust, Newcastle upon Tyne, England

Rosemary Toomey — Department of Psychiatry, Massachusetts Mental Health Center, Harvard Medical School, Boston, Massachusetts

Ming T. Tsuang
Department of Psychiatry
Massachusetts Mental Health Center
Harvard Medical School
Boston, Massachusetts

Peter Tyrer
Academic Unit of Psychiatry
St. Charles Hospital
London, England

Ehud Ur
Division of Endocrinology
Queen Elizabeth II Health Sciences Centre
Dalhousie University
Halifax, Nova Scotia, Canada

Philip Winn
School of Psychology
University of St. Andrews
St. Andrews, Scotland

Malcolm Wiseman
Department of Child Psychiatry
Guy's Hospital
London, England

Athanasios P. Zis
Department of Psychiatry
University of British Columbia
Vancouver, British Columbia, Canada

PREFACE

It is now widely recognized that biological psychiatry is rapidly coming into its own. For over the last three decades dramatic advances in this young discipline have been made, all of which attest to the staying power of the experimental method. Those who made this revolution in knowledge happen are a breed of investigators availing themselves of the tools of molecular biology, pharmacology, genetics, and perhaps, above all, the technology of neuroimaging. The introduction of the interdisciplinary method of approach to the study of psychopathology had made it very clear that neuroimaging, as a set of techniques, is unique in that it is gradually providing us with evidence supporting Kraepelin's original view that mental illness is closely associated with abnormal changes in the brain.

Broadly speaking, there are presently two structural techniques in neuroimaging—computed tomography and magnetic resonance imaging (MRI)—and three functional techniques—single photon emission tomography (SPECT), positron emission tomography (PET), and functional magnetic resonance imaging (fMRI). Through PET technology, for example, we have learned that, in early brain development, the primitive areas, mostly the brain stem and thalamus, are the first to show high activity in an infant. This is followed by the development of cortical areas by year one. Between the ages of four to 10, the cortex is almost twice as active in the child as in the adult. This information alerts us to what might happen in the way of trauma in abused children, especially those under the age of three. Child abuse increases the risk of physical changes, not only in the stress systems, but also in brain development (Glaser and Weissman). In addition to the difficult problem of post-traumatic stress disorder (PTSD), we have to take into account the possibility of other types of mental illness as the consequences of child abuse. These include depression, eating disorders, and drug and alcohol problems.

The combination of PET and fMRI represents a more remarkable example of the power of neuroimaging since the two have made it feasible to map accurately *in vitro* identifiable cortical fields, or networks. In a landmark NIH investigation of human cortical reorganization (plasticity), persuasive evidence was brought forward showing that the process of learning as a motor task involves a specific network of neurons. These neurons occur in the cortical field that is responsible for that particular task. Such findings are important partly because they provide evidence supporting the current notion that labor in the cortex is divided among *ensembles* of specialized neurons that cooperate in the performance of complex tasks. Cooperation, then, in this sense, implies crosstalk among *ensembles* and that signals are both processed and retransmitted to neighboring *ensembles.* To understand the workings of these *ensembles,* much better spatial and temporal resolution in functional brain mapping is required. This can be achieved with an NMR instrument whose magnet is 4.1 Tesla or more.

At this writing, it is, perhaps, safe to say that the cause of schizophrenia is still largely a mystery. However, there is considerable agreement that it involves complex genetics as well as environmental factors that interact with, and independently of, the genes (Winn). Strictly, neither the dopaminergic endophenotype, nor the sensory gating endophenotype, nor reduced neuropil models, as they presently stand, provide an adequate explanation. That schizophrenia represents a neurodevelopmental disorder is an old view, but it has been revived by Weinberger in the light of evidence that is compatible with reduced expression of plasticity in CA 3–4 neurons of the hippocampal formation. Very briefly, he makes the case that the neuronal phenotype findings are mostly localized to the CA 3–4 region. They are not an epiphenomenon but, rather, more likely the result of neonatal abnormalities related to defective genes. The impetus for checking whether the lesion is of retroviral origin may have come from such thinking, but there is all the more reason to attach special weight to the recent finding of high reverse transcriptase activity in the spinal fluid of schizophrenics (Yolken). It is a matter of more than passing interest that human endogenous retroviruses (HERV-W) are a set of viruses commonly found in the human genome, a textbook example being neurological diseases such as multiple sclerosis. One crucial question now is whether or not the lesion develops *in utero* during the first four to six weeks and whether its pathogenesis involves multiple defective proteins, rather than a single protein presumably falling in the class of molecules that guide the migration of growth cones.

It is important at this stage to decide whether there is any validity to the conclusion drawn by McCarley, Slentor, and colleagues that there are structural NMR changes in schizophrenics. By distinguishing between specific and nonspecific structural changes in the brain, they were able to obtain ample evidence of distinct alterations in brain structure in schizophrenics. These changes are particularly prominent in the cortical and medial temporal areas. McCarley and colleagues also emphasize the point that these structural changes are not the same as those seen in patients with bipolar psychosis. Conceivably, with some improvement of the two functional techniques of PET and fMRI to map the CA3-4 region of the hippocampal formation more carefully, it will become possible to verify the presence of abnormal activity and follow its time-course. Interest in this particular approach is heightened by the discovery that manganese can be used as an MRI tracer over time to track both neuronal activity and connectivity (Koretsky).

As a stimulus to fresh thinking, we must bear in mind that the dopamine theory of schizophrenia has been weakened by two pieces of evidence. One is the finding of excessive sensitivity of the postsynaptic membrane to dopamine that is accompanied by reduced dopamine release by the presynaptic membrane (retrograde cycle). The other piece of evidence is that the new, atypical antipsychotic drugs, such as risperidone, are also powerful inhibitors of the 5-HT_2 family of serotonin receptors. To this, we have to add another complicating observation, namely, that the knock-out mouse without its dopamine transporter (DAT) is associated with dopaminergia. In this instance, however, the behavior of the mouse bears a more striking resemblance to that of the patient with attention deficit hyperactivity disorder (ADHD) than to the schizophrenic (Caron).

Studies of the knock-out mouse lacking the gene encoding the 5-HT1B subtype of receptor show the development in this mouse of increased aggression toward intruders (wild-type mouse). The mere fact that a serenic drug, such as Eltoprazine, which acts as an antagonist of the 5-HT1B receptor, reduces the level of aggression is taken as evidence that the effectiveness of this new class of antipsychotic drugs is largely due to its action on the central serotonergic system rather than on the dopaminergic system. So far, the atypical antipsychotic drugs, have been found effective not only in the treatment of schizophrenia (Colasanti) but also in the treatment of several conditions, including rage in autistic children, pathological violence in adults and, those with borderline personality disorder who have brief psychotic episodes, and in certain patients with mental retardation and so forth.

Hitherto, individual neuromodulators of rage have not been identified, although a wide variety of endogenous substances, notably, ACh, vasopressin, substance P, and glutamate are known to evoke both anger and aggression. Based upon mapping through electrical stimulation in human and animal brain, the circuits of RAGE have been pinpointed. They run from medial areas of the amygdala to regions of the hypothalamus and then down into the periaqueductal grey of the midbrain (Panksepp). These well-defined and discrete areas have higher functions that depend on the lower ones and that interact with higher cognitive functions. This is borne out by the seminal work of LeDoux who has identified more fully the amygdala as being the main site of fear processing and fear learning. In this view, the organization of this region is determined at the level of the subnuclei instead of the nuclei. Although primate data apply to the human brain, very little is known of the way in which the transition from emotional reaction to emotional action occurs. There is no doubt that this kind of knowledge would shed much light on how the fear system works in anxiety disorders.

Fear and anxiety, for example, as emotions, are processed within the amygdala circuits, thereby giving rise to outputs from the central nucleus. Thus, it is not surprising that damage to the amygdala dramatically changes the way in which both humans and animals act in threatening situations. Rats lose their fear of cats and are prevented from learning about stimuli warning them of danger. Now, it is also realized that the two cortical areas, the hippocampus and the medial prefrontal cortex, play a key role in the response to fear and fear conditioning. This is underscored by the fact that damage to the hippocampus, in turn, leads to dysregulation of the stress response system which is marked by the release of the key peptide, corticotropin releasing factor (CRF). Thus, according to the modern outlook, stress not only impairs the anatomy, physiology, and behavioral functions of the hippocampus (McEwen and Sapolsky) but also affects the functions of the prefrontal cortex.

The principal neuromodulator of the nucleus of the amygdala is CRF, which is essentially anxiogenic in its effect (Davis), whereas its antagonists are anxiolytic. Here there is a distinction to be made between stimulus-specific fear and the less specific stimulus, which we call anxiety. This can be measured with the aid of the acoustic startle reflex. A prime example that illustrates this in humans is the patient diagnosed as having post-traumatic stress disorder (PTSD) (Ur). Although normal fear reactions are seen in PTSD, the levels of anxiety are abnormally high. By analogy, and to elaborate on the distinction just made, rats showing specific signs of fear also show activation of the central nucleus of the amygdala and, in turn, activation of hypothalamic and brain stem regions. Rats displaying less specific signs of fear (i.e., anxiety) in response, for example, to a threatening environment, show activation of the bed nucleus of the stria terminalis. This, in turn, leads to activation of hypothalamic and brain stem regions. The importance of this pathway is brought out by observations that the intraventricular administration of CRF potentiates the startle reflex and mimics anxiety seen in patients with generalized anxiety disorder (GAD). It is thus easy to see why it would be vital, in terms of therapy, to have nonpeptide CRF antagonists made available that are also free of side-effects.

Apart from the fact that stress results in raised levels of both glucocorticoids and cortisol (Ur), evidence is available that serotonin also contributes to hippocampal atrophy by indirectly leading to reduced synthesis of brain-derived neurotrophin factor (BDNF) in the hippocampus. This has clinical implications because atrophy, as measured by MRI, is often found in Vietnam veterans with combat-related PTSD (Bremner). This is a situation closely associated with deficits in shortterm memory. As is expected, an extraordinary parallel exists between Bremner's findings of hippocampal atrophy and memory deficits and those found in patients with major depression. More than half those with clinical depression also show cortisolemia, but whether those with atrophy are the same patients that have raised cortisol levels is not yet known. Bearing in mind that the prefrontal cortex receives input from the amygdala, the recent work of Rajkowska and colleagues has shed new light on the question of whether changes in the prefrontal cortex occur in patients with major depression. These workers have been able to demonstrate the presence of neuronal and glial changes in the prefrontal cortex of postmortem brains of these patients—changes characterized by reductions in the density of both neuronal populations. Moreover, imaging studies are confirmatory in that they show reduced metabolic activity and volume in several subregions of the prefrontal cortex.

During the 1960s, two schools of thought concerning the etiology of depression emerged. One was the low catecholamine school led by Kety, and the other the low serotomin school led by Coppen. This was followed by a school that upheld the concept that depression is closely connected to the ratio of serotomin-to-catechols in certain regions of the brain. This ratio concept, however, was soon overshadowed by the locus ceruleus theory stating that this structure is unusually well positioned to receive information and influence fear-related neural structures. This theory soon gave way to the idea of crosstalk among the various transmitter systems, including that of GABA and, more surprisingly, two presumably less important systems: cholecystokinin (CCK), a gut hormone that acts as a panicogen, and neuropeptide Y(NPY), an anti-anxiety principle. This had wide appeal partly because much had been learned already about transduction signaling. It thus made it more convenient, but not necessarily more realistic, to view anxiety disorders as resulting from a breakdown in crosstalk. Nonetheless, mounting evidence pointed to the fact that presynaptic 5-HT reuptake inhibitors (SSRIs) were, and still are, the firstline drugs in the successful treatment of depression. In addition, they proved to be the drugs of choice in the treatment of different types of anxiety disorders, sleep–wake disorders, panic disorders, obsessive–compulsive disorder, social phobia, and post-traumatic stress disorder. Thus the only reasonable conclusion that could be drawn is that the extraordinary benefits obtained with SSRIs is in keeping with the fact that the serotonin system is not only phylogenetically the oldest, but also the most elaborate among the major transmitter systems. Such a conclusion agrees with Grove's model, which assumes that serotonin input to the forebrain comes wholly from the amygdala where the highest density of serotonergic cell bodies occurs. It is, therefore, of special significance that LeDoux has recently shown that lesions of the medial prefrontal cortex result in extinction of conditioned fear.

From these considerations it becomes more apparent that the serotonin system serves as a behavior inhibition system in which the circuits of fear, anxiety, panic, and so forth, are modulated by the central levels of serotonin. Expressed in a more generalized way, it may be stated that decrements in brain serotonin are associated with impulsive behavior, various forms of aggression, or even suicide, whereas increments have the opposite effect, mostly that of social interaction and assertiveness. From this standpoint, it is hardly surprising that the ambition of the pharmaceutical industry is to develop a second line of drugs that, by design, are not only more specific and selective in their actions but also free of side-effects.

It should be emphasized in passing that, besides neuropeptide Y(NPY), oxytocin and prolactin behave as anti-anxiety systems. Oxytocin, for example, is known to exert anti-stress effects in lactating women, each suckling episode being followed by a fall in blood pressure and cortisol levels. Though oxytocin promotes bonding between mother–infant or female–male, we really know very little about its role in higher-order functions.

In sharp contrast, considerable information about mental retardation has accrued over the last few years, mainly because of the enormous progress made in molecular genetics—for example, in positional cloning—in addition to the recognition of behavioral phenotypes associated with mental retardation (Stern). Although runs of trinucleotide repeats vary in the normal population and may be harmless, there is a clear correlation between the number of repeats and the age of onset of the illness. The numbers of repeats can increase in the next generation to a point that is identified with disease in successive generations, appears earlier and is more severe. A dramatic example is the patient with the Huntington's disease gene in whom CAG repeats fall in the 37-to-121 range, while the normal range is nine to 37 copies. A parallel situation is that of the patient with spinalbulbar muscular atrophy in whom CAG repeat of the androgen receptor gene falls in the 30-to-62 range, which is twice the normal range. In the fragile X syndrome, the CGG repeats of the FMR-1 gene fall in the 50-to-1500 range. After Down syndrome, the fragile X syndrome is the most common cause of mental retardation. It is significant that half the genes ascribed to the X-chromosome are disease related. In the case of the fragile X syndrome, the chromosome locus is Xq27.3, and in the spinalbulbar muscular atrophy, the locus is Xq21.3.

Moreover, mental retardation is recognized as being closely associated with an enormous number of amino acidurias of which phenylketonuria is a textbook example. It essentially involves a metabolic block in the hydroxylation of phenylalanine to tyrosine. Hence, the risk to the fetus in the untreated mother with phenylketonuria is great; these include microcephaly, mental retardation, and multiple malformations. Furthermore, among the 17 human disorders linked to peroxisomal dysfunction, 15 are found to involve neuropsychiatric deficits. The majority of these disorders are related to single peroxisomal enzyme defects, meaning that they are attributable to a single gene defect such as that found in Refsum's disease, a phytanic acid storage disease. Those with the hereditary type of lactic acidosis of mitochondrial origin have a poor prognosis since the intracellular acidosis is intractable to treatment. Stern's useful schema outlining the examination and management of patients with mental retardation indicates that every detail is important.

Mental retardation is found in 65% to 90% of autistic children. This is seen three times more often in boys than in girls. One-third have epilepsy. In sharp contrast, mental retardation is hardly seen in children with Asperger's syndrome. The Rett syndrome is only seen in girls in whom there is mental retardation, absence of speech, crying spells, night laughing, and stereotypic hand movement. This is representative of the behavioral phenotype. In the absence of a satisfactory explanation of the Rett syndrome, Asperger's syndrome, and early infantile autism, there is general agreement that all three represent a neural developmental lesion of complex genetic origin that has its onset in the first trimester of intrauterine life.

The problem of alcoholism in the field of biological psychiatry centers by and large on four key transmitter systems in the brain; these are the serotonin, dopamine, GABA, and NMDA systems (Overstreet and Karpov-Polevoy). The current working hypothesis is that alcohol stimulates the release of 5-HT, thus leading to its interaction with the 5-HT_3 receptor. The interaction with 5-HT_3, in turn, is followed by the release of dopamine. Although antagonists of 5-HT_3 are accompanied by reduced alcohol intake in animals, they are relatively ineffective in humans. However, this is not the case with the dopamine antagonist, tiapride, which so far has given more promising results. There is another therapeutic approach to the vexed problem of alcohol abuse and dependence. This involves the use of the opiate antagonist Naltrexone or Nalfamene, which is based on the rationale that alcohol increases the release of brain opioids. Both antagonists have so far been shown to have benefial effects in humans. However, it could be more profitably argued that the central problem in alcoholics is intimately related to anxiety and depression which arise because of a deficit in brain 5-HT and dopamine. With more than 7% of the nation being diagnosed as suffering from alcohol abuse and/or dependence, one fifth of whom have anxiety disorders and half of whom have depression disorders, including mania, it seems very likely that SSRIs would end up as the drug line of choice. The rationale behind this is quite straighforward, since SSRIs act by reversing deficits in serotonin and dopamine in the brain (Cheetham and Heal). With respect to alcohol's action on the GABAergic system, γ-hydroxybutyrate, a metabolite of GABA, is found to reduce alcohol intake, both in humans and experimental animals. As might be expected, benzodiazepines (BZP) mitigate alcohol's GABAergic action, since they enhance the function of GABA by binding to a modulatory site on the GABA-A receptor to increase chloride ion flux. Since the antidepressant buproprion is a weak inhibitor of adrenalin, serotonin, and dopamine uptake, it might turn out to be the first choice in the treatment of alcoholism, as it has already been found effective in the treatment of nicotine addiction in smokers.

Although half a century has elapsed since Cade of Australia introduced lithium for the treatment of bipolar psychosis, its underlying mechanism of action remains somewhat obscure. Lithium remains the treatment of choice for acute mania and prophylaxis in bipolar disorder. It also stabilizes recurrent depression when depression is associated with unipolar disorder, and it proves effective in the treatment of major depression if the disorder is refractory to anti-depressant therapy. Manji and co-workers have more recently demonstrated that lithium reduces protein kinace C activity in rat frontal cortex and hippocampus. They have suggested that this critical enzyme, when inhibited, exerts an anti-manic effect, particularly since the enzyme regulates neuronal excitability and transmitter release. Though structurally dissimilar to lithium, valproate is able to mimic the effect of lithium on protein kinase C. Additionally, Manji and co-workers have produced evidence suggesting that chronic administration of lithium or valproate produces several changes in gene expression. These lithium-genes, as they are now called, have not yet been cloned.

In editing this book, we have not tried to cover everything. What we have attempted is to outline to a large extent the biological basis of psychiatry and those aspects of classical psychiatry that cannot yet be discussed in molecular terms. A typical example is child psychiatry. Although it is clear that there is currently a rapidly growing interest in biological psychiatry, we are conscious of how much more knowledge is still needed to achieve an understanding of mental illnesses. We cannot deny the fact that our perception of biological psychiatry is that it is essentially an experimental discipline. It is also our perception that the methods of brain imaging, in combination with the new technology of gene chips, will revolutionize psychiatry as we know it today.

As editors, we are most obliged to those who have contributed chapters to make this book possible. Indeed, we owe them a special debt for their enthusiasm, patience, and scholarly commitment.

E. EDWARD BITTAR,
NEVILLE BITTAR

Chapter 1

Consciousness

TERENCE W. PICTON and DONALD T. STUSS

Biological Psychiatry, pages 1-25.

ISBN: 1-55938-819-6

INTRODUCTION

General Characteristics of Human Consciousness

Long considered beyond the scope of science, human consciousness is now being viewed from a biological perspective (Flanagan, 1992; Revonsuo and Kamppinen, 1994; Lyons, 1995). This presumes that consciousness is part of the real world and therefore accessible to scientific investigation. Those aspects of consciousness, such as its personal and private quality, that make it so interesting to philosophy and so difficult to comprehend, should not be omitted from a scientific view of consciousness. Biology may help us to understand how consciousness occurs, but such understanding must be deeper than propositions that consciousness is "nothing but" a pattern of active neurons or a mixture of synaptic transmitters. Consciousness plays an important role in human biology. It should not be considered as inexplicable nor dismissed as epiphenomenal. It is the defining factor of our biological existence.

The word "consciousness" has multiple meanings that vary from the collective knowledge of a group, through different types of awareness of the world and the self, to the state of normal wakefulness (Natsoulas, 1978). Consciousness is closely related to many other concepts—awareness, sentience, self, soul, thought, knowledge, will, conscience—each with its own multiple meanings. Even the word "awareness" has many levels since an individual can be aware of sensory data, of remembered or imagined information, of the self, and of the self's intentions. Different views of consciousness consider it as either a state of the brain or some particular process of the brain. In this chapter, we shall concentrate on consciousness as a process or function. William James (1890; 1912) described consciousness as the function of knowing and proposed that this function was performed by thought. He considered the "stream of thought" as

1. *Personal.* This subjective quality of consciousness makes it inaccessible to public scrutiny unless the individual describes his or her experience to others.
2. *Transient.* As the momentary present between past and future, consciousness is always changing.
3. *Continuous.* Within consciousness, there are no gaps either in space or time.
4. *Representational.* Consciousness deals with objects in a way that makes them separate from the conscious self.
5. *Selective.* Our consciousness attends to some things and ignores others.

Of these the crucial characteristic is perhaps the fourth. A conscious individual considers the world—the past, present, and future world as well as the internal and external world—as something separate.

Some Principles Underlying Consciousness

The first principle that we shall use in presenting a biological view of consciousness is that the brain constructs a model of the world in order to understand it, to remember how it was, and to predict what might happen. The generation of the model is the process of consciousness. Information coming from the senses enters into a feedback system wherein the brain builds a neuronal model of the world. Consciousness also participates in our response to the world by modeling goals to be attained and generating behavior to fit these goals.

A second principle is that the model-making of the human brain occurs at a hierarchy of levels and is related to the different levels of experienced consciousness. On the sensory side, the hierarchy goes from simple sensations like "red" and "pain" to the experience of the self; on the motor side, from simple reflex responses to planned behavior.

A third principle is that the brain also works in a parallel manner. Even within one sensory modality, various aspects of the sensory information are analyzed in separate pathways. Nevertheless, the information from different pathways does come together at higher levels in the hierarchy. The brain is balanced between connection and independence.

Cerebral Systems Underlying Consciousness

We therefore propose that consciousness is that particular pattern of cortical neuronal activity that uses feedback circuits to construct, maintain, and operate a model of the world (Picton and Stuss, 1994). A major consequence of this view of the brain is that sensation is not considered as intrinsically different from motor activity. It is an active process rather than a passive viewing of the world. Feedback is widespread in many biological systems (Riggs, 1970) and has long been used to explain the processes of human cognition (Miller et al., 1960). The concept of modeling as the way in which conscious thought operates has been widely discussed in philosophy (Craik, 1943), artificial intelligence (Johnson-Laird, 1983), and cognitive psychology (Yates, 1985). In neurophysiology, the concept of the brain's model of the world largely derives from studies of the orienting reflex, the "what is it?" response invoked when the world is not what the brain expects (Sokolov, 1963).

Most neurophysiology of the cortex has looked at cortical function as the transfer of information from one region of the cortex to the next. This conceptualization cannot easily explain the extensive recurrent connections both within a cortical region and between cortical regions or the spontaneous neuronal activity that occurs in cortex in the absence of input. We suggest that cortical tissue can be considered as an intrinsic modeling system along the lines of Figure 1.

The different boxes in the figure represent populations of neurons. Within any region of cortex, these populations are intermixed. Whether each population is recognizable in terms of anatomy or chemistry is not known. The structure is probably essentially similar from one region of cortex to another, although particular regions of cortex can concentrate on particular aspects of incoming information or outgoing behavior. Connections between regions of cortex can be arranged in both hierarchical and parallel systems.

The center of such a concept of the cortex is a set of neurons that serve to "generate" the model. These cells can be active spontaneously or in response to signals from higher levels. Most commonly they are excited by incoming information from the "input" neurons. Once active, they initiate the firing of "model" cells, which encode both a perceptual model and a model for action. The modeling process continues until the model "compares" well with the input. A lack of fit of the model to the input brings about further generator activity. Once set up, the activity

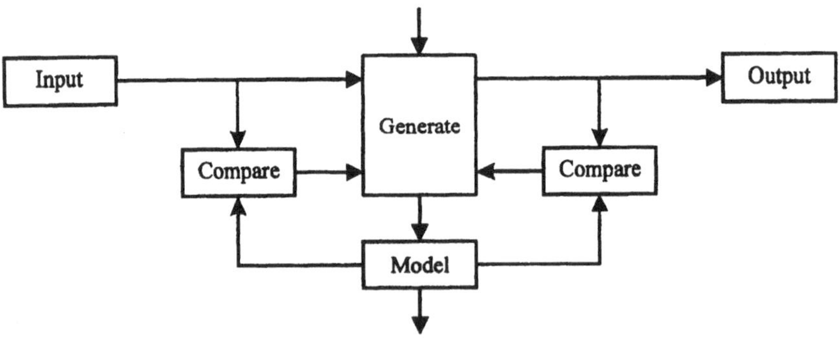

Figure 1. Diagrammatic representation of a modeling system of cortical activity

patterns of the model cells can become dormant until initiated again by the generator cells. The generator neurons also initiate "output" and this is fit to a model of the desired action using a feedback comparison process similar to that used in fitting the perceptual model to the incoming information. Since the model has much in common with what we consider memory, this way of looking at brain function supports the idea that memory is mainly located in those areas of the cortex that process the information to be remembered.

There is little in the way of direct evidence for such a pattern of cortical activity, although some cortical recordings can be construed along these lines. Logothetis and Schall (1989) have studied the neuronal activity in monkeys viewing moving stimuli. When the stimuli to the two eyes were different ("binocular rivalry"), the monkeys acted in much the same way as human subjects who fluctuate between two contradictory perceptions. In the superior temporal sulcus, some neurons responded specifically to the movement of stimuli in one eye independently of how it was perceived (input neurons), and other neurons responded to the perceived movement (model neurons). In another set of experiments, Maunsell and Ferrera (1995) studied neurons in the V4 region of cortex during a match-to-sample task wherein the monkey had to respond when the orientation of a test stimulus was the same as the orientation of a sample stimulus. The responses of some neurons to the test stimulus varied with its orientation whereas the responses of other neurons varied with the orientation of the sample stimulus (regardless of the orientation of the test stimulus evoking the response). The former would be the input neurons and the latter would be the model neurons that are activated in expectancy.

Comparisons between incoming information and expectancy can also show up in cortical recordings. Desimone and his colleagues (1995) have found neurons in the inferotemporal cortex that fire more rapidly when a test stimulus does not match a sample than when it does. These firing patterns could represent the feedback loop that drives the generator neurons when the model (or expectancy) does not fit the incoming information. Recent studies of the electrical and magnetic activity generated by the human auditory cortex have shown a "mismatch negativity" that occurs when a deviant stimulus occurs in a train of standard stimuli (Näätänen, 1992). This indicates that neuronal activity in the cortex compares incoming information to some model of what is expected.

We also propose that the human brain is arranged in a hierarchy of feedback systems. Lower levels are concerned with sensory experience and motor activity whereas higher levels are concerned with self-awareness. In this view higher and lower can be considered in terms of the distance (or the number of synapses) from sensory receptors or muscle effectors. A model active at one level of the hierarchy can control the modeling at lower levels through a process of "subsumption", to use a word from philosophy that means the bringing of one concept under a broader concept. Consciousness occurs at the level of the highest active generator system, and the contents of consciousness are the model at that particular level.

The contents of the lower models subsumed in the higher may form what has been called the "fringe" of consciousness (James, 1890)—a background of dimly perceived ideas surrounding what is the focus of present consciousness. This view of consciousness has some similarities to the "dominant action system" proposed by Shallice (1978) and the "multiple drafts model" of Dennett (1991). In both, the content of consciousness is the brain activity that wins the competition for the control of behavior.

The process of subsumption allows the whole brain to function in a manner similar to what we are suggesting for particular regions of cortex. In this view, the generator activity is largely in the prefrontal areas and the models largely in more posterior cortical regions. Studies of cerebral blood flow during thought and attention show activity in the prefrontal cortex concomitantly with posterior regions specific to the information being attended to or thought about (Roland, 1982; Roland and Friberg, 1985). Usher and Niebur (1996) have proposed a model of neruonal activity in the brains of monkeys performing match-to-sample tasks. Activity in the prefrontal cortex ("working memory") activates neurons in the inferotemporal cortex ("sensory memory") so as to predispose this region to respond to stimulus information coming from primary visual cortex ("input layer") when it matches the sample. Goldman-Rakic (1990) has suggested that the prefrontal regions of the brain are important to the organization of what is in the working memory of our conscious minds. Stuss and colleagues (1995) have proposed that a model of prefrontal cortex might activate, maintain, or change "schemata" in other cortical areas.

SENSATION AND PERCEPTION

Cortical Activity Evoked by Sensory Stimuli

A sensory stimulus elicits many different patterns of neuronal activity from the cerebral cortex. These patterns can be recorded as poststimulus time histograms from microelectrodes that pick up the activity of single neurons and as evoked potentials from macroelectrodes that record the activity of populations of neurons. Some of the responses reflect the transfer of activity through the different regions of sensory cortex. Others reflect the synchronization or desynchronization of spontaneous rhythmic activity. In a conscious subject, the cortical activity evoked by a sensory stimulus persists for several hundreds of milliseconds. If the stimulus is improbable and relevant to what the subject is doing, the sensory evoked potential leads into a late positive wave called the P300 on the basis of its typical latency in milliseconds (Picton, 1992). This large positive wave is related to consciousness in that a subject is almost certainly conscious of a stimulus that elicits a P300 wave, although not all stimuli that reach consciousness elicit a P300. In relation to our proposed modeling system, the P300 has been associated with the updating of a model of sensory input held in working memory (Donchin et al., 1983).

The human brain contains specialized neuronal systems to analyze different aspects of a perceived object. The nervous system thus arrays the sensory data in many disparate areas of the sensory cortex. For example, color might be encoded in the responses of neurons in the inferior occipital lobe, motion in lateral tempero-occipital neurons, and location in parieto-occipital neurons. How a particular object, such as a red ball moving toward me from my right, is perceived from this disparate array of data is the "binding problem" (Crick, 1994). There are several approaches to this problem. One proposes that the neurons encoding each of the aspects of the object all project to some "grandmother" neuron (i.e., a neuron that responds to an object as specific as one's grandmother). This runs into combinatorial problems because of the many possible objects and views. A second approach is that the various features of the object participate in some specific pattern of neuronal activity. For example, the neurons active in response to the different features may be able to synchronize their firing so that a rhythm with a frequency near 40 Hz can both integrate the processing of a perceived object and act as its perceptual signature (Singer and Gray, 1995). These "gamma" rhythms, which can be recorded in the human electroencephalogram and specifically driven by stimuli presented at rates near 40 Hz, are enhanced by attention and decreased during sleep and anesthesia (Plourde, 1993). A third approach to the binding problem is to propose that the nervous system constructs a model to account for the activity in the different neurons activated by the senses. The gamma rhythms recorded in the cortex might then represent the fitting of this model to the sensory input.

Electrical Stimulation of the Brain

Some insight into conscious perception can be obtained by directly stimulating the cortex rather than by observing the response of cortex to sensory input. These studies can occur in patients who have electrodes implanted in the nervous system for the control of pain or movement-disorders and in patients who have their sensory cortex exposed during neurosurgery. Electrical stimulation of different levels of the somatosensory system can lead to conscious perception. At the thalamus and cortex, near-threshold stimuli must be repeated over a period of several hundred milliseconds before conscious experience occurs (Libet et al., 1964). Furthermore, electrical stimulation of the cortex can mask the conscious perception of a previous stimulus, presented either peripherally or cortically, if the interval between the stimulus and the electrical stimulation is less than about 200 ms. These results suggest that sensory awareness requires sustained cortical activity lasting several hundred milliseconds. It appears that some level of "neuronal adequacy" must be achieved before conscious perception can occur, and that this process can be disrupted if additional unrelated activity occurs in the cortex.

These studies have also provided some information about the timing of conscious awareness. Libet and colleagues (1979) performed experiments wherein the timing of a train of electrical stimuli presented to the cortex or medial lemniscus is perceived in relation to the timing of a single peripheral stimulus presented to the skin. If the peripheral stimulus occurs at the same time as the train is initiated, a train of stimuli to the medial lemniscus is perceived as having occurred at the same time as the peripheral stimulus, even though it takes up to 500 ms for such a train of stimuli to reach neuronal adequacy. Libet suggests that the central stimulus is perceptually antedated to the time of the initial cortical-evoked potential, which occurs after the first stimulus in the train. A train of cortical stimuli that, unlike lemniscal stimuli, does not elicit any normal cortical-evoked potential is perceived as occurring much later than a simultaneously presented peripheral stimulus. The train of cortical stimuli has no cortical-evoked potential to provide itself with a timing, and perception occurs (after neuronal adequacy is reached) without any antedating.

Imagery and Illusions

Consciousness independent of sensory input occurs during imagery. What regions of the brain are active during the maintenance of a visual image? Some studies have suggested that visual imagery is associated with activity in primary visual cortex (Kosslyn et al., 1993), although most studies only show activity as far back as the extrastriate visual areas (Roland and Gulyas, 1994). Whatever the resolution of this debate, it is still clear that imagery involves activity in the areas of the cortex subserving the normal perception. Concomitant with this activity is the fact that there is increased activity in the frontal regions of the brain. In terms of our proposed modeling system, imagery involves activating the model neurons without fitting the model to external input (i.e., with the comparison process turned off). However, the model must be made consistent with whatever use for which the image is created. Thus, activity freely modeled at one level will be compared with other models at higher levels where the comparison process may function quite accurately.

Some patients with lesions to the visual areas of the brain show relatively preserved imagery despite severely abnormal perception in the form of object agnosia (Moscovitch et al., 1994b). This means that imagery must involve different activities than perception. It is possible that both imagery and perception involve the same cortical areas but in different ways. For example, the binding together of information into a unified object and the separation of this as a "figure" from the "ground" of other incoming information is essential for normal perception but unnecessary in imagery when one starts with the object.

Illusions are disorders of perception wherein what is perceived does not truthfully represent reality. The location of cerebral activity leading to illusionary perception appears to be similar to the activity subserving veridical perception.

Prolonged viewing of a stimulus moving in one direction (as when staring at a waterfall) causes stationary objects to appear as though moving in the opposite direction when the movement stops. During this illusory perception, functional MRI studies showed selective activation of the midtemporal (MT) area of the human cortex (Tootell et al., 1995). These results also indicated that the awareness of the illusion was associated with relatively more neuronal activity in the MT region than in more posterior areas of visual cortex. The persistent neuronal activity in the MT region is probably caused by a release from inhibition. Neurons activated by the moving stimulus inhibit neurons that normally respond to motion in the opposite direction. Once the movement stops these neurons would be released from inhibition and might fire at a high rate for a brief period. If so, the illusion is veridical in reference to the cortical activity but not in reference to reality.

Dreams and Hallucinations

Sensory experience independent of sensory input exists on a continuum that goes from imagery to daydreams to sleeping dreams. Sleep is the most common alteration in consciousness. During sleep, consciousness is usually considered present but separated from external input, from logical constraint, and from ongoing memory (Llinás and Paré, 1991): separated but not divorced because some external stimuli are incorporated into dreams, some dreams run according to a logic, and sometimes dreams are remembered. In human subjects, dreams occur most vividly in sleep stages associated with rapid eye movements (REM-sleep). This stage of sleep is initiated and maintained by brainstem systems (Hobson, 1988). During REM sleep, the electroencephalogram and the evoked potentials are similar to those recorded during wakefulness. In REM-sleep, the cortex probably functions as it does during wakefulness except that the generator system is more spontaneously active, the comparison process less precise, and the model neurons less persistent.

During hallucinations, a patient is vividly conscious of events that have no basis in reality and are often inconsistent even within themselves. Hallucinations occur for a variety of reasons. In some cases, they may represent the inappropriate occurrence of dream activity during wakefulness (Fisher, 1991). This may be triggered by toxic effects or by direct lesions to the brainstem systems that initiate REM sleep. In our schema of consciousness, these hallucinations represent the modeling system acting without any constraint from reality (i.e., with no comparison to incoming information). In other cases, hallucinations may result from disordered activity in focal areas of the brain subserving perception. These hallucinations may represent the modeling system trying to interpret abnormal input.

Blindsight

One of the most striking dissociations between consciousness and behavior occurs in patients with blindsight (Weiskrantz, 1990). Patients with lesions to the occipital visual cortices suffer from a loss of vision in the visual field contralateral to the lesioned cortex. Some of these patients may, however, respond to visual information within their blind fields. Indeed, experiments have shown that these patients can respond to such precise aspects of the visual stimuli as their location, colour, and texture. Although they may point to the stimuli or press buttons in response to these stimuli, the patients do not report any conscious experience of the stimuli and often have to be coaxed to guess. Occasionally they may report some consciousness of the stimuli, but this is attenuated and limited to only some characteristics (Barbur et al., 1993). The visually directed behavior of patients with blindsight is probably mediated by connections through the superior colliculus, pulvinar and parietal cortex. The major conclusion of these studies is that awareness within a particular perceptual domain requires that the normal functioning of the cortical systems processing that domain; in this case, normal visual awareness requires an intact occipital cortex.

Kolb and Braun (1995) have induced a state akin to cortical blindness in normal subjects by superimposing complementary visual textures on a target stimulus. For example, one eye may be activated by stimuli oriented oppositely from the other eye, and the subject using binocular perception does not see the target that is clearly visible to just one eye. The subject responds significantly above chance in a forced-choice localization of the target but reports no conscious awareness of the target. These results lead perhaps to a somewhat different conclusion than those deriving from clinical blindsight. Perceptual awareness of the total stimulus is integrated at a stage of processing that combines the complementary textures into one view of a stimulus. This perceptual awareness overrides the information available at lower levels that can nevertheless control behavior through unconscious mechanisms.

Electrical stimulation of the somatosensory thalamus at levels below those necessary for a subject to report awareness can still influence behavior (Libet et al., 1991). If the stimulus occurs during either the first or second of two lights and the subject has to choose which, the response is more accurate than chance even when the subject is not consciously aware of the stimulus. This is a clear demonstration that information can reach the cortex and yet not reach consciousness even when the subject is specifically attending to the stimuli. However, it is unlikely that this is the usual outcome. In general, the thresholds for conscious experience are very similar to the thresholds for the activation of cortical neurons. In the case of electrical stimulation, the unnatural stimulus is probably difficult to model other than as "paresthesiae", or abnormal sensations, and these may not be convincing at low intensities.

Agnosia

Agnosia is a disorder wherein certain aspects of perception are selectively deficient. In the original formulations of the problem, the patient was assumed to be able to experience something but not to understand what—to have a percept stripped of its meaning (Teuber, 1968). Visual form agnosia commonly occurs following bilateral damage to the extrastriate regions of the inferior occipital lobes. Patients can detect an object in their field of vision but are unable to identify what it is despite being able to manipulate it correctly and sometimes even make an accurate drawing of it. Milner and Goodale (1995) have interpreted the syndrome as a dissociation of visual input from a ventral stream of cortical analysis that leads to the conscious perception of an object. Visual agnosias can involve all aspects of perceptual processing or just particular aspects, with selective impairments for color, faces, and motion (Young, 1994).

Spatial agnosia represents an inability to evaluate the spatial aspects of stimuli. This typically results from lesions to the parietal regions of the brain and represents disruption of the dorsal stream of cortical analysis (Milner and Goodale, 1995). Interestingly, this type of perceptual problem is less closely associated with a clear loss of consciousness. These patients are conscious of objects but unaware of their spatial relations.

Some cortical lesions that cause neurological deficits are accompanied by anosognosia—an unawareness of the deficit (Schacter, 1990; Young, 1994). This is usually part of a syndrome of neglect or inattention. Anosognosia is superficially similar to denial except that it is not susceptible to argument. A patient with anosognosia may deny blindness or hemiplegia even when presented with incontrovertible evidence of such deficits. Young (1994) has suggested that anosognosia may represent a disorder of monitoring in addition to the specific sensory or motor disorder. Similarly, Stuss (1991) has proposed that anosognosia may be caused by some disconnection of the prefrontal cortex from the posterior cortical systems that mediate specific types of knowledge. It is as though the high-level modeling systems of the frontal lobe build a world that omits some information without admitting that it is missing. The model only attends to what has access to it.

ATTENTION AND MEMORY

Selective Attention

Attention defines the contents of our consciousness: we are aware of what we are attending to. Attention is also crucial to memory in that we are best able to remember those experiences to which we have paid (or which have demanded) the most attention. Attention is concerned both with selecting what will be attended to and with providing the mental effort to process the attended information.

Attention facilitates the processing of particular aspects of the available information. Neurophysiologically, attention shows up as an increased neuronal activity (Richmond and Sato, 1987; Hillyard et al., 1995) and increased blood flow (Roland, 1982) in those regions of the brain that evaluate the attended information. The neuronal mechanisms of attention are not clear. The activity in particular regions of the brain may be facilitated either by some thalamocortical gating mechanism (Yingling and Skinner, 1977) or by some dynamic routing of corticocortical connections (Olshausen et al., 1993). One simple mechanism to increase the cortical processing might be to alter the precision of the cortical comparisons. Making these more precise would lead to more intensive modeling and a more accurate model. It is not clear how this might show in neuronal recordings. In monkey cortex, attention alters the receptive fields of neurons processing the attended information independently of what stimuli occur (Desimone et al., 1990).

Attention also activates regions of the brain other than those specifically processing the attended information: the parietal cortex, the anterior cingulate, and the prefrontal cortex (Posner and Rothbart, 1992). The parietal cortex may be involved in allocating attention to particular regions of space (spotlight), binding information into figures or objects (highlight), and setting the scale of evaluation (zoom). The anterior cingulate appears to be involved in any task that requires effort. Different regions of the prefrontal cortex may be involved in various procedures that control attention. The prefrontal cortex receives input from posterior brain regions and sends feedback connections to these areas (Pandya, 1987). The prefrontal cortex may therefore control attention by selecting what sensory or remembered information to process and which responses to facilitate (Stuss et al., 1995).

Remembering

Human memory is subserved by multiple brain systems, some of which are conscious (explicit) and some that are unconscious (implicit). The consciousness associated with remembered information is quite different from the consciousness of present experience. Distinguishing between remembered and present experience has been termed "reality monitoring" (Johnson, 1991). Two types of memory are clearly related to consciousness since the subject is conscious of the recalled information (Tulving, 1985; Wheeler et al., 1997): episodic memories are remembered clearly as events in the subject's personal life; semantic memories are recalled to consciousness without any specific personal context.

A series of experiments measuring cerebral blood flow has suggested that the frontal lobes are involved in the encoding and retrieval of episodic memories. The left frontal lobe appears to be involved in encoding and the right frontal lobe in retrieval (Tulving et al., 1994; Nyberg et al., 1996). Exactly what role the right frontal lobe plays in retrieval is not yet known. The activity does not relate specifically to the conscious experience of remembered information since it occurs equally when the retrieval process brings back a memory and, in the case of novel stimuli, when the retrieval process is unsuccessful (Kapur et al., 1995). It may be mainly related to some conscious monitoring of the retrieval process.

Regions of the left frontal cortex are involved in many tasks as well as memory-encoding. The lateral prefrontal regions may participate in multiple tasks that share the requirement for "generative" processes. Patients with frontal lobe lesions typically display reduced "fluency" as tested by the ability to produce lists of words within a particular category (e.g., words beginning with a particular letter or words denoting animals). Consciousness is a creative process. It must be able to come up with possible interpretations to explain reality. The prefrontal regions of the brain seem to be particularly involved in this generative activity and in monitoring how well the models fit with reality. This is consonant with the multiple reciprocal connections of the prefrontal areas to other regions of the brain. The relationship to memory is that those interpretations that fit well with reality are entered into memory.

Confabulation and Delusions of Memory

Disruption of the normal linkages between memory and consciousness can lead to delusions of memory. Temporal lobe dysfunction, particularly transient dysfunction related to epilepsy, can produce states of *jamais vu* wherein a patient has no conscious recollection of something previously experienced and *déjà vu* wherein a patient experiences something new as though familiar from prior experience.

Confabulation is a disorder wherein a patient consciously remembers something that did not happen. Most patients with amnesia do not confabulate but simply fail to remember. However, some patients fill in the memory gaps with invented and sometimes fanciful detail. These patients usually have a frontal lobe dysfunction in addition to the more posterior disorders causing the amnesia (Shapiro et al., 1981; Moscovitch and Melo, 1997).

Some unusual delusional syndromes may also represent dissociations between memory and consciousness (Benson and Stuss, 1990; Stuss, 1991). Reduplicative paramnesia occurs when a patient considers his present location in two ways. Although the correct location is supported by other knowledge, the patient remains obstinately aware of being in a different but previously familiar location. In Capgras syndrome, a patient declares a close relative to be an imposter. Again, there are two interpretations, one logically correct and another that the patient consciously believes. The only resolution of these two interpretations is that the perceptually correct but consciously unfamiliar person is an imposter. Both of these delusions appear to result from dysfunction between frontal and posterior regions of the brain with disruption of the normal linkages between consciousness and memory. Normally, personal memories are associated in our consciousness with an experience of warmth and intimacy (James, 1890). In these delusions of place and person, the experience of warmth and intimacy is present in two contradictory memories, and the patient cannot resolve the contradiction because of impaired functioning in the frontal cortex.

Conscious and Unconscious Processing

Memories that occur without any relationship to consciousness include memories for motor skills and priming. Priming is the improvement in perceptual processing caused by prior experience that occurs even when the subject cannot consciously remember having experienced the priming stimulus (Schacter et al., 1993). Priming for visually presented words can be demonstrated by a task such as fragment completion in which a subject completes a word for which only some of the letters are given (A--AS-I-, ASSASSIN). Priming causes words that were seen prior to the task to be more readily completed than words that were not seen. Priming is distinguished from explicit memory processes by not being affected by the depth of encoding during initial stimulus presentation.

Another way to look at conscious and unconscious processes is from the point of view of automaticity. Practice with a motor or cognitive skill can make it extremely efficient and automatic in the sense that it no longer reaches consciousness or requires conscious effort. A child learning to read does with intense conscious effort what by adulthood has become automatic. Practice on a task can significantly alter the patterns of cerebral blood flow during task-performance (Raichle et al., 1994). During practiced performance, there was much less activation of prefrontal cortex, anterior cingulate, and cerebellum than during naïve performance. An input–output transformation could be made automatic by strengthening the response generation and attenuating the modeling and comparison activities that accompany normal conscious processing. Responses are then made without need of top-down activation and there is little or no conscious content in terms of active model neurons.

Much of what goes on in the human brain is unconscious. Although we are conscious of our thoughts, we are generally unconscious of how they came about. Fodor (1983) has proposed that the brain contains multiple "modules" that automatically process information and make it available to conscious awareness in a central cognitive processor. These modules are characterized by domain specificity and cognitive impenetrability: they work on specific aspects of information and do so independently of conscious awareness or control. Most of the sensory analyzers operate in modular fashion. A more complicated example would be a cerebral module that looks up the meaning of a word. Fodor suggests that such a module acts rapidly and efficiently to look up all the possible meanings of a word. The central cognitive processor then selects whichever meaning is appropriate to the present context. Cognitive modules appear to be mainly determined by some inherited neural architecture that becomes competent after experience of the appropriate information during development. The extent to which modules are cognitively impenetrable is open to some debate (Van Petten and Kutas, 1991), but the idea of separate neuronal systems working efficiently on specific processes fits well with our knowledge of the anatomy and physiology of the brain.

Much of our behavior is therefore affected by unconscious processes (Kihlstrom, 1987). These can be the implicit memories of prior experience, the overlearned procedures for motor or cognitive skills, or the modular processes that are innate in the brain. An example of how behavior results from a complex interplay between conscious and unconscious processes is the retrieval of information from memory (Moscovitch, 1994a). Once inititated, memory retrieval occurs rapidly and independently of conscious control. However, one can consciously manage memory retrieval by selecting what triggers to consider and by organizing the information retrieved. For example, when trying to remember someone's name, one might deliberately recall various times when one was with that person, in the hope that the associated name will also pop into mind. Moscovitch has suggested that this "working with memory" largely depends on activity in the prefrontal cortex, whereas the automatic retrieval process involves the hippocampus.

PERSON AND PLANNING

The Self

Essential to any understanding of conscious experience is the concept of the person undergoing the experience. Some higher order model in the brain is aware of the external world and controls behavior. This can be the self, the soul, the homunculus or the "center of narrative gravity" (Dennett, 1991). The concept of self has engendered much recent discussion in neuropsychology. Lesions to the prefrontal regions of the brain cause a multiplicity of symptoms, many of which (such as apathy and impulsivity) are apparently contradictory. These can be interpreted as a disorder of personality, that is, a change in the stable response patterns that define an individual (Stuss et al., 1992). A continuity of person is necessary for any memory that is organized according to a personal viewpoint. It is also necessary for knowledge, since knowledge involves prediction and this requires that the person be around to experience what is predicted. Personal continuity seems attenuated in patients with frontal lobe damage. They can recall memories and make intellectual predictions of the future but they do not apply these memories or predictions to themselves: they live only in the present (Hutton, 1947).

Psychology often postulates some conscious executive process to manage the brain's unconscious systems: James' "pilot" (1890) or Shallice's "supervisory attentional system" (1978). This executive notices unexpected occurrences, allocates attentional resources, makes decisions, considers possibilities, and plans ahead. Unlike the multiple parallel procedures being managed, the executive works in a serial manner, considering one thing at a time. Much evidence in patients with lesions to the frontal lobe suggests that the prefrontal cortex is essential to such an executive (Stuss and Benson, 1986; Knight and Grabowecky, 1995). Our recent thinking suggests that the prefrontal cortex may not act in an undifferentiated manner but rather be composed of separate modules that work together in different ways to perform executive functions (Stuss et al., 1995).

Cognitive Development and the Theory of Mind

The human brain appears designed to find better and better ways of representing the world. The cognitive development of children can be considered in terms of "representational redescription" (Karmiloff-Smith, 1992). A powerful way to gain knowledge is to keep recasting the knowledge we have already (either innate or learned) until it can be used more efficiently. Thus children learn better laws of physics and psychology. This appears to be an explicit or conscious process, even though it derives largely from implicit knowledge and leads largely to automatic procedures.

The developing human infant quickly grasps important concepts about the world (Spelke, 1988; Leslie, 1988; Dasser et al., 1989). Interactions between innate tendencies and worldly experience bring about four main concepts. The first is the concept of *object*: those pieces of sensory information that correlate with each other, and in particular that move together, can be considered as features of an object. A second level of conceptualization has to do with *causation*. This probably derives initially from objects bumping into one another and transferring momentum. A third concept concerns *animism*: some objects are alive in the sense that they do things that are not externally caused. A final level concerns the fact that some living objects (myself included) have a *mind*. Young children develop a sense of their own minds at about the same time as they develop a sense of what other people think and believe (Gopnik, 1993). Our sense of our own minds may therefore be as much a theoretical construct as our sense of the minds of others. Continued experience of our own mind provides an expertise that makes the awareness of self more "direct" than the awareness of others.

One clear demonstration of the ability to understand the state of mind of another occurs when an individual recognizes that another individual may hold a false belief. This is the basis of a test that has been used to study how children perceive the minds of others (Baron-Cohen, 1995). The test involves a simple puppet play. One puppet (Sally) places her marble in a basket, and then leaves the scene. Another puppet transfers Sally's marble from the basket to a box. Sally returns and the child is asked where Sally will look for the marble. Normal children and children with Down's syndrome correctly state that Sally will look for the marble in the basket where she placed it rather than in the box. Children with autism often answer that Sally will look for the marble in the box. They are unable to recognize Sally's false belief: they are blind to the minds of others.

Divided Minds

In certain patients with otherwise intractable epilepsy, the corpus callosum can be cut to decrease the frequency and intensity of seizures. The neuropsychological evaluation of such "split brain" patients has led to a greater understanding of how the brain works and in particular how consciousness occurs (Gazzaniga, 1985). A major initial finding was that some split-brain patients appeared to act as if their consciousness as well as their hemispheres had been divided, with two conscious individuals vying for the control of action. This does not easily fit with the basic idea that consciousness is a unified process. However, it does fit with the common experience of being of two minds about something. Consciousness seems to be set up so that it seeks some single interpretation but sometimes, by pathology or by perplexity, different interpretations coexist.

Other studies of split-brain subjects are particularly important to ideas of conscious and subconscious control of behavior. The subjects were shown one slide in the left visual field (going only to the right hemisphere) and a different slide in the right visual field (going only to the left hemisphere) and then asked to choose from a series of pictures those which were most related to the slides. When a patient was presented with a chicken-foot on the right and a snow-scene on the left, the patient pointed to a picture of a chicken with his right hand and a picture of a shovel with his left hand. When asked why he picked these two pictures, the patient said "the chicken claw goes with the chicken and you need a shovel to clean out the chicken shed". The patient was only able to speak with the left hemisphere. The left hemisphere was unable to explain why the left hand (controlled by the right hemisphere) was choosing the shovel and invented a theory that made sense given the information available to the left brain. Gazzaniga has therefore proposed that the left hemisphere contains an "interpreter" system that seeks to understand why we are doing what we do. These findings fit well with a large psychological literature (Nisbett and Wilson, 1977) that shows that we are often not aware of what causes our behavior. After the fact, we come up with reasonable but not necessarily true explanations for our actions. The need for such interpretations only occurs when we have to make inferences about something beyond our knowledge. It is unnecessary when we have direct knowledge, and verbal reports of conscious experience are demonstrably reliable (Ericsson and Simon, 1980). Even our interpretations of why we act the way we do are more often correct than not, unless we refuse to consider or, like the split-brain patient, are unable to consider certain possibilities.

Volitional Behavior

Preparation to make a motor response is associated with activity preceding the movement in the frontal regions of the brain. This activity can be recorded from the human scalp in the form of a Bereitschafts potential or "readiness potential".

This negative wave begins several hundred milliseconds before a motor act. As the time before the act decreases, the readiness potential increases in amplitude and becomes more focal in its distribution over the motor cortex responsible for the act. Recent experiments have suggested that the midline regions of the frontal cortex are involved with the motor cortex in the conscious initiation of an act (Shibasaki et al., 1993; Toro et al., 1993).

Libet and colleagues (1983) investigated the relationship between the onset of the readiness potential and the conscious initiation of the movement. The results suggested that the readiness potential had already begun by the time that subjects perceived that they were consciously initiating the movement. Libet and colleagues therefore concluded that the brain initiates volitional activity independently of consciousness, with consciousness only playing a possible veto role in the control of behavior. This certainly fits with some behaviors: the sprinter only becomes aware of leaving the blocks after it has already happened. However, we clearly differentiate between such reflex responses and voluntary acts. Unfortunately, in the Libet and colleagues experiments, it is very difficult to understand either what the subject was mentally timing in terms of the initiation of the motor act or what the readiness potential represented. One prepares to initiate a motor act by first getting into a generally responsive state and then specifically initiating the motor act. When during this preparatory period is the act initiated? When does the readiness potential change from general preparation to specific motor activation? Furthermore, as demonstrated in the other experiments concerning the timing of electrical stimulation to the brain, a subject's perception of time can be projected away from exact physical time. Given these problems with the experiment, it seems unnecessary to accept that consciousness plays no role in controlling those behaviors that we specifically consider voluntary.

One rare but intriguing disorder of conscious control that occurs with lesions to the frontal lobe is the "alien hand" syndrome (Goldberg and Bloom, 1990). The patient is unaware of the movements of his own hand, which moves in a groping or grasping manner independently of conscious control. The syndrome is usually associated with a lesion of the supplementary motor area or some disconnection of this area from the motor cortex activating the muscles of the alien hand. It appears as if the motor control mechanisms are active without any feedback control from the model of planned action.

Science often has difficulty considering our consciousness of free will. Perhaps, free will is illusory and our actions occur like those of the alien hand without any conscious control. Perhaps some random uncertainty in synaptic potentials might make our brains sufficiently unpredictable that they can be considered free. We think that this difficulty is a throwback to the times when consciousness was considered immaterial and therefore unable to affect a material brain. If consciousness is real, it can cause behavior. Although many of our actions are performed unconsciously, those that are consciously energized and directed need to be understood in terms of the brain.

CONCLUDING COMMENTS

Culture and Evolution

Any biological view of consciousness must consider consciousness in terms of evolution. In terms of where on the evolutionary scale consciousness develops, one would have to consider conscious any nervous system capable of constructing neuronal models of input. The particular level of consciousness that can postulate a self and exercise a theory of mind is probably only found in higher primates.

One particular aspect of the model-making approach to consciousness where evolution may provide some explanatory justification concerns the fitting of the model to reality. If we are only conscious of reality in terms of a model that we create, how can we tell whether this model is accurate? Indeed, how can we tell that there is a real world independently of our model? Evolutionary theory would allow the survival of only those individuals who can construct reasonably truthful models of reality and who can therefore accurately predict comfort or danger.

In terms of human consciousness, evolutionary theory requires expansion to consider language and culture (Donald, 1991). These developments have caused the human animal to evolve more rapidly than any other. Most human activity now has only faint ties to the actions of lower animals. Human memory is organized completely differently and exists in books and in computers, as much outside of the brain as within it.

The role of culture in human evolution makes it very difficult to consider consciousness as "epiphenomenal". Culture involves the transmission of conscious information among individuals. That such transmission should occur with consciousness only being aware of but not actively directing the process leads to a communication process divorced from the codes of both transmitter and receiver. At the present stage of human development, the accumulation and sharing of conscious knowledge is what makes the individuality of each person and guides the evolution of our species.

Intentionality and Folk Psychology

Our present theories divide into two main groups on the basis of how they consider the "intentional" aspects of human consciousness (Lyons, 1995; Dennett, 1996). Intentionality is the "aboutness" of something—what it signifies or points to. Intrinsic to intentionality is an attitude or perspective: thus an individual may believe something to be true or desire that something may happen. In one camp (e.g., Fodor, 1987), the intentional aspects of our mental life are real—beliefs and desires are in the brain and control behavior. We act to fulfil our desires in the light of our beliefs. In the other camp (e.g., P.S. Churchland, 1986; P.M. Churchland, 1995), beliefs and desires are but hypothetical interpretations of what is going on in the brain. Although they can be used to describe brain functions, they are probably unrelated to the actual transformations of neuronal activation patterns that occur in the brain. If we fully understood these cerebral processes, we would find our present view of intentions as misplaced as our ideas of an earth-centered universe. We do not have a true science but a "folk psychology", a set of ideas about the mind believed by common folk but not proven by experiment (Goldman, 1993). In this second view of intentionality, the procedures of psychotherapy, based as they are on our ideas about how beliefs and desires control behavior, would fade to insignificance. However, like Fodor (1987), we feel that our present view of intentionality, although far from exact, is probably not radically different from what actually occurs in the conscious brain. Why else would we be able to communicate with each other? Communication acts to change the beliefs and desires of others. The main problem with psychotherapy is not so much that it is determined by an intentional view of the brain as that this view is culturally dependent. However, since culture is so great a part of human consciousness, this may be a necessary caveat rather than an essential defect.

Directions

Although consciousness is difficult to understand, its importance to human life demands that it be studied. Our intuition is that it is not beyond our intellectual grasp. Observation, hypothesis and experiment are the tools to examine consciousness. Our present understanding is based more on observation and hypothesis than on experiment and this imbalance needs redress. Multidisciplinary studies are almost certainly necessary; looking at the physiology of consciousness without simultaneously examining its psychology can easily lead to difficulties. Determining the basis of human consciousness should be one of the main goals of our next century, much as the discoveries of the structure of the atom and the nature of the gene have been in our present century.

SUMMARY

Human consciousness is based upon the ability of the brain to construct models of the past, present, and future world. The cerebral cortex generates a model of what we experience in patterns of neuronal activity and compares this model to information coming from the senses. As well, the cerebral cortex can make models of planned behavior and generate motor activity to fit within these plans. The modeling process can occur both within a region of cortex and within a hierarchy of different cortical areas. In the hierarchical version of the modeling process, the prefrontal regions of the brain are responsible for generating models of what we have perceived, for predicting what might happen in the future, and for organizing our behavior. This view of consciousness is supported by physiological measurements of neuronal activity and cerebral blood flow during conscious processing.

Disruptions of the modeling process might explain some of the abnormalities of consciousness that occur in neurological and psychiatric disorders.

REFERENCES

Barbur, J.L., Watson, J.D.G., Frackowiak, R.S.J., & Zeki, S. (1993). Conscious visual perception without V1. Brain, 116, 1293-1302.

Baron-Cohen, S. (1995). *Mindblindness. An Essay on Autism and Theory of Mind.* MIT Press, Cambridge, MA.

Benson, D.F. & Stuss, D.T. (1990). Frontal lobe influences on delusions: A clinical perspective. Schizophr. Bull. 16, 403-411.

Churchland, P.M. (1995). *The Engine of Reason, the Seat of the Soul. A Philosophical Journey into the Brain.* MIT Press, Cambridge, MA.

Churchland, P.S (1986). *Neurophilosophy: Toward a Unified Science of the Mind-Brain.* MIT Press, Cambridge, MA.

Craik, K. (1943). *The Nature of Explanation.* Cambridge University Press, Cambridge, MA.

Crick, F. (1994). *The Astonishing Hypothesis. The Scientific Search for the Soul.* Charles Scribner's Sons, New York.

Dasser, V., Ulbaek, I., & Premack, D. (1989). The perception of intention. Science 243, 365-367.

Dennett, D.C. (1991). *Consciousness Explained.* Little, Brown & Company Ltd., Boston, MA.

Dennett, D.C. (1996). *Kinds of Minds.* BasicBooks, New York.

Desimone, R., Wessinger, M., Thomas, L., & Schneider, W. (1990). Attentional control of visual perception: Cortical and subcortical mechanisms. Cold Spring Harbor Symposia on Quantitative Biology. 50, 963-971.

Desimone, R., Miller, E.K., Chelazzi, L., & Lueschow, A. (1995). Multiple memory systems in the visual cortex. In: *The Cognitive Neurosciences.* (Gazzaniga, M.S., Ed.), pp. 475-486, MIT Press, Cambridge, MA.

Donald, M. (1991). *Origins of the Modern Mind.* Harvard University Press, Cambridge, MA.

Donchin, E., McCarthy, G., Kutas, M., & Ritter, W. (1983). Event-related brain potentials in the study of consciousness. In: *Consciousness and Self Regulation.* (Davidson, R.J., Schwartz, G.E., Shapiro, D., Eds.), pp. 81-121, Plenum Press, New York.

Ericsson, K.A., & Simon, H.A. (1980). Verbal reports as data. Psychol. Rev. 87, 215-251.

Fisher, C.M. (1991). Visual hallucinations on eye closure associated with atropine toxicity. A neurological analysis and comparison with other visual hallucinations. Can. J. Neurol. Sci. 18, 18-27.

Flanagan, O. (1992). *Consciousness Reconsidered.* MIT Press, Cambridge, MA.

Fodor, J.A. (1983). *The Modularity of Mind.* MIT Press, Cambridge, MA.

Fodor, J.A. (1987). *Psychosemantics. The Problem of Meaning in the Philosophy of Mind.* MIT Press, Cambridge, MA.

Gazzaniga, M.S. (1985). *The Social Brain. Discovering the Networks of the Mind.* Basic Books, Inc., New York.

Goldberg, G. & Bloom, K.K. (1990). The alien hand sign: Localization, lateralization and recovery. Am. J. Physical. Med. Rehab. 69, 228-238.

Goldman, A.I. (1993). The psychology of folk psychology. Behav. Brain Sci. 16, 15-28, 1993.

Goldman-Rakic, P.S. (1990). The prefrontal contribution to working memory and conscious experience. In: *The Principles of Design and Operation of the Brain,* (Eccles, J.C. & Creutzfeldt, O., Eds.), pp. 389-407. Springer-Verlag, New York.

Gopnik, A. (1993). How we know our minds: The illusion of first-person knowledge of intentionality. Behav. Brain Sci. 16, 1-14.

Hillyard, S.A., Mangun, G.R., Woldorff, M.G., & Luck, S.J. (1995). Neural systems mediating selective attention. In: *The Cognitive Neurosciences.* (Gazzaniga, M.S., Ed.), pp. 665-681, MIT Press, Cambridge, MA.

Hobson, J.A. (1988). *The Dreaming Brain.* Basic Books, New York.

Hutton, E.J. (1947). Personality change after leucotomy. J. Ment. Sci. 93, 31-42.

James, W. (1890). *The Principles of Psychology.* Original: Henry Holt & Co 1890; Reprint (1950): Dover, New York.

James, W. (1912). Does consciousness exist? In: *Essays in Radical Empiricism.* Longmans Green, New York; Reprint (1976), Harvard, Cambridge, MA.

Johnson, M.K. (1991). Reality monitoring: Evidence from confabulation in organic brain disease patients. In: *Awareness of Deficit after Brain Injury.* (Prigatano, G. & Schacter, D.L., Eds.), pp. 176-197. Oxford University Press, New York.

Johnson-Laird, P.N. (1983). *Mental Models: Towards a Cognitive Science of Language, Inference, and Consciousness.* Harvard University Press, Cambridge, MA.

Kapur, S., Craik, F.I.M., Jones, C., Brown, G.M., Houle, S., & Tulving, E. (1995). Functional role of the prefrontal cortex in memory retrieval: A PET study. NeuroReport 6, 1880-1884.

Karmiloff-Smith, A. (1992). *Beyond Modularity: A Developmental Perspective on Cognitive Science.* MIT Press, Cambridge MA.

Kihlstrom, J.F. (1987). The cognitive unconscious. Science 237, 1445-1451.

Knight, R.T. & Grabowecky, M. (1995). Escape from linear time: Prefrontal cortex and conscious experience. In: *The Cognitive Neurosciences.* (Gazzaniga, M.S., Ed.), pp. 1357-1371, MIT Press, Cambridge, MA.

Kolb, F.C. & Braun, J. (1995). Blindsight in normal observers. Nature 377, 336-338.

Kosslyn, S.M., Alpert, N.M., Thompson, W.L., Malijkovic, V., Weise, S.B., Chabris, C.F., Hamilton, S.E., Rauch, S.L., & Buonanno, F.S. (1993). Visual mental imagery activates topographically organized visual cortex: PET investigations. J. Cog. Neurosci. 5, 263-287.

Leslie, A.M. (1988). The necessity of illusion: Perception and thought in infancy. In: *Thought without Language.* (Weiskrantz, L., Ed.), pp. 185-210. Clarendon Press, Oxford, England.

Libet, B., Alberts, W.W., Wright, Jr., E.W., Delattre, L.D., Levin, G., & Feinstein, B. (1964). Production of threshold levels of conscious sensation by electrical stimulation of human somatosensory cortex. J. Neurophysiol. 27, 546-578.

Libet, B., Gleason, C.A., Wright, E.W., & Pearl, D.K. (1983). Time of conscious intention to act in relation to onset of cerebral activity (readiness-potential). The unconscious initiation of a freely voluntary act. Brain 106, 623-642.

Libet, B., Pearl, D.K., Morledge, D.E., Gleason, C.A., Hosobuchi, Y., & Barbaro, N.M. (1991). Control of the transition from sensory detection to sensory awareness in man by the duration of a thalamic stimulus. Brain 114, 1731-1757.

Libet, B., Wright, E.W. Jr., Feinstein, B., & Pearl, D.K. (1979). Subjective referral of the timing for a conscious sensory experience. A functional role for the somatosensory specific projection system in man. Brain 102, 193-224.

Llinás, R.R. & Paré, D. (1991). Of dreaming and wakefulness. Neurosci. 44, 521-535.

Logothetis, N. & Schall, J.D. (1989). Neural correlates of subjective visual perception. Science 245, 753-761.

Lyons, W. (1995). *Approaches to Intentionality.* Clarendon Press, Oxford, England.

Maunsell, J.H.R. & Ferrera (1995). Attentional mechanisms in visual cortex. In: *The Cognitive Neurosciences.* (Gazzaniga M.S., Ed.), pp. 451-461, MIT Press, Cambridge, MA.

Miller, G.A., Galanter, E., & Pribram, K. (1960). *Plans and the Structure of Behavior.* Holt, Rinehart and Winston, New York.

Milner, A.D. & Goodale, M.A. (1995). *The Visual Brain in Action.* Oxford University Press, Oxford, England.

Moscovitch, M. (1994a). Memory and working with memory: Evaluation of a component process model and comparisons with other models. In: *Memory Systems.* (Schacter, D.L. & Tulving, E., Eds.), pp. 269-310. MIT Press, Cambridge, MA.

Moscovitch, M. (1994b). Do PETs have long or short ears? Mental imagery and neuroimaging. Trends Neurosci. 17, 292-294.

Moscovitch, M. & Melo, B. (1997). Strategic retrieval and the frontal lobes: Evidence from confabulation and amnesia. Neuropsychologia 35, 1017-1034.

Näätänen, R. (1992). *Attention and Brain Function.* Erlbaum, Hillsdale, NJ.

Natsoulas, T. (1978). Consciousness. Amer. Psychologist 33, 906-914.

Nisbett, R.E. & Wilson, T.D. (1977). Telling more than we can know: Verbal reports on mental processes. Psychol. Rev. 84, 231-259.

Nyberg, L., Cabeza, R., & Tulving, E. (1996). PET studies of encoding and retrieval: The HERA model. Psychonom. Bull. & Rev. 3, 135-148.

Olshausen, B.A., Anderson, C.H., Van Essen, D.C. (1993). A Neurobiological Model of Visual Attention and Invariant Pattern Recognition Based on Dynamic Routing of Information. J. Neurosci. 13, 4700-4719.

Pandya, D.N. & Barnes, C.L. (1987). Architecture and Connections of the Frontal Lobe. In: *The Frontal Lobes Revisited.* (Perecman, E., Ed.), pp. 41-72. IRBN Press, New York.

Picton, T.W. (1992). The P300 wave of the human event-related potential. J. Clin. Neurophysiol. 9, 456-479.

Picton, T.W. & Stuss, D.T. (1994). Neurobiology of conscious experience. Curr. Opinion Neurobiol., 4, 256-265.

Plourde, G. (1993). The clinical use of the 40 Hz auditory steady state response. Int. Anaesthesiol. Clin. 31, 107-120.

Posner, M.I. & Rothbart, M.K. (1992). Attentional mechanisms and conscious experience. In: *The Neuropsychology of Consciousness.* (Milner, A.D. & Rugg, M.D., Eds.), pp. 91-111. Academic Press, London.

Raichle, M.E., Fiez, J.A., Videen, T.O., MacLeod, A-M.K., Pardo, J.V., Fox, P.T., & Petersen, S.E. (1994). Practice-related changes in human brain functional anatomy during nonmotor learning. Cereb. Cortex 4, 8-26.

Revonsuo, A. & Kamppinen, M. (1994). *Consciousness in Philosophy and Cognitive Neuroscience.* Lawrence Erlbaum, Hillsdale, NJ.

Richmond, B.J. & Sato, T. (1987). Enhancement of inferior temporal neurons during visual discrimination. J. Neurophysiol. 58, 1292-1306.

Riggs, D.S. (1970). Control Theory and Physiological Feedback Mechanisms. Williams and Wilkins, Baltimore.

Roland, P.E. (1982). Cortical regulation of selective attention in man. J. Neurophysiol. 48, 1059-1077.

Roland, P.E. & Friberg, L. (1985). Localization of cortical areas activated by thinking. J. Neurophysiol. 53, 1219-1243.

Roland, P.E. & Gulyás, B. (1994). Visual imagery and visual representation. Trends Neurosci. 17, 281-297.

Schacter, D.L. (1990). Toward a cognitive neuropsychology of awareness: Implicit knowledge and anosognosia. J. Clin. Exp. Neuropsychol. 12, 155-178.

Schacter, D.L., Chiu, C.Y.P., & Ochsner, K.N. (1993). Implicit memory: A selective review. Ann. Rev. Neurosci. 16, 159-182.

Shallice, T. (1978). The dominant action system: An information-processing approach to consciousness. In: *The Stream of Consciousness. Scientific Investigations into the Flow of Human Experience.* (Pope, K.S. & Singer, J.L., Eds.), pp. 117-157. Plenum, New York.

Shapiro, B.E., Alexander, M.P., Gardner, H., & Mercer, B. (1981). Mechanisms of confabulation. Neurology 31, 1070-1076.

Shibasaki, H., Sadato, N., Lyshkow, H., Yonekura, Y., Honda, M., Nagamine, T., Suwazono, S., Magata, Y., Ikeda, A., Miyazaki, M., Fukuyama, H., Asato, R., & Konishi, J. (1993). Both primary motor cortex and supplementary motor area play an important role in complex finger movement. Brain 116, 1387-1398.

Singer, W. & Gray, C.M. (1995). Visual feature integration and the temporal correlation hypothesis. Ann. Rev. Neurosci. 18, 555-586.

Sokolov, E.N. (1963). Higher nervous functions: the orienting reflex. Ann. Rev. Physiol. 25, 545-580.

Spelke, E.S. (1988). The origins of physical knowledge. In: *Thought Without Language.* (Weiskrantz, Ed.), pp.168-184. Clarendon Press, Oxford, England.

Stuss, D., Gow, C.A., & Hetherington, C.R. (1992). "No longer Gage": Frontal lobe dysfunction and emotional changes. J. Consult. Clin. Psychol. 60, 349-359.

Stuss, D.T. (1991). Self, awareness, and the frontal lobes: A neuropsychological perspective. In: *The Self: Interdisciplinary Approaches.*(Strauss, J. & Goethals, G.R., Eds.), pp. 255-278. Springer-Verlag, New York.

Stuss, D.T. & Benson, D.F. (1986). *The Frontal Lobes.* Raven Press, New York.

Stuss, D.T., Shallice, T., Alexander, M.P., & Picton, T.W. (1995) A multidisciplinary approach to anterior attentional functions. In: *Structure and Function of the Human Prefrontal Cortex.* (Grafman, J., Holyoak, K.J., & Boller., F., Eds), Annals of the New York Academy of Sciences, 769, 191-211.

Teuber, H.L. (1968). Alteration of perception and memory in man. In: *Analysis of Behavioral Change.* (Weiskrantz, L., Ed.), pp. 274-328, Harper and Row, New York.

Tootell, R.B.H., Reppas, J.B., Dale, A.M., Look, R.B., Sereno, M.I., Malach, R., Brady, T.J., & Rosen, B.R. (1995). Visual motion aftereffect in human cortical area MT revealed by functional magnetic resonance imaging. Nature 375, 139-141.

Toro, C., Matsumoto, J., Deuschl, G. Roth, B.J., & Hallett, M. (1993). Source analysis of scalp-recorded movement-related electrical potentials. Electroenceph. clin. Neurophysiol. 86, 167-175.

Tulving, E. (1985). Memory and consciousness. Can. Psychol. 26, 1-12.

Tulving, E., Kapur, S., Craik, F.I.M., Moscovitch, M., & Houle, S. (1994). Hemispheric encoding retrieval asymmetry in episodic memory: Positron emission tomography findings. Proc. Nat. Acad. Sci. 91, 2016-2020.

Usher, M. & Niebur, E. (1996). Modeling the temporal dynamics of IT neurons in visual search: A mechanisms for top-down selective attention. J. Cog. Neurosci. 8, 311-327.

Van Petten, C. & Kutas, M. (1991). Electrophysiological evidence for the flexibility of lexical processing. In: *Word and Sentence.* (Simpson, G., Ed.), pp129-174. Elsevier, Amsterdam, The Netherlands.

Weiskrantz, L. (1990). Outlooks for blindsight: Explicit methodologies for implicit processes. Proc. R. Soc. Lond. B. 239, 247-278.

Wheeler, M.A., Stuss, D.T., & Tulving, E. (1997). Towards a theory of episodic memory: The frontal lobes and autonoetic consciousness. Psych. Bull. 121, 331-354.

Yates, J. (1985). The content of awareness is a model of the world. Psychol. Rev. 92, 249-2848.

Yingling, C.D. & Skinner, J.E. (1977). Gating of thalamic input to cerebral cortex by nucleus reticularis thalami. In: *Attention, Voluntary Contraction, and Event-related Cerebral Potentials. Progress in Clinical Neurophysiology.* (Desmedt, J.E., Ed.), vol. 1, pp. 70-96. Karger, Basel, Switzerland.

Young, A.W. (1994). Neuropsychology of awareness. In: *Consciousness in Philosophy and Cognitive Neuroscience* (Revonsuo, A. & Kamppinen, M., Eds.), pp. 173-203. Lawrence Erlbaum, Hillsdale, NJ.

NOTE

1. Recent reviews that the reader might find helpful in understanding the present state of research into consciousness are those of Black (1998) and Frith and colleagues (1999). The relationship between consciousness and the human EEG has been recently investigated in terms of γ rhythms (Tallon-Baudry et al., 1997) and signal complexity (Coenen, 1998). The role of acetylcholine in modulating cortical activity has been reviewed by Woolf (1997). The function of the prefrontal cortex in conscious awareness has been further highlighted in studies of cerebral blood flow (McIntosh et al., 1999) and focal lesions (Levine et al., 1998).

RECOMMENDED READINGS

Black, I.B. (1998). Genes, brain, and mind: The evolution of cognition. Neuron 20, 1073-1080.

Coenen, A.M. (1998). Neuronal phenomena associated with vigilance and consciousness: From cellular mechanisms to electroencephalographic patterns. 7, 42-53.

Frith, C., Perry, R. & Lumer, E. (1999). The neural correlates of conscious experience: An experimental framework. Trends Cognit. Sci. 3, 105-114.

Levine, B., Black, S.E., Cabeza, R., Sinden, M., McIntosh, A.R., Toth, J.P., Tulving, E., & Stuss, D.T. (1998). Episodic memory and the self in a case of isolated retrograde amnesia. Brain 121, 1951-1973.

McIntosh, A.R., Rajah, M.N., & Lobaugh, N.J. (1999). Interactions of prefrontal cortex in relation to awareness in sensory learning. Science 284, 1531-1533.

Tallon-Baudry, C., Bertrand, O., Delpuech, C., & Permier, J. (1997). Oscillatory γ-band (30-70 Hz) activity induced by a visual search task in humans. J. Neurosci. 17, 722-734.

Woolf, N.J. (1997). A possible role for cholinergic neurons of the basal forebrain and pontomesencephalon in consciousness. Conscious. Cogn. 6, 574-596.

Chapter 2

Emotional Circuits of the Mammalian Brain:

Implications for Biological Psychiatry

JAAK PANKSEPP

Biological Psychiatry, pages 27-58.

ISBN: 1-55938-819-6

ON THE NATURE OF EMOTIONS

The subjective power of emotionality makes it one of the most difficult and one of the most intriguing topics for psychiatry and neuroscience. Most people (especially children) are intensely aware when emotions enter their lives, while many others (e.g., self-confident, highly intellectualized adults living in safe environments) have comparatively little appreciation of the powerful emotions that could be elaborated in their brains. Indeed, most normal adults have difficulty talking about their emotional feelings straightforwardly. There are many reasons for this:

1. The brain areas that mediate emotions are more active in early childhood than in adulthood (Chugani, 1994).
2. Most people cultivate self-restraint and are hesitant to reveal their true feelings, for self-exposure can weaken self-esteem and make one vulnerable to psychic injury.
3. The prison-houses of language and propositional logic, which are so highly valued in society, impose enormous constraints on our ability to communicate clearly about emotional matters.
4. Our ability to express, decode, and feel emotions is substantially different in our left and right hemispheres.
5. Finally, males and females typically excel in different emotional skills.

Males generally experience more difficulty in articulating their feelings than females. Partially this may be due to the fact that language functions are specialized in the left hemisphere while the higher cognitive aspects of most emotions, especially the negative ones, arise more from the right side of the brain (Davidson, 1992; Robinson, 1996). Since females exhibit better interhemispheric communication than males (Shaywitz, 1994), they may be constitutionally more adept at explicit emotional communication. This adaptation probably reflects the special importance of such skills in the rearing of children and in language acquisition. Males often lag behind in such social abilities, but are more skilled in assuming dominant social stances, which may be enhanced by the ability to control one's emotions. Social dominance skills may come easier if one remains cognitively remote from their deeper social feelings, and males are more adept in distancing themselves from certain emotions, especially the nurturant ones.

Although subcortical systems govern our spontaneous facial expressions and our affective feelings, cortical ones control our voluntary facial gestures (Rinn,

1984). Our linguistically adept left hemisphere accomplishes most of our social communication, while our right hemisphere elaborates our deeper, more egocentric feelings (Ross et al., 1994). Because of our tendency to put forward a confident and friendly social front, we commonly smile more from the right side of our face, which is controlled by the linguistically skilled left hemisphere. The left half of our face, controlled by the right cerebral cortex, often conveys our deeper, internally focused feelings (Borod, 1993a,b). Neurological studies have also affirmed that the right side of the brain contributes more to the reception of emotional information and the expression of emotional intonations (i.e., prosody) than does the left hemisphere.

The receptive and expressive aspects of prosody are organized in brain regions in the right hemisphere that correspond to Wernicke's and Broca's areas on the left side of the brain (Ross, 1982). However, individuals who exhibit such expressive deficits still seem to experience emotions fairly normally. The cortex only helps decode emotional information. It is not the ultimate source of emotional feelings. Emotional feelings primarily reflect subcortical functions that typically reside at the edges of adult awareness. During development, however, emotional feelings are probably among the first types of neural functions to emerge into consciousness. This may have been as true in phylogeny as it is in ontogeny.

Thus, is it any wonder that most people, including neuroscientists and mental health professionals, commonly experience difficulty in dealing clearly and openly with emotional issues? The most emotionally deficient adults, those who are especially likely to experience psychosomatic disorders, are called *alexithymic* (Greek: *a* = lack, *lexis* = word, *thymos* = emotion)—a psychic condition characterized by the profound inability to talk about emotional matters (Lumley, Stettner, and Wehmer, 1996; Sifneos, 1966). This disorder is not accompanied by an inability to be expressive in language (i.e., to have prosody) or by the inability to elaborate the bodily signs of emotions. It is the language that is deficient. Sometimes, this deficit arises from traumatic events, which lead one to become too externally oriented. Although one still exhibits the bodily signs of emotions, the internally experienced emotional view of life has become deficient. Such symptoms may reflect extreme cortical inhibition over emotional circuits as well as a shift to extreme left hemisphere dominance, which promotes dealing with the world in more calculated, unfeeling ways (Henry, 1993).

Indeed, one of the adaptations that promoted cerebral evolution was the massive emergence of inhibition over primary-process subcortical processes. Higher cognitive activities can proceed more effectively when the primitive brain functions are restrained. When intense, primary-process emotional feelings break through regulatory defenses, which commonly become tattered by sustained stress, the road to other psychiatric disorders is paved. The most well documented, stress-induced tear in the neural fabric is the failure of negative feedback over the pituitary adrenal stress response (which transpires in a healthy paraventricular nucleus of the hypothalamus). This marker of abnormal brain regulation of the stress response, as evaluated through the Dexamethasone Suppression Test, commonly accompanies depression but also several other disorders (Ribeiro, et al., 1993; Rush, et al., 1996). Since it is adaptive in most situations to have well-regulated emotional responses, alexithymia notwithstanding, it is much less common for emotional underarousal to be deemed psychiatrically significant than is overarousal of emotional processes.

Because of the complexity of the underlying mechanisms, psychiatrists, psychologists, and neuroscientists have not been able to agree upon a unitary conceptualization of emotions. There are many problems with the concept of emotions (Griffith, 1997), but the three main stumbling blocks are detailed below.

The Problem of Content. Emotional systems were forged comparatively early in brain evolution, and it is difficult for us to agree upon the nature of such ancient processes. Unlike cognitive systems, which are tightly linked to information arising from the senses, the existence of emotional systems is not as unambiguously dependent on environmental events. Emotional systems are so ingrained in the brain that they can be aroused without external precipitants. Although people typically attribute their emotional feelings to external causes—after all, that is the way the underlying neural systems are designed—feelings ultimately reflect genetically provided potentials for certain types of neurodynamics within the brain.

The Problem of Complexity. Because of their importance early in brain organization, many additional layers of neural control have evolved in more recent brain areas to provide regulation over the core functions of ancient emotional systems. Because the basic emotional systems can interact with practically all other brain systems, it is difficult to distinguish clearly emotional and cognitive systems of the brain. Since they intermesh so extensively in certain areas such as frontal and temporal cortices, it may indeed be impossible to distinguish them at higher levels of the neuroaxis.

The Problem of Emotional Consciousness. Scientists have found it difficult to conceptualize how certain brain states, such as subjectively experienced affective feelings, could ever emerge from brain activities. That would require acceptance of some type of neural entity such as "the self" in brain matter. Such a possibility has not been a welcome concept in neuroscience. Likewise, because most neuroscientists have yet to come to terms with psychological processes that can be generated by neural activities, it has been especially difficult to specify what role such seemingly ephemeral subjective processes as affective states can have in the control of behavior. Of course this problem could be solved rather simply if we accept that emotional feelings mediate learning and the control of future behaviors as opposed to simply controlling immediate, unconditioned actions. We might even wish to consider how emotional feelings may emerge from the interactions of specific emotional circuits with a primitive form of neurosymbolic bodily "self-representation" within the brain (Panksepp, 1998).

In any event, we can now be confident that many specific neural systems do exist in the brain for the elaboration of a variety of distinct emotional processes. A serious scientific confrontation with the nature of the emotional organization of the brain has begun, and a credible neuroscience of emotions is finally emerging (Damasio, 1994; LeDoux, 1996; Panksepp, 1998). Accordingly, I will dispense with the older historical issues succinctly. For present purposes, suffice it to say that the serious study of brain emotional systems originated with the rejection of earlier theories (e.g., the James–Lange perspective) that suggested emotions arose indirectly from commotion within the viscera (Cannon, 1931). The recognition that specific brain circuits may be essential for emotions (Papez, 1937) was followed by a great deal of research suggesting that emotions emerge from specific midline, visceral regions of the brain that we are now accustomed to calling the Limbic System (MacLean, 1990). Along the way, it has been detailed how changes in the activity of the autonomic nervous system are essential for the manifestations of emotions (Smith and DeVito, 1984).

Although there is still considerable debate over how many core emotional systems exist in the brain and how precisely they are organized, there is growing consensus that fundamental emotional expressions and feelings arise largely from the intrinsic, genetically-dictated subcortical functions of the brain interacting with certain higher areas such as anterior cingulate and frontal and temporal cortices. There are distinct neural systems for the basic emotional responses that have been traditionally recognized down through the ages. To the best of our knowledge, we share these core systems homologously with all other mammals. In humans, however, these core systems are surrounded by a variety of cognitive controls and other higher regulatory mechanisms that permit us to respond in more sophisticated ways to the simple-minded dictates of the emotional systems we share with the other creatures. Through the study of these shared systems in carefully selected animal models, we can finally construct a deep understanding of the fundamental sources of human emotions, and thereby the nature of our primitive value systems.

To the best of our knowledge, the higher cognitive layers of the human brain cannot generate emotionality without the participation of subcortical emotional circuits. Of course the evolution of higher brain functions has imposed a new order over the ancient emotional systems. Because of the new layers of complexity, several distinct approaches can be taken to analyzing emotionality in the brain, including complex social constructivist approaches that seek to clarify the many cognitive and cultural reflections of emotional impulses. At the other extreme, neuroscientists, using various animal models, can describe the fine details of the ancient neural mechanisms for emotionality and provide new biological ways to rebalance pathological emotional impulses and moods. More recently, new brain imaging technologies are finally revealing the interactions of the various levels of control, allowing us to conceive how specific higher brain functions become imbalanced in psychiatric disorders (George et al., 1996). These technologies offer new and powerful diagnostic and outcome measures (Harris and Hoehn-Saric, 1995).

Despite all these spectacular advances, the most reasonable starting place for our discussion of the specific emotional systems is at a folk-psychological level that all people understand. When powerful waves of affect overwhelm our sense of ourselves in the world, we say that we are experiencing an emotion. When similar feelings are more tidal—weak but persistent—we say we are experiencing moods. The basic emotions come in various dynamic forms, which certainly include anger, fear, eager anticipation, joy, sadness, playfulness, and various other social emotions ranging from shame to envy to pride. Some of these states are fundamental to the organization of the nervous system (reflecting "basic emo-

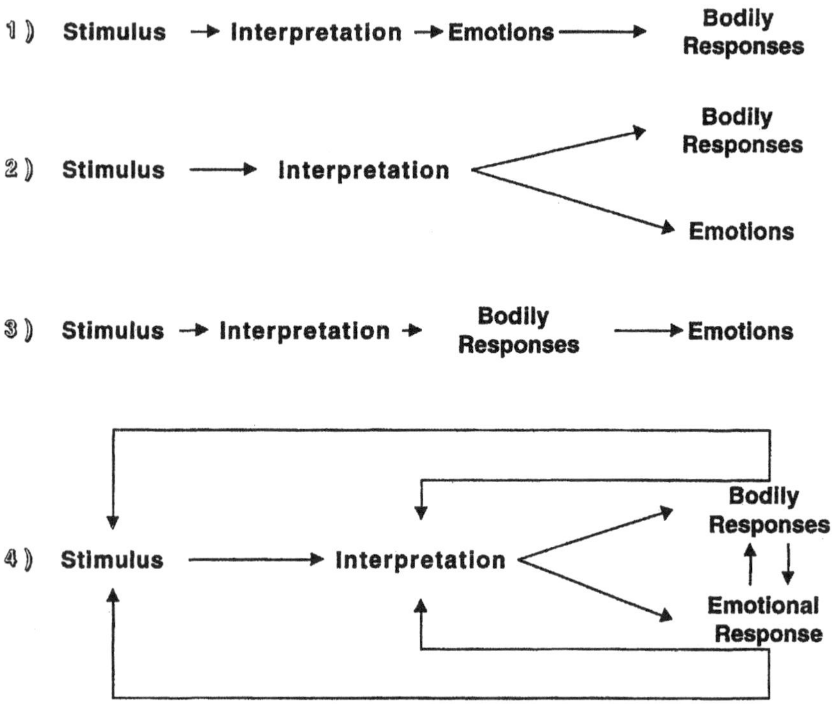

Figure 1. Four possible ways of viewing the role of affective consciousness in the generation of adaptive behavior in emotional situations: (1) the "commonsense" view that emotions cause bodily responses; (2) the possibility that the two are independently but concurrently organized; (3) the counterintuitive James-Lange type of view, that emotions arise by the way we bodily respond in emotional situations; and (4) a more realistic view which suggests that al levels of information processing in the generation of emotional responses interact with each other. Although these schemes suggest that stimulus "interpretation" is important for emotions, it is also possible to evoke emotions directly by artificially activating certain brain circuits.

tions") while others are derived via social learning. Some claim that an understanding of emotional feelings is essential for us to understand the organizations of the brain (Panksepp, 1998), while others assert that a focus on feelings in emotion research is a counterproductive distraction (LeDoux, 1996). If one takes the former view, there is a possibility of understanding how affect is elaborated in the brain; if one takes the latter, the issue is deemed an unproductive line of inquiry. Accordingly, there is still considerable debate over these issues as well as many other controversies (see Ekman and Davidson, 1994). The various conceptual ways in which one can conceive of conscious emotional feelings in the flow of how emotions are regulated are summarized in Figure 1.

ON THE DEFINITION OF EMOTIONS

One traditional barrier to progress in this area has been the absence of an adequate scientific definition of "emotions." Although we all have a basic instinctive understanding of what it means to be under the sway of emotional storms, this offers only marginal guidance for systematic, empirical investigations of the brain. Of course the importance of definitions in the analysis of the various basic (genetically-provided) psychobehavioral systems of the brain can be overstated. Adequate definitions for such evolutionarily constructed brain processes can only emerge toward the end of our empirical analyses rather than at the beginning, and they are more likely to be composed of lengthy treatments as opposed to short dictionary-type descriptors. At present, the basic emotional systems of the brain need to be provisionally recognized as certain types of dynamic neural representations, and we only need objective indicator variables (operational definitions) to proceed with the needed empirical clarification of the underlying neural system (Panksepp, 1991).

A reasonable working definition for emotional systems would consist of a specified set of shared neural attributes such as those summarized in Table 1. Virtually all investigators agree that emotions are powerful and coherent functions of the nervous system that tend to compel our actions, modify the activities of our internal organs, modify our judgments and perceptions about events, and produce a variety of strong internal feelings, both positive and negative. This last property—the subjective feeling state of emotions—remains difficult to capture in the butterfly net of objective measurements. Human self-reports, as systematized through various mood scales, and object/place preference measures in animals are the best we can do. This supreme attribute of emotions is also, of course, the main motivator for many individuals to seek psychiatric help. Although it has commonly been claimed that first-person experiences cannot be clarified through the third-person methodologies of science, psychobiology and biological psychiatry have now matured as experimental disciplines to a point where such neurodynamic processes of the nervous system can be approached empirically with the guidance of inductive observations and theoretical inference (Panksepp, 1998).

Table 1. Neurally Based Definition of Emotional Processes in the Brain

1. The underlying circuits are genetically prewired and designed to respond unconditionally to stimuli arising from major life-challenging circumstances.
2. The circuits organize diverse behaviors by activating or inhibiting motor subroutines (and concurrent autonomic-hormonal changes) that have proved adaptive in the face of such life-challenging circumstances during the evolutionary history of the species.
3. Emotive circuits change the sensitivities of sensory systems relevant to the behavior sequences that have been aroused.
4. Neural activity of emotive systems outlast the precipitating circumstances.
5. Emotive circuits can come under the conditional control of emotionally neutral environmental stimuli.
6. Emotive circuits have reciprocal interactions with brain mechanisms that elaborate higher decision-making processes and consciousness.

The neural complexity of these systems is finally being addressed, but we still need metaphoric images to guide our thinking. One that helps visualize the sources of complexity is the image of clusters of trees—neuronal trees with spreading axonal canopies and dendritic roots that interact both with the soil of the body and the vagaries of the external environment. For instance, emotional operating systems have deep roots in the lower brainstem and spinal cord, coherent trunks of neural pathways in the mesencephalon and diencephalon, and a spreading canopy of processes that innervate the limbic system and certain cortices, especially the frontal, motor-planning areas of the brain. However, information does not just go in one direction in these systems. Everything is bidirectional, and each emotional system is hierarchically organized, with higher functions being more dependent on lower functions than vice versa. Because of such levels of control, we can have many different views concerning the nature of emotions. The higher ramifications of each system interact with specific world events, while the lower roots provide a varied but integrated orchestral output for each emotional response. The middle level of analysis provides the clearest view of the various basic emotions that have discrete trunk lines—distinct neural circuitries to impose a symphonic coherence on both higher and lower functions. An analysis at this middle level is beginning to reveal the command neurochemistries for discrete emotions, and they appear to be peptidergic (Panksepp, 1993). That type of knowledge will eventually yield a new generation of medications for the various emotional disorders. For instance, most investigators presently believe that the a nonpeptide corticotrophin release hormone antagonist will be a remarkably effective drug in helping regulate various stress and distress disorders (Chalmers et al., 1996).

This arboreal image can easily incorporate other conceptualizations of emotions such as *dimensional* ones, where the various feelings we can experience lie dispersed along several orthogonal processes, such as approach–avoidance or the degree of arousal on one axis and the type of emotional valence (positive to negative) on the other (Lang, 1995; Lang et al., 1993). These dimensions reflect various types of neurodynamics that can be elaborated by underlying neural circuits, and they may reflect interactions with nonspecific neurochemical systems such as those summarized in Figure 1. Cognitive approaches can also be incorporated through a focus on the complexities of the arboreal (cortical) canopies of each emotional system; this allows us to appreciate how emotional states fluctuate as a function of time, as a function of minor changes in events, and especially as a function of our changing appraisal of these events (Mandler, 1975; Lazarus, 1991).

Although certain categorical emotional processes do exist as real systemic entities within the brain, in the absence of a complete and satisfactory knowledge, it remains rather difficult to speak about those processes unambiguously. Accordingly, the lexical approach that I have favored is the use of a universally accepted folk-psychological classifications approach that includes anger, fear, grief, and desire as major systems. One key research aim should be to try to match up such conceptual entities with the emerging neuroscience evidence (Panksepp, 1998). Since the match-up must initially be imperfect, to facilitate communication, I have recommended the use of capitalized designators for the underlying neural systems (e.g., FEAR, RAGE, and PLAY circuits). Such a convention not only helps highlight the probable existence of specific types of brain operating systems that are essential for emotions, but it also helps remind us that our present knowledge is not sufficient to explicate all the major attributes of such emotional processes in either humans or other animals. Fortunately for young scientists, most of the exciting scientific work still lies ahead.

A HISTORICAL SKETCH OF PROGRESS IN THE NEUROSCIENTIFIC UNDERSTANDING OF EMOTIONS

A thumb-nail history of emotion research has already been sketched, but let me briefly summarize the specific milestones that have brought us to our present understanding.

1. The first breakthrough—the observation of intense emotional displays in decorticate animals—led to the recognition that basic brain mechanisms for certain emotions are subcortically situated. However, since decorticate animals did not always direct their temperamental energies correctly (to appropriate targets), their emotional displays were commonly deemed to be "pseudoaffective". This terminology reflected the widespread belief that such animals did not actually experience internal states that corresponded to their outward emotional behaviors. The issue of whether, and to what extent, the cortex is important for generating affective experience remains unresolved (see Gainotti and Caltagirone, 1989), but

the weight of evidence suggests that the primal integrative forces for most affective experiences emerge directly from subcortical circuits.

2. A second set of seminal findings was that surgical removal of several brain areas, including the ventromedial hypothalamus (VMH), septal area, frontal lobes, and temporal lobes, can dramatically modify emotionality in animals and humans (Aggleton, 1992). Temporal lobectomy makes animals less fearful, hypersexual, and hyperoral (the Kluver Bucy Syndrome). Frontal lobe lesions can make animals more placid (they generally seem to live in the present moment, without much thought about past or future) but also promote simple-minded emotional outbursts when the animal is thwarted. Septal lesions produce hyperemotional/hyperaggressive animals as do VMH lesions, but in the latter case, individuals remain persistently savage while the rage of septal animals diminishes gradually over time (indeed, the animals eventually seem to become friendlier than normal).

3. An especially influential series of findings (starting with the 1949 Nobel laureate Walter Hess' work in Zurich) was the highly replicable observations that restricted electrical stimulation of a variety of brain sites concentrated in the hypothalamus and midbrain could produce coordinated behavior sequences in awake animals. This suggested the animals are experiencing emotional states. Animals could be induced to act angry, fearful, curious, hungry, or nauseous, by stimulation of specific sites within these parts of the brain, which came to be called the "head-ganglia of the autonomic nervous system." This eventually led to the observation that animals come to crave stimulation of certain brain sites (e.g., they self-stimulate many brain sites, especially those along the course of the medial forebrain bundle, which courses through the dorsolateral hypothalamus), while they dislike stimulation of other sites (e.g., they would learn to terminate stimulation applied to brain sites concentrated especially in anterior and ventrolateral hypothalamus as well as the mesencephalic periaqueductal gray (PAG)). However, some of the more social emotions, such as separation-distress and playfulness are mediated more by medial thalamic circuits than hypothalamic ones. One can also obtain different forms of self-stimulation and aversion from stimulating these circuits. Overall, such work clearly indicated that "pleasure -approach" and "distress-avoidance" are elaborated by distinct brain circuits. Brain stimulation studies have provided some of the clearest insights into the nature of intrinsic emotional systems in the mammalian brain (i.e., they suggested there were executive "command" structures, or "trunk lines", for certain emotions within fairly ancient regions of the brain).

4. A more recent breakthrough has been the observation that animals, like humans, may express desire for certain pharmaceutical agents, especially opiates and amphetaminelike psychostimulants. These studies are now bringing us close to an understanding of the neurochemistry of human and animal pleasure and cravings. A detailed understanding of the brain chemistries that permit these drugs to produce their effects (e.g., the ascending dopamine systems that arise from A-10 cell groups of the ventral tegmental area, Figure 2), has opened up a

Pandora's Box of knowledge concerning ways to modify the moods and emotions of humans by pharmaceutical means.

5. Closely related to the previous achievements, neuroscientists discovered a variety of discrete biogenic-amine and acetylcholine pathways in the brain (Figure 2) that appear to mediate generalized brain functions such as arousal, attention, and stress, which are important for controlling all behaviors and all moods (Puglisi-Allegra and Oliverio, 1990). While norepinephrine promotes the processing of information in the brain by reducing background noise, acetylcholine promotes arousal and attention by increasing the meaninful signals that get through; dopamine promotes psychomotor arousal, while serotonin increases background "noise" and hence diminishes and constrains the impact of all types of information in the brain. Likewise, GABA reduces the arousability of practically all parts of the brain. It is remarkable that most of the prominent successes of biological psychiatry have arisen from our ability to manipulate these chemical systems (Leonard, 1992; Schatzberg and Nemeroff, 1995). Broadly speaking, the

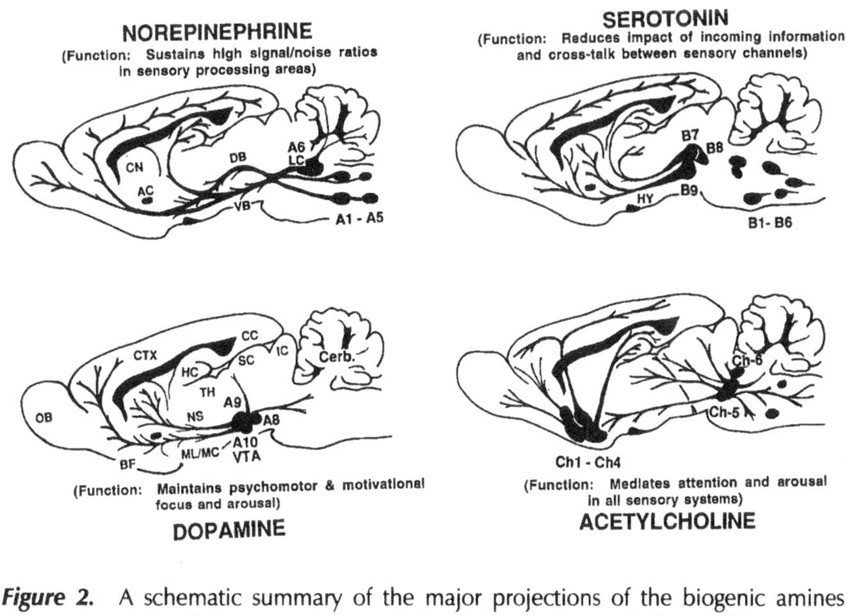

Figure 2. A schematic summary of the major projections of the biogenic amines systems (norepinephrine, serotonin, and dopamine) and acetylholine pathways on a saggital section of the rat brain. Abbreviations: CN: caudate nucleus; AC: anterior commissure; DB: Dorsal norepinephrine bundle; VB: Ventral norepinephrine bundle; LC: Locus Coeruleus; OB: Olfactory bulb; BF: Basal Forebrain; CTX: Cortex; HC: Hippocampus; TH: Thalamus, CC: Corpus Callosum; SC: Superior Colliculus; IF: Inferior Colliculus; Cerb.: Cerebellum; VTA: Ventral Tegmental Area; NS: Nigrostriatal pathway; ML/MC: Mesolimbic & Mesocortical Pathways; HY: Hypothalamus.

various antidepressants facilitate norepinephrine, serotonin, or dopamine activity, while antimanic agents may stabilize activity in these same systems. Antipsychotics dampen dopamine activity, while cognitive enhancers are being developed to facilitate acetylcholine activity. Antianxiety agents, on the other hand, generally facilitate GABA activity.

More recently a large number of neuropeptide pathways that can mediate specific psychobehavioral tendencies have been identified (e.g., Figure 3). The recent revolution in neuropeptide research suggests that the brain has specific chemical codes for various emotional tendencies. For example, cholecystokinin (CCK) can precipitate panic, oxytocin can promote nurturance, and as mentioned, CRF in the brain integrates a coherent and psychologically powerful distress response. Many other neuropeptides seem to have very discrete psychobehavioral effects. (For a detailed summary, see Panksepp, 1993.) Such findings have opened the door to the development of a new generation of psychiatric medicines, which may be able to modify distinct mood states quite specifically.

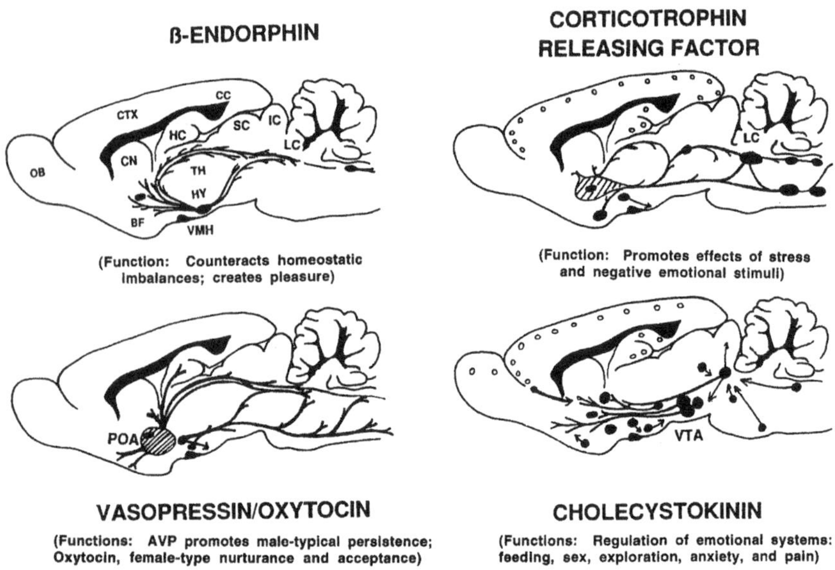

Figure 3. A schematic summary of four neuropeptide systems that have been implicated in the control of distinct emotional responses. CN: Caudate Nucleus; LC: Locus Coeruleus; OB: Olfactory bulb; BF: Basal Forebrain; CTX: Cortex; HC: Hippocampus; TH: Thalamus, CC: Corpus Callosum; SC: Superior Colliculus; IF: Inferior Colliculus; VTA: Ventral Tegmental Area; HY: Hypothalamus; VMH: Ventromedial Hypothalamus; POA: Preoptic Area.

In sum, the above work has led to a recognition that emotionality is governed largely by subcortical circuits, which can be differentiated from those that harvest the sensory information destined for thalamic and cortical processing. In other words, there is a visceral-emotional axis in the brain (PAG–hypothalamic–limbic–cingulate/frontal/temporal cortical circuits) that, besides controlling our visceral and hormonal patterns, can generate a variety of internally experienced affective feeling states and corresponding action plans (Depaulis and Bandler, 1992). This emotional brain may be contrasted to the sensory–cognitive (exteroceptive thalamic–neocortical) axis, which processes information from the external world, and hence mediates cognitive abilities. In the highest reaches of the brain, it is clear that emotionality is linked more strongly to motor systems in front of the central sulcus than to sensory systems behind it. The hypothalamic–limbic–frontal axis contains slowly firing neural systems with a great abundance of neuropeptide chemistries, while the thalamic–cortical axis exhibits much higher neuronal firing rates and, with a few prominent exceptions (e.g., somatostatin and CCK), is comparatively poor in neuropeptide systems.

Since many neuropeptides are also manufactured in various visceral organs (in the enteric nervous system), there are several potential avenues for brain–body interactions besides the classic sensory and hormonal systems. For instance, gastric hormones such as pentagastrin and CCK fragments may figure heavily in the genesis of certain emotional changes (Harro et al., 1995). The key to future progress will be the anatomical, neurophysiological and neurochemical delineation of the various basic emotional systems of the brain. The broad outlines of these systems in brain organization have now been recognized, and the detailed analysis of certain systems such as FEAR have begun in earnest (LeDoux, 1996). In general, it appears that basic emotional systems facilitate the synchronization and coordination of a variety of brain and bodily systems, including prominently various autonomic and hormonal changes. It is important to recognize that these changes enable organisms to behave in effective ways in response to the world. In other words, these systems control autonomic processes (which will not be detailed in this chapter) in order to facilitate behavioral adaptations. For example, when an individual is angry (because their freedom to act or resources have been threatened), it is useful to prepare the body for increased muscular exertion in order to compete effectively for available resources. Thus the increases in heart-rate and redistribution of blood flow from gastric organs to muscles that are triggered by arousal of anger circuits in the brain, should be seen not as the central hallmark of an emotion but only as one component of an intrinsic neurobehavioral strategy that can be aroused by a widely ramifying emotional command-circuit of the brain. Precisely, how many such circuits exist in the mammalian brain remains to be resolved.

EMOTIONAL OPERATING SYSTEMS OF THE BRAIN

According to current estimates, there are anywhere from six to a dozen core emotional circuits within the brain. Of course the number will be determined by one's willingness to include or exclude certain affective processes. The number increases as we include bodily related hungers such as thirst and sex (and their attending pleasures and displeasures, which are neuropsychic signals to help signal deviations from ideal bodily homeostasis), and the number decreases if we are unwilling to consider those as well as a variety of ambiguous categories such as disgust and surprise and other, more subtle higher processes such as guilt and shame, which may be relatively unique to humans. Here, the guiding principle for defining basic emotions will be the existence of executive circuits in the brain with certain characteristics (e.g., as summarized in Table 1), which can simultaneously synchronize a broad spectrum of coherent psychobehavioral tendencies in response to appropriate life-challenging situations. With this restriction in mind, only a handful of command systems for emotionality have presently been reasonably well identified.

As already indicated, many behavioral neuroscientists presently prefer dimensional schemes that acknowledge only two types of emotive systems in the brain—those which mediate approach and avoidance—but we can now be certain that there are several distinct types of emotional systems that can be placed under the approach and avoidance categories. The best current guide is the ability of localized brain stimulation to evoke coherent emotional displays in experimental animals (i.e., in the footsteps of the research program initiated by Hess; see "milestone" #3 of the previous section). For brain stimulation to activate the various coordinated behavior patterns (which are accompanied by affective states as indicated by behavioral approach and withdrawal tests), electrodes have to be situated in very specific parts of the visceral/limbic brain. Once an electrode is in the correct brain location, essentially the same emotional tendencies can be evoked in all mammals, including humans.

Four primal emotional circuits that mature soon after birth can presently be so designated with confidence; others for more subtle social emotions such as nurturance and playfulness are also coming to be well accepted. The most well-studied systems are (1) a FEAR system that mediates freezing and flight (which I will discuss in some greater detail in Chapter 8 of this volume; also see Burrows et al., 1990; Denny, 1991); (2) an appetitive motivation SEEKING system that helps elaborate energetic search, foraging, and a variety of goal-directed behaviors in behalf of a variety of distinct motivational systems, such as those that mediate hunger and sexual craving (Panksepp, 1982); (3) a RAGE (or defensive aggression) system that can probably modulate several forms of aggression (as discussed in Chapter 5 as well as by Bandler, 1990 and Siegel and Brutus, 1990); and finally, (4) a separation distress or PANIC system, which is especially important in the elaboration of social emotional processes related to attachment and which may help generate human grieving, sadness, depression, and other mood processes related to loss (see Panksepp, 1981; Reite and Fields, 1985; Newman, 1988). Because of the limited space available here, I will briefly focus on the first three of these systems and then spend a bit more time on the fourth—the social emotional system—which may have especially important implications for a biological psychiatry of the future. The brain systems for the mediation of sexuality, maternal nurturance, and playfulness, will also be briefly discussed. For a detailed coverage of all these systems as well as comprehensive citations to the primary research literature, see Panksepp (1998).

A FEAR System in the Brain

Although several systems for negative affect/trepidation may exist in the brain, a specific FEAR circuit was surely designed over the course of evolution to help animals reduce pain and the possibility of destruction. When artificially stimulated via implanted electrodes, this circuit leads animals to run away as if they are extremely frightened. With very weak stimulation, animals exhibit just the opposite motor tendency of freezing, a response that is common when animals are placed in circumstances where they have previously been hurt or frightened. Humans stimulated in these same brain areas report being engulfed by intense anxiety. Animals seem to experience comparable feelings, since they avoid environments where such brain stimulation has been applied. If forced to be in those environments (even in the absence of brain stimulation), they exhibit the type of "up-tight" immobility that is characteristic of frightened animals.

This circuit arises from specific subareas of the lateral and central amygdala and courses through the medial hypothalamus and down through the medial core of the brain stem known as the periaqueductal gray (for details, see Chapter 20 in the volume). Various human anxiety disorders may arise from overactivity of this brain system. Minor tranquilizers, such as chlordiazepoxide and diazepam, which can effectively dissolve anticipatory anxiety, achieve their success, in part, by reducing activity within this circuitry by increasing GABA inhibition. We cannot yet be sure which synaptic transmitters activate fear, but an endogenous anxiety-producing chemical called DBI (Diazepam Binding Inhibitor), which produces an effect opposite to that of the minor tranquilizers, may still be a key player. Other transmitters such as glutamate, acetylcholine, α-melanocyte stimulating hormone (αMSH), and corticotrophin releasing factor (CRF) can also stimulate fearlike behaviors in animals when applied directly into the brain (see Figure 3). As mentioned, intravenous application of CCK can precipitate panic attacks in humans, but it is likely that the response is distinct from the one mediated by the FEAR circuit discussed above. Indeed, it is likely that several distinct anxiety-related systems exist in the brain, but a proper taxonomy of such systems remains to be achieved.

The Appetitive Motivation SEEKING System

Another basic emotional system is the one that arouses animals to explore their world and causes them to become excited when they are searching for rewards and about to get what they desire. This system allows animals to find and eagerly anticipate the many things they need for survival, including food, water, warmth, and their ultimate evolutionary survival need, sex. Animals are especially willing to voluntarily activate this system (i.e., to show self-stimulation behavior). In humans, it may be one of the main brain systems that generates the intensity of curiosity and the eagerness of desire. When it is underactive, one result is a form of depression. An understanding of this system will have important implications for treating schizophrenia, mania, and various forms of craving, from food and drugs to gambling. When it becomes spontaneously overactive or poorly regulated, especially as reflected in elevated D_2 receptors, schizophrenia may follow—especially "functional" forms of schizophrenia in which positive symptoms (delusions and hyperemotionality) can be readily treated with many existing antipsychotic medications. This can be contrasted to more chronic forms of schizophrenia in which negative symptoms predominate (withdrawal and psychomotor retardation) that may result from brain degeneration (as indexed by ventricular enlargement). These forms of schizophrenia are more resistant to treatment (even though atypical antipsychotics such as clozapine and risperidone do counteract some negative symptoms).

A key neurochemical in the SEEKING system is dopamine, especially the dopaminergic mesolimbic and mesocortical dopamine circuits that emanate from the Ventral Tegmental Area situated at the very back of the hypothalamus (see Figure 2). These dopamine circuits tend to energize many higher brain areas that mediate planning and foresight (such as the amygdala, nucleus accumbens, and frontal cortex) and promote states of eagerness and focused plans and purposes in both humans and animals. Many peptide-containing neural circuits that secrete molecules such as neurotensin, opioids, cholecystokinin, Substance P, and other neurokinins converge on this key brain area, allowing diverse neuropsychic influences to control exploration and anticipatory eagerness. We now know a great deal about the properties of this system, especially since animals vigorously activate this circuit if given an opportunity. The phenomenon of rewarding electrical self-stimulation of the brain has led investigators to do a great deal of detailed work on the underlying circuitry in hopes of elucidating the concept of reinforcement. However, it seems certain that this circuit mediates not simply pleasure but rather an anticipatory incentive state (i.e., appetitive behavior) that normally precedes consummatory behavior. Indeed, the reinforcement itself may arise from a sudden reduction of activity in this circuit, since cessation of the appetitive phase of behavior (i.e., when consummatory behavior begins) signals biological relevance. Many psychostimulant drugs, especially amphetamine and cocaine, derive their appeal and potential to produce psychosis by temporarily overarousing this emotional system of the brain. Other drugs of addiction, such as opiates, nicotine, and alcohol, exert similar effects.

An Anger-Promoting RAGE System of the Brain

Working in opposition to the curiosity/anticipation system is one that is aroused by irritation and frustration and that mediates the anger response. It has long been known that one can enrage both animals and humans by stimulating very specific parts of the brain, which in fact parallel the trajectory of the fear system, from the medial amygdala to the PAG. This system not only helps animals defend themselves by provoking fear in other animals, but it also energizes behavior when an animal is irritated or restrained. Human anger probably gets much of its psychic "energy" from this brain system. Brain tumors that irritate this circuit often cause pathological rage, while brain damage along the trajectory of this system promotes serenity.

We know much about the anatomical details of this system, but its neurochemistries remain ambiguous. Acetylcholine, acting on muscarinic-type receptors, is certainly important in helping trigger this emotional system, and glutamate, the most abundant excitatory transmitter in the brain, facilitates rage (as it does every other emotional behavior that has been studied). A distinct anger transmitter remains to be identified, but the possibility that Substance P is a key component has recently emerged. Arousal of many neurochemical systems, including serotonin, norepinephrine, GABA and oxytocin, can make animals more peaceful, and there is now a new class of drugs, not yet approved for clinical practice, called "serenics"(the main example of which is *eltoprazine*) that selectively reduces aggression by apparently increasing brain serotonin 1A- and 1B-receptor activity.

The Separation Distress (PANIC) and Social Bonding Systems of the Brain

To be a mammal is to be born socially dependent. Brain evolution has provided safeguards to assure that parents (especially mothers) take care of the offspring, while the offspring have powerful emotional systems to indicate that they are in need of care. One of the main behavioral outputs of this system is a form of crying, the emission of separation calls. I will discuss these systems in greater detail because they are the most recent ones to receive substantial attention from psychobiologists, and this knowledge should have especially profound implications for biological psychiatry, because so many emotional disorders are ultimately related to feelings of social loss as well as to flaws in the ability to relate socially.

Furthermore, the psychotherapeutic enterprise is ultimately a social process. The personal qualities of a therapist, especially his or her sense of self-mastery and faith in the clients' unawakened potentials, are as important as the specific behavioral and cognitive techniques used in promoting positive personal change. A therapist who knows how to approach another person at a deeply empathetic level, and is able to engage in a dialogue that is firmly rooted in affective affirmation, is most likely to provide substantive help. Indeed, the phenomena of faith-healing and placebo-effects in humans may tap similar processes (since they appear to be accompanied by opioid release in the brain). The neurochemical dynamics that control social affect probably reflect the generalized nature of social emotional systems that we share with other animals. This may also explain why pet-assisted therapy is often quite effective in mild cases of emotional distress.

Neuroscience work on social-affect systems began with the assumption that the heart of such a system lies in the circuitry that mediates the separation-call. It was assumed that this system originally evolved from those that modulated pain, for the social bond is first established with caretakers who alleviate distress. Since it was assumed that the perception of social loss carried the messages of pain within its deep neurochemical structure (indeed the semantics of social loss are similar to the semantics of pain—it hurts to lose someone), it was hypothesized that brain opioids would inhibit separation distress as effectively as they controlled pain. In fact, opioids turned out to be remarkably effective in reducing such distress.

There were other reasons to believe that opioids might be important in the construction of social bonds. The dynamics of social emotions suggest that there are underlying similarities between the brain substrates that support narcotic dependence and those for social dependence. Social bonding and habitual opiate use share three main features: (1) an initial addiction euphoria-attraction phase, (2) a tolerance-habituation phase whereby the effects of the narcotics diminish spontaneously as a function of time, as does the attractive impact of social interactions (which may promote weaning and other forms of species dispersal, including potential human practices such as divorce), and (3) a powerful withdrawal phase when the object of attachment is lost, reflecting a background of endogenous neurochemical "dependence" that characterizes the social bond. Key predictions that have been experimentally confirmed are that opiate receptor agonists should diminish social motivation while opiate antagonists should intensify social motivations. β-Endorphin is the most powerful endogenous opioid that can do this.

Evidence indicates that brain opioids can inhibit separation distress in both young and mature animals and that brain opioids participate in the gratifications that arise from many social interactions. Opioid systems of young animals are quite active in the midst of rough-and-tumble play. Likewise, when adult monkeys share friendly time grooming each other, their brain opioid systems are aroused. The auditory system is also rich in opioid receptors, which may help mediate the attractive quality of certain familiar sounds (perhaps the specific sound of a loved one). In addition, as discussed more fully below, the reward of sexual gratification is, at least in part, due to opioid release within the brain. From all this, it is tempting to hypothesize that one reason certain people become addicted to opiates is because they are able to replace with drugs the pleasures normally derived from social interactions. Indeed, opiate addiction in humans is most common in environments where social isolation and alienation are endemic.

Experimentally, it is possible to increase opiate self-administration in animals simply by separating them from companionship.

In sum, it is now clear that positive social emotions and social bonds are, to some extent, mediated by opiate-based addictive processes in the brain, and this knowledge should have profound implications for understanding certain psychiatric disorders. For instance, it is commonly believed that autistic children have a constitutional inability to establish emotional bonds with other people. Indeed many of the symptoms of autism resemble behavioral changes produced by exogenous opiates—including diminished pain sensitivity, decreased responsivity to social isolation, potentiation of various stereotyped behaviors, and other abnormalities, leading to the first psychopharmacological therapy for humans developed on the basis of preclinical work that sought to understand emotional systems in the mammalian brain (for a review, see Panksepp et al., 1991).

Although opioids are exquisitely effective in alleviating separation-distress in all species that have been tested, there are many other chemistries that can modulate this emotional system (Panksepp, 1991), and presumably each has implications for understanding psychiatric disorders. Various chemistries can amplify the separation response, including corticotrophin releasing factor (CRF), which has a gross anatomical trajectory that resembles the neuronal circuits containing β-endorphin (Figure 3). Indeed, both of these chemistries course through major brain areas where localized electrical stimulation can activate the separation call (including the bed nucleus of the stria terminalis, the dorsomedial thalamus, and the central gray of the mesencephalon). Separation distress can also be activated by intraventricularly administered glutamate analogs, especially those that activate kainate and NMDA receptors, as well as by centrally administered curare, which may also exert its effect via interactions with the glutamate receptor system. In addition, other neurochemistries can powerfully inhibit the separation call, including somatostatin and oxytocin. Let us briefly focus on oxytocin, since it has recently emerged as a preeminent social neurohormone in the brain (see Pedersen et al., 1992).

Considering the importance of oxytocin for maternal nurturance (see below), it was anticipated that this molecule might also inhibit the negative emotions that rise from separation. Perhaps both mother and child derive physical and psychological pleasure from their ability to promote a mutual release of oxytocin in each others brains during the act of nursing. Indeed, oxytocin and its ancestral molecule, vasotocin (but not vasopressin, which differs from vasotocin by a single amino acid), are both extremely powerful inhibitors of the separation call, further affirming that social comfort was produced by the same brain chemistries that help mediate maternal behaviors. In this context, it also becomes noteworthy that the Proline-Leucine-Glycine (PLG) tail of the oxytocin molecule can modulate the sensitivity of brain opioid systems. As is well known, organisms typically exhibit tolerance to opiates; thus, addicts need to administer ever increasing amounts to obtain the same psychological response. Both oxytocin and PLG can inhibit opioid tolerance, a dynamic that may provide a way for the maternal experience to sustain social reward from the release of endogenous opioids within the brain. It would be disastrous for the future of any mammalian species if mothers lost their ability to experience intense social reward when their offspring were still quite young.

Additional Social Emotions: Maternal Nurturance and PLAY

Clearly, a momentous evolutionary passage occurred when animals capable of caring for their young evolved on the face of the earth, but how could nurturance have evolved from a state of nonnurturance? We cannot go back in evolutionary history, but we now know the script was written with ancient chemistries, such as vasotocin, which mediated egg-laying in reptiles. The pattern of hormone changes in the body that precedes parturition (decreasing progesterone, with rapidly increasing estrogen and prolactin) arouses brain systems that control maternal urges. At this point, it should come as no surprise that one system genetically aroused by this pattern of hormone change is the oxytocin system (through both the increased transcription from the oxytocin gene and a proliferation of oxytocin receptors in specific parts of the brain). Thus, at birth, oxytocin promotes the delivery of infants by its effects on the proliferation of oxytocin receptors on uterine smooth muscles, and concurrently, the mother's psyche is given a nurturant boost, making childcare a more attractive proposition for the mother than it may have been before. If brain oxytocinergic synapses are blocked at the onset of the first delivery (a manipulation that does not impair the ongoing process of birth), maternal behavior is weak and inconsistent, at least in rats. However, oxytocin receptor antagonists only function when administered during the first delivery, before the mother has developed maternal habits. Likewise, oxytocin administered into the brain can precipitate nurturant behavior in virgin female rats—if testing conditions are optimal. Specifically, virgin rats must be primed with estrogen (to proliferate oxytocin receptive fields in the brain) and the oxytocin works best if the potential step-mothers cannot smell the pups (since nonmaternal females normally dislike the smell of rat pups and oxytocin apparently is not the endogenous factor that counteracts this aversion at the time of birth).

At the present time, both oxytocin and opioid systems appear to be prime movers in the construction and maintenance of social bonds. Indeed, animals do prefer to spend more time with others in whose presence they have received either oxytocin or opioid infusions. It seems almost as if friendships are cemented by the same chemical systems that mediate maternal urges and alleviate separation distress. Perhaps this is one of the primitive emotional reasons why we are more likely to help family and friends than strangers. Not only do we feel better about them than strangers, but we also have their subtle ways of being, their faces, and their voices engraved in our memories.

How does this engraving process occur? Again, there is evidence that low levels of oxytocin (as well as vasopressin), can facilitate social recognition in animals. It is also known that animals are less likely to harm relatives than strangers, especially when it comes to their offspring. In rats, administration of oxytocin reduces infanticide, as well as practically all other forms of aggression. Since it has been found that sexual experiences promote oxytocin synthesis within the male brain, it would be predicted that availability of sex, at least within their own family situation, might make males less aggressive. Indeed, it has now been shown that sexual activity can diminish the tendency of male rats to exhibit infanticide (especially at the time the offspring would be born—three weeks after impregnation), but it remains to be proved that this effect is mediated by oxytocin. Likewise, it is reasonable to hypothesize that the elevated tendency of sexually experienced males to exhibit maternal nurturance in various species may be due to the elevated oxytocin activity within their brains. In short, the tendency of mammals to exhibit cooperative behaviors and to develop friendships probably reflects, at least in part, the neurochemical dynamics of such intrinsic social emotional systems of the brain.

Even within the context of friendship, however, there is always competition, and there appear to be social emotional systems, specifically social play systems, which facilitate vigorous competition and may help generate joy and feelings of exhilaration. During childhood, specific brain circuits for playful/joyful behaviors are especially strongly expressed in the brain. Although the importance of play has long been recognized in human development, the basic brain substrates for play have only recently received attention from neuroscientists (MacDonald, 1993). Because of newly developed animal models, which allow for rigorous, well-controlled experiments on rough-and-tumble social play, new insights into this important process are gradually emerging. Touch is an especially important sensory stimulus for triggering play, and it is clear that playful animals have specific sensory areas on the body that resemble "tickle skin" in humans (i.e., around the rib-cage), stimulation of which facilitates playful moods. Indeed, the brain areas that receive and process touch messages (e.g., the parafascicular area of the posterior thalamus and somatosensory cortex) are presently the ones that have been most clearly implicated in the organization of play. Although a variety of neurochemistries, including brain opioids and acetylcholine, control the instigation of playful activities, our understanding of the underlying neural systems remains rudimentary. However, perhaps we can now understand why one cannot physically tickle oneself--tickling is an innate social response that is tuned to the presence of a playful interaction with another individual. Indeed, we have recently discovered that even laboratory rats exhibit a tickling-induced chirping response (that may be functionally related to human laughter; Panksepp, 1998).

In addition to these social emotional systems, several others may exist, but the evidence is even more sketchy. For instance, the influence of PLAY systems on brain activity appears to diminish as a function of age, and it is possible that these impulses gradually come to be expressed in dominance urges, which are prevalent in nature and certainly figure in the power politics and sport-loving tendencies of human beings. It may also be that there are dominance-promoting systems other than those that arise from early playfulness. Considering the importance of material resources in evolutionary fitness, it is likely that there are specific brain systems for hoarding and possessiveness. Such behaviors are easily measured in animals. For instance, when one rat is given a special treat and another rat tries to take it away, the owner shows a rapid pivoting response—of turning a "cold-shoulder"—to that kind of approach. Does human greed emerge partially from such circuits? We do not presently know, but a full understanding of these types of behavior in animals is bound to have important implications for the understanding of normal as well as abnormal human behaviors.

SEXUALITY SYSTEMS OF THE BRAIN

One dimension of emotionality that emotion researchers commonly ignore, although no discussion can be complete without it, is sexuality. Sexuality is the source for some of the most powerful human feelings, and we finally have a good understanding of these systems in the animal brain (Crews, 1987, Levay, 1992). There has been great resistance to accepting the implications of this work for the human condition; that may largely represent a well established societal stance of artificially separating the present cultural condition of humans from our animal past. Of course the variety of sexual strategies in nature is vast, and the underlying brain details will vary accordingly. For instance, the neural timing of sexual receptivity is bound to be different between seasonal breeders who must advertise their receptive status and those species who, through the social benefits of hidden estrus, remain receptive throughout the year. However, we can still anticipate that general principles of sexuality will still be conserved despite such differences. The notion that human sexual behavior is largely a matter of choice as opposed to biology, is an unfortunate exaggeration preferred by many psychologists (e.g., Bem, 1996) that is preventing a natural integration of biological knowledge derived largely from animals into our thinking about the underlying matters.

Male and female sexuality are subservient to distinct brain controls, although they also share some influences. If all biochemical events go according to schedule during the neonatal organizational phase of gender determination, the male brain is masculinized *in utero* by the timed secretion of testosterone after its conversion to estrogen via aromatization. The female fetus is protected from masculinization by prophylactic molecules, such as α-fetoprotein, which help sequester the influence of maternal estrogens that would otherwise tend to masculinize the brain. To be masculinized means that certain areas of the brain, especially specific nuclear groups in anterior hypothalamus, grow larger than is typical in females (through the slowing of selective neuronal death as well as via the neurotrophic effects of estrogen). In rats, the major specific nuclear groups that are masculinized are called the sexually dimorphic nuclei of the preoptic area (SDN-POA), and in humans they are called the Intermediate Nuclei of Anterior Hypothalamus (INAH). The magnitude of hypertrophy of these brain areas in male rats is much greater than human males, which suggests that behavioral sex differences are magnified in rodents as compared with humans. In both species, however, these hypertrophied hypothalamic circuits surely participate in the elaboration of male-typical sex behaviors. Thus, preoptic hypothalamic damage has a much greater deleterious effect on male sexual behavior than female behavior. As we will see, neural changes in the ventromedial hypothalamus are essential for female receptivity but not for that of males. In other words, the male brain is also defeminized by early testosterone secretions, a process that results in active suppression of female sex potential (Becker et al., 1992).

The brain organizational effects of early hormone secretions go a long way toward explaining homosexual tendencies, for the hormones that ultimately trigger the organization of the male brain (testosterone aromatized to estrogen) are distinct from those that trigger the organization of the male body (testosterone 5α-reduced to dihydrotestosterone (DHT)). Due to this branching of control factors for brain and body organization, it is quite possible for the male body to contain a female-type brain, and for a female body to contain a male-type brain. Indeed, it has been repeatedly shown in animal models that maternal stress can hinder the normal process of brain masculinization by desynchronizing the underlying physiological processes (neonatal testosterone is secreted too early, before receptors are available to receive the message), and stress also impairs aromatase activity, retarding conversion of testosterone to estrogen.

These different gender potentials in the brain, laid down during fetal development, are brought to life (activated) by the maturation of gonadal steroid activity during puberty. To have a male brain means many things, but one of the best established effects in rats is the higher prevalence of arginine-vasopressin (AVP) circuits in males than in females. These circuits are under the tonic influence of testosterone, for the genetic expression of brain AVP and brain estrogen receptors diminish dramatically following castration, although these neurochemical deficits can be restored by testosterone injections. AVP appears to intensify male sexual arousal in many ways, leading to increased behavioral persistence, especially in matters of sex, as diverse as the marking of territories to sex-related aggression. Without brain AVP, which is naturally low in certain genetic strains of animals (e.g., the Brattleboro rat), male sexual behavior is sluggish. In human plasma, AVP is elevated during the arousal phase of masturbation but it declines rapidly at orgasm. Oxytocin, on the other hand, remains low during sexual arousal but is released vigorously during orgasm. As discussed below, oxytocin in the brain can facilitate both male and female sexual behavior, while AVP only promotes male behavior. From the perspective that male and female sexual behavior is differently organized in the brain, it is noteworthy that female sexual behavior is selectively diminished when AVP is infused into their brains while oxytocin can increase sexual behavior in both males and females.

Although female sexual behavior is also partially controlled by the preoptic area, such control appears to emerge from nearby zones other than the SDN-POA that are larger in females than males, and where a key chemistry is Luteinizing Hormone–Releasing Hormone (LH–RH), which can selectively increase female libido. In addition, the area of the brain just caudal to this zone, namely the ventromedial hypothalamus, is critical for sexual receptivity in females. Lesions of these areas dramatically diminish female sex behavior without much effect on male behavior. Indeed, hormonally induced receptivity (i.e., estrogen injections for several days followed by progesterone a few hours before behavioral testing) leads to dramatic anatomical and neurochemical changes in the medial hypothalamus. The prevailing neurochemical principle appears to be oxytocin. Hormone priming (just like normal estrus) leads to a proliferation of oxytocin receptors in the medial hypothalamus as well as an expansion of the dendritic fields that physically expand toward the incoming oxytocinergic innervation. The genetic expression of oxytocin is also under the positive control of estrogen. Female receptivity can be markedly increased by administering oxytocin into various brain areas which normally contain oxytocin circuits (but only if the females have been adequately primed beforehand with estrogen) and sexual receptivity is compromised by administration of oxytocin antagonists. Male sexual behavior is also strongly diminished with these antagonists.

Thus, the distinct male and female sexual urges appear to come together with oxytocin secretion. As already mentioned, male ejaculation and orgasm is accompanied by peripheral (and perhaps central) oxytocin secretion, although it is not yet clear whether this mediates the actual experience of orgasm or the "after-glow", when sexual activity is inhibited. It may participate in both, for oxytocin administered into the brain of male rats can provoke erections, but it can also prolong the sexual refractory period following ejaculation. There have also been occasional reports that intranasal oxytocin is able to facilitate sexual performance in humans. Although much more needs to be learned about these fascinating systems before useful interfaces to psychiatric issues can be formulated, it is likely that basic sexual urges are controlled by these neurochemistries in both rats and humans.

Finally, it should be emphasized that the location of maternal behavior circuits remains closely intermeshed with those that control sexuality in many areas of the brain, especially the anterior hypothalamus. It is possible that the gratification of maternal behavior arose from preexisting circuits that already mediated sexual attraction and pleasure prior to the evolution of the "social bond" that came to characterize mammalian species. This confluence may lend some credence to widely-debated Freudian notions of infantile sexuality, but those types of connections require exceedingly large inferential leaps. Although future progress in deciphering the ancient neurosymbolic functions of the brain will require daring theoretical inferences, to be useful, they should be constructed in ways that are empirically testable.

To what extent do the animal studies summarized above have implications for the human condition? It is fair to say that we can not know for sure until the corresponding evidence is obtained from humans (which in most cases will prove to be exceedingly difficult). However, if one ascribes to evolutionary principles (e.g., that evolution can only build on preexisting solutions), then it seems unlikely that the foundation principles for the basic emotions and motivations could have been discarded or replaced in the short evolutionary time that separates existing mammalian species. The fine details of the underlying mechanisms are bound to be different, and the massive human cortex is bound to put a new twist on the old solutions that arise from the subcortex, but until demonstrated otherwise, it is wiser to assume that the general principles abide. While new levels of cortical control that exist in the human brain can surely provide overriding principles, the power of subcortical emotional circuits surely remains decisive in the emotional quality of human lives (LeVay, 1993; Hamer and Copeland, 1994).

HOW DOES THE CORTEX PARTICIPATE IN THE GENERATION OF EMOTIONS?

Although subcortical areas of the brain appear to contain the executive systems for the basic mammalian emotions, to appreciate the full complexity of human emotions, we must also consider the role of higher cortical processes. Unique emotional complexities arise from the massive cortical evolution in hominids, and the utility of lower animal models in clarifying those processes is limited.

Emotional subtleties, ranging from the appreciation of art to the need to express oneself through creative acts, arise largely from neocortical functions, but the key question is whether the cortex acts as more than an initial information processor in the elaboration of emotions (e.g., as the initiator of appraisal processes that have access to subcortical emotional circuits). A great number of studies employing localized electrical stimulation of the neocortex have found no clear affective changes during such experiences. On the other hand, the more ancient paleocortical areas (in cingulate, frontal, and temporal limbic areas that receive strong inputs form brainstem circuits) do help encode affective states. For instance, seizures emerging from temporal areas are commonly accompanied by a variety of affective auras and behaviors (MacLean, 1990).

The neocortex probably participates in emotions in at least three general ways. Firstly, in order to function effectively in the elaboration of cognitive/rational processes, the neocortex exerts tonic inhibition over lower circuits that might otherwise trigger excessively impulsive and unproductive behaviors. The relative lack of such downward inhibition may help explain the hyperkinetic symptoms of children with attention deficit disorders, as well as the impulsivity of young children in general. Secondly, various subcortical circuits (Figure 2), including emotional ones (Figure 3), can exert powerful effects over neocortical functions. How each emotional system actually modifies and disrupts cognitive processes, however, remains an open area of inquiry. Thirdly, it seems likely that the neocortex learns from emotional experiences. There is bound to be considerable functional localization in the cortex for the processing of emotion-specific cues and cognitive appraisals that modulate emotional processes. For instance, emotional perceptions and feelings related to positive and negative social interactions appear to find a locus of control within the cingulate cortex, and this helps to explain the importance of such brain areas for depressive disorders (Drevets et al., 1997; Mayberg. et al., 1997). Thus it is understandable how psychic tensions that lead to panic disorders and agoraphobia can be markedly diminished following cingulate cortex damage.

A reasonable working hypothesis is that specific emotional processes have "centers of gravity" in specific cortical areas. In other words, different cortical modules may preferentially process specific types of environmental representations (e.g., angry faces, sad voices, looming objects, etc.) and thereby help generate the appraisals (the decisions and judgments) that precipitate different emotions. Thus, neural computations that lead to fear and anger appear to be heavily represented in the temporal lobe. It is to be expected that various perceptions that are especially effective in arousing these emotions are processed in temporal cortical areas with strong excitatory connections (learned as well as instinctual) to specific regions of the amygdala. Likewise, it is generally accepted that the frontal lobes help to anticipate events and hence help to generate expectancies and foresights about the world. People with frontal lobe damage exhibit deficits in planning, and consequently live at a "lower" level of psychic existence. It is noteworthy that the left frontal lobes participate more in positively valenced expectancies, while right frontal areas are more involved in negatively valenced expectancies (Davidson, 1992).

The newer brain imaging techniques, including modern computational analysis of EEG tracings, will eventually provide better answers to such localization issues, but such experiments will first require the development and implementation of new procedures for the generation and control of emotions in laboratory settings (see Clynes and Panksepp, 1988). The utilization of music may be an especially effective way to achieve this, but practically nothing is known yet about the manner in which music comes to arouse emotions in people as effectively as it does. Attempts to find specific higher areas that mediate the effects of happy and sad music have not yet been found (Panksepp and Bekkedal, 1997).

ON THE COMPLEXITY OF HUMAN EMOTIONS AND PSYCHIATRIC IMPLICATIONS

Obviously human emotional subtleties are more complex than can be captured in animal models. However, that complexity may not emerge from any fundamental difference in the principles by which the basic emotional systems operate but rather from the interactions of basic emotional systems within the higher cognitive reaches of the human brain (see Gray, 1990; Wagner and Manstead, 1990). In newborn humans, emotional responsivity is comparatively simple (as in "lower" animals), but through life experience, our emotional repertoire can become as complex as our most subtle psychological processes.

Accordingly, several levels of analysis need to be pursued in the laboratory as well as in the clinic: At the most primitive level, one will need to approach emotions as basic systems of the brain, while at a higher level, we need to recognize that many different components of the nervous system can be coordinated by emotional experiences. At still higher levels, we need to recognize that emotional manifestations in artistic and cultural expressions can be constructed according to sociocultural and individual existential standards. Although these levels of analysis have traditionally vied for primacy as mutually exclusive ways of approaching the question of emotions, we should recognize that they can all work together (Zahn-Waxler et al., 1986). While psychosocial approaches can help describe the diversity of emotional manifestations in the everyday world, the neural level of analysis can provide useful new biomedical approaches for the modification and regulation of emotions with pharmacological and other somatic interventions such as rapid Transcranial Magnetic Stimulation (rTMS) (George et al., 1996). We can only hope that future generations of investigators will continue to bring these levels of analysis closer together rather than farther apart.

Indeed, the complex interactions are such that even discarded Freudian notions of how the psyche is organized may end up being compatible with the findings of modern neuroscience. For instance, the concept of childhood sexuality is beginning to make some sense as we recognize that infantile social bonding and adult sexual behaviors and social attachments are cemented, in part, by common systems such as brain oxytocin circuits (Insel, 1992; Nelson and Panksepp, 1996). This does not mean that such a concept can be translated readily and precisely, but it should be recognized that the subcortical localization of emotional circuits resembles quite well the Freudian notion of subconscious psychic energies that control cognitive development. One of Freud's staunchest associates and best biographers, Ernest Jones, highlighted such a view long ago when he suggested that Freud's approach

> "...has permitted access to a hitherto veiled part of the mind, designated the Unconscious, and the explorations thus carried out have yielded information of very considerable value about the unsuspected significance of this more emotional region of the mind. It would appear from these investigations that man is endowed with a far more intense emotional nature than is

> generally imagined, and that powerful barriers exist, the function of which is to restrain its manifestations. All the emotions of which we become aware, either in ourselves or in others, represent only trickling through from the volcanic reservoir that is pent up in the unconscious regions of the mind... The dams that impede a freer flow of emotion are the restrictions against uncurbed action that have been painfully acquired during the civilization of the race and the training of the individual, and the reason for their existence is the fact that the pent-up or 'repressed' emotional life is of a rude and savage character incompatible with the demands of civilized standards." (Jones, 1915)

Although it is no easy task to study the dynamic circuit properties of evolutionarily ancient parts of the human brain that mediate emotionality, animal models have provided a provisional outline of the hierarchical brain substrates whereby our basic animal passions are constructed. Many external experiences can trigger emotions, but only neuroscience can reveal the nature of the underlying mechanisms that ultimately allow us to feel the way that we do. Since this knowledge has to be collected in a piecemeal way, it is often difficult to envision how an individual, as a whole, fits into this body of knowledge. Because of such difficulties, many are prone to remain skeptical about empirical approaches that seek to isolate distinct brain processes from the wholes in which they are embedded, especially when it comes to matters as close to our heart as human feelings. Although there are few empirical alternatives, we should never forget that in all biological systems, the whole is always more complex than its parts. Still, the knowledge that has already been collected has profound implications for understanding human nature and its emotional disorders. We are finally in a position to do what the young Freud wished to do—to decode, in a scientifically credible way, the ancient neurodynamic structures and molecules of the psyche that create our passions and personalities. Through the continuation of such efforts, Freud's early search for a neurobiological science of the human psyche (as documented in his posthumously published *Project for a Scientific Psychology)* will gradually materialize. Once we understand the nature of the underlying circuitries, we will be in a much better position to understand the longterm consequences of positive and negative emotions on human mental and physical health (see Ryff and Singer, 1998).

SUMMARY

Evidence for basic brain systems that mediate affective-emotional processes is summarized. The key to understanding the fundamental sources of human emotions lies in unraveling the nature of a variety of subcortical neurochemical circuits that mediate spontaneous emotional behaviors in animals: freezing and flight (for fear), separation distress vocalizations (for grief), fighting and biting (for anger), rough-and-tumble play (for joy), and appetitive approach and other anticipatory behaviors (for interest/expectancy). There are also basic brain circuits that generate various affective experiences, including male and female sexuality, dominance, and perhaps greed. These systems are modulated by a variety of nonspecific neurochemical systems, including acetylcholine, dopamine, norepinephrine, and serotonin, which control arousal and attention. The arousal of specific emotions appears to be related more to discrete neuropeptide systems. Just as a detailed understanding of the former circuits served as a foundational pillar for the biological psychiatry of the past, new knowledge of how neuropeptide systems arouse distinct emotions will be a pillar for the biological psychiatry of the future.

REFERENCES

Aggleton, J.P. (Ed.)(1992). *The Amygdala.* Wiley-Liss, New York.

Bandler, R. (1988). Brain mechanisms of aggression as revealed by electrical and chemical stimulation: Suggestion of a central role for the mdibrain periaqueductal gray region. In: *Progress in Psychobiology and Physiological Psychology.* (Epstein, A.N. & Morrison, A.R., Eds.), vol. 123, pp. 67-154. Academic Press, New York.

Becker, J.B., Breedlove, S.M., & Crews, D. (Eds.)(1992). *Behavioral Endocrinology.* MIT Press, Cambridge, Massachusetts.

Bem, D.J. (1996). Exotic becomes erotic: A developmental theory of sexual orientation. Psych. Rev. 103, 320-335.

Borod, J.C. (1993a). Cerebral mechanisms underlying facial, prosodic, and lexical emotional expression: A review of neuropsychological studies and methodological issues. Special Section: Neuropsychological perspectives on components of emotional processing. Neuropsychology 7, 445-463.

Borod, J.C. (1993b). Emotion and the brain--anatomy and theory: An introduction to the Special Section. Special Section: Neuropsychological perspectives on components of emotional processing. Neuropsychology 7, 427-432.

Burrows, G.D., Roth, M., & Noyes, R., Jr. (Eds.)(1990). *Handbook of Anxiety, Vol. 3, The Neurobiology of Anxiety.* Elsevier, Amsterdam.

Cannon, W.B. (1931). Again the James-Lange and the thalamic theories of emotion. Psychological Review 38, 281-295.

Chalmers, D.T., Lovenberg, T.W., Grigoriadis, D.E., Behan, D.P., & De Souza, E.B. (1996). Corticotrophin-releasing factor receptors: From molecular biology to drug design. Trends Neurosci. 17, 166-172.

Chugani, H.T. (1994). Development of regional brain glucose metabolism in relation to behavior and plasticity. In: *Human behavior and the developing brain.* (Dawson, G. & Fischer, K.W., Eds.), pp. 153-175. Guilford Press, New York.

Clynes, M. & Panksepp, J. (Eds.) (1988). *Emotions and Psychopathology.* Plenum Press, New York.

Crews, D. (Ed.)(1987). *Psychobiology of Reproductive Behavior: An Evolutionary Perspective.* Prentice Hall, New York.

Damasio, A.R. (1994). *Descartes' Error.* G.P. Putnam's Sons, New York.

Davidson, R.J. (1992). Anterior cerebral asymmetry and the nature of emotion. *Brain and Cognition,* 20, 125-151.

Denny, M.R. (Ed.)(1991). *Fear, Avoidance, and Phobias: A Fundamental Analysis.* Lawrence Erlbaum Associates Pubs. Hillsdale, New Jersey.

Depaulis, A. & Bandler, R. (Eds.) (1991). *The Midbrain Periaqueductal Gray Matter: Functional Anatomical and Neurochemical Organization.* Plenum Press, New York.

Drevets, W.C., Price, J.L., Simpson, J.R. Jr., Todd, R.D., Reich, T., Vannier, M., & Raichle, M.E. (1997). Subgenual prefrontal cortex abnormalities in mood disorders. Nature. 386, 824-827.

Gainotti, G. & Caltagirone, C. (Eds.)(1989). *Emotions and the Dual Brain.* Springer-Verlag, Berlin.

George, M.S., Ketter, T.A., Kimbrell, T.A., Speer, A.M., Steedman, J.M., & Post, R.M. (1996). What functional imaging has revealed about the brain basis of mood and emotion. *Advances in Biological Psychiatry, Vol. 2.* (Panksepp, J., Ed.), pp. 63-113. JAI Press, Greenwich, CT.

Gray, J.A. (Ed.)(1990). *Psychobiological Aspects of Relationships Between Emotion and Cognition. A special issue of Cognition and Emotion.* Lawrence Erlbaum Associates Pubs., Hillsdale, N.J.

Griffiths, P.E. (1997). *What Emotions Really Are.* University of Chicago Press, Chicago.

Hamer, D. & Copeland P. (1994). *The science of desire: The search for the gay gene and the biology of behavior.* Simon & Schuster, New York.

Harris, G.J. & Hoehn-Saric, R. (1995). Functional neuroimaging in biological psychiatry. In: *Advances in Biological Psychiatry, Vol. 1.* (Panksepp, J., Ed.), pp. 113-160. JAI Press Inc., Greenwich, CT.

Harro, J., Vasar, E., Koszycki, D., & Bradwejn, J. (1995) Cholecystokinin in panic and anxiety disorders. In: *Advances in Biological Psychiatry, Vol 1.* (Panksepp, J., Ed.), pp. 235-262, JAI Press, Greenwich, CT.

Henry, J.P. (1993). Psychological and physiological respones to stress: The right hemisphere and the hypothalamo-pituitary-adrenal axis, and inquiry into problems of human bonding. Integrative Physiological & Behavioral Science, 28, 369-387.

Insel, T.R. (1992). Oxytocin: A neuropeptide for affiliation-evidence from behavioral, autoradiographic, and comparative studies. Psychoneuroendocrinology 17, 3-35.

Jones, E. (1915). War and individual psychology. Sociol. Rev. 8, 167-187.

Lang, P.J. (1995). The emotion probe. Amer. Psychol. 50, 372-385.

Lang, P.J., Greenwald, M.K., Bradley, M.H., & Hamm, A.O. (1993). Looking at pictures: Affective, facial, visceral, and behavioral reactions. Psychophysiology 30, 261-273.

Lazarus, R.S. (1991). *Emotion & Adaptation.* Oxford University Pess, New York.

LeDoux, J. (1996). *The Emotional Brain.* Simon & Schuster, New York.

Leonard, B.E., (1992) *Fundamentals of Psychopharmacology.* Wiley, New York.

LeVay, S. (1993). *The Sexual Brain.* MIT Press, Cambridge, Massachusetts.

Lewis, M. & Haviland, J. (Eds.) *Handbook of Emotions.* Guilford Press, New York.

Lumley, M.A., Stettner, L., & Wehmer, F. (1996). How are alexithymia and physical illness linked? A review and critique of pathways. *J. Psychosom. Res.* 41, 505-518.

MacDonald, K. (Ed.)(1993). *Parent-Child Play: Descriptions & Implications.* SUNY Press, Albany, NY.

MacLean, P.D. (1990). *The Triune Brain in Evolution,* Plenum Press, New York.

Mandler, G. (1975). *Mind and Emotion.* Wiley, New York.

Mayberg, H.S., Brannan, S.K., Mahurin, R.K., Jerabek, P.A., Brickman, J.S., Tekell, J.L., Silva, J.A., McGinnis, S., Glass, T.G., Martin, C.C., & Fox, P.T. (1997). Cingulate function in depression: A potential predictor of treatment. Neuroreport 8, 1057-1061.

Metherate, R., Tremblay, N., & Dykes, R.W. (1988). The effects of aceylcholine on response properties of cat somatosensory cortical neurons. J. Neurophysiol. 59, 1231-1252.

Nelson, E. & Panksepp, J. (1996). Oxytocin mediates acquisition of maternally associated odor preferences in preweanling rats pups. Behav. Neurosci. 110, 583-592.

Newman, J.D. (Ed.)(1988). *The Physiological Control of Mammalian Vocalization.* Plenum Press, New York.

Panksepp, J. (1981) Brain opioids—A neurochemical substrate for narcotic and social dependence. In: *Theory in Psychopharmacology* (Cooper, S.J., Ed), pp. 149-175, Academic Press, New York.

Panksepp, J. (1982). Toward a general psychobiological theory of emotions. *Behav. Brain Sciences,* 5, 407-467.

Panksepp, J. (1991). Affective neuroscience: A conceptual framework for the neurobiological study of emotions. In: *International Review of Studies on Emotion, Vol. 1.* (Strongman, K.T., Ed.), pp. 59-99. Wiley, Chichester, England.

Panksepp, J. (1993). Neurochemical control of moods and emotions: Amino acids to neuropeptides. In: *Handbook of Emotions.* (Lewis, M. and Haviland, J.M., Eds.), pp. 87-107. Guilford Press, New York.

Panksepp, J. (1998). *Affective Neuroscience: The Foundations of Human and Animal Emotions.* Oxford University Press, New York.

Panksepp, J. & Bekkedal, M.Y.V. (1997). The affective cerebral consequences of music: Happy vs. sad effects on the EEG and clinical implications. *Intl. J. Arts Medicine,* 5, 18-27.

Panksepp, J., Lensing, P., Leboyer, M., & Bouvard, M.P. (1991). Naltrexone and other potential new pharmacological treatments of autism. *Brain Dysfunction* 4, 281-300.

Papez, J.W. (1937). A proposed mechanism of emotion. *Archiv. Neurol. Psych.* 38, 725-744.

Pedersen, C.A., Caldwell, J.D., Jirikowski, G.F., & Insel, T.R. (Eds.)(1992). Oxytocin in maternal, sexual, and social behaviors. Ann. N.Y. Acad. Sci. 652, .

Plutchik, R. & Kellerman, H. (Eds.) (1986). *Emotion: Theory, Research, and Experience, Vol. 3, Biological Foundations of Emotion.* Academic Press, New York.

Price, J.L., Carmichael, S.T., & Drevets, W.C. (1996). Networks related to the orbital and medial prefrontal cortex; a substrate for emotional behavior? Progress in Brain Research. 107, 523-536.

Puglisi-Allegra, S., & Oliverio, A. (Eds.) (1990). *Psychobiology of Stress.* Kluwer Academic Publishers, Dordrecht, The Netherlands.

Reite, M. & Fields, T. (Eds.)(1985). *The Psychobiology of Attachment and Separation.* Academic Press, New York.

Ribeiro, S.C., Tandon, R., Grunhaus, L., & Greden, J.F. (1993). The DST as a predictor of outcome in depression: A meta-analysis. Am. J. Psych. 150, 1618-1629.

Rinn, W.E. (1984). The neuropsychology of facila expression: A review of the neurological and psychological mechanisms for producing facial expressions. Psychological Bulletin 95, 52-77.

Robinson, R.G. (1996). Emotional and psychiatric disorders associated with brain damage. In: *Advances in Biological Psychiatry.* (Panksepp, J., Ed.), Vol. 2, pp. 27-62, JAI Press, Greenwich, CT.

Ross, E.D. (1981). The aprosodias: Functional-anatomic organization of the affective components of language in the right hemisphere. Arch. Neurol. 38, 561-569.

Ross, E.D., Homan, R.W., & Buck, R. (1994). Differential hemispheric lateralization of primary and social emotions. Neurops., Neuropsych. Behav. Neurology 7, 1-19.

Rush, A.J., Giles, D.E., Schlesser, M.A., Orsulak, P.J., Parker, C.R. Jr., Weissenburger, J.E., Crowley, G.T., Khatami, M., & Vasavada, N. (1996). The dexamethasone suppression test in patients with mood disorders. J. Clin. Psychiat. 57, 470-484.

Ryff, C.D. & Singer, B. (1998). The contours of positive human health. Psych. Inquiry 8, (In press.)

Schatzberg, A.F. & Nemeroff, C.B. (Eds.)(1995). *The American Psychiatric Press Textbook of Psychopharmacology.* American Psychiatric Press, Washington, DC.

Shaywitz, B.A., Shaywitz, S.E., Pugh, K.R., Constable, R.T., Skudlarski, P., Fulbright, R.K., Bronen, R.A., Fletcher, J.M., Shankweller, D.P., Katz, L., & Gore, J.C. (1995). Sex differences in the functional organization of the brain for language. Nature 373, 607-609.

Siegel, A. & Brutus, M. (1990). Neural substrates of aggression and rage in the cat. In: *Progress in Psychobiology and Physiological Psychology.* (Epstein, A. & Morrison, A., Eds.), Vol. 14, pp., 135-223, Academic Press, New York.

Smith, O.A. & DeVito, J.L. (1984). Central neuronal integration for the control of autonomic responses associated with emotions. Ann. Rev. Neurosci. 7, 43-65.

Sifneos, P.E. (1996). Alexithymia: Past and present. Am. J. Psych. 153, 137-142.

Wagner, H. & Manstead, A. (Eds.)(1990). *Handbook of Social Psychophysiology*. Wiley, Chichester, England.

Waterhouse, B.D., Azizi, S.A., Burne, R.A., & Woodward, D.J. (1990). Modulation of rat cortical area 17 neuronal responses to moving visual stimuli during norepinephrine and serotonin microiontophoresis. Brain Res. 514, 276-292.

Zahn-Waxler, C., Cummings, E.M., & Iannotti, R. (Eds.)(1986). *Altruism and Aggression: Biological and Social Origins*. Cambridge University Press, Cambridge, England.

Chapter 3

The Sexual Brain

WILLIAM BYNE and EILEEN KEMMETHER

Biological Psychiatry, pages 59-86.

ISBN: 1-55938-819-6

INTRODUCTION

As the organ of psychology, the brain is clearly the organ of sexuality, but it also can be considered a sex organ in the true physiological sense. Not only does it modulate the autonomic aspects of sexual arousal, but by regulating the pituitary gland, it influences the maturation and function of the gonads and the timing of puberty. In addition to maintaining the sex organs and secondary sex characteristics, the hormones secreted by the gonads have several actions on the brain. In laboratory animals, these actions cause certain portions of the brain to develop differently in males and females. Over the past two decades, much has been learned about the cellular and molecular mechanisms by which those sex differences develop. The hormonally mediated sexual differentiation of the brain is responsible, in part, for the subsequent differences in the behavioral repertoires of males and females. Whether or not the human brain exhibits sex differences comparable to those described in animals remains a matter of ongoing research and debate. This chapter will examine the various roles of the brain in sex and sexuality and review the research on sexual differentiation of the human brain particularly as it pertains to current theories of sexual orientation and gender identity.

THE BRAIN AND GONADAL ACTIVITY

The testes and ovaries secrete several hormones that are frequently referred to as the "sex hormones" because of their roles in sexual development, fertility, and the maintenance of the secondary sex characteristics. In addition, they act on the brain where they influence sexual behavior and motivation. The sex hormones are all steroids and thus similar to one another in chemical structure. While the androgens (e.g., testosterone) are sometimes called the "male hormones" and estrogens (e.g., estradiol) and progestins (e.g., progesterone) are called the "female hormones", all of these hormones are manufactured by both testes and ovaries and present in the bloodstreams of males and females, albeit in differing amounts (Goy and McEwen, 1980; Wilson and Foster, 1985).

The Hypothalamic-Pituitary-Gonadal Axis

Of all brain regions, the hypothalamus is the most intimately associated with the endocrinological aspects of sex. The hypothalamus is named for its position

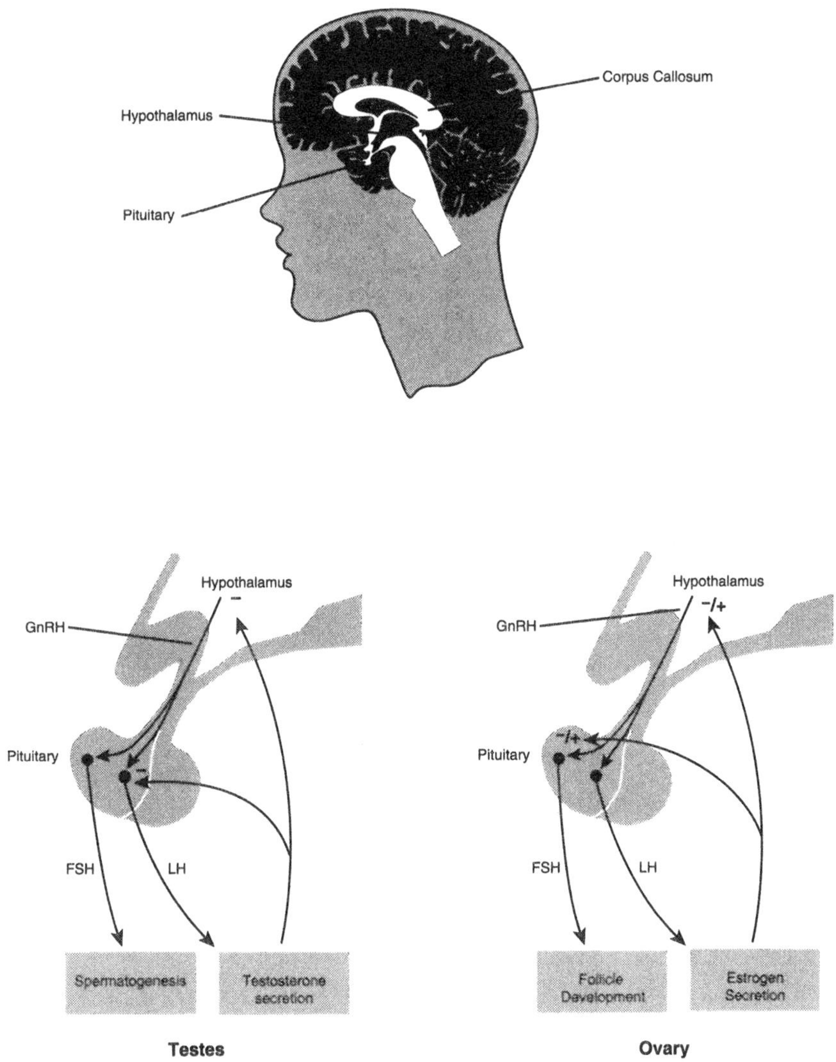

Figure 1. The hypothalamic-pituitary-gonadal axis. Testosterone exhibits only negative feedback effects at the level of both the hypothalamus and the pituitary gland. Low levels of estrogen exert negative feedback while high levels of estrogen exert positive feedback resulting in increased secretion of both LH and FSH.

below (i.e., hypo), the thalamus. It is connected to the pituitary gland by a structure called the infundibulum (Figure 1). Within the hypothalamus are many specialized cell groups called *nuclei*. Some of these nuclei produce substances called releasing hormones that travel to the anterior pituitary gland through specialized blood vessels in the infundibulum and regulate the secretion of the anterior pituitary hormones. Some of these pituitary hormones in turn regulate the secretion of other endocrine glands. Two anterior pituitary hormones, luteinizing hormone (LH) and follicle stimulating hormone (FSH), regulate the function of the gonads in both males and females (Wilson and Foster, 1985). Together, LH and FSH are called gonadotropins and the hypothalamic hormone that regulates their secretion is called gonadotropin-releasing hormone (GnRH).

Sex Hormone Systems in the Male

In the male, LH controls testosterone production while FSH and testosterone act together to regulate sperm production. Testosterone levels are held at relatively constant levels because the hypothalamus, the pituitary gland, and the testes operate in a negative feedback loop. High levels of testosterone act at the brain to inhibit GnRH secretion and at the pituitary to inhibit LH secretion. As testosterone levels fall, the inhibition of GnRH and LH secretion is reduced. This feedback loop is similar to that between a thermostat and a furnace which acts to hold room temperature constant. Testosterone levels show a daily variation (higher in the morning than at night) and they may also vary in response to psychological factors including stress (Wilson and Foster, 1985; Banks and Gartrell, 1995).

Sex Hormone Systems in the Female

Gonadal function in the female is also regulated by the pituitary hormones, LH and FSH. FSH acts to stimulate the growth of ovarian follicles, and LH stimulates the follicles to secrete estrogen. As the follicles grow larger in response to FSH, they secrete more and more estrogen. The low levels of estrogen secreted by immature follicles exert negative feedback on the brain and pituitary to inhibit both LH and FSH secretion, but when the follicles are mature and ready for ovulation they secrete high levels of estrogen.

When estrogen reaches a high threshold level in the bloodstream, it no longer exerts negative feedback effects. Instead, it does just the opposite: It exerts positive feedback effects that result in an abrupt increase in LH secretion. This abrupt and transient rise in LH is referred to as the *LH surge*. The LH surge has two effects. One is to cause the follicular membrane to rupture and release its oocyte; the other is to stimulate the remnants of the follicle (called the *corpus luteum*) to secrete progesterone (Wilson and Foster, 1985).

Development of the Hypothalamic-Pituitary-Gonadal Axis

In the human fetus, the gonads become functional before the hypothalamus and pituitary gland do. The activity of the fetal testis that initiates the differentiation of the external genitalia is not regulated by GnRH from the hypothalamus. Instead it is regulated by a product of the placenta called *human chorionic gonadotropin* (Moore, 1982). GnRH-producing cells in the hypothalamus appear by the tenth week of gestation, and the gonadotropin-secreting cells of the pituitary gland appear between the tenth and fifteenth weeks. The portal plexus that carries releasing factors from the hypothalamus to the pituitary matures around the middle of gestation and stimulates increased gonadal activity (Styne, 1994).

Gonadotropin secretion decreases toward the end of gestation presumably due to the development of inhibitory inputs to the hypothalamus from other brain regions in addition to the onset of responsiveness of the negative feedback system described above. Following a transient increase in gonadotropin secretion at birth (possibly due to release from the negative feedback effects of maternal estrogen), a sharp reduction occurs in both pituitary and gonadal activity until 10 or 12 years of age. The precise mechanism responsible for this *juvenile pause* is unknown, but it is believed to be mediated by the central nervous system (Styne, 1994).

As puberty approaches, the hypothalamus begins to secrete more GnRH, particularly at night; the pituitary becomes increasingly sensitive to the effects of GnRH and the levels of gonadal hormones in the bloodstream rise. In the later stages of puberty, daytime gonadotropin secretion increases, while the nocturnal secretion typical of early puberty decreases. The onset of positive feedback occurs later in puberty. This allows high levels of estrogen to trigger the LH surge which induces ovulation (Kulin and Reiter, 1976; Styne, 1994).

SEXUAL DIFFERENTIATION OF THE BRAIN

Rodent Studies

Most of what is known about the neurobiology of sex has been learned from experiments carried out in laboratory rodents. While findings in one species may sometimes apply to another, often they do not. One reason for this may be that females of many species will copulate only when they are physiologically ready to conceive. In such species, the hypothalamus must coordinate the physiological and behavioral aspects of reproductive behavior to maximize the odds of conception. Because the reproductive strategies of different species vary tremendously as a function of their ecological niche, it should not be surprising to find a corresponding variation in the hormonal and brain mechanisms that underlie those strategies. One must therefore be very cautious when trying to extrapolate findings from one species to another.

Rodents display sex differences in a variety of behaviors and physiological functions, not all of which seem to be directly related to sex or reproduction (Goy and McEwen, 1980). One of the more obvious sex differences is in copulatory behaviors. Males mount females, and if the females are sexually receptive (i.e., in heat) they display a posture called lordosis which allows the male to achieve intromission. While mounting and lordosis are often referred to as male and female behaviors, respectively, it is not unusual for an animal to display a copulatory behavior more typically associated with the other sex. Thus, sex differences in the display of particular behaviors tend to be quantitative rather than qualitative.

Because males and females have different amounts of particular sex hormones in their bloodstreams, one might wonder if those hormonal differences account for sex differences in behavior. That is, would giving testosterone to a female rat cause her to show male levels of mounting behavior; or would an adult male rat show lordosis if he were injected with estrogen and progesterone? The answer to those questions is, "No." The primary role of the sex steroids in adult rodents is to activate sexual behaviors, but whether or not a particular hormone will activate a specific behavior depends largely on how the brain was organized in early development. Thus, we may speak of the *activational effects* of hormones which bring preexisting brain circuits into operation, and the *organizational effects* of hormones which were responsible for wiring those brain circuits to begin with (Phoenix et al., 1959).

As we shall see, male and female rodents have differently organized brains. Organizational effects that result in the absence of female-typical behaviors, responses, or structures are referred to as *defeminizing*. Organizational effects that instate male-typical behaviors or structures are referred to as *masculinizing* (Goy and McEwen, 1980). Estrogen will activate female-typical behaviors and physiological responses more readily from animals with brains that have not been defeminized. Similarly, testosterone will activate male-typical behaviors more readily from animals whose brains have been masculinized. The organizational effects of hormones are relatively enduring compared to the activational effects, which are temporary. Nevertheless, activational effects may persist for a while after the hormone has been eliminated from the bloodstream. For example, in males of most species, mounting behavior will not stop abruptly upon castration but will gradually decrease in frequency.

Sexual Differentiation of the Rodent Brain

Although the brains of male and female rodents are organized differently in many respects, studies of sexual differentiation have concentrated largely on the origins of three differences: (1) sex differences in copulatory behaviors as discussed above; (2) sex differences in the positive feedback effects of estrogen on LH secretion (high levels of estrogen exert positive feedback on LH release in normal female rodents but not in male rodents); and (3) sex differences in brain structure. A variety of structural sex differences have been described in the rodent brain (Bleier et al., 1982). The best studied of these involves a cell group in the hypothalamus of the rat that has come to be known as the sexually dimorphic nucleus of the preoptic area (SDN-POA). This nucleus is much larger in males than in females (Gorski et al., 1978).

Table 1. Effects of Early Testosterone Exposure on Sexual Differentiation of the Brain in Genetic Male and Female Rats

Genotype	*Perinatal treatment*	*Mounting frequency*[a]	*Lordosis frequency*[b]	*LH response (-)feed back*	*LH response (+)feed back*[b]	*SDN-POA*
XY (Male)	none	high	low	yes	no	large
	castration	low	high	yes	yes	small
XX (Female)	none	low	high	yes	yes	small
	ovariectomy	low	high	yes	yes	small
	testosterone	high	low	yes	no	large

Notes: [a]In response to testosterone given in adulthood
[b]In response to estrogen given in adulthood

The rat has been a particularly useful animal to employ in studies of sexual differentiation of the brain because it is born at a very immature stage in which sexual differentiation is still in progress. This allows researchers to easily observe the impact of various experimental manipulations on sexual differentiation. As summarized in Table 1, whether a rat has male-typical or female-typical brain organization depends not on its genetic sex but on its pattern of early exposure to testosterone or its metabolites, although other factors may also be involved (Fausto-Sterling, 1992). Under normal circumstances, of course, hormonal exposure is genetically regulated.

Time Course of Sexual Differentiation

In the rat, the testes begin to secrete testosterone around the 16th day of gestation, and birth occurs at approximately 22 days. Testosterone acts on the developing brain to masculinize behavior (e.g., increase mounting) between the 18th day after conception and the fourth day after birth. Complete masculinization of hypothalamic structure requires the presence of testosterone from the 18th day of conception through the first week after birth. Defeminization of behavior (suppression of the lordosis response) and of estrogen feedback (suppression of positive feedback) occur primarily during the first 3 days after birth. Because the various organizational effects of testosterone are exerted during different (but overlapping) periods of development (Figure 2), they can be manipulated somewhat independently by differently timed hormonal manipulations (Goy and McEwen, 1980; Byne and Bleier, 1987).

Note that, in the rat, masculinization of behavior and of hypothalamic anatomy begin prenatally and extend into the early postnatal period, while defeminization of behavior and estrogen feedback occur postnatally. Thus, the brain of a male rat that was castrated at birth will be partially masculinized, but it will not be defeminized. Due to the partial masculinization, testosterone injections in adulthood would activate mounting behavior in such a male, but due to the lack of definization such a male would readily exhibit an LH surge and the lordosis response if injected with estrogen (Goy and McEwen, 1980).

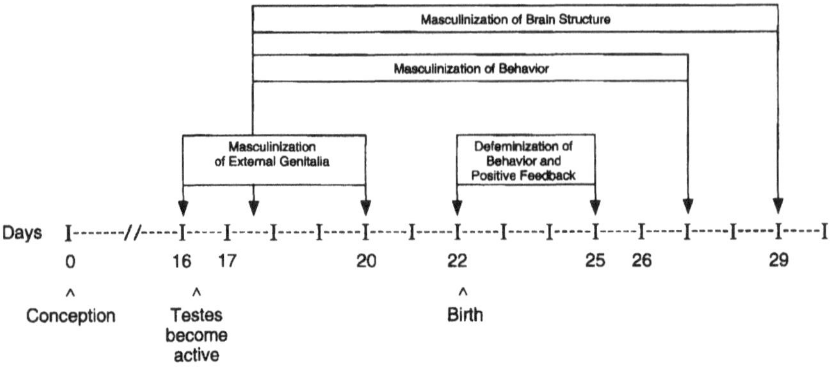

Figure 2. Approximate time course of sexual differentiation in the rat. Days refer to time after conception.

Metabolic Requirements of Sexual Differentiation

As shown in Figure 3, testosterone may influence sexual differentiation of the rodent brain by two different pathways (Goy and McEwen, 1980; Olsen, 1983). In one of these pathways, which we refer to as the *androgen pathway*, testosterone (or its derivative, 5-α dihydrotestosterone) interacts with target neurons that contain androgen receptors. In the second pathway, which we refer to as the *aromatase pathway*, testosterone is converted to estrogen by the enzyme aromatase after it enters the brain. This brain-derived estrogen exerts its effects by interacting with estrogen receptors in target cells. In some of the more commonly studied laboratory species, it appears that defeminization of the brain requires the aromatase pathway. The metabolic requirements of masculinization show more variability across species. While the aromatase pathway may be required for brain masculinization in some species including the rat, masculinization in other species occurs independent of this pathway or involves the activation of both androgen and estrogen receptors (Goy and McEwen, 1980; Olsen, 1983).

If estrogen can cause defeminization and masculinization of the brain, one might wonder why the brains of developing female rodents are not defeminized and masculinized by estrogen of ovarian origin. Part of the answer to this may be that estrogen that enters the bloodstream becomes bound to a protein called α-fetal protein, which prevents it from entering the brain (McEwen et al., 1977).

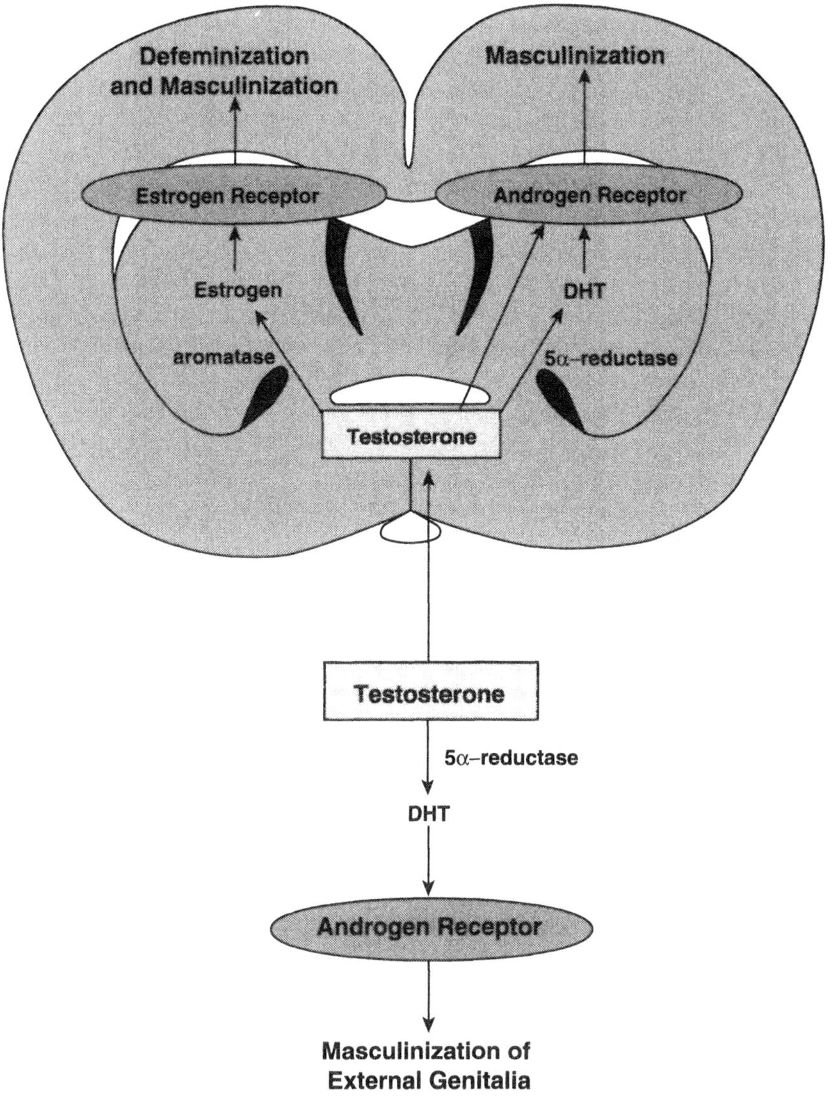

Figure 3. Metabolic pathways by which testosterone influences sexual differentiation in laboratory rodents.

Human Studies

Men and women exhibit differences in a large variety of behaviors, cognitive traits, and social roles (McGlone, 1980; Fausto-Sterling, 1992; Hyde, 1994). Most of these differences are of a statistical nature. That is, the trait may be observed to some degree in both genders, but it is more common in one than the other. For example, boys tend to do better on tests of visuospatial ability than girls, but many girls do very well on these tests and many boys do poorly. Although gender is a poor predictor of visuospatial ability and vice versa, statistically speaking, there is a gender difference in performance on tests of visuospatial ability.

The origins of psychological and behavioral gender differences are poorly understood. Three general hypotheses have been advanced to account for them. Of course, all differences need not be explained by the same hypothesis. According to the *biological* hypothesis, the human brain, like that of rats, undergoes a process of sexual differentiation in early development. Behavioral and psychological gender differences would then follow from the sexually differentiated state of the brain. According to the *psychosocial* hypothesis, these differences would be determined primarily by social forces. Advocates of psychosocial explanations emphasize that boys and girls are socialized differently from the moment of birth. The *interactional* hypothesis holds that behavioral and psychological differences are shaped by a dynamic interplay between both biological and psychosocial factors. Advocates of the interactional hypothesis emphasize that experience can produce measurable changes in brain anatomy and chemistry (Bhide and Bedi, 1984; Kraemer et al., 1984; Turner and Greenough, 1985). One can thus imagine an ongoing cycle of mutual influence in which experience alters the brain, which in turn influences behavior and the future experiences that alter the brain and so on.

We have already discussed some of the well-documented sex differences in the brains of laboratory animals. We will now examine the evidence for differences between the brains of men and women.

Functional Studies

A popular hypothesis holds that men tend to use predominantly the left hemisphere of their brains for language functions and the right hemisphere for visuospatial functions, while women tend to use both hemispheres more symmetrically (McGlone, 1980). In other words, men's brains are held to be more lateralized than those of women. Extensive reviews of the literature, however, suggest that the issue of gender differences in brain lateralization is a complicated matter that has yet to be resolved (McGlone, 1980; McKeever, 1981; Hier et al., 1994).

If language functions are indeed more lateralized to the left hemisphere in men than in women, one would predict that men would suffer more pronounced language deficits following left-sided strokes. While a number of studies have found more male than female stroke patients in language disorder clinics, most failed to adequately account for gender differences in the incidence and severity of strokes. A study that controlled for these variables found no evidence for gender differences in the cerebral lateralization of language; however, another gender difference was suggested. Compared to women with language disturbances men tended to have more extensive brain damage and the damage tended to be located more posteriorly within the left hemisphere than in women (Hier et al., 1994).

Recently developed techniques have allowed brain function to be visualized in living subjects. One study described differences in the regional activity of the brain between resting men and women (Gur et al., 1995), and another study found gender differences in subjects performing a language task (Shaywitz et al., 1995). During the language task, brain activity was more lateralized to the left hemisphere in men than in women. While functional brain imaging studies may be used to demonstrate differences in patterns of brain activity between men and women, they do not say much about the reason for those differences. Gender differences in patterns of brain activation are consistent with the hypothesis that men and women have different brains, but they are equally compatible with the hypothesis that men and women have identical brains, which they use differently. For example, a visuospatial approach to solving a particular problem would result in a different pattern of brain activity than a verbal approach to solving the same problem.

Neuroanatomical Studies

Since the middle of the last century, many reports have described sex differences in the structure of the human brain. Historically, however, such reports have had a poor track record for reliability owing in part to the difficulty of obtaining sizeable numbers of autopsied human brains that have been adequately preserved for quantitative anatomical studies (Fee, 1979; Gould, 1981; Fausto-Sterling, 1992; Byne, 1995). On the basis of this track record, it has been suggested that reports of sex differences in the human brain should be viewed skeptically until they have been replicated by at least three independent laboratories—provided there are no intervening failures of replication (see Byne, 1996). To date, no report of structural sex differences in the human brain meets that criterion except for the finding that men tend to have slightly larger brains than women. This does not mean that other structural sex differences do not exist—just that to date none has been unequivocally demonstrated. It would be surprising if no other structural sex differences exist in the human brain given that there are sex differences in virtually every other organ system, that various regions of the brain contain receptors for the gonadal steroid hormones, and that men and women have different concentrations of these hormones in their circulations throughout much of their lives.

Recent studies have described sex differences in several brain regions, including the corpus callosum, the planum temporale, the anterior commissure, the massa intermedia, two subcomponents of the bed nucleus of the stria terminalis (BNST), and three interstitial nuclei of the anterior hypothalamus (see Byne, 1995). These structures are shown schematically in Figure 4. The corpus callosum is the largest bundle of fibers that connects the right and left hemispheres of the brain. A few studies have found portions of the callosum to be larger or thicker in

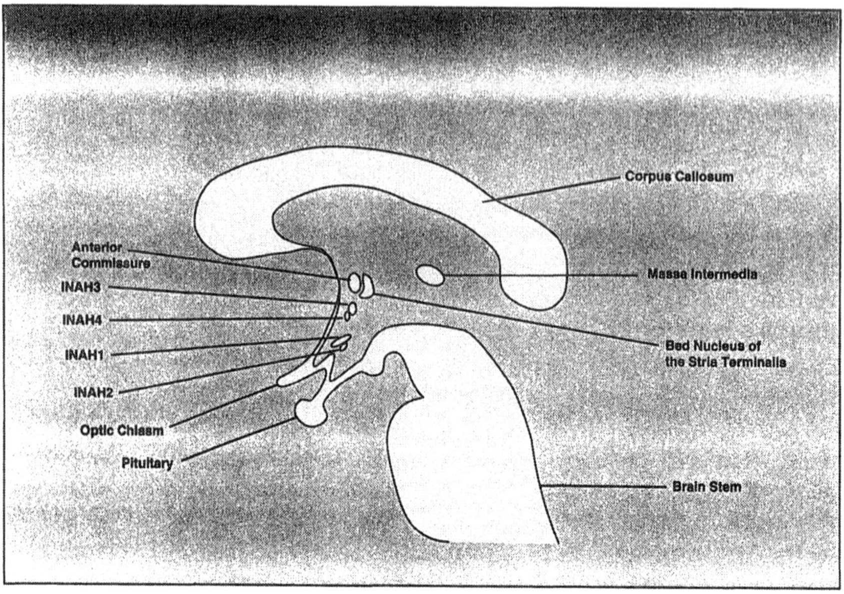

Figure 4. A profile view of several brain structures that have been reported to differ between men and women.

women than in men. If correct, those studies might suggest that more fibers interconnect the right and left hemispheres in women than in men. This increased connectivity between the hemispheres might then explain why the two hemispheres are sometimes observed to function more symmetrically in women than in men. However tantilizing, such conjectures must be viewed cautiously in light of the lack of consensus that has emerged among the large number of studies (more than 40) that have examined the corpus callosum for gender differences (Wahlsten and Bishop, 1997).

The anterior commissure is a small bundle of fibers that interconnects portions of the left and right temporal and frontal lobes. Two studies from one laboratory found it to be larger in women than in men, whereas another laboratory found it to be larger in men. The massa intermedia is a small group of nerve cells that usually crosses the midline of the brain to interconnect the left and right thalamus. This structure that is missing in perhaps 20% of individuals has been reported to be larger in women than in men. The planum temporale is a portion of the temporal lobe that is believed to play a role in language. A sex difference in the left-to-right asymmetry of this structure has been described but has yet to be replicated. The stria terminalis is a collection of fibers that connects portions of the amygdala with portions of the hypothala-

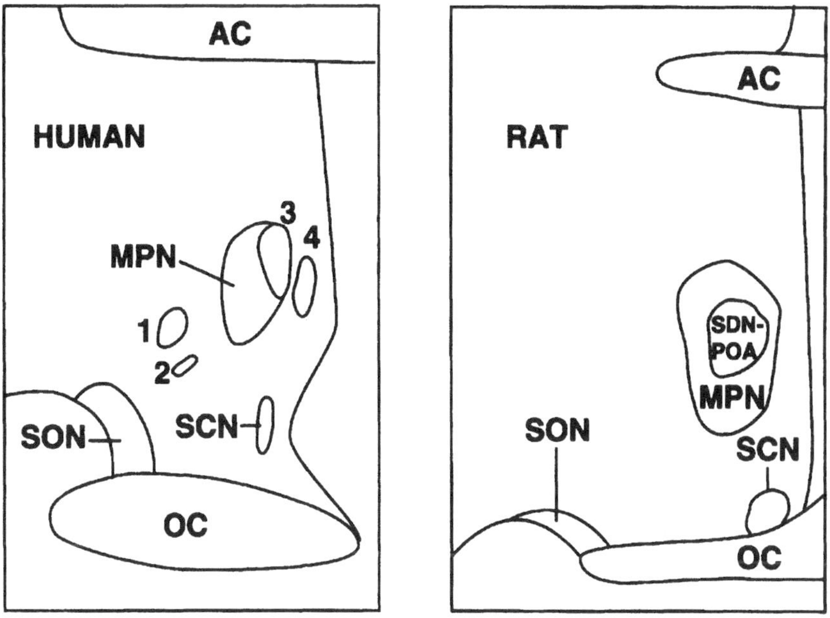

Figure 5. Comparison of the interstitial nuclei of the anterior hypothalamus (INAH) in the human with the sexually dimorphic nucleus of the preoptic area in the rat. The four interstitial nuclei in the human are indicated by the corresponding numbers. Note that INAH3, like the SDN-POA, is contained within the medial preoptic nucleus (MPN).

mus. It is surrounded by collections of nerve cells known as its bed nucleus. Portions of the BNST have been reported to vary according to sex in laboratory animals and humans. In rodents the BNST is thought to play a role in reproductive behavior such as orchestrating behavioral responses to salient olfactory stimuli. Its role in humans remains unknown. The first, second, and third interstitial nuclei of the anterior hypothalamus (INAH1, INAH2, and INAH3, respectively) have been reported to be larger in men than in women. Nothing is known regarding the function of these nuclei, although it has been suggested that one or more of them may correspond to the SDN-POA of the rat. Positional criteria suggest that INAH3 may be the best candidate for homology since, like the SDN-POA, it is situated within the medial preoptic nucleus. In contrast, INAH1 and INAH2 are situated more laterally (Figure 5).

Neurohormonal Studies

One of the most-studied sex differences in the rat brain pertains to its role in the positive feedback effect of estrogen on LH release. As discussed above, the brain of the normal female rat will support the positive feedback effects of estrogen on LH secretion but the brain of the normal male will not. This is because in the course of normal development testosterone defeminizes the feedback mechanism in male rats.

Many textbooks and popular accounts suggest that defeminization of the positive feedback mechanism also occurs in human male development; however, several lines of evidence suggest that it does not. In fact, laboratory work carried out on nonhuman primates suggests that defeminization of the positive feedback mechanism may not occur in any primates (see Byne and Parsons, 1993). Prolonged developmental exposure to testosterone defeminizes the feedback mechanism in neither genetic female monkeys (Goy and Resko, 1972) nor in human females with congenital virilizing adrenal hyperplasia (Wilkins et al., 1952). Moreover, ovarian tissue continues its cyclic pattern of hormonal secretion when transplanted into male monkeys that were castrated as adults (Norman and Spies, 1986). Developmental studies suggest that the positive feedback system matures during puberty in boys as well as in girls (Kulin and Reiter, 1976).

SEXUAL ORIENTATION AND GENDER IDENTITY

From the turn of the century into the 1970s a popular hypothesis held that the amounts of androgens or estrogens in the bloodstream of adult men and women might influence or determine their sexual orientation. This hypothesis is no longer viewed favorably because an overwhelming majority of studies failed to demonstrate a correlation between sexual orientation and adult hormone levels (Meyer-Bahlburg 1977; 1984). In fact, androgens have been found to increase libido (interest in sex) in both sexes with no effect on sexual orientation (Sherwin, 1991; Glass and Johnson, 1944). Moreover, sexual orientation has not been shown to shift in adults as a consequence of alterations in hormone levels resulting from gonadal malignancies, trauma, or surgical removal (Gooren, 1990).

Currently, the major impetus for research into a hormonal basis of sexual orientation focuses on the role of prenatal hormones. The prenatal hormonal hypothesis of sexual orientation posits that heterosexual men and homosexual women were exposed to high levels of androgens during a critical period of early development and consequently have brains that are defeminized and masculinized. Conversely, heterosexual women and homosexual men are held to have been exposed prenatally to low androgen levels and consequently have brains that retain their intrinsic female organization (Dorner et al., 1975; Gladue et al, 1984; LeVay, 1991).

It has also been proposed that prenatal hormonal exposure influences subsequent gender identity (Zhou et al., 1995). Gender identity refers to one's concept of being a man or a woman and should not be confused with or conflated with sexual orientation. Because sexual orientation and gender identity can vary independently of one another, one must propose that gender identity and sexual orientation would be sensitive to the organizing effects of androgens during different periods of development. Men who identify as women are called male to female transsexuals (MTF); women who identify as men are female to male transsexuals (FTM). Among MTF who have sex-change surgery, the majority are sexually attracted primarily to men, but a substantial minority (perhaps 40 percent) are attracted either exclusively to women or to both women and men. Similarly, a majority, but not all, FTM are sexually attracted to women (Coleman et al., 1993; Zucker, 1995).

The prenatal hormonal hypothesis of sexual orientation draws upon the observation that in rodents the balance between male and female patterns of mating behaviors is strongly influenced by the amount and timing of early testosterone exposure (Dorner et al., 1975). Extrapolating from observable behaviors in rodents to psychological phenomena in humans is problematic. The neonatally castrated male rat that shows lordosis when mounted by another male is sometimes considered to be homosexual as is the prenatally androgenized female rat that mounts others (Dorner et al., 1975). The male that mounts another male is sometimes considered to be heterosexual as is the female that displays lordosis when mounted by another female. Thus, in this particular laboratory paradigm, sexual orientation is defined in terms of specific behaviors and postures. In contrast, sexual orientation in humans is defined not by the motor patterns of copulation but by one's pattern of erotic responsiveness and the gender of one's preferred sex partner (Byne and Parsons, 1993).

Because of the problems in equating behaviors in rodents with sexual orientation, researchers have begun to employ a variety of strategies to assess partner preference in animals (Paredes and Baum, 1995). This is sometimes done by seeing whether a test animal chooses to approach a male or a female stimulus animal placed in opposite arms of a T-maze. While some unaltered laboratory animals will spontaneously direct most of their sexual behaviors toward their own sex, such studies are usually carried out on animals that have been experimentally manipulated. For example, a genetically male rodent may either be castrated as a neonate (Goy and McEwen, 1980), depriving his developing brain of androgens, or particular androgen-responsive regions of his brain may be destroyed (Hennessey et al., 1986; Paredes and Baum, 1995). In order to activate the display of female-typical behaviors and preferences in such animals, estrogen injections are required in adulthood (Goy and McEwen, 1980; Hennessey et al, 1986, Paredes and Baum, 1995). Because homosexual men and women have hormonal profiles that are indistinguishable from those of their heterosexual counterparts

(Meyer-Bahlberg, 1977; 1984), it is not clear how findings based on these hormonally abnormal animals relate to human sexual orientation.

Human Intersexes

If the prenatal hormonal hypothesis were correct, one might expect that a large proportion of men with medical conditions known to involve prenatal androgen deficiency or insensitivity would be homosexual or transsexual, as would a large proportion of women exposed prenatally to excess androgens. Because androgens are required for the development of the external male genitalia, such individuals may be born with genitals that are intermediate in morphology between those of normal males and females. Such individuals are referred to as *intersexes* because their sexual differentiation is in some respects intermediate between male and female.

Regardless of their genetic sex or the nature of their prenatal hormonal exposure, it has been reported that intersexes often become heterosexual in accordance with the gender they are assigned—provided that the gender assignment is made early and that rearing is unambiguous with respect to that assignment (Meyer-Bahlburg, 1993). It seems unlikely, however, that the rearing of individuals born with genital ambiguity is unambiguous with regard to gender (Zucker, 1995). This may be especially so when the individual has had multiple surgeries in attempt to construct more normal appearing genitalia. Even if the surgeries were performed so early that the individual has no memory of them, they may leave scars or other anomalies which give rise to concerns about gender. Moreover, parents may remain ambivalent about the gender of the child, and this ambivalence may be unwittingly and nonverbally communicated to the child. In a related case, which did not involve an intersex condition, the penis of an eight month old XY infant was damaged during circumcision and it was subsequently decided to ablate the penis and to raise the individual as a girl. Initially, this individual was described as developing into a normally functioning female; however, subsequently it was reported that the gender of rearing was rejected at puberty and that this individual switched to living as a male. While this case was cited initially in support of the theory that gender identity is determined by socialization and independent of biology, it has more recently been cited as evidence for the reverse—that gender identity is relatively independent of socialization. In actuality, however, the realities of this case are far more complex than suggested by either of those simple interpretations. Although the article describing the switch to a male gender identity is titled, "Sex Reassignment at Birth: Long Term Review and Clinical Implications", the sex reassignment was not done at birth but many months later. Moreover, the mother's ambivalence about the reassignment is evidenced by the following quotation, "As soon as he had the surgery, the doctor said I should start treating him as a girl.... But that was a disaster. I put this beautiful little dress on him and he [immediately tried] to rip it off; I think he knew it...was for girls, and he wasn't a girl" (Diamond and Sigmundson, 1987). In the mother's mind, if not in reality, at least the rudiments of male gender identity were in place prior to the reassignment. The relative contributions of biological and social factors in the outcome of this case are anything but clear.

Three intersex conditions are often discussed in the context of hormonal theories of sexual orientation and gender identity. These will be briefly reviewed here.

Congenital Virilizing Adrenal Hyperplasia. In this genetically recessive condition, the adrenal glands enlarge and over-secrete androgens beginning in fetal life. A number of studies have suggested an increased incidence of homosexual fantasy or behavior in affected women. At present, however, these findings are inconclusive due to problems in experimental design such as of finding appropriate comparison groups (Friedman and Downey, 1993). Affected individuals require ongoing medical management to control the adrenal over-secretion, and many receive repeated genital examinations. It is therefore possible that, compared to most control subjects, they would have lowered thresholds for describing sexual behaviors and fantasies to medical researchers. Moreover, even if such an increase were to be unequivocally demonstrated, it could not automatically be attributed to a hormonal effect on the brain. Psychosocial factors such as those discussed above would need to be considered.

5-α Reductase Deficiency. The enzyme, 5-α reductase, converts testosterone to dihydrotestosterone, the hormone responsible for masculinizing the external genitalia. Human males with this deficiency are born with severe genital ambiguity and are sometimes assumed to be females. If the condition is untreated, masculinization occurs at puberty: The voice deepens, the phallus enlarges, the testes descend into the labialike scrotum, and a muscular habitus develops. Despite having been raised as girls, during puberty many untreated individuals with this condition develop the gender identity and sexual orientation of heterosexual men (Imperato-McGinley et al., 1979). Taken at face value, these observations might suggest that the sex of rearing is not as important as prenatal exposure of the brain to testosterone; however, some have questioned whether these individuals were truly reared unambiguousy as girls (Zucker, 1995). If the condition is diagnosed prior to puberty and the individuals are castrated to prevent masculinization of the body, they may be more likely to develop the gender identity and sexual orientation of heterosexual women (Johnson et al., 1986). While these observations allow no definitive conclusions to be drawn concerning the relative importance of hormone exposure and sex of rearing, one possibility is that the outcome depends largely on whether or not the individual was castrated prior to puberty. It has been suggested that androgen exposure, both prenatally and at the time of puberty, is important for the evolution of male gender identity and heterosexual orientation (Imperato-McGinley et al., 1979).

Androgen insensitivity. A number of syndromes have been described in which target cells are unable to fully respond to normal or even elevated levels of androgens. This insensitivity, which may be partial or complete, may result from either a deficiency of androgen receptors or a defect in other intracellular mechanisms (see Byne and Parsons, 1993). The complete form of androgen insensitivity is sometimes referred to as the syndrome of *testicular feminization* (Tfm). In this sex-linked condition, the testes are normal but there is a congenital reduction of androgen receptors. Although affected individuals are genetically male and have abdominal testes they have normal appearing female external genitalia and are reared as girls. In the absence of androgen receptors, breast development and female fat distribution occur at puberty in response to estrogen of testicular origin. Unless the testes become prominent within the labia, they are usually left in place until after puberty so that secondary sex characteristics can develop without the need for hormonal replacement therapy. Sometimes these individuals do not come to medical attention until after puberty when they present with complaints of amenorrhea (failure to menstruate) and infertility (Moore, 1982). In addition to having normally functioning testes, humans with complete androgen insensitivity are believed to have normal levels of aromatase enzymes and estrogen receptors in their brains. Recalling that, in rats, the majority of testosterone's organizational effects on the brain are exerted through the aromatase pathway and do not require androgen receptors (refer back to Figure 3), we might predict that the brains of individuals with complete androgen insensitivity would be defeminized and masculinized. In fact, a syndrome comparable to complete androgen insensitivity has been described in mice. Genetically male mice with this syndrome have female genitalia, but defeminized and partially masculinized brains. That is, these mice look like normal females but respond behaviorally and endocrinologically more like males (Olsen, 1983).

In contrast, psychosexual assessments of humans with androgen insensitivity suggest that they are indistinguishable from heterosexual genetic females (Money et al., 1984). Some authors have conjectured that the "complete absence of masculine tendencies" in these individuals suggests that defeminization and masculinization of the human brain are independent of the aromatase pathway and are mediated via the androgen pathway instead (Goy and McEwen, 1980). However, an equally plausible hypothesis is that the absence of masculine tendencies in these individuals is a byproduct of their female appearance and rearing. Precisely because androgen-insensitive individuals are usually reared unambiguously as girls (due to the delay in diagnosing the condition), this syndrome cannot provide unequivocal evidence favoring a major role of hormones in the development of either gender identity or sexual orientation.

Hormonal Feedback Studies

Two laboratories have published evidence that homosexual men exhibit feminized positive feedback responses when injected with estrogen (Dorner et al., 1975; Gladue et al. 1984). However, other studies have found no correlation between sexual orientation and the amount of LH secreted in response to estrogen (Gooren 1986a, 1986b; Hendricks et al., 1989). The suggestion that homosexual men might have feminized feedback responses to estrogen injections is highly suspect on theoretical grounds since, as discussed above, the brain mechanism regulating LH appears to be the same in both men and women rather than taking two sexually distinct forms as in rodents (Gooren, 1986). While men and women do have different patterns of LH secretion, that is because they have different gonads and different levels of androgens and estrogens in their circulations—not because they have differently organized brains. If there is no sex difference in the human feedback mechanism to begin with, then it cannot be argued that the mechanism should be feminized in male homosexuals.

Neuroanatomical Studies

Speculation that the SDN-POA may be involved in the regulation of mounting behavior in male rats has stimulated considerable interest in finding a comparable nucleus in humans. Four candidates have been identified and designated as the interstitial nuclei of the anterior hypothalamus (INAH1, INAH2, etc.)(see Figure 4)(Allen et al., 1989). As shown in Table 2, measurement of these nuclei by different laboratories has generated inconsistent results. Nevertheless, two studies are in agreement in finding INAH3 to be larger in men than in women (Allen et al., 1989; LeVay, 1991). One of these studies also reported that in homosexual men, INAH3 was as small as in women (LeVay, 1991).

Although that study has been widely interpreted as strong evidence that the size of the INAH3 determines or influences sexual orientation, it has yet to be corroborated and a variety of problems render its evidence inconclusive (Byne

Table 2. Summary of Studies of Hypothalamic Interstitial Nuclei

	INAH1	*INAH2*	*INAH3*	*INAH4*
Swaab & Fliers (1985)	Larger in men	Not studied	Not studied	Not studied
Allen et al. (1989).	No sex difference	Larger in men	Larger in men	No sex difference
LeVay (1991).	No sex difference	No sex difference	Larger in heterosexual men than in women or gay men	No sex difference

and Parsons, 1993; Byne, 1995). Furthermore, all of the brains of gay men came from persons with AIDS. This poses an interpretive problem because at the time of death virtually all men with AIDS have decreased testosterone levels as the result of AIDS itself or of the side effects of particular treatments (Croxon et al., 1989). In some laboratory animals, the size of a structure comparable to the SDN-POA of rats is influenced by testosterone levels in adulthood as well as in early development (Commins and Yahr, 1984). Thus, it is possible that the effects on the size of the INAH3 that were attributed to sexual orientation were actually due to the hormonal abnormalities associated with AIDS. The inclusion of a few brains from heterosexual men with AIDS did not constitute an adequate control to rule out that possibility.

In addition to INAH3, single uncorroborated studies have reported that the size of the anterior commissure (Allen and Gorski, 1992) and suprachiasmatic nucleus (Swaab and Hoffman, 1990) vary with sexual orientation in men. Both of these studies relied on the brains of homosexual men who died from causes related to AIDS and are subject to some of the same interpretive problems the study of INAH3.

SEX CIRCUITS IN THE BRAIN

The sexual circuitry of the brain is exceedingly complex and poorly understood. It may be best conceptualized as involving a number of "centers" that receive sexually salient inputs from diverse brain regions and process or relay those inputs to other components of the circuit. In this sense, a "center" is merely a node within a complex circuit where connections with other brain regions converge. In order for sex behavior to occur in an appropriate context and in response to appropriate social and sensory cues, the sex "centers" must, at a minimum, receive inputs from most sensory modalities and some higher cortical centers. Because sex behavior involves motoric responses and activity of the both the sympathetic and parasympathetic divisions of the autonomic nervous system (Jacobs, 1984) at a minimum, the sex "centers" must communicate with the nuclei of the autonomic nervous system as well as those brain regions involved in the coordination and execution of movement.

Animal Studies

The medial preoptic area of the hypothalamus has been consistently implicated in the regulation of mounting behavior by a variety of experimental approaches. Destruction of portions of this region reduce or abolish mounting behavior (Arendash and Gorski, 1983), while electrical stimulation of electrodes implanted in this region has the opposite effect (Perachio et al., 1979). Recording the activity of single neurons in this region reveals that they change their firing rates during sexual activity and in response to sexually salient stimuli (Oomura et al., 1988).

The medial preoptic area contains a number of nuclei that differ in size between males and females (Gorski et al., 1978; Bleier et al., 1982; Commins and Yahr, 1984; Hines et al., 1985). It is likely that these sexually dimorphic nuclei have some sexual function. They are rich in receptors for the sex hormones, and the sex differences in their volumes depend on sex differences in early exposure to testosterone during the period of behavioral masculinization and defeminization (see Byne and Bleier, 1987). For the most part, however, the precise functions of these nuclei remain to be elucidated. The size of the much-studied SDN-POA correlates positively with the amount of mounting behavior displayed by rats: the larger the nucleus, the more mounting is displayed (Anderson et al., 1986). This correlation has generated much speculation regarding a role for the SDN-POA in the regulation of mounting. Correlation, however, does not prove causation as evidenced by the fact that the SDN-POA can be destroyed on both sides of the rat's brain without decreasing mounting behavior (Arendash and Gorski, 1983).

The precise region of the medial preoptic area that must be destroyed to impair mounting behavior is situated slightly above the SDN-POA in the vicinity of a structure called the bed nucleus of the stria terminalis (BNST). Like the SDN-POA, portions of the BNST are larger in males than in females and are rich in sex hormone receptors (Bleier et al., 1982; Hines et al., 1985). The SDN-POA may function to block the expression of lordosis in males. Following relatively large lesions that included the SDN-POA, male rats were found to exhibit lordosis in response to appropriate tactile stimulation if they were injected with estrogen and progesterone (Hennessey et al., 1996).

Another brain region consistently implicated in male-typical sex behavior is the medial amygdala (MA). This structure, which is located to the side of the hypothalamus, communicates with the BNST as well as with other regions of the medial preoptic area. Like the SDN-POA and portions of the BNST, the posterodorsal portion of the MA (MApd) is larger in male rats than in females and is rich in steroid hormone receptors (Hines et al., 1992). The BNST and MApd are components of a circuit believed to relay sexually salient olfactory information to the medial preoptic area.

The same experimental approaches that have established the importance of the medial preoptic area in male-typical sex behavior have implicated the ventromedial nucleus (VMN) of the hypothalamus in the regulation of female-typical sexual behaviors (Pfaff, 1980). In addition to impairing the lordosis response, damage to the ventrolateral portion of the VMN in rats also impairs the expression of female-typical courtship behaviors. Conversely, electrical stimulation of the VMN elicits these behaviors. Recording experiments have found that the activity of some neurons in the VMN is strongly correlated with sexual activity.

The ventrolateral VMN is rich in receptors for both estrogen and progesterone. Estrogen acts in part by inducing the VMN cells to manufacture receptors for progesterone. Following the induction of progesterone receptors, injections of progesterone into the VMN will bring female rats into heat (Cohen and Pfaff, 1992). As noted above, male rats that were castrated in adulthood can be stimulated to display lordosis if their medial preoptic area is damaged (Hennessey et al., 1986). Because the VMN receives projections from the medial preoptic area, it is possible that neurons in the medial preoptic area (possibly the SDN-POA) of normal males prevent the display of lordosis by inhibiting the VMN.

Another important nucleus for sexual functioning in female rats is the anterior periventricular nucleus (AVPVN). This nucleus is situated in the very front of the medial preoptic area, is larger in females than in males, and is rich in estrogen and progesterone receptors (Bleier et al., 1982 [AVPVN is designated MPN]; Byne and Bleier, 1987). In rats this nucleus is believed to be the primary site where estrogen exerts its positive feedback effects on LH release (Terasawa et al., 1980). Estrogen will not induce an LH surge in rats whose AVPVN has been destroyed. The site where estrogen exerts its positive feedback effects varies considerably from one species to the next. In guinea pigs, positive feedback seems to be exerted at the ventrobasal portion of the hypothalamus (Terasawa and Weigand, 1978), whereas in most primates, the pituitary gland may be the primary site of positive feedback (Karsch et al., 1973).

Human Studies

Sexual arousal is dependent upon the integrity and interaction of the neural, hormonal, vascular, and psychological systems. Studies pertaining to sexual circuitry in humans are by and large limited to observations of changes in sexuality or sexual functioning following brain damage (Miller et al., 1986). Hyposexuality is common following brain injury, but lesions seen with this condition are sufficiently variable to prevent the identification of any particular brain region responsible for the diminished sexual interest. Hypersexuality is seen less commonly and is usually seen in the context of lesions, such as those of the frontal lobes, that impair judgment and result in a general disinhibition of behavior. Consequently, it is not clear whether such lesions increase sex drive or merely decrease the inhibitions that regulate its expression. Hypersexuality may be especially common following lesions that involve or are adjacent to the medial septal area. Interestingly, patients have described experiencing sexual sensations following electrical stimulation of this region (Heath, 1964).

The new onset of overt homosexual interests has also been reported to accompany brain pathology (Cummings, 1985; Miller et al., 1986). Again, however, it is possible that these changes were the result of decreased social inhibition. For example, one description told of a married female patient who began to show sexual interest in other women following bilateral temporal lobe damage. She reportedly lost sexual interest in her husband but would comply with his sexual advances (Miller et al., 1986). It is not possible, however, determine whether the temporal lobe damage changed the woman's sexual orientation or merely produced disinhibition that allowed her to express preexisting sexual desires.

The importance of the hypothalmus in sexual functioning across species, including primate species, suggests that it is also important in human sexual behavior. While there has been much speculation regarding the role of particular portions of the hypothalamus in sexual orientation (Dorner et al., 1975; LeVay, 1991) and gender identity (Zhou et al., 1995), existing evidence allows no definitive conclusions to be drawn (Byne, 1996). At a minimum, the hypothalamus plays a role in the autonomic nervous system and the hormonal systems integral to sexual arousal and functioning. The most common sexual effect of hypothalamic damage in humans is hyposexuality (Bauer, 1958; Miller et al., 1986).

SUMMARY AND CONCLUSIONS

In addition to its psychological role, the brain plays a major role in the physiology of sex. It modulates the autonomic aspects of sexual arousal and, by regulating the pituitary, influences the maturation and function of the gonads as well as the timing of puberty. In addition to maintaining the sex organs and the secondary sex characteristics, the hormones secreted by the gonads exert many influences on the brain. To a first approximation, these influences can be described as organizational effects that determine how particular brain circuits are wired during development and activational effects that subsequently put those circuits into action.

In laboratory animals, the organizational effects of testosterone and its derivatives result in differences between the brains of males and females through the process of sexual differentiation. The extent to which the human brain undergoes a similar process of sexual differentiation remains to be determined. Similarly, whether or not sexual orientation and gender identity are influenced by the sexually differentiated state of the brain has yet to be established. Because all psychological phenomena rely on brain activity, behavioral gender differences, gender identity, and sexual orientation require a biological substrate. The fact that they require a biological substrate, however, does not suggest that they are necessarily biologically determined. The structure and function of the brain are influenced not only by biological factors but also by learning and experience. An integrated biopsychosocial perspective is necessary to fully appreciate the complexities of the human brain and its roles in sex and sexuality.

REFERENCES

Allen, L.S. & Gorski, R.A. (1990). Sex differences in the bed nucleus of the stria terminalis of the human brain. J. Comp. Neurol. 302, 697-706.

Allen, L.S. & Gorski, R.A. (1992). Sexual orientation and the size of the anterior commissure in the human brain. Proc. Natl. Acad. Sci. U.S.A. 89, 7199-7202.

Allen, L.S., Hines M., Shryne, J.E., & Gorski R.A. (1989). Two sexually dimorphic cell groups in the human brain. J. Neurosci. 9, 497-506.

Allen, L.S., Richey, M.F., Chai, Y.M., & Gorski, R.A. (1991). Sex differences in the corpus callosum of the living human being. J. Neurosci. 11, 933-942.

Anderson, R.H., Fleming, D.E., Rhees, R.W., & Kinghorn, E. (1986). Relationships between sexual activity, plasma testosterone, and the volume of the sexually dimorphic nucleus of the preoptic area in neonatally stress and non-stressed rats. Brain Res. 370, 1-10.

Arendash, G.W. & Gorski, R.A. (1983). Effects of discrete lesions of the sexually dimorphic nucleus of the preoptic area or the medial preoptic regions on the sexual behavior of male rats. Br. Res. Bull. 10, 147-154.

Banks, A. & Gartrell, N.K. (1995). Hormones and sexual orientation: A questionable link. J. Homosex. 28, 247-268.

Bauer, H.G. (1958). Endocrine and metabolic conditions related to pathology in the hypothalamus: A review. J. Nerv. Ment. Dis. 128, 323-338.

Bhide, P.G. & Bedi, K.S. (1984). The effects of a lengthy period of environmental diversity on well-fed undernourished rats. II. synapse to neuron ratios. J. Comp. Neurol. 227, 305-310.

Bleier, R. (1984). *Science and Gender: A Critique of Biology and Its Theories on Women*. Pergamon Press, New York.

Bleier, R., Byne, W., & Siggelkow I. (1982). Cytoarchitectonic sexual dimorphisms of the medial preoptic and anterior hypothalamic areas in guinea pig, rat, hamster and mouse. J. Comp. Neurol. 212, 118-130.

Byne, W. (1995). Science and Belief: Psychobiological research on sexual orientation. J. Homosexual. 28, 303-344.

Byne, W. (1996). Biology and sexual orientation: Implications of endocronological and neuroanatomical research. In: *Comprehensive Textbook of Homosexuality*. (Cabaj, R. & Stein, T., Eds.), American Psychiatric Press, Washington, DC.

Byne, W. & Bleier R. (1987). Medial preoptic sexual dimorphisms in the guinea pig. I. An investigation of their hormonal dependence. J. Neurosci. 7, 2688-2696.

Byne, W. & Parsons, B. (1993) Sexual orientation: The biological theories reappraised. Arch. Gen. Psychiatry 50, 228-239.

Cohen, R.S. & Pfaff, D.W. (1992). Ventromedial hypothalamic neurons in the mediation of long-lasting effects of estrogen on lordosis behavior. Prog. Neurobiol. 38, 423-453.

Commins, D. & Yahr, P. (1984). Adult testosterone levels influence the morphology of a sexually dimorphic area in the Mongolian gerbil brain. J. Comp. Neurol. 224, 132-140.

Croxson, T.S., Chapman, W.E., Miller, L.K., Levit, C.D., Senie, R., & Zumoff, B. (1989). Changes in the hypothalamic-pituitary-gonadal axis in human immunodeficiency virus–infected men. J. Clin. Endocrinol. Metab. 89, 317-321.

Cummings, J.L. (1985). Sexual Behavior. In: *Clinical Neuropsychiatry*. (Cummings, J., Ed.), Grune and Stratton, New York.

Demeter, S. Ringo, J.L., & Doty, R.W. (1988) Morphometric analysis of the human corpus callosum and anterior commissure. Hum. Neurobiol. 6, 219-226.

Diamond, M. & Sigmundson, H.K. (1997). Sex reassignment at birth. Long-term review and clinical implications. Arch. Pediatr. Adolesc. Med. 151, 298-304.

Dorner, G., Rhode, W., Stahl, F., Krell, L., & Masius, W.G. (1975). A neuroendocrine predisposition for homosexuality in men. Arch. Sex. Behav. 4, 1-8.

Fausto-Sterling, A. (1992). *Myths of Gender: Biological Theories about Women and Men*. Basic Books, New York.

Fee, E. (1979). Nineteenth century crainiology: The study of the female skull. Bull. Hist. Med. 33, 415-433.

Friedman, R.C. & Downey, J. (1993). Neurobiology and sexual orientation: Current relationships. J. Neuropsychia Clin. Neurosci. 5, 131-153.

Gladue, B.A., Green, R., & Hellman, R.E. (1984). Neuroendocrine response to estrogen and sexual orientation. Science 225, 1496-1499.

Glass, S.J. & Johnson, R.W. (1944). Limitations and complications of organotherapy in male homosexuality. J. Clin. Endocrinol. 4, 550-554.

Gooren, L. (1986a). The neuroendocrine response of luteinizing hormone to estrogen administration in heterosexual, homosexual and transsexual subjects. J. Clin. Endocrinol. Metab. 63, 583-588.

Gooren L. (1986b). The neuroendocrine response of luteinizing hormone to estrogen administration in the human is not sex-specific but dependent on the hormonal environment. J. Clin. Endocrinol. Metab. 63, 589-593.

Gooren, L. (1990). Biomedical theories of sexual orientation: A critical examination. In: *Homosexuality/ Heterosexuality: Concepts of Sexual Orientation.* (McWhirter, D.P., Sanders, S.A., & Reinisch, J.M., Eds.), pp. 71-87. Oxford University Press, New York.

Gorski, R.A., Gordon, J.H., Shryne, J.E., & Southam, A.M. (1978). Evidence for a morphological sex difference in the medial preoptic area of the rat brain. Brain Res. 148, 333-346.

Gould, S.J. (1981). *The Mismeasure of Man.* W.W. Norton & Company, New York.

Goy, R.W. & McEwen, B.S. (1980). *Sexual Differentiation of the Brain.* MIT Press, Cambridge, Massachusetts.

Goy, R.W. & Resko, J.A. (1972). Gonadal hormones and behavior of normal and pseudohermaphroditic nonhuman female primates. Recent Prog. Horm. Res. 28, 707-733.

Gur, R.C., Mozley, L.H., Mozley, P.D., Resnick, S.M., Karp, J.S., Alavi, A., Arnold, S.E., & Gur, R.E., (1995). Sex differences in regional cerebral gulcose metabolism during a resting state. Science 267, 528-531.

Hamer, D.H., Hu, S., Magnuson, V.L., Hu, N., & Pattatucci, A.M.L. (1993). A linkage between DNA markers on the X chromosome and male sexual orientation. Science 261, 321-327.

Heath, R.G. (1964)(Ed.) Pleasure response of human subjects to direct stimulation of the brain: Physiologic and psychocynamic considerations. In: *The Role of Pleasure in Behavior.* pp 219-243. Harper and Row, New York.

Hennessey, A.C., Wallen, K., & Edwards, D.A. (1986) Preoptic lesions increase display of lordosis by male rats. Brain Res. 370, 21-28.

Hier, D.B., Yoon, W.B., Mohr, J.P., Price, T.R., & Wolf, P.A. (1994). Gender and aphasia in the stroke data bank. Brain and Language 47, 155-167.

Hendricks, S.E., Graber, B., & Rodriquez-Sierra, J.F. (1989). Neuroendocrine responses to exogenous estrogen. No differences between heterosexual and homosexual men. Psychoneuroendocrinology 14, 177-185.

Hines, M., Allen, L., & Gorski, R.A. (1992) Sex differences in subregions of the medial nucleus of the amygdala and the bed nucleus of the stria terminalis of the rat. Brain Res. 579, 321-326.

Hines, M., Davis, F.C., Coquelin, A. Goy, R.A., & Gorski, R.A. (1985). Sexually dimorphic regions in the medial preoptic area and the bed nucleus of the stria terminalis of the guinea pig brain: A description and an investigation of their relationship to gonadal steroids in adulthood. J. Neurosci. 5, 40-47.

Imperato-McGinley, J., Peterson, R.E., Teofilo, G., & Sturla, E. (1979). Androgens and the evolution of male-gender identity among pseudohermaphrodites with 5-alpha-reductase deficiency. N. Eng. J. Med. 300, 1233-1237.

Hyde, J. (1994). *Understanding Human Sexuality.* McGraw Hill, New York.

Jacobs, J.A. (1984). Physiology of male sexual arousal. In: *Sexual Arousal.* (Ficher, M., Fishkin, R.E., & Jacobs, J.A., Eds.), Charles C. Thomas, Springfield.

Johnson, L., George, F.W., Neaves, W.B., Rosenthal, I.M., Christensen, R.A., Decristoforo, A., Schweikert, H., Sauer, M.V., Leshin, M., Griffin, J.E., & Wilson, J.D. (1986). Characterization of the testicular abnormality in 5-alpha-reductase deficiency. J. Clin. Endocrin. Metab. 63, 1091-1099.

Karsch, F.J., Dierschle D.E., & Knobil, E. (1973). Sexual differentiation of pituitary function: Apparent difference between primates and rodents. Science 179, 484-486.

Kinsey, A.C., Pomeroy, W.B., & Martin, C.C. (1948). *Sexual Behavior in the Human Male.* W.B. Saunders Co., Philadelphia.

Kraemer, G.W., Ebert, M.H., Raymond, L., & McKinney, W.T. (1984). Hypersensitivity to d-amphetamine several years after early social deprivation in rhesus monkeys. Psychopharmacology 82, 266-271.

Kulin, H. & Reiter, E.O. (1976). Gonadotropin and testosterone measurements after estrogen administration to adult men, prepubertal and pubertal boys, and men with hypogonadism: Evidence for maturation of positive feedback in the male. Pediatric Res. 10, 46-51.

LeVay, S. (1991). A difference in hypothalamic structure between heterosexual and homosexual men. Science 253, 1034-1037.

McEwen, B.S., Lieberburg, I., Chaptal, C., & Krey, L.C. (1977). Aromatization: Important for sexual differentiation of the neonatal rat brain. Horm. Behav. 9, 249-263.

McGlone, J. (1980). Sex differences in human brain asymmetry: A critical survey. Behav. Brain Sci. 3, 215-263.

McKeever, W.F. (1981). Sex and cerebral organization: Is it really that simple? In: *Sex Differences in Dyslexia.* (Ansara, A., Geschwind, N., Galaburta, A., Albert, M., & Gartrell, N., Eds.), pp. 97-108. The Orton Dyslexia Society, Towson, Maryland.

Meyer-Bahlburg, H.F.L. (1977). Sex hormones and male homosexuality in comparative perspective. Arch. Sex. Behav., 6, 297-325.

Meyer-Bahlburg, H.F.L. (1984). Psychoendocrine research on sexual orientation: Current status and future options. Prog. Brain Res. 71, 375-397.

Meyer-Bahlburg, H.F.L. (1993). Gender development in intersex patients. Child. Adoles. Clin. N. Am. 2, 501-512.

Miller, B.L., Cummings, J.L., McEntyre, H., Ebers, G., & Grodes, M. (1986). Hypersexuality or altered sexual preference following brain injury. J. Neurol. 49, 867-873.

Money, J. & Ehrhardt, A.A. (1972). *Man and Woman, Boy and Girl.* Johns Hopkins University Press, Baltimore, Maryland.

Money, J. (1987). Sin, sickness, or status: Homosexual gender identity and psychoneuroendocrinology. Am. Psychol. 42, 384-389.

Money, J., Schwartz, M., & Lewis, V.G. (1984). Adult erotosexual status and fetal hormonal masculinization and demasculinization: 46,XX congenital virilizing adrenal hyperplasia and 46,XY androgen-insensitivity syndrome compared. Psychoneuroendocrinology 9, 405-414.

Moore, K.L. (1982). The Developing Human. W.B. Saunders Co., Philadelphia.

Norman, R.L. & Spies, H.G. (1986). Cyclic ovarian function in a male macaque: Additional evidence for a lack of sexual differentiation in the physiological mechanisms that regulate the cyclic release of gonadotropins in primates. Endocrinology 118, 2608-2610.

Olsen, K.L. (1983). Genetic determinants of sexual differentiation. In: *Hormones and Behavior in Higher Vertebrates.* (Balthazart, J., Prove, E., & Gilles, R., Eds.), Springer-Verlag, Berlin.

Page, D.C. et al. (1987) The sex-determining region of the human Y chromosome encodes a finger protein. Cell 51, 1091-1104.

Paredes, R.G. & Baum, M.J. (1995). Altered sexual partner preference in male fettets given excitotoxic lesions of the preoptic area/anterior hypothalamus. J. Neurosci. 15, 6619-6630.

Perachio, A.A., Marr, L.D., & Alexander, M. (1979). Sexual behavior in male rhesus monkeys elicited by electrical stimulation of preoptic and hpothalamic areas. Brain Res. 177, 127-144.

Oomura, Y., Aou, S., Koyama, Y., & Yoshimatsu, H. (1988). Central control of sexual behavior. Br. Res. Bull. 20, 863-870.

Parsons, B., MacLusky, N.J., Krey, L., Pfaff, D.W., & McEwen, B.S. (1980). The temporal relationship between estrogen-inducible progestin receptors in the female rat brain and the time course of estrogen activation of mating behavior. Endocrinology 107, 774-779.

Pfaff, D.W. (1980). *Estrogens and Brain Function*. Springer-Verlag, New York.

Phoenix, C.H., Goy, R.W., Gerall, A.A., & Young, W.C. (1959). Organizational action of prenatally administered testosterone propionate on the tissues mediating behavior in the female guinea pig. Physiol. Behav. 12, 1045-1055.

Shaywitz, B.A., Shaywitz, S.E., Pugh, K.R., Constable, R.T., Skudlarski, P., Fulbright, R.K., Bronen, R.A., Fletcher, J.M. et al. (1995). Sex differences in the functional organization of the brain for language. Nature 373, 607-609.

Sherwin, B.B. (1991) The psychoendocrinology of aging and female sexuality. Ann. Rev. Sex. Res. 2, 181-198.

Styne, D.M. (1994). Physiology of puberty. Horm. Res. 4 (Suppl. 2), 3-6.

Swaab, D.F. & Fliers, E. (1985). A sexually dimorphic nucleus in the human brain. Science 228, 1112-1114.

Swaab, D.F. & Hoffman, M.A. (1984). Sexual differentiation of the human brain: A historical perspective. Prog. Brain Res., 61, 361-383.

Swaab, D.F. & Hoffman, M.A. (1990). An enlarged suprachiasmatic nucleus in homosexual men. Brain Res. 537, 141-148.

Terasawa, E. & Wiegand, S.J. (1978). Effects of hypothalamic deafferentation on ovulation and estrous cuclicity in the female guinea pig. Neuroendocrinol. 26, 229-248.

Terasawa, E., Wiegand, S.J., & Bridson, W.D. (1980). A role for medial peroptic nucleus on the afternoon of proestrus in female rats. Am. J. Physiol. 238, E533-E539.

Turner, A.M. & Greenough, W.T. (1985). Differential rearing effects on rat visual cortes synapses. I. Synaptic and neuronal density and synapses per neuron. Brain Res. 329, 195-203.

Wilkins, L., Crigler, J.F., Silverman, S.H., Gardner, L.I., & Migeon, C.J. (1952). Further studies on the treatment of congenital adrenal hyperplasia with cortisone. II. The effects of cortisone on sexual and somatic development with an hypothesis concerning the mechanism of feminization. J. Clin. Endocrinol. Metab. 12, 277-295.

Willmott, M. & Brierley, H. (1984). Cognitive characteristics and homosexuality. Arch. Sex. Behav. 13, 311-319.

Wilson, J.D. & Foster, D.W. (1991). *Williams Textbook of Endocrinology*. Saunders, Philadelphia.

Zhou, J., Hoffman, M.A., Gooran, J.G., & Swaab, D.F. (1995). A sex difference in the human brain and its relation to transexuality. Nature 378, 68-70.

Zucker, K.J. (1995). *Gender Identity Disorder*. Gilford Press, New York.

RECOMMENDED READINGS

Andersen, N.C. (1997). What shape are we in? Gender, Psychopathology, and Brain. Am. J. Psych. 154, 1637-1639.

Arpels, J.C. (1996). The Female Brain: Hyperestrogenic continuum from the premenstrual syndrome to menopause. J. Reprod. Med. 41, 633-639.

Byne, W. (1998). The medial preoptic and anterior hypothalamic areas of the rhesus monkey: Comparison with the human and evidence for sexual dimorphism. Brain Research 793, 346-350.

de Courten-Meyers, GM (1999) The human cerebral cortex: Gender differences in structure and function. J. Neuropathol. Exp. Neurol. 58, 217-226.

Fausto-Sterling, A (1995). Animal models for the development of human sexuality: A critical evaluation. J. Homosex. 28, 217-236.

LeVay, S. (1993). Sexual Brain. MIT Press, Cambridge, Massachusetts.

Schober, J.M. (1998). A surgeon's response to the intersex controversy. J. Clin. Ethics 9, 393-397.
Wilson, B.E. & Reiner, W.G. (1998). Management of intersex: A shifting paradigm. J. Clin. Ethics. 9, 360-369.

Chapter 4

The Biology of Personality

ROGER T. MULDER

Biological Psychiatry, pages 87-101.

ISBN: 1-55938-819-6

INTRODUCTION

The relative importance given to environmental and biological factors in determining personality has fluctuated throughout history. The western medical tradition has tended to emphasize biological factors in explaining human behavior. Many of the theories presented a simplistic and even prejudiced view of human behavior. However, there now exists an extensive body of data from disparate fields of research that supports the concept that biology contributes significantly to personality.

This chapter will briefly review the history linking biology to personality. It will then discuss the evidence for the importance of biological factors influencing behavior. Current biological theories will be examined and research on these theories presented. Finally, an attempt will be made to link biology of personality with current medical and psychiatric practice.

DEFINITION

The word "personality" has been described as one of the most abstract words in our language. Its exact meaning generally varies within the context in which it is used. The concept of personality discussed in this chapter might be more properly labeled temperament. Temperament may be seen as simple, nonmotivational, noncognitive, stylistic characteristics. Such behaviors appear early in childhood, show substantial stability over time, and reflect a significant heritable bias. This contrasts with the "higher nervous system", which through language, thought, memory, and other cognitive processes, shapes the expansion of the underlying temperamental dispositions.

MODELS OF PERSONALITY

The history of formal personality characterization can be traced to the Greeks. The two major theories to explain personality were based on the concepts of constitution and temperament. Both have continued in one form or other until the present. Constitution is equivalent to the "conformation of the body" and relates behavioral style to physical attributes. Physiognomy or "character writing" was possibly the first constitutional classification system and was brought to its finest form by Theophrastus. He presented brief descriptions delineating a particular trait to highlight major features in an individual. Many of these descriptions are often based on physical appearance of various animals (hence the expression "sly as a fox") as well as on facial expressions or racial stereotypes.

The concept of temperament has been even more influential. The Hippocratic view, later elaborated by Galen, classified behaviors into four temperaments—sanguine, phlegmatic, choleric, and melancholic. These were related respectively to excessive blood, phlegm, yellow bile, or black bile. Galen elaborated the behavioral manifestations so that sanguine temperament had a leaning toward optimism, the phlegmatic temperament was associated with an apathetic disposition, the choleric temperament was related to irascibility, and the melancholic temperament was inclined to sadness.

Attempts to appraise others by observing facial expressions, postural attitudes, and style of movement have continued until the present. The first widely practiced systematic attempt to relate psychological functioning to external morphology was phrenology, introduced by Gall in the late eighteenth century. Contending that the brain is the central organ of thought and emotion, Gall concluded that the intensity and character of thought and emotion should correlate with variations in the size and shape of the brain's encasement, the skull. Despite the transparent weakness of Gall's system, it was widely accepted for several decades. Earlier this century, Kretschmer and Sheldon developed systems of describing personality based on body shape. Sheldon's typology postulated on three basic types, endomorphy (a predominance of roundness of the body), mesomorphy (predominance of bone and connective tissue), and ectomorphy (fragile body build).

The humoral theory of temperament has had a more profound impact on medical tradition, and the principle of psychophysical substances being seen as the determinants of behavior survives in a modified form in our current era of empirical research.

While the dominant paradigm within the literate medical tradition has been that biology and personality are linked, there have always existed folk and magical views, which ascribed human behavior to external malevolent agents. By the nineteenth century, the medical literature had become reoccupied with the relative weight to be attributed to physical and to moral (roughly psychological) determinants of behavior and insanity. Some believed that moral causes, centering on an individual's lack of moderation and on excesses of all kinds were important. Others favored physical causes, especially hereditary endowment. A philosophical movement, Associationism, proposed by Hobbs, Locke and others, argued that the mind was an empty slate and that behavior were acquired due to what was seen and heard and which habits were practiced or encouraged.

These ideas were given a further boost by the growth of psychoanalysis around the turn of the century. Personality was believed to be related to the frustrations and indulgences of instinctual drives associated with specific stages of psychosexual maturation. Early behaviorists also appeared to exclude the concept of temperament altogether—the most dogmatic arguing that even human sexual behavior has no basis in biology. There was therefore no need to measure or explain concepts such as temperament or personality.

By the middle decades of this century, the medical tradition widely endorsed an overwhelmingly environmental view of personality, particularly in the United States. Neonates were considered psychological nonentities, a bundle of reflexes and therefore totally malleable. "Heredity" was widely misinterpreted to mean determined, or fated, as it sometimes is now. In addition, the distortion of racial and behavioral characteristics by Nazi ideology and the horrific consequences this brought about further strengthened the case of those who put forward an overwhelmingly environmental view of personality. Genetics steered clear of behavior even as they found gene after gene that was intimately involved in most other aspects of human metabolism and functioning. Differences in children's moods and behaviors were seen primarily the result of early family experiences. Excessive timidity or aggressiveness in a child was a sign that the family had done something wrong.The dangers inherent in a primarily psychosocial model of personality should not be underestimated. Assuming that any child can become schizophrenic and that all cases of panic or antisocial behavior are psychologically traceable to parental failures 20 years earlier is just as flawed and simplistic as the racial eugenics arguments. These theories generated a great deal of suffering and undeserved guilt in patients and their families.

THE BIOLOGY OF PERSONALITY

The change from antipathy to acceptance of biological factors with regard to personality has occurred rapidly. Today's scientific and popular presses are fueled with claims that biology plays a role in everything from sexual attraction to shyness. The reason for this shift in attitude is related to a number of factors. These include the resurgence in interest in behavioral genetics, studies of child temperament, neurochemical studies, and a number of sociopolitical factors.

Genetics

As discussed previously, the eugenics movement and Nazi Germany had meant that, for a long time after World War II, geneticists avoided the search for genes linked with behavior. Eventually the evidence became too much to ignore. The huge accumulation of data about hereditary influences in animal behavior could not be overlooked. Nor could the larger and more sophisticated twin and adoption studies of human beings that were consistently reporting that between 40 and 60% of observed variance in personality was due to genetic factors. Behavioral genetics is now a major field of research and there is general acceptance that genetics play at least some role in determining personality and behavior.

Temperament

The modern study of child temperament began with the New York Longitudinal Study initiated in the 1950s by Thomas and Chess. Rather than focusing on parent–child interactions, the study focused on individual differences in the child as an active agent in its own environment with an emphasis on emotional rather cognitive aspects of development. This study and several others made it increasingly clear that individual children differ from each other in a host of ways that can be reliably measured. These measures include concepts such as activity level, autonomic activity, behavioral inhibition, sociability, and impulsivity. These differences are relatively stable and matter in terms of later psychological development: More recent studies have shown that infants have relatively sophisticated sensory discriminations at birth, that they have learning abilities, and that they already display substantial interindividual variability.

Neurobiology

The third influence results from the study of central neurotransmitters and their relationship to observable behavior. Forty years ago, mapping behavior onto existing models of the brain was impossible. Noradrenaline and acetylcholine were the only chemicals known to act at the synapse. There is now a large literature on the chemistry of the brain that is full of apparent contradictions and lacks compelling heuristic models. Nevertheless, there is some consensus on the behavioral functions of the central neurotransmitters and explanations for the apparent contradictions are becoming clearer. Methodology is becoming more sophisticated and there is now copious evidence that altering neurochemical pathways, either pharmacologically or anatomically, affect behavior in quite specific ways.

Sociocultural

Finally, the influence of sociopolitical factors in augmenting acceptance of the importance of biology and personality should not be underestimated. There is arguably less need at present to promote a view of humankind as being shaped almost exclusively by culture. This is not to say that some social scientists and others do not still argue forcefully for the basic similarity of all human groups and the capacity of humans to change their personal qualities through experience. What is often not appreciated is that temperamental differences are malleable by experience. Temperament does not imply a rigid determinism. While major brain pathways are specified in the genome, neural connections and levels of neurotransmitters reflect socially mediated experiences as well.

THE CURRENT ERA

We have now reached a point where some matters are less controversial. Today, few would dispute that children are temperamentally different from one another and that some of this difference is genetically determined. It is also generally accepted that brain pathways affect behavior. The major challenges lie in measuring these neurophysiological changes and constructing a model of temperament with which to test possible relationships.

The logical place to look for biological correlates of temperament is in brain anatomy and brain physiology. At this point, it appears most likely that temperament will be related to physiological differences, particularly those involving neurochemistry. However, there is some evidence, in animals at least, that neuroanatomy may be linked with behavior. Domesticated strains of horses and pigs, for example, have smaller brains than wild varieties and are much less fearful. Animals who are reluctant to explore novel environments have larger hippocampus projections than those who are highly exploratory.

Neurophysiological Studies

Exploring brain physiology in humans presents a number of problems. A fundamental difficulty is that the activity of the central nervous system (CNS) pathways can only be measured indirectly. Using the electroencephalogram (EEG), early studies related electrical activity to arousal level or used evoked potentials and event-related potentials to study the nervous system's reaction and response to environmental stimuli. The studies particularly focused on measures of introversion and neuroticism and they also often used "psychopaths" (contrasting their responses to nonpsychopathic individuals). Unfortunately, despite a considerable number of studies, the results have been inconsistent. Similarly, studies of autonomic functioning, including electrodermal activity and heart rate, which have attempted to relate these to measures of psychopathy and various personality scales, have been disappointing. Although there appears to be a trend for decreased electrochemical activity and lower tonic heart rate in psychopaths, the studies remain inconclusive. Theories have been advanced to explain the variable results, but little consensus is emerging.

Neurochemical Studies

Another area of study is CNS neurochemical measures, particularly monoamine oxidase systems. Studies use cerebrospinal fluid (CSF) measures, probes that exploit monoamine-mediated functions such as hormone release, receptor studies, and timed urine collections. These indirect measures have significant shortcomings. Lumbar puncture is invasive and difficult to justify in normal subjects, and the trauma of the procedure may bias measures. Measurement of the concentration of catecholamines and their metabolites in humans has not yet yielded uniform results across laboratories.

Seasonality, time of day, diet, age, sex, and weight may also affect measures. Despite this, some results are beginning to emerge with reasonable consistency. Most studies, until now, have centered on the three main monoamines and their role as neurotransmitters.

Serotonin

Reasonable agreement has been reached on the role of the serotonergic system. Most regard it as a behavioral inhibition system, which acts in opposition to the two catecholamine systems, noradrenaline and dopamine. Human studies have linked the 5-HT metabolite (5-hydroxindol acetic acid, 5-HIAA), with measures of aggression, impulsivity, and disinhibition in humans. Low CSF 5-H1AA has been linked to violent suicide, impulsivity, arson, and aggressive behavior. This evidence has been supported by more recent studies using serotonin probes, such as prolactin release, with several studies reporting that indices of serotonergic function are negatively correlated with impulsivity and aggression.

Noradrenaline

The role of noradrenaline is less clear. Animal studies suggest that noradrenergic neurones play a reinforcement role assisting in reward-associated learning. Several studies have demonstrated impaired learning in animals with noradrenaline depression. Recent studies in rats and moles report that highly affilative animals showed distress in response to social isolation, a phenomenon be reduced by the administration of oxytocin. There have also been hypotheses that noradrenaline mediates general arousal of all motor systems. Human studies looking at noradrenaline blood levels are uncommon. Some have shown that compulsive gamblers may have significantly higher levels of noradrenaline metabolites than controls, and alcohol craving has been linked with low levels of the neurotransmitter.

Dopamine

There is more consensus on the role of dopamine. This monoamine appears to be related to behavioral activation. Dopamine-depleting drugs reduce active behavior, whereas facilitation of dopaminergic neurotransmission increases motor activity, exploratory behavior, and approach to novel stimuli. Studies on rodents link highly active behavior with independent hippocampal fiber terminal fields containing high levels of dopamine. Human studies indicate that low basal dopaminergic activity is associated with suicidal behavior. There is also some evidence that decreased dopamine β-hydroxylase has been associated with unsocialized conduct disorder in boys, and this may reflect a disorder of CNS dopaminergic transmission.

A recent study reported that striatal dopamine re-uptake site densities were markedly lower in nonviolent alcoholics than in those who were violent.

In summary, human studies show some degree of consistency with animal studies. There is considerable evidence that serotonin and behavioral inhibition are linked. Depletion of this neurotransmitter leads to general disinhibition of behavior. Dopamine has been associated with activating an incentive function in animals, and it is possible that it is related to exploratory and possibly antisocial behavior in humans. The role of noradrenaline in both animals and humans is much less clear.

MODELS OF TEMPERAMENT

As we have seen, there is now strong evidence that individual differences appear to be relatively stable, genetically influenced temperamental traits that are related to CNS neurotransmitter functioning. A major problem impeding further research is a lack of an agreed upon cohesive psychobiological model.An ideal model of temperament would be useful at a number of levels. Firstly, at the descriptive level, it would situate individuals in a unique multidimensional personality space. Ideally the measures obtained would also relate to current concepts of psychiatric illness. Secondly, the measured dimensions would reflect the underlying genetic structure, and thirdly, the dimensions would reflect the underlying biological causal mechanisms.

A number of descriptive models have been proposed. It is probable that a model involving three to seven measurable dimensions would be most useful. If, as seems likely, multiple genes are responsible for genetic influences on behavioral dimensions, a continuum of genetic risk is likely to extend from normal to abnormal behavior. Therefore, such dimensions should be reasonably normally distributed in a general population. In addition, neurochemical measures that are related to behavior appear to follow a dimensional pattern and cut across categorical boundaries. Although there have been some interesting links between categorical classes of personality disorder and biological measures, this is unlikely to be fruitful in the long run.

Models using factor analytic techniques reduce down to two to five major independent dimensions of variation (besides intelligence) in personality space. The most influential are Eysenck's model of personality and the so-called "big-five" or five-factor model (FFM), which overlap to some degree. The FFM has received extensive psychometric evaluation and has arguably the most compelling empirical support as a dimensional model of normal personality functioning. The five factors are called neuroticism (N), extroversion (E), openness to experience (O), agreeableness (A), and conscientiousness (C). (See Table 1 below for further explanation of what these terms mean). Unfortunately, there has been very little theoretical or empirical work relating these measures to biological factors. This does not seem to have been a priority for those using the model to study personality. There have

Table 1. Five Main Determinants of Personality

Extraversion	
Is outgoing, decisive, persuasive, and enjoys leadership roles	Is retiring, reserved, withdrawn, and does not enjoy being the center of attention
Neuroticism	
Is emotionally unstable, nervous, irritable, and prone to worry	Quickly gets over upsetting experiences and is stable and not prone to worries and fears
Conscientiousness	
Is planful, organized, responsible, practical, and dependable	Is impulsive, careless, irresponsible, and cannot be depended upon
Agreeableness	
Is sympathetic, warm, kind, good-natured, and will not take advantage of others	Is quarrelsome, aggressive, unfriendly, cold, and vindictive
Openness	
Is insighful, curious, orginal, imaginative, and open to novel experiences and stimuli	Has narow interest, is unintelligent, unreflective, and shallow

Note: Modified from Bouchard, 1994.

been some studies looking at the heritability of the five factors and this is reported to be between 40% to 50%, which is similar to most other dimensional behavioral measures. The FFM is difficult to relate to current psychiatric categories. Neuroticism has been well studied and is increased in most psychiatric illnesses, but the other dimensions are less closely linked. There have been attempts to relate the dimensions to personality disorders, but the results are mixed.

Eysenck's model has three dimensions: extroversion/introversion (E), neuroticism (N), (which is similar to but not identical to the FFM neuroticism scale), and psychoticism (P). The central core of Eysenck's work was his hypothesis that the ease of acquiring a conditioned response was related to an individual's location on the axes of extroversion and neuroticism. Introverted (low E) subjects conditioned more rapidly and extinguished more slowly than extroverts; these effects were exaggerated if they were also neurotic (high N). A number of biological measures—galvanic skin response, salivation, eye blinks, and so forth—were related to these differences. However, intercorrelations were modest and the effects of these measures were small. Eysenck's attempts to relate his dimensions to the CNS, particularly the ascending reticular activating system and the limbic system, were less successful .

The Tridimensional Model of Temperament

A more recent biosocial model of temperament was the tridimensional model of personality proposed by Cloninger. This was a theoretical model specifically designed to encompass current knowledge about the biology of behavior.

Cloninger began by postulating brain systems and then worked outward to their behavioral manifestations. There are three temperament dimensions named novelty seeking (NS), harm avoidance (HA), and reward dependence (RD). Variation of each of these dimensions was held to be associated with activity in a specific monoamine pathway. Novelty seeking (NS) was hypothesized to effect heritable individual differences in the behavioral activation system. High novelty seeking manifested itself in exploratory and active avoidance of monotony and was associated with low basal firing rates of dopaminergic neurons. In contrast, individuals with high spontaneous firing rates would be expected to be relatively placid and content regardless of external circumstances. Harm avoidance (HA) reflects heritable individual differences in the activity of the behavioral inhibition system. High harm avoidance manifests itself as a tendency to respond intensely to aversive stimuli and to learn to avoid punishment, and it is associated with high basal firing rates of serotonergic neurones. Individuals with low spontaneous firing rates would be expected to be confident, bold, risk-taking, and disinhibited.

Finally, reward dependence (RD) is supposed to reflect heritable individual differences in the behavioral maintenance system. High reward dependence manifests itself with an intense response to signals and reward, as well as with maintaining or resisting extinction of behavior. It is associated with low basal noradrenergic activity. Reward-dependent individuals may be described as dedicated, committed, loving, and emotionally dependent.

There is a pleasing simplicity about the way Cloninger's model relates each dimension to a single monoamine, but it is probably preferable to think of three monoamines as "emblematic" neurotransmitters that might reflect overall CNS activity. Multiple other substances including amino acids, neuropeptides, second

Table 2. The Tridimensional Model of Temperament

Brain System (Related Temperament Dimension)	*Principal Monoamine Neuromodular*	*Behavior*	
		High Scorers	*Low Scorers*
Behavioral Activation (Novelty Seeking)	Dopamine	exploratory & curious; impulsive; extravagant & enthusiastic; disorderly	indifferent; reflective; frugal & detached; orderly & regimented
Behavioral Inhibition (Harm Avoidance)	Serotonin	worrying & pessimistic; fearful & doubtful; shy; fatigable	relaxed & optimistic; bold & confident; outgoing; vigorous
Behavioral Maintenance (Reward Dependence)	Noradrenaline	sentimental & warm; dedicated & attached; dependent	practical & cold; withdrawn & detached; independent

messengers, and so on, are also important in regulating behavior. Cloninger's model, like Eysenck's, has been made operational by the construction of a hundred-item self-report—the tridimensional personality questionnaire (TPQ), which generates scores for novelty seeking, harm avoidance, and reward dependence.

Research Using the TPQ

At a descriptive level, results using the TPQ have been promising. Internal consistency and factor analysis suggest that there are three normally distributed independent measures. The findings were problematic for reward dependence and this has resulted in a subsequent fourth dimension—persistence, which was extracted from reward dependence. Personality disorders are related to the temperament extremes in a logical way. DSM 111 cluster A personality disorders (paranoid, schizoid, and schizotypal) have low reward dependence, cluster B personality disorders (antisocial, borderline, narcissistic, and histrionic) have high novelty-seeking, and cluster C personality disorders (avoidant, dependent, and obsessive compulsive) have high harm avoidance. Recent revision has seen the addition of three character measures to the original four temperament measures, but their relevance to biological measures is less obvious and they will not be discussed here.

Genetic analysis of data from 2680 twin pairs demonstrated that the genetic contribution accounts for 54% to 61% of the stable variation in the traits of novelty seeking, harm avoidance, and reward dependence. Further analysis suggests that the TPQ is measuring other (although still incomplete) descriptors of heritable personality that are different from those measured by Eysenck or the FFM. There have been two reports of a modest association between the dopamine D4 receptors and the trait of novelty seeking, but these were not replicated.

At a biological level, results using the TPQ have been mixed. Studies using traditional methods of measuring CNS monoamines have been unable to show any significant relationship between such measures and the TPQ dimensions. This was the case whether the biological examples were obtained using imipramine binding, prolactin response to alcohol, CSF monoamine metabolite concentration, or sensitivity to diazepam. On the other hand, there have been some more promising findings. One study showed a significant correlation between dopamine uptake in the left caudate and TPQ novelty seeking as would be predicted. It also reported that patients with Parkinson's disease have less novelty-seeking behaviors, an observation consistent with expectations since patients with Parkinson's disease have low dopamine turnover. High plasma GABA levels, which are related to anxiety proneness, have been correlated with low harm avoidance. Low levels of urinary 3-methoxy-4-hydroxyphenylglycol (MPGH), the metabolite of noradrenaline, have been reported in alcoholic patients with high reward dependence.

Which Model of Temperament

It emerges that the major difficulty in constructing a model of temperament is that there is an infinite number of alternative ways of summarizing temperamental traits and placing them in a three-dimensional space. Factor analysis can help determine the minimum number of measured dimensions but cannot decompose the underlying causal structure. Any useful biological model needs to be validated at the genetic and biological level as well as at the descriptive level. Presently, it seems fair to say that models are based on authority or tradition. Cloninger's tri-dimensional model is a qualified success at the descriptive level since the dimensions are independent, stable, conceptually attractive, able to be related to animal models, and reasonably easy to integrate with psychiatric diagnoses. At the genetic level, the TPQ measures appear genetically homogeneous and independent. However, it remains limited at the biochemical level.

A fundamental question is whether there is a significant problem using questionnaires to measure temperament. There may be serious differences between an adult self-evaluation and direct observation of that person. For example, many more adults answer to being shy than those who inherit a temperamental bias to be unusually shy. The items in questionnaires are restrictive and reflect the beliefs of the individuals who designed them. Factors that emerge depend on the questionnaire containing several items that reflect the same quality.Unfortunately, the alternative strategy of prolonged close observation of infant (or adult) behavior means that only small samples can be used. This may result in subtle differences in temperament being missed or in exaggerated importance being placed on a few unusual cases. Computer technology has begun to provide a middle road. In these systems, the way a questionnaire is scored as well as the answer to individual items are considered. For example, if individuals answer questions rapidly, then this is taken into account when scoring dimensions such as novelty-seeking or impulsivity.

There have also been new animal models for studying the biology of temperament, and this gives a further strategy to tackle the problem. Rather than measure behavior and relate this to biological variables, genetic manipulations of transporter genes allow highly specific alterations of CNS neurochemistry, and the resultant behavioral change can then be carefully observed. For example, a recent study produced a strain of mice in which the gene encoding the dopamine transport gene (DAT) had been disrupted—so called "DAT knockout mice". This resulted in CNS dopamine persisting at least 100 times longer in the extracellular space. The mice were extremely hyperactive and exploratory, gained weight more slowly than the control mice, and had impaired maternal behavior—something akin to extreme novelty seeking.

Another study produced mice lacking monoamine oxidase A (MAO-A), an enzyme that degrades serotonin and norepinephrine. In pup brains, serotonin concentrations were increased and pup behavioral alterations including trembling and fearfulness (resembling increased behavioral inhibition or harm avoidance) were reversed by a serotonin synthesis inhibitor. Adult mice had a different pattern of behavioral alterations consisting of offensive, aggressive behavior. Deficiency in MAO-A was shown to be associated with aggressive behavior in men in a Dutch family, so this parallel is interesting. The problems in these models include the common one of translating animal behavior to human behavior and also the fact that the genetic manipulations result in extreme biochemical abnormalities rather than a variance on normal CNS biochemistry. Such extremes may result in categorical behavioral differences rather than representing one end of dimensional continuum.

THE IMPORTANCE OF THE BIOLOGY OF TEMPERAMENT

An obvious question is that, given the serious problems in measuring both temperament and biology, why bother to pursue this area of research? The obvious answer is that this research is important. Most models of medical and psychiatric research have assumed specificity from a nosological/categorical perspective. Researchers have studied symptoms and biological measures within diagnostic categories in an attempt to guide treatment and predict the course of illness. Measuring temperament affords an additional strategy that proposes that behavioral traits are present across categories. It is these traits that have biological relationships rather than symptom measures. For example, serotonin disturbances have been reported in a variety of illnesses including major depression, antisocial personality disorder, obsessive–compulsive disorder, schizophrenia, and disruptive behavior disorder in children. The reason for this may be that such disturbances are nonspecific from the illness perspective but specific from a functional/dimensional perspective in that they correlate with particular temperament dimensions (in this case, impulsivity and aggression) across diagnoses.

The evidence also suggests that temperament differences within diagnostic categories may help explain the contradictory biological findings and diverse treatment responses. Cortisol hypersecretion, for example, which is present in many depressed patients, may be more related to measures of temperament than depressive symptoms. Hence, this biological marker is not especially specific for depression and may not be reported in some samples of depressed patients. Patients within categories may also respond differently to specific treatments. Depressed individuals with extreme temperaments often respond less well to tricyclic antidepressants. One study has found that temperament measures were much more significant in predicting the response to antidepressants than were any clinical symptoms or signs.

The finding from many studies that different psychiatric illnesses exist in one family might reflect a pedigree based on vulnerable temperamental traits rather than implying that these illnesses are directly related in some way. It may be more fruitful to map genes contributing to temperament, which has a relatively simple genetic architecture, than to attempt to map genes to a disease such as schizophrenia, which is probably caused by multiple susceptibility dimensions, each of which may be oligogenic. If it is possible to successfully map temperament, subsequent susceptibility to complex disorders like schizophrenia and alcoholism could be evaluated in the terms of risk from heritable temperament traits and possible disease-specific factors.

CONCLUDING REMARKS

This chapter offers a selected review of the literature in the field of biology and personality. It has done little more than hint at the confusion inherent in the large quantity of pertinent data that exists. However, this confusion should not obscure the fact that there is now overwhelming evidence to suggest that biology and personality are linked. Temperament traits appear to be relatively stable, measurable, genetically influenced, and related to CNS neurotransmitter functioning. Classification is a necessary step for this science to proceed. It is currently not clear whether research should proceed with animal models using genetic manipulation of CNS neurochemistry and analysis of resultant behavior, with human models using simple questionnaires, or with prolonged observation of small children. All have significant problems related to measurement and the meaning of the measured behavior.

It is hoped that all methods of research continue. New theoretical models are needed to help link biology and temperament. These models will overlap to some extent with the proposed current models but, until then, the existing models must serve for the coherent formulation of available data and the generation of testable hypotheses. It is now clear that it is no longer tenable to ignore the biological aspects of temperament in research or clinical practice.

SUMMARY

Historically, models of personality have generally postulated a link with biology. This century has witnessed a major revision of these ideas, with behavioral and psychoanalytic theorists emphasizing life experiences as being mainly responsible for behavior as adults. Challenges to this assumption have come from genetic studies, infant temperament studies, and neurobiological research. An extensive body of data now exists that suggests that biology contributes significantly to individual variability. This biological contribution occurs at a relatively low level in the central nervous system and is better defined as temperament. This biological contribution has suffered from the lack of a cohesive psychobiological model. A number of models are reviewed; all of them have significant shortcomings.

Alternative strategies of observing infant behavior or using genetically modified animals are discussed. The importance of continuing research into the biology of personality for psychiatry is reviewed. Biological abnormalities in psychiatric illness may be related to temperament traits instead of diagnostic categories. Such traits may predict treatment response. Genetic vulnerability to psychiatric illness may be via heritable temperamental traits rather than disease-specific factors. It is no longer reasonable to ignore the biological aspects of personality in research or clinical practice.

RECOMMENDED READING

Bauchard, T.J. (1997). Experience producing drive theory: How genes drive experience and shape personality. Acta. Ped. 422 (Suppl.), 60-64.

Benjamin, J., Ebstein, R.P., & Belmaker, R.H. (1997). Personality and Temperament. Israel J. Personality Genetics 70(2), 387-412.

Cloninger, C.R. (1987). A systematic method for clinical description and clarification of personality variants. A proposal. Arch Gen Psych. 44, 573-588.

Eysenck, H.J. (1967). *Biological Basis of Personality*. Charles C. Thomas, Springfield, Illinois.

Kagan, J. (1994).*Galen's Prophecy: Temperament in Human Nature*. Basic Books, New York.

Nigg, J.T. & Goldsmith, H.H. (1998). *Developmental Psychopathology*. Amer. Psych. Press, Washington, D.C.

Thomas, A. & Chess, S. (1997). *Temperament and Development*. Brunner/Mazel, New York.

Silk, K.R (Ed)(1998). *Biology of Personality Disorders*. American Psychiatric Press, Washington, D.C.

Chapter 5

The Biology of Aggression

PAUL F. BRAIN

INTRODUCTION

The main problem with specifying the biology of "aggression" is knowing which phenomena the biological variables should be related to. Aggression is clearly a heterogeneous concept (Brain, 1984) in which biological variables, environmental factors, and social learning all play complex, intertwined roles. Violence is a com-

Biological Psychiatry, pages 103-113.

ISBN: 1-55938-819-6

plex phenomenon and one *must* specify the importance (and limitations) of genes, diet, neural systems, hormones, and drugs in its expression. Furthermore, as well as forming the basis of some clinical treatments, a knowledge of biological correlates may have some predictive value in assessing the risk of aggression in populations of humans. It must also be maintained that individuals with greatly disordered biologies (e.g., brain tumors or endocrine disorders) cannot be regarded as wholly responsible for their actions, but the legitimate use of such claims in legal circles is often contentious. It should be self-evident that there are considerable dangers in seeking to equate *all* examples of human aggression to medical disorders. If a disordered biology has a role, it is only likely to be of relevance to *some* forms of interpersonal behavior, not group activities such as riots and war, which clearly have a largely sociocultural basis. It should be admitted at the outset that even focusing on particular expressions of aggression, such as assault, homicide, or rape, does *not* lead to simple associations with biological factors.

This review broadly looks at some of the more recent writings on the role(s) of biological factors in aggression with the intention of illustrating the modern approaches and concerns.

AGGRESSION AND GENES

Behavior genetics (e.g., Royce and Mos, 1979) is currently a growth area of research. In spite of the public misconception that scientists are looking for *the* gene involved in aggression, it is obvious that genes interact, influence developing anatomical and physiological systems, and are involved in interplay with the environment. For many years, people have been intrigued about suggestions of links between genetic endowment and human hostility (e.g., Christiansen, 1968; Owen, 1972; Cadaret, 1978; Gottesman et al., 1983; Mednick et al.,1987). This area has been recently substantially reviewed and re-evaluated by Carey (1994) and a *Symposium on Genetics of Criminal and Antisocial Behaviour* (edited by Bock and Goode, 1996). Carey (1994) concludes that "there is a trend in most studies (of violence and human genetics), albeit not always a statistically significant one, consistent with the hypothesis of a genetic effect on adult and perhaps adolescent antisocial behaviour." (p. 42). He points out, however, that it is not easy to integrate this literature into contemporary criminological research on violence in the U.S. In addition, the genetics of antisocial behavior do not fit any simple additive model.

Carey (1994) points out that "the joint effects of marital assortment, temporal trends over time, nonadditive genetic variance, special twin effects, etc., must be considered in studying the genetics of antisocial behaviour" (p. 42). The evidence for a genetic effect primarily on offenses involving physical aggression is not impressive. This is partially due to the relatively low base rate for violent offenses, which consequently require large samples to give meaningful and replicable results. There is no real evidence that violent crime aggregates in twins or in biological relatives of violent adoptees despite a genetic liability to crime. Rutter (1996) supports the view that the genetic influence on violent crime is less than that in petty property crime. Although there is good evidence that the Y chromosome contributes to some forms of aggressive behavior in the mouse (Maxson, 1996), there is no evidence (Baker et al., 1989) for a similar effect on human antisocial behavior. The 47 XYY chromosomal anomaly (with double the "male" chromosome), although not leading to violent behavior as initially maintained, is associated with behavioral factors that increase the likelihood of antisocial acts (Ratcliffe, 1994).

There is evidence (Carey, 1994) for a genetic association between antisocial behavior and alcohol (as well as possibly other substance) abuse. There is a strong suspicion (Cloninger, 1987) that alcohol abuse is genetically heterogeneous and that one meaningful subgroup is related to antisocial personality. Rutter (1996) claims that "adoptee data also point to the need to differentiate criminality associated with alcohol abuse from that unassociated with alcohol problems" (p. 266). Virkkunen and colleagues (1996) are of the view that "it is likely that several new gene markers linked to and associated with impulsive violence, alcoholism, and related behaviors will be characterized within the next few years" (p. 174). One should note that alcohol abuse (Brain, 1986a) has very wide-ranging effects on physiology and may impair the cognitive skills required to solve potentially violent social interactions (Brain et al., 1993). Rutter (1996) concludes that "nevertheless, it is evident that genetic factors do play a significant role in antisocial behaviour and that their investigation is likely to be useful, with respect to both theory and practice" (p. 265). He suggests that "...it is apparent that researchers in this arena are constructively self-critical, with a deep awareness of the methodological hazards to be avoided, the inconsistencies across studies to be explained, and the marked limitations in the inference that can be drawn from the findings so far" (p. 265). Genes seem most influential in early-onset pervasive hyperactivity, which is an excellent predictor of adult antisocial behavior (Farrington et al., 1990), and they have less effect on adolescent-onset delinquency nonpreceded by hyperactivity.

Although genetic research concentrates on individual differences in liability for violence (i.e., the predictive value), it may be used to look at changes in rates of crime with time or differences between different geographical locations. Rutter (1996) maintains that there is a requirement to consider person–environment correlations and interactions and to establish whether part of the genetic influence on antisocial behavior involves modifying the ways that people select and shape their environments.

AGGRESSION AND DIET

Brain (1984) claimed that "there are many reports, both anecdotal and scientific, suggesting that diet influences aggression in populations" (p. 105). Substances such as glucose, caffeine, alcohol, tryptophan (an amino acid), environmental lead, food additives, natural salicylates, vitamins, and minerals have all at various times been investigated as potential dietary influences on antisocial behavior. Benton (1996) has reviewed the evidence and concludes that "...there is little doubt that in a minority of people food intolerance can induce antisocial behaviour" (p. 140) including extreme violence. Such effects are, however, only revealed with careful use of double-blind studies. Cases of food intolerance have involved red dye, peanuts, wheat, sugar, milk, and potatoes, and the defence has been used successfully in both the U.S. and U.K.

Benton (1996) has also looked at the claimed links between low blood sugar levels and aggression expressed in many diverse situations and concludes that modest declines in blood glucose levels can cause increased irritability. Whether this mood change leads to aggressive behavior depends on the degree of provocation, the individual's social skills, and other features of the situation. This area is also examined by Kanarek (1994) who concludes that neither violent behavior nor attention deficit hyperactivity disorder (ADHD) can be clearly related to reduced blood sugar levels. Violent behavior is not easily related to changes in carbohydrate metabolism because studies are often confounded by uncontrolled variables (such as overcrowding or drug use), food intake is rarely measured, and the antisocial behaviors are inadequately specified. Neither dietary challenge studies nor contrasting hyperactive and normal children support the view that blood sugar level plays a major role in ADHD. Notwithstanding this lack of scientific evidence, the belief that there is such a connection persists in sections of the parenting and teaching communities.There is some support for the claim (Kanarek, 1995) that low blood cholesterol levels can be linked to violent behavior (e.g., accidents, suicide, or homicide). Although such relationships are not revealed in all studies, such behavioral changes have been seen in cases where blood cholesterol values are reduced by dietary manipulation as well as by drug treatment. There is sometimes a failure to record levels of alcohol ingestion in the populations studied—a problem because this activity influences blood cholesterol levels.

Exposure to lead (especially in early life) has been associated with increased antisocial behavior (Kanarek, 1994). This toxic metal increases the incidence of ADHD, which we have already noted is an established risk factor for adult antisocial behavior.Benton (1996) has also looked at the claimed links between vitamin and mineral intake and antisocial behaviors in institutionalized (i.e., in prisons and orphanages) individuals. Recent adequately controlled studies have confirmed earlier impressions that correcting inadequate diets with multivitamin and mineral supplements decreases the incidence of rule-breaking activity.

Neither Kanarek (1994) nor Benton (1996) regard diet as being especially important in influencing the incidence of aggressive behaviors. Both seem convinced, however, that it may change the predisposition for producing such responses.

AGGRESSION AND NEURAL SYSTEMS

There has been a long tradition of considering the roles of parts of the brain in aggression. Animal experiments with cats suggested that activities such as "sham rage" could be manipulated by stimulating parts of the hypothalamic region of the brain or ablating other areas. The simplistic view that there are "on" and "off" centers for aggression in the brains of higher vertebrates has been challenged by the following: (a) the recognition that different forms of attack have very different physiologies (Brain, 1981a); (b) the acceptance that the brain functions via "constellations" of linked areas acting on a common process by modulating subprocesses and each other (Delgado, 1981), and (c) demonstrations that precisely the same input will have different outcomes when the same animal is in different social contexts (Plotnik et al., 1971).

The initial technologies of "correcting" human aggression via psychosurgery (as advocated by, for example, Heath, 1981) were challenged by Valenstein (1980). The whole area has been re-evaluated by Mirsky and Siegel (1994) who conclude that, so far as patients with neuropsychiatric disorders are concerned, there is tentative evidence of a connection between aggression and alcohol abuse that seems mediated via alcohol-induced brain damage. The links between alcohol ingestion and violence are complex (Brain et al., 1993), in some cases being produced by alcohol-induced deficiencies in cognitive and social skills (this may be the genesis of the "battered alcoholic" syndrome). The links between epilepsy and aggression have also been reviewed (Mirsky and Siegel, 1994). Ictal violence (i.e., seizure-concomitant) is unusual, being found in around 0.25% of epileptics, is only rarely associated with angry mood (fear is more common), and can hardly be viewed as "deliberate" attack on another person. Interictal violence (i.e., occurring between seizures) is much more difficult. Violence associated with this condition has been regarded as an occult, or hidden, seizure or as evidence of damaged neural mechanisms associated with emotions. There is a debate concerning whether one can abolish interictal violence by surgical removal of a seizure focus. There are also some provocative claims concerning the incidence of subtle brain abnormalities in sexual offenders, including the view that the actual abnormality may differ in the varied types of offender. Mirsky and Siegel (1994) conclude, however, that it is uncertain whether human aggression can be unequivocally linked to disordered brain mechanisms. There is the difficulty of controlling for the effects of lifestyle, and brain injuries are likely to be more common in certain groups of individuals.

AGGRESSION AND HORMONES

Brain (1981b, 1994) and Brain and Susman (1997) have reviewed the wide-ranging literature, which has attempted to link the secretions of the endocrine system (notably the hormones of the hypothalamus-pituitary-gonadal [HPG] and the hypothalamus-pituitary-adrenal axes along with the thyroid and the adrenal medulla) to various manifestations of aggression in human and infra-human animals. Some hormones are implicated in this behavior, but the effects may be properly categorized as the following:

1. the organizing effects of perinatal hormones on the potential for adult behavior
2. direct activational effects in adulthood presumably mediated via hormonal influences on the central nervous system
3. indirect signaling effects by altering the cues used in social interactions (e.g., anatomy, odors, etc.)
4. effects of social experiences on hormone production

In terms of their organizing effects, sex steroids in prenatal life (humans) and/or postnatal life (rats and mice) produce subtle structural changes in the brain that masculinize and defeminize the behavioral potential (Meyer-Bahlberg et al., 1995). A wide range of reproductive and nonreproductive behaviors appear to be influenced by the sexual dimorphisms induced by gonadal steroids. For example, Hines (1990) has reviewed the evidence that gonadal hormones influence human cognitive development. These include sexual orientation, cognitive characteristics concerned with language and visuospatial abilities, patterns of childhood play, and *propensities for physical aggression*. The humans involved in such studies are generally subjects in which the early hormones are elevated endogenously (by an endocrine disorder such as congenital adrenal hyperplasia, which increases maternal androgens) or exogenously by direct hormonal treatment (e.g., to prevent a miscarriage). Interestingly, "males" may be individuals who are exposed at key developmental stages to sex steroids (the testis is active earlier than the ovary), estrogens may be more strongly implicated in masculinization than androgens, and sociocultural effects can augment or override hormonal influences on behavior at many points in development.

Adolescence is associated with dramatic changes in hormonal secretions, and these have been linked with the "rebellious attitude" said to characterize this phase of development. Testosterone and adrenal androgens can influence antisocial behavior in adolescents, but these effects are complicated by factors noted by Hays (1981). She attaches a series of provisos to this area:

a. Mood changes produced by hormones (including factors unrelated to the gonadal system) are perhaps consequences of the instigation of drives with few socially acceptable outlets in young people.
b. Development can involve changes in behavioral sensitivities to hormones as well as changes in hormones *per se*.
c. Circadian rhythms are important considerations when relating hormones and behavior at puberty as hormonal concentrations seem to covary with moods and behaviors.
d. Interactions between hormones may be more important than titers of single hormones.

Truscott (1992) has provided support for the view that violence is transmitted intergenerationally from parents to their adolescent offspring and that psychological mechanisms (perhaps including hormones) have a role in this phenomenon. Early established aggressive tendencies are predictive factors in adult unprovoked aggression in a laboratory test (Hammock and Richardson, 1992).

Brain and Susman (1997) have reviewed the use of manipulations of the HPG axis hormones to control human aggression. Certainly, castration, estrogen therapies, and anti-hormone treatments have all been utilized clinically. For a variety of reasons (e.g., the view that not all forms of aggression are amenable to such treatments, development of side-effects, ethical considerations), they conclude that, currently, there is an apparent reduction in the enthusiasm for treating human violence by castration or by anti-androgens, estrogens, or progesterone derivatives.

It has also been noted (Brain and Susman, 1997) that there were many early attempts to link plasma levels of testosterone with human aggression in prison populations and others with a known history of anti-social behavior. Brain's (1994) conclusion was "it seems unlikely that androgens have a simple causal effect on human aggression and violence but the patterns of sex steroids do appear to alter several factors (e.g., "aggressive feelings", self image, and social signalling) that predispose individuals towards (sic.) carrying out actions that can receive this label" (p. 221). The reasons for this conclusion are as follows:

1. Meta analysis of studies using the Buss-Burkee Hostility Inventory score and plasma testosterone titers shows a low but positive relationship between these factors (N.B. this links aggressive feelings, i.e. angry mood, rather than actions).
2. Studies on sexually aggressive men show that certain categories of offender evidence altered levels of particular hormones (notably the adrenal androgen dehydroepiandrosterone).
3. Investigations of "winners" in a variety of circumstances reveal that altered mood and apparent status increase plasma testosterone levels.
4. Saliva concentrations of testosterone have been positively correlated with self-rated spontaneous aggression.

5. Males (but not females) show links between plasma levels of testosterone and estradiol and a variety of types of self-reported aggression.
6. "Free" testosterone in saliva has been correlated with the degree of violence involved in the crimes for which young male U.S. prisoners were convicted.

It is evident that there are some complex effects of environmental and developmental factors on the associations between hormones and the plethora of forms of human aggression.

AGGRESSION AND DRUGS

This association between psychoactive compounds and violence has been looked at in a variety of ways. There have been numerous attempts to link human aggression to the use of a range of legal (e.g., alcohol; Brain, 1986) and illegal (e.g., amphetamines, heroin, hallucinogens, and cocaine; Reiss and Roth, 1993) drugs. Suffice it to say that the links are not simple here. Many psychoactive drugs can influence neural development in the offspring of mothers who take the material while pregnant (e.g., fetal alcohol syndrome and "crack babies") but these effects are induced in a variety of ways (Brain et al., 1994). These same drugs will alter certain emotions or impair, in some cases, cognitive and social skills, which could both make the escalation of conflict more common. The links to violence can be less direct, in other cases, with the drug use altering relevant circumstances such as the perceived reliability of witnesses, the ease of capture of offending individuals, or the need to resort to crime to maintain an expensive habit. The Reiss and Roth (1993) volume, for example, concludes "there is fairly strong evidence that individual differences beyond the biological processes discussed above intervene in the relationships between violent behavior and the use of alcohol and other psychoactive drugs" (p. 195).

The other way in which psychoactive drugs have been linked with human aggression has been in the clinical use of these compounds to control such activities. Itil (1981) suggested that drugs had an important role in the acute emergency situation when sedation was effectively required. He suggested that barbiturates, neuroleptics (major tranquilizers), and anxiolytics (minor tranquilizers) all could be useful in controlling acute aggressive states. The "anti-aggressive" effects in such cases are simply due to CNS depression. The treatment of persistent or episodic aggression was, according to Itil (1981), much more complicated, with "the classical CNS depressants, CNS stimulants, anticonvulsants, hormones, narcotics, and compounds without classical pharmacological properties" (p. 497), all being advocated for the treatment of aggression largely on empirical grounds. Itil maintains that "...the aggressive symptomatology may be better controlled with a drug that can "normalize" a deviated human brain function as determined by electroencephalogram" (p. 497). There have, notably, been recent concerns about the appropriateness of using "chemical coshs" to treat human violence and there certainly have been cases when drug treatments have been used inappropriately. Miczek and colleagues (1994) confirm that, "in emergency situations, injections of benzodiazepines effectively calm violent individuals" (p. 280) but suggest that extended treatment of aggressive subjects within clinical populations typically results in a development of tolerance to the sedative actions without reducing the therapeutic influences on violent behavior. Miczek and colleagues (1994) maintain that the selective serotonin 1A anxiolytics show considerable promise as anti-aggressive agents that lack sedative effects. Some patients, however, show "paradoxical rage" following benzodiazepine treatment, and there is an urgent need to identify factors that predispose individuals to react in this way. Attempts to develop other specific anti-aggressive compounds such as fluprazine or DU 27716 have had only limited success (e.g., Olivier et al., 1984), probably because it is not really possible to influence the wide range of aggressive behaviors without changing other behavioral elements (Benton et al., 1983).

CONCLUSIONS

It should be evident from the detailed considerations of the individual sections on biological factors that there is considerable interaction and overlap between them. Our current theme is that "aggression" is such a diverse concept that maintaining that any one biological variable will be a unique predictor of violence or even of a potential specific therapy is a deeply flawed view. Biology can certainly be implicated in aggression, but biology is expressed differently in different forms of behavior, operates at different stages of the individual's life, and *cannot* easily be divorced from environmental and learned responses in our species. Perhaps even more problematic is a gradual recognition that even defined classes of aggressive acts (e.g., homicide, rape, and assault) are markedly heterogeneous such that biology will vary its degree of involvement on a case-by-case basis. This means that one cannot make sweeping generalizations about the biology of aggression. Although such a view may cause some initial distress in individuals who desire general statements, I feel that this growing appreciation of the complexities of biological effects on aggression has some real benefits. It should improve our understanding of a complex (and essentially biological) phenomenon and counter the unhelpful, overly simplistic ideas that have surrounded some areas of study.

REFERENCES

Baker, L.A., Mack, W., Moffitt, T.E., & Mednick, S (1989). Sex differences in property crime in a Danish adoption cohort. Beh. Genetics 19, 355-370.

Benton, D. (1996). *Food for Thought*. Penguin, Harmondsworth, England.

Benton, D., Brain, P.F., Jones, S., Colebrook, E., & Grimm, V. (1983). Behavioural examinations of the anti-aggressive drug Fluprazine. Beh. Brain Res. 10, 325-338.

Bock, G.R. & Goode, J.A. (1996). *Genetics of Criminal and Antisocial Behaviour*. John Wiley and Sons, Chichester.

Brain, P.F. (1981a). Differentiation of attack and defense in rodents. In: *Multidisciplinary Approaches to Aggression Research.*(Brain, P.F. & Benton, D., Eds), pp.53-78. Elsevier, Amsterdam, North Holland.

Brain, P.F. (1981b). Hormones and aggression in infra-human vertebrates. In: *The Biology of Aggression.* (Brain, P.F. & Benton, D., Eds), pp.181-213. , Sijthoff-Noordhoff, Alphen aan den Rijn, The Netherlands.

Brain, P.F. (1984). Human aggression and the physical environment. In: *Mental Health and the Environment.*(Freeman, H.L., Ed.), pp. 97-120. Churchill-Livingstone, London.

Brain, P.F. (1986a). Multidisciplinary examinations of the "causes" of crime: The case of the link between alcohol and violence. Alcohol and Alcoholism, 21, 237-240.

Brain, P.F. (Ed.)(1986b). *Alcohol and Aggression.* Croom-Helm, London.

Brain, P.F. (1994). Hormonal aspects of aggression and violence. In: *Understanding and Preventing Violence Vol. 2, Biobehavioral Influences* (Reiss, A.J., Jr., Miczek, K.A., & Roth, J.I., Eds.), pp.173-244, National Academy Press, Washington D.C.

Brain, P.F. & Susman, E.J. (1997). Hormonal aspects of aggression and violence. In: *Handbook of Antisocial Behavior.* (Stoff, D.M., Breiling, J., & Maser, J.D., Eds.) pp. 314-323, John Wiley and Sons, New York.

Brain, P.F., Miras, R.L., & Berry, M.S. (1993). Diversity of animal models of aggression: Their impact on the putative alcohol/aggression link. J. Studies on Alcohol 11(Suppl.), 140-145.

Brain, P.F., Kurishingal, H., Whiting, K., & Restall, C.J. (1994). An ethopharmacological approach to behavioural teratology. In: *Ethology and Psychopharmacology.* (Cooper, S.J. & Hendrie, C., Eds.), pp. 225-239. John Wiley and Sons, Chichester, England.

Cadoret, R.J. (1978). Psychopathology in adopted-away offspring of biologic parents with antisocial behavior. Archiv. Gen. Psych. 35, 176-184.

Carey, G. (1994). Genetics and violence. In: *Understanding and Preventing Violence, Vol. 2, Biobehavioral Influences.* (Reiss, A.J., Miczek, K.A., & Roth, J.A., Eds.), pp. 21-58. National Academy Press, Washington D.C.

Christiansen, K.O. (1978). The genesis of aggressive criminality: Implications of a study of crime in a Danish twin study. In: *Origins of Aggression.* (Hartup, W.W. & de Wit, J., Eds.), pp. 99-120. Mouton, The Hague, The Netherlands.

Cloninger, C.R. (1987). Neurogenetic adaptive mechanisms in alcoholism. Science 236, 410-416.

Delgado, J.M.R. (1981). Neuronal constellations in aggressive behavior. In: *Aggression and Violence: A Psycho/Biological and Clinical Approach.* (Valzelli, L. & Morgese, L., Eds.), pp. 82-98. Edizioni Saint Vincent, Saint Vincent, Italy.

Farrington, D.P., Loeber, R., & van Kammen, W.B. (1990). Long-term criminal outcomes of hyperactivity-impulsivity–attention deficit and conduct problems in childhood. In: *Straight and Devious Pathways from Childhood to Adulthood.* (Robins, L.N. & Rutter, M., Eds.), pp. 62-81. Cambridge University Press, Cambridge, England.

Gottesman, I.I., Carey, G., & Hanson, D.H. (1983). Pearls and perils in epigenetic psychopathology. In: *Childhood Psychopathology and Development.* (Guze, S.B., Earls, E.J., & Barrett, J.E., Eds.), pp. 287-300. Raven Press, New York.

Hammock, G.S. & Richardson, D.R. (1992). Predictors of aggressive behavior. Aggressive Behavior 18, 219-229.

Hays, S.E. (1981). The psychoendocrinology of puberty and adolescent aggression. In: *Biobehavioral Aspects of Aggression.* (Hamburg, D.A. & Trudeau, M.B., Eds.), pp. 107-119. Alan R. Liss Inc., New York.

Heath, R.G. (1981). The neural basis for violent behaviour: Physiology and anatomy. In: *Aggression and Violence: A Psycho/Biological and Clinical Approach.* (Valzelli, L. & Morgese, L., Eds.), pp. 176-194. Edizioni Saint Vincent, Saint Vincent, Italy.

Hines, M. (1990). Gonadal hormones and human cognitive development. In: *Hormones, Brain and Behaviour in Vertebrates.* (Balthazart, J., Ed.), pp.51-63. Karger, Basel, Switzerland.

Itil, T.M. (1981). Drug therapy in the management of aggression. In: *Multidisciplinary Approaches to Aggression Research.* (Brain, P.F. & Benton, D., Eds.), pp. 489-502. Elsevier Biomedical Press, Amsterdam, The Netherlands.

Kanarek, R.B. (1994). Nutrition and violent behavior. In: *Understanding and Preventing Violence, Vol. 2, Biobehavioral Influences.* (Reiss, A.J., Miczek, K.A., & Roth, J.A., Eds.), pp. 515-539. National Academy Press, Washington, D.C.

Maxson, S.C. (1996). Issues in the search for candidate genes in mice as potential modelsof human aggression. In: *Genetics of Criminal and Antisocial Behaviour.* (Bock, G.R. & Goode, J.A., Eds.), pp.21-35. John Wiley and Sons, Chichester, England.

Mednick, S.A., Gabrielli, W.F., & Hutchings, B. (1987). Genetic factors in the etiology of criminal behavior. In: *The Causes of Crime: New Biological Approaches.* (Mednick, S.A., Moffitt, T.E., & Stack, S.A., Eds.), pp. 74-91. Cambridge University Press, New York.

Meyer-Bahlberg, H.F.L., Ehrhardt, A.A., Rosen, L.R., Gruen, R.S., Veridiano, N.P., Van, F.H. & Neuwalder, H.F. (1995). Prenatal estrogens and the development of homosexual orientation. Devel. Psych. 31, 12-21.

Miczek, K.A., Mirsky, A.F., Carey, G., DeBold, J., & Raine, A. (1994). An overview of biological influences on violent behavior. In: *Understanding and Preventing Violence, Vol. 2, Biobehavioral Influences.* (Reiss, A.J., Jr., Miczek, K.A., & Roth, J.A., Eds.), pp. 1-20. National Academy Press, Washington, D.C.

Mirsky, A.F. & Siegel, A. (1994). The neurobiology of violence and aggression.In: *Understanding and Preventing Violence, Vol. 2, Biobehavioral Influences.* (Reiss, A.J., Jr., Miczek, K.A., & Roth, J.A., Eds.), pp. 59-172. National Academy Press, Washington, D.C.

Olivier, B., Mos, J., van de Poel, A.M., Krijzer, F.N.C., & Kruk, M.R. (1984). Effects of a new psychoactive drug (DU 27716) on different models of rat agonistic behavior and EEG. In: *Biological Perspectives on Aggression.* (Flannelly, K.J., Blanchard, R.J., & Blanchard, D.C., Eds.), pp. 261-279. Alan R. Liss Inc., New York.

Owen, D.R. (1972). The 47, XYY male: A review. Psych. Bull. 78, 209-233.

Plotnik, R., Mir, D., & Delgado, J.M.R. (1971). Aggression, noxiousness, and brain stimulation in unrestrained rhesus monkeys. In: *The Physiology of Aggression and Defeat.* (Eleftheriou, B.E., and Scott, J.P. Eds.), pp. 143-228. Plenum Press, New York.

Ratcliffe, S.G. (1994). The psychosocial and psychiatric consequences of sex chromosomal abnormalities in children based on population studies. In: *Basis Approaches to Genetic and Molecular Biological Developmental Psychiatry.* (Poutska, F., Ed.), pp. 99-122. Quintessenz, Berlin.

Reiss, A.J. & Roth, J.A. (Eds.)(1993). *Understanding and Preventing Violence.* National Academy Press, Washington, D.C.

Royce, J.R. & Mos, L.P. (Eds.)(1979). *Theoretical Advances in Behavior Genetics.* Sijthoff and Noordhoff, Alphen aan den Rijn, The Netherlands.

Rutter, M. (1996). Concluding remarks. In: *Genetics of Criminal and Antisocial Behaviour.* (Bock, G.R. & Goode, J.A., Eds.), pp. 265-271. John Wiley and Sons, Chichester, England.

Truscott, D. (1992). Intergenerational transmission of violent behavior in adolescent males. Aggressive Behavior 18, 327-335.

Valenstein, E.S. (1980). *The Psychosurgery Debate.* W.H. Freeman and Co., San Francisco.

Virkkunen, M., Goldman, D., & Linnoila, M. (1996). Serotonin in alcoholic violent offenders. In: *Genetics of Criminal and Antisocial Behaviour.* (Bock, G.R. & Goode, J.A., Eds.), pp. 168-182. John Wiley and Sons, Chichester, England.

Chapter 6

Psychological Aspects of Hypothalamic–Pituitary–Adrenal Axis Activity

EHUD UR

Biological Psychiatry, pages 115-133.

ISBN: 1-55938-819-6

INTRODUCTION

The hypothalamo–pituitary–adrenal (HPA) axis plays a central role in the integration of the response of the organism to stress. This is achieved through various homeostatic mechanisms controlling intermediary metabolism, blood pressure, and behavior. The neuroendocrine cascade triggered by stress begins with central perception of a stressor, thus leading to the release of corticotropin releasing hormone (CRH), arginine vasopressin (AVP), and other secretagogues for adrenocorticotrophin (ACTH). ACTH is released from the corticotrophic cells of the anterior pituitary, following posttranslational modification of its parent molecule proopiomelanocortin (POMC), and this peptide promotes the release of glucocorticoids from the adrenal gland.

For technical reasons of accessibility, early work on the physiology of the HPA axis mostly concentrated on the distal effector limb of this cascade. Thus cortisol has long been known to be secreted in response to a wide range of environmental stimuli. Cortisol activates lipolysis, gluconeogenesis, and a whole host of additional metabolic pathways aimed at increasing the supply of circulating energy substrates. It has also been known that glucocorticoids inhibit numerous "energy-expensive" processes, including growth, reproductive function, and immunological and inflammatory responses. More recent findings of widespread glucocorticoid receptor immunoreactivity within the central nervous system indicate that cortisol plays a role in behavioral modulation through its effects on neurophysiological function.

Technical inaccessibility has meant that less is known about the rostral connections of the HPA axis and, in particular, about the regulatory role played by the hippocampus, a major component of the limbic system and the dominant site of glucocorticoid receptor activity in the brain. In this chapter, we set out to review the evidence linking the hippocamus with stress responses and particularly with activation of the HPA axis.

Since the isolation and characterization of CRH-41 (Vale et al., 1981), interest has developed in the central effects of this compound aside from those involved in the promotion of corticotropin release. Direct intracerebroventricular instillation of CRH has resulted in increased sympathetic-nervous-system activity, suppression of luteinizing hormone (LH) release, and a number of behavioral changes unrelated to ACTH and steroid activation.

Other factors with significant roles in the modulation of ACTH secretion are also known to exhibit profound neurotropic activity. AVP, an important (if not preeminent) secretagogue for corticotropin, is thought to play a role in memory. In view of their putative role in the neurophysiology of affect, it is of interest that there is compelling morphological and pharmacological evidence for the involvement of monoamines and especially central catecholaminergic modulation of HPA axis function. The role of endogenous opiates in the control and function of the HPA axis remains unclear, as does the physiological significance of opioid analogues co-released with ACTH in response to acute stressors. Nevertheless it is noteworthy that differential processing of POMC takes place and this appears to be determined by qualitative factors in the stressful situation. In more recent years, there has been intense speculation about the role of cytokines, excitatory amino acids, and nitric oxide in the control of the HPA axis and in the functions these novel modulators may have in mediating the biochemical substrates of behavioral effects.

Ever since Cushing's original description of the pituitary, significant disturbances of psychological function have been recognized to be among the cardinal manifestations of several conditions of pituitary–adrenal dysfunction of primarily endocrine etiology. Recently, consistent disturbances in HPA axis function have been found in a number of disorders of psychiatric origin, the best characterized example of which is the hypercortisolemia of depression. Obviously, as our knowledge of HPA axis physiology has improved, it has become possible to construct and test a number of hypotheses that seek to explain these findings.

STRESS

It has been a hotly debatable issue as to whether stress represents the principle paradigm for HPA axis activation. The psychobiological concept of stress continues to defy definition. Thus, stress is variously described as either a characteristic of the environment, of the response of the individual, or of the interaction of the individual's perception of the environment with his response. Researchers have emphasized differences between acute and chronic stress, the importance of the psychological aspects of the provocative stimulus, and the ability of the organism to respond appropriately and effectively.

Selye's (1936) notion of a specific "general adaptational syndrome" has fallen out of favor with current workers who prefer to define a variable constellation of adaptive changes, which are brought into play by a number of neurohumoral effectors. Sokolov (1966) has provided a model for a system in which the organism matches immediate events with a neural representation of prior events. In this model, the stressed organism has a set of prior expectations. If the environment contains new contingencies that defy the preexisting conditions, a physiological response is evoked. In this context, Gray (1984) introduced a further psychological concept—that of coping—which is defined as an ability to predict and control stressors, a process that results in modulation of physiological activation and the damping of autonomic and hormonal excursions.

STRESS AND THE HYPOTHALAMO-PITUITARY-ADRENAL (HPA) RESPONSE

Over the past fifty years, numerous studies have been carried out looking at cortisol responses to stressful circumstances generated by a wide variety of experimental paradigms. These have been reviewed by Mason (1968), Rose (1980), and Ur (1991). Comparisons must be made cautiously in view of the wide variation in operational definitions for stress, in subjects, and in experimental design. Nevertheless, it has become increasingly clear through the demonstration of differential sensitivity, that Selye's (1936) notion of an undifferentiated and nonspecific stress response is no longer tenable. What is recognized is that an individual's psychological perceptions of a stressor are more important than the physical characteristics of the stress itself in determining adrenocortical responses. Factors that modulate this response can be ascribed to parameters specific to the stimulus or to the individual (Table 1).

Stimulus Parameters

With regard to environmental parameters, those most readily studied involve stimulus intensity. Although it is not possible to devise a protocol to look at strict

Table 1. Parameters Modulating Cortisol Responses

Stimulus Parameters	*Physiological Parameters*	*Psychological Parameters*	*Social Parameters*
Stimulus Intensity	Gender	Personality	Position in hierarchy
Stimulus Duration	Age	Behavior	
Novelty	Physical Fitness	Affect	
Control		Coping Mechanisms	
Predictability		Locus of Control	

dose–response relationships in humans, a number of studies looking at the problem from a qualitative perspective have demonstrated that corticosteroid responses are proportional to the intensity of stress. Plumpton and colleagues (1969) have shown that the cortisol response to major surgery is larger and of longer duration than that occurring with minor procedures. Other stimulus parameters are even less amenable to precise definition; for example, the duration of a stress may be prolonged, despite its relatively discrete temporal onset, by its psychological sequelae (e.g., bereavement).

Studies of cortisol responses looking at paradigms involving persistent threat or constant high levels of demand, have generally shown rapid adaptation to the stimulus such that the individual's baseline response to chronic stimulation shows little if any hormonal difference from the unstimulated state. As it turns out, it is apparent that psychobiological stress responses are most intense when the provoking stimulus is a novel one, and this is particularly true of glucocorticoid responses. Levine's (1983) classical study of parachutists and other studies of individuals engaged in similarly dangerous pursuits have demonstrated maximal glucocorticoid and catecholamine responses on the first exposure to the stimulus, with rapid diminution over repeated exposure. Such progressive insensitivity may be a consequence of psychological modulation (e.g., coping; see below) or may reflect physiological adaptation built into many neurobiological mechanisms. It certainly is not a manifestation of Selye's (1936) notion of "exhaustion", as adequate adrenal reserve may be readily demonstrated through the application of an appropriate novel stimulus.

Behavioral control is a psychological notion that refers to the availability of a behavioral response that can be used to switch off or modulate an aversive stimulus. Animal studies have shown that individuals lacking such control typically have higher corticosteroid responses.

Individual Parameters

Even if controls for variation in stimulus parameters are maintained, there are still significant interindividual differences in stress responses. Constitutional factors, including gender, age, and physical fitness have all been cited as important contibutors to such variations. Moreover, it has become clear that an individual's psychological perception of a stressor is just as important as the physical characteristics of the stress itself in determining adrenocortical responses. For example, a number of studies have demonstrated that anticipation of a stressful event (e.g., surgery, exercise, examination) is as effective a stimulus as the stressful event itself.

Attempts have been made to ascribe interindividual differences to a variety of psychological factors including variations in personality type as well as the use of different coping mechanisms. Several personality variables have been shown to modulate psychobiological stress responses. In the majority of studies, personality traits such as anxiety, neuroticism/emotional lability, and type-A behavioral characteristics have been highlighted as important factors accounting for major interindividual differences in stress-induced adrenocortical responses; even so, some workers have reported no relationship between cortisol responses to stress and psychological variables. Herbert (1986), looking at endocrine and psychological function in 38 male medical students, found that increased cortisol was predicted by several trait scales. The "lie" scale of the Eysenck personality inventory and "debilitating" anxiety as defined by the Alpert-Heber scale correlated negatively with changes in cortisol. Type-A coronary-prone behavior has been evaluated in relation to glucocorticoid response in a number of studies; the data obtained from these are rather conflicting, with evidence both for and against differences in cortisol responses between groups defined according to the A/B dichotomy. This is probably because of the wide variety of populations studied, paradigms invoked, and techniques used, for assessing type-A behavior. Lambert and colleagues (1987) looked at 149 young Swedish adults using a Jenkin's activity survey in order to assess for clustering of type-A behavior (irritability, competitiveness, hurried behavior, and work achievement) among a checklist of parameters. Their results showed that only certain aspects of the type-A "syndrome" correlated with stress responses, whereas other variations were gender dictated. Specifically, in males, higher irritability and increased competitiveness were predictive of lower cortisol levels, but in females, it was higher irritability and lower competitiveness.

Coping

Coping is a psychological concept that is used to denote those mechanisms invoked by an individual in order to avert or reduce the impact of potentially stressful events once they have taken place. It involves both cognitive processes (appraisal of the stressor) and behavioral strategies (denial, displacement, and disengagement). Coping styles are important modulators of the adrenal stress response, although it is probable that it is the effectiveness rather than the type of the coping strategy applied by an individual that is of prime significance. Vickers (1988) has suggested that effectiveness of defense, a clinical assessment based on emotional reaction to stress, disruption of psychological and social functioning, and the ability to mobilize additional defenses to deal with acute superimposed stress correlates well with cortisol responses and that ineffective defenses are associated with higher cortisol excursions. Thus, psychological conditions such as depression and anxiety, which interfere with and impair coping strategies, may result in unattenuated cortisol stimulation.

Disturbances in affect, such as depression or anxiety, are undoubtedly associated with abnormal cortisol responses. Whether these come about as a consequence of disrupted psychological homeostasis through impairment of coping strategies or represent derangements at a more fundamental level (neurotransmitter control of HPA axis activity—*vidé infra*) remains a matter for speculation.

Social Parameters

The final set of factors that have a bearing on glucocorticoid responses relate to the individual's position in a social hierarchy. Thus, it has been shown in human and in animal studies that individuals with a low level of control over their environment (i.e., external locus of control)—those who lie at the lowest echelons of a social structure—tend to display the most dramatic adrenocortical responses in reply to a given stimulus (Sapolsky, 1989).

In conclusion, it may be stated that maximal cortisol responses may be invoked by stressors that are life-threatening, unpredictable, and novel in individuals who are socially nondominant, behaviorally submissive, and in possession of poor coping strategies and a disturbed affect.

LIMBIC SYSTEM AND HIPPOCAMPUS

The role played by psychological elements in the modulation of adrenocortical responses to stressors points to the importance of higher centers in the brain in the regulation of hormonal output. At the center of such modulation lies the mechanism of neurohumoral transduction whereby cognitive processes are transformed into hormonal ones.

Although our structural understanding of the limbic system is firmly grounded upon decades of neuroanatomical work, the functional significance of its various constituents and the role they play in the neurophysiology of emotion remain substantially unaltered since Papez's (1937) hypothesis. While appreciating the fact that cognitive elements perceived at cortical levels acquire "emotional" attribution and humoral sequelae by the interposition of limbic and hypothalamic influences, evidence for specific functional/structural correlation remains scanty.

Theories that attempt to explain the neurophysiology underlying mechanisms that control the neurohumoral transduction center on the integrative role of the hippocampus. This region is of particular interest and importance because of its highly organized structure and its recognized role in memory. Studies of the accumulation of radiolabeled glucocorticoid have shown that the hippocampus is the principal uptake site for cortisol in the brain. The significance of this finding remains unclear.

A model has been proposed whereby the hippocampus, bearing a representation of the external environment based on previous experience and stored as neural engrams, acts as a comparator looking for mismatches between expected and actual events. When a mismatch is found, an effector sequence is initiated that results in the activation of a neuroendocrine cascade. Constant modulation (coping) dampens this output unless the mismatch is perceived as a threat to the organism in which case this system takes control of behavior. Thus the hippocampus functions in two modes: checking or scanning for most of the time, and then on appropriate occasions, transforming itself into an activating role.

CORTICOTROPIN-RELEASING HORMONE (CRH)

Geoffrey Harris (1948) hypothesized that the hypothalamus acts as the key mediator between the central nervous system and the pituitary gland, effecting a wide range of adaptive responses to both physical and psychological stimuli. Soon after that, Saffran and Schally (1955) demonstrated the presence in hypothalamic extracts of humoral factors capable of stimulating the secretion of ACTH from the pituitary gland. Sequencing and synthesis of the 41-amino-acid, corticotropin-releasing hormone was achieved in 1981.

There are several major systems of CRH neurons. The best characterized of these is the paraventricular nucleus–median eminence (PVN–ME) pathway, which is responsible for most of the ACTH regulatory activity associated with CRH. CRH neurons are also found in the limbic system (particularly the amygdala and the nucleus of the stria terminalis) and as scattered interneurons within the cerebral cortex. CRH immunoreactivity has also been localized to a number of peripheral sites including the lungs, liver, gastrointestinal tract, adrenal medulla, and testis, but the role played by CRH at these sites is unclear. The human placenta is particularly rich in CRH immunoreactivity. The CRH gene has been isolated from a variety of species and shows considerable interspecies homology. The rat and human sequences share 94% nucleotide homology, while the ovine sequence shares approximately 80% with each of these. In humans, the gene is located on chromosome 8 (Arbiser et al., 1988). Tissue-specific expression of the rat CRH gene has been reported (Thompson et al., 1987) and closely parallels the immunohistochemical distribution described above. The hCRH gene has cAMP, glucocorticoid, and estrogen response elements (Vamvakopolous and Chrousos, 1994). The demonstration of direct estrogen effects implicates the CRH gene as a mediator of gender-related differences in the stress response. As far as the gene product is concerned, the rat and human peptides are identical while the ovine molecule varies by only seven amino-acid substitutions. CRH stimulatory action causes a dose-related increase in the synthesis and secretion of ACTH; this effect is brought about by the binding of the peptide to specific cell membrane receptors on corticotrophs. In addition to the pituitary, CRH receptors are also present in most areas of the cerebral cortex, limbic system, and brain stem which are known to contain CRH neurons (Aguilera et al., 1987).

Most studies indicate that there is only one class of receptor: this is coupled to guanylyl nucleotide proteins and adenylyl cyclase. Activation induces a rise in intracellular cAMP, transmembrane Ca^{2+} flux, ACTH secretion and proopiomelanocortin (POMC) synthesis (Insel et al., 1988).

Hypothalamic CRH

CRH motoneurons are concentrated in the paraventricular nucleus of the hypothalamus and project to the median eminence, thereby effecting the activation of the HPA axis. These neurons receive a variety of inputs. Chief among these are relays via the *striae terminalis* from limbic structures, which include the amygdala and the hippocampus formation, as well as ascending visceral sensory information (α_2-noradrenergic from the *nucleus solitarius* and α_1-noradrenergic neurons from the ventrolateral medulla). There is also a wealth of connections from nearby hypothalamic regions—no doubt acting as a mechanism for mediating local hormonal interactions.

Behavioral Effects of CRH

In the rat, direct intracerebroventricular (ICV) administration of CRH (Koob and Bloom, 1985) results in activation of the HPA axis as well as of the sympathetic nervous system, with consequent elevation in plasma catecholamines and suppression of the hypothalamopituitary gonadal axis. Distinct behavioral responses have also been observed including dose-dependent locomoter stimulation, general arousal, decreased feeding and sexual behavior, increased exploratory maneuvers, hostility, enhancement of conditioned fear responses, and improved acquisition of the learned response in a visual discrimination task. These effects are independent of factors at lower levels of the HPA axis as they occur in hypophysectomized and dexamethasone pretreated animals. In contradistinction to the effects of the agonist, ICV administration of a CRH antagonist increases exploratory behavior and diminishes the acquisition of conditioned fear responses (Britton et al., 1986).

As there is no evidence to support the suggestion that peptides can cross the bloodbrain barrier, it is interesting to note that a number of observers have described neurotropic effects of CRH administered peripherally. These include behavioral changes in rhesus monkeys after peripheral intravenous administration of ovine CRH. In humans, hCRH has been shown to augment selective attention as indicated by an increased difference between evoked potential waveforms to attended and unattended stimuli (Fehm, 1987). Circumventricular organs (the subfornical organ, the subcommisural organ, and the *area postrema*) may be subject to bloodborne peptides, and a number of hormones are known to influence receptors located there.

ARGININE VASOPRESSIN (AVP)

AVP-Containing Pathways in the Hypothalamus

There are several distinct AVP-containing pathways in the hypothalamus. A pathway originating in the magnocellular neurons of the PVN and SON, passing through the *zona interna* of the median eminence, and concluding at the neurohypophysis is responsible for the circulating peptide with its classical renal and vascular actions. Another pathway originates in the medial parvocellular nucleus of the PVN and passes to the *zona externa* of the median eminence at the portal capillary plexus. AVP secreted into the portal blood acts as a secretogogue for ACTH, potentiating the action of CRH.

Behavioral Effects of AVP

Apart from its well known effects on the renal collecting ducts and arteriolar smooth muscle, AVP plays an important role in the control of ACTH release and also manifests significant behavioral effects. There is an extensive literature on the role of vasopressin in memory. This stems from the primary observation that the removal of the posterior pituitary from the rat was found to interfere with escape behavior, a deficit that was restored using pitressin. Subsequent studies have shown that AVP and endocrinologically inactive analogues enhance consolidation and retrieval of memory and that these effects are long-lived, lasting for several days or even weeks. The posterior thalamus and, in particular, the parafascicular nucleus, appear to be the main site of this action. Bilateral lesions to this area have abolished the behavioral effects of AVP. Interestingly, the effect of AVP on passive avoidance has been shown to be blocked by lesions to the dorsal noradrenergic system made in the *locus ceruleus* (Kovacs et al., 1979). In human subjects, desmopressin given daily for one week has been shown to improve both long- and short-term episodic memory (Nebes et al., 1984). Studies looking at patients with senile dementia of the Alzheimer type (SDAT) who were given desmopressin show significant enhancement of semantic memory (Weingartner et al., 1981). Therapeutic responses have also been seen in patients with retrograde amnesia and Korsakoff's syndrome, although there also are data that conflict with these findings.

ADRENOCORTICOTROPIC HORMONE (ACTH)

Regulation of ACTH

Adrenocorticotropin (ACTH) derives from a larger precursor molecule, proopiomelanocortin (POMC), which also gives rise to a number of other peptides including β-lipotrophin (βLPH) and its carboxy terminal opioid β-endorphin. ACTH and βLPH are stored in the same secretory granules and are released simultaneously. The release of POMC products is under the control of a number of mechanisms. The most straightforward of these is the regulatory role played by cortisol through the mechanism of inhibitory feedback. A rate-sensitive feedback system maintains fine control over basal secretion of ACTH in situations of acute stress, modulating release of ACTH from secretory granules. A second delayed system acts on POMC gene transcription and mRNA translation. In addition to these mechanisms, ACTH release has also been shown to be modulated by a number of hypothalamic factors. These include AVP and CRH, which have been shown to act as effective secretagogues both individually and in synergy. The functional significance of an α_1-noradrenergic stimulatory mechanism is unknown. It seems to be mediated through AVP and is known to play a part in the ACTH response to ingestion of a meal, and the cortisol secretory pattern during waking hours (Aldamluj et al., 1987). ACTH release is known to be inhibited by opiates, though this effect would seem to be mediated through CRH and atrial natriuretic peptide. The roles played by other monoamine neurotransmitter mechanisms remain unclear.

Behavioral Effects of ACTH

There is substantial evidence showing that peptides derived from POMC exert neurotropic effects. De Wied has shown that hypophysectomized rats, which are thus rendered deficient in ACTH, show impaired acquisition of avoidance conditioning. This can subsequently be restored by the administration of ACTH or indeed by a number of the ACTH molecule's fragments including αMSH and $ACTH_{4\text{-}10}$. βLPH, which shares the amino acid sequence 4-10, has even more potent effects. Smotherman and Levine (1978) have shown that, in rats, ACTH delays the extinction of learned responses such as taste aversion induced with lithium and alleviates the amnesia produced by CO_2 inhalation, electroconvulsive shock, or intracerebral administration of puromycin, an inhibitor of protein synthesis. Lesions to the *nucleus parafascicularis* in the thalamus and the anterior hippocampus block these effects. Administration of $ACTH_{4\text{-}10}$ to human subjects has been shown to increase selective attention to the exclusion of other environmental clues but does not alter consolidation or retrieval of memory (d'Elia and Frederikson, 1980). These behavioral effects are generally short-lived lasting from one to four hours. Thus, ACTH and subfragments such as $ACTH_{4\text{-}10}$, which lack endocrine capability, appear to exert profound behavioral effects. They do not have a direct effect on memory but accentuate selective attention allowing the individual to focus on the task at hand, and eliminating extraneous influences.

CORTISOL AND CEREBRAL GLUCOCORTICOID RECEPTORS

In recent years, there has been great interest in the molecular biology of corticosteroid action on the brain, particularly with regard to the interaction between the hormone molecule and receptor systems. Studies in rats reveal the presence of two distinct populations among glucocorticoid receptors (De Kloet, 1991). It appears that the CR (corticosterone-preferring receptor) has its principal localization in neurons of the hippocampus and mediates tonic influences of corticosterone (the dominant glucocorticoid in the rat) on hippocampus-associated functions. On the other hand, the glucocorticoid receptor (GR) mediates feedback action of corticosterone on stress-activated brain processes. A number of behavioral effects of corticosteroids have been attributed to their action via GR receptors. These include effects on sleep, the detection and perception of sensory stimuli, food intake, and affect. In contrast to CR receptors, GR receptors are subject to autoregulation. Thus, adrenalectomy results in the upregulation of GR receptors (increased activity or numbers of receptors) but has no effect on CR receptors.

PSYCHOLOGICAL ASPECTS OF ENDOCRINE DISORDERS OF THE HPA AXIS

Cushing's Syndrome

Since Cushing's original observation published in the American Journal of Insanity in 1913, psychiatric disturbances have been recognized as a central feature of hyperadrenocorticalism. Despite the heterogenous nature of Cushing's syndrome, most investigations of behavioral change in these disorders have made no distinction between pituitary and ectopic ACTH-dependent or adrenal diseases. Mental changes are usually diagnostically nonspecific, although depressive symptoms predominate in Cushing's disease whereas euphoria is generally believed to be more common in exogenous hypercortisolism. Risk factors for depression, including an excess of adverse early life events and a family history of suicide and affective disorders, are common in patients with Cushing's disease, prompting the (not–so–far-fetched) suggestion that psychiatric factors may even be etiological in pituitary-dependent diseases. In Trethowan and Cobb's (1952) original study of 25 patients with Cushing's syndrome, depression was the most common complaint. In Cohen's (1980) study of 29 patients, 86% displayed depressive symptoms that included insomnia, tearfulness, irritability, and somatic preoccupation; 50% had a family history of depression. Kelly and colleagues (1980), using a present state examination, found that 50% of their patients were depressed. Symptoms were more prominent in patients with active as compared with inactive disease. Treatment resulted in a significant decline in the severity of their depression as measured by the Hamilton Rating Scale.

Acute administration of corticosteroids at doses above 30 mg prednisolone per day is commonly associated with the onset of hyperactivity, hyperphagia, insomnia, and other euphoric symptoms within several days (Ling et al., 1981). Their sudden withdrawal may precipitate depressive symptoms. There has been considerable speculation about the relative contributions of ACTH and cortisol in the etiology of these behavioral changes. A number of investigators have found a much higher rate of depression in patients with pituitary-dependent disease suggesting that ACTH levels may be significant. Interestingly, however, treatments that elevate ACTH, such as metyrapone, mitotane, and adrenalectomy, have been used effectively to eliminate mental symptoms. In the case of Nelson's syndrome, the exceedingly high levels of ACTH that arise as a consequence of bilateral adrenalectomy are not known to be associated with any psychiatric complications. The increased propensity of ACTH-dependent sources to invoke mental changes may be related to some other as yet unexplored characteristic. For example, it has been suggested that the critical factor in these mental disturbances is not absolute levels of corticosteroid but rather sudden changes in output against a background of cerebral corticosteroid receptor up- or downregulation (Ur et al., 1992b). In this respect, the increasingly recognized tendency of pituitary-dependent and ectopic sources to cycle may be a significant factor. Further studies are warranted.

Addison's Disease

The occurrence of psychological symptoms in Addison's disease is well recognized and has formed the basis for a number of studies. In Addison's original study, patients with adrenal insufficiency were described as having "an inability to concentrate, drowsiness, restlessness, insomnia, irritability, apprehension, and disturbed sleep". Engel and Margolin's (1941) study of 25 Addisonian patients revealed 16 with significant psychiatric symptoms. Of these three were psychotically depressed, and a further six, were severely depressed. Symptomatic improvement has depended upon hydrocortisone replacement. Ur and colleagues (1992b) describe an acute self limiting manic response to treatment with physiological doses of glucocorticoids, which they suggest is due to upregulation of hippocampal glucocorticoid receptors as a consequence of prolonged hypocortisolemia.

HPA AXIS DISTURBANCES IN NEUROPSYCHIATRIC DISORDERS

Major Depression

Over the past 30 years, there has been extensive research into the relationship between the hypothalamic–pituitary–adrenal axis activity and psychiatric disorders. Studies involving the measurement of plasma cortisol concentrations have shown increased secretion of this hormone in some depressed subjects (Butler and Besser, 1968). Large numbers of studies have shown that a certain proportion of patients with major depression (MD) demonstrate resistance to pituitary–adrenal suppression with the administration of dexamethasone, and this phenomenon has been clinically formalized in the dexamethasone suppression test (DST) as described by Carroll (1982). In this paper and in subsequent publications, a range of 40% to 50% of DST nonsuppressors has been found within selected populations of patients diagnosed as depressed according to various criteria. Pfohl and colleagues (1985) have shown consistently high ACTH levels in dexamethasone nonsuppressors with significant differences compared to suppressors and normals. Further studies have shown enhanced adrenal sensitivity to exogenous synacthen stimulation and radiographic evidence of adrenal hyperplasia in depressed subjects relative to normal controls.

It has been suggested that the hypercortisolemia of depression is a consequence of excessive hypothalamic CRH drive. Nemeroff and colleagues (1984) have shown raised levels of immunoreactive CRH in the cerebrospinal fluid of depressed subjects. Studies examining hormonal responses to an intravenous bolus dose of oCRH have shown characteristic rises in ACTH levels in normal subjects and exaggerated responses in patients with Cushing's disease (CD). In contrast, patients with ED have shown attenuated ACTH (though normal cortisol) responses (Gold et al., 1984). Continuous 24-hour infusions of oCRH in normal subjects have resulted in a modest hypercortisolemia, which shares some of the features of that found in MD. These include the preservation of circadian rhythms in cortisol and ACTH secretion and an attenuated ACTH response to a further bolus dose of oCRH (Schulte et al., 1985). It has been suggested that the attenuated ACTH response to a bolus injection of oCRH in depressives is a consequence of the functionally robust negative feedback inhibition exerted by raised levels of cortisol. More recent studies have shown that with the abolition of this negative feedback inhibition using metyrapone—an 11-β-hydroxylase inhibitor—hypercortisolemic depressives show exaggerated rises in ACTH as compared with normocortisolemic controls (Ur et al., 1992a). Similar responses are seen in normal subjects infused with CRH over 48 hours, suggesting that CRH hyperactivity may be an underlying characteristic of major depression.

Although the hypercortisolemia of depressive illness and the underlying CRH overdrive appear to be consistently demonstrable, it remains unclear as to whether these biological markers reflect primary derangements in neural mechanisms that modulate HPA axis activity or whether they exist purely as epiphenomena. Mullen and colleagues (1986) have argued that HPA axis activation is a consequence of sleep disturbance and weight loss alone and have replicated dexamethasone resistance in normal volunteers by fasting and sleep deprivation. On the other hand, animal studies have consistently shown that central administration of CRH produces behavioral and neuroendocrine changes characteristic of severe stress, many of which are also seen in major depression. These include behavioral activation in nonstressful surroundings and anxiogeniclike effects in novel environments that appear to augment the effects of stress. It has therefore been proposed that activation of CRH, both hypothalamic and possibly extrahypothalamic, plays a crucial role in the central and peripheral manifestations of depressive illness. The development of lipid-soluble, orally active analogues of CRH-41, both agonists and antagonists, should advance our understanding of the pathogenesis of depression considerably.

Posttraumatic Stress Disorder

Posttraumatic stress disorder (PTSD) is another psychiatric condition that is associated with significant alterations in neuroendocrine regulation. The syndrome is characterized by three sets of symptoms: (i) recurrent and intrusive recollections of a traumatic event, (ii) avoidant symptoms (feelings of detachment and estrangement), and (iii) persistent symptoms of hyperarousal (e.g., hypervigilance, exaggerated startle response).

Evidence of HPA dysregulation in these patients includes lower mean 24-hour urinary cortisol secretion, lower baseline plasma cortisol concentrations, increased glucocorticoid binding in lymphocytes, increased sensitivity to the HPA-suppressive effects of dexamethasone, and lower levels of pituitary ACTH secretion in response to exogenous challenge with CRH. Taken together, these abnormalities suggest *enhanced* negative-feedback sensitivity in these patients.

There is also evidence for hyperactivity in the sympathetic nervous system. Twenty-four–hour urinary excretion of catecholamines is significantly elevated in patients with PTSD and this probably underlies observed increases in resting heart rate, systolic blood pressure, and conditioned response. Moreover, catecholamine levels appear to correlate with specifically intrusive PTSD symptoms such as flashbacks (Southwick et al., 1993). It has been suggested by several experts, notably Yehuda and his colleagues that the development of PTSD may be facilitated by an atypical biological response in the immediate aftermath of a traumatic event, which in turn leads to a maladaptive psychological state.

Chronic Fatigue Syndrome

A chronic fatigue syndrome (CFS) has been described following various infections. This syndrome is associated with hypersomnia or sleep disturbance and may be closely related to the vegetative form of atypical depression. The Centers for Disease Control (CDC) have defined chronic fatigue syndrome as a persistent or relapsing debilitating fatigue for at least six months in the absence of any definable medical diagnosis. Symptom criteria include abrupt onset of fever, arthralgias, myalgias, painful adenopathy, neuropsychological complaints, and sleep disturbances. A number of hypotheses have been advanced in order to account for the etiology of this condition. Recent findings of abnormalities in cell-mediated and humoral immunity, along with atypical profiles of antibody responses to viral antigens, have led to the suggestion that persistent viral infection is of pathogenetic significance. Chronic and excessive production of cytokines after viral infection or other antigenic challenge has been proposed as the pathophysiologic basis of CFS. This suggestion has been supported by the fact that a syndrome resembling CFS results from the administration for therapeutic purposes of recombinant α-interferon and of interleukin-2 (IL-2). Several pieces of evidence suggest that the HPA axis is of interest as a possible site of pathology in patients with CFS. Many of the symptoms of the syndrome, and fatigue in particular, are characteristic of glucocorticoid insufficiency. Demitrack and colleagues (1991) have found evidence for impaired activation of the HPA axis in a large group of patients with CFS and have proposed a putative deficiency of CRH as being fundamental to the pathophysiology of the condition.

CONCLUSIONS

While stress is difficult to define and even harder to quantify, activation of the HPA axis has long been considered one of its most crucial hallmarks. Most studies have concluded that activation of the HPA axis during psychological stress will occur either as a specific type of coping response, particularly passive coping, or as subjective perception of inability to cope. Activation of the HPA axis under these circumstances gives rise to a cascade of neuropeptide events including release of central peptides such as CRH-41, vasopressin, and ACTH, and peripheral activation of ACTH and cortisol. The latter seems to have as its principal target the glucocorticoid receptors in the hippocampus, the hippocampus itself being a site where cognitive and motivational systems interact. The neuropsychological effects of corticosteroids at these sites remain obscure, expecially as data in humans arise principally from pathological states of chronic excess in which serious defects in effective mental functioning are apparent. However, extrapolation from animal studies might suggest that steroid receptor activation in the hippocampus modifies information-processing at this site, enabling the organism to deal more effectively with changes in its environment. While changes in this system clearly occur in states of depressive illness, the relation between cause and effect in this situation remains unelucidated.

REFERENCES

Aguilera, G., Millan, M., Hauger, R.I., & Catt, K.J. (1987). Corticotropin releasing factor receptors: Distribution and regulation in brain, pituitary, and peripheral tissues. Ann. N. Y. Acad. Sci. 512, 48-66.

Aldamluji, S.A., Perry, L., Tomlin, S., Bouloux, P.M.G., Grossman, A.B., Rees, L.H. & Besser, G.M. (1987) Alpha-adrenergic stimulation of corticotrohin secretion by a specific central mechanism in man. Neuroendocr. 45, 68-76.

Arbiser, J.L., Morton, C.C., Bruns, G.A., & Majzoub, J.A. (1988). Human corticotropin releasing hormone gene is located on the long arm of chromosome 8. Cytogenet. Cell. Genetics 47, 113-116.

Britton, K.T., Lee, G., Vale, W., Rivier, J., & Koob, G.F. (1986). Corticotropin releasing factor (CRF) receptor antagonist blocks activating and anxiogenic actions of CRF in the rat. Brain Res. 369, 303-306.

Butler, P.W.P. & Besser, G.M. (1968). Pitutary adrenal function in severe depresive illness. Lancet 1, 1234-1236.

Carroll, B.J. (1982). The dexamethasone suppression test for melancholia. Br. J. Psychiatry 140, 292-304.

Cohen, S.I. (1980). Cushing's syndrome: A psychiatric study of 29 patients. Br. J. Psychiatry 136, 120-124.

d'Elia, G. & Frederikson, S.O. (1980). ACTH410 and memory in ECT-treated and untreated patients. Acta Psych. Scand. 62, 418-428.

De Kloet, E.R. (1991). Brain corticosteroid receptor balance and homeostatic control. Front. Neuroendocr. 12, 95.

Demitrack, M.A., Dale, J.K., & Straus, S.E. (1991). Evidence for impaired activation of the hypothalamo–pituitary–adrenal axis in patients with chronic fatigue syndrome. J. Clin. Endocrinol. Metab. 73, 1224-1234.

Engel, G.I. & Margolin, S.G. (1941). Neuropsychiatric disturbances in Addison's disease and the role of impaired carbohydrate in the production of abnormal cerebral function. Arch. Neurology and Psychiatry 45, 881-884.

Gold, P.W., Chrousos, G., Kellner, C., Post, R., Roy, A., Augerinos, P., Schulte, H., Oldfield, E., & Loriaux, D.L. (1984). Psychiatric implications of basic and clinical studies with corticotropin-releasing factor. Am. J. Psych. 141, 619-627.

Gray, J.A. (1984). The hippocampus as an interface between cognition and emotion. In: *Animal Cognition.* (Roitblat, H.I., Bever, T.G., & Terrace, H.S., Eds.), pp. 607-626. Erlbaum, Hillsdale, NJ.

Harris, G.W. (1948). Neural control of the pituitary gland. Phys. Rev. 28, 134.

Herbert, J., Moore, G.F., De La Riva, C., & Watts, F.N. (1986). Endocrine responses and examination anxiety. Biolog. Psych. 22, 215-226.

Insel, T.R., Battaglia, G., Fairbanks, D.W., & De Souza, E.B. (1988). The ontogeny of brain receptors for corticotropin-releasing factor and the development of their functional association with adenylate cyclase. J. Neurosci. 8, 4151-4158.

Kelly, W.F., Checkley, S.A., & Bender, D.A. (1980). Cushing's syndrome, trytophan and depression. Br. J. Psych. 136, 125-132.

Koob, G.F. & Bloom, F.E. (1985). Corticotrophin-releasing factor and behaviour. Fed. Proc. 44, 259-263.

Kovacs, G.l., Bohus, B., & Versteeg, D.H. (1979). Facilitation of memory consolidation by vasopressin: Mediation by terminals of the dorsal noradrenergic bundle? Brain Res. 172, 73-85.

Lambert, W.W., Macevoy, B., Klackenberglarsson, I., Karlberg, P., & Karlberg, J. (1987). The relation of stress hormone excretion to type A behaviour and to health. J. Human Stress 13, 128-135.

Levine, S. (1983). Coping: An overview. In: *Biological and Psychological Basis of Psychosomatic Medicine.* (Urswin, H. & Murison, R., Ed.), pp. 15-26. Pergamon, Oxford, England.

Ling, M.H., Perry, P.J., & Tsuang, M.T. (1981). Side effects of corticosteroid therapy. Arch. Gen. Psych. 38, 471-477.

Mason, J.W. (1968). A review of psychoendocrine research on the pituitaryadrenal cortical system. Psychosom. Med. 30, 576-607.

Mullen, P.E., Linsell, C.R., & Parker, D. (1986). Influence of sleep disruption and calorie restriction on biological markers for depression. Lancet 2, 1051-1055.

Nebes, R.D., Reynolds, C.F., & Horn, L.C. (1984). The effect of vasopressin on memory in the healthy elderly. Psych. Res. 11, 49-59.

Nemeroff, C.B., Widerlov, E., Bissette, G., Walleus, H., Karlsson, I., Eklund, K., Kilts, C.D., Loosen, P.T., & Vale, W. (1984). Elevated concentrations of CSF corticotropin-releasing factor like immunoreactivity in depressed patients. Science 226, 1342-1344.

Papez, J.W. (1937). A proposed mechanism of emotion. Arch. Neur. Psych. 38, 725-743.

Pfohl, B., Sherman, B., Schlecte, J., & Stone, R. (1985). Pituitary–adrenal axis rhythm disturbances in psychiatric depression. Arch. Gen. Psych. 42, 897-903.

Plumpton, F.S., Besser, G.M., & Cole, P.B. (1969). Corticosteroid treatment and surgery. Anaesthesia 24, 3-11.

Rose, R.M. (1980). Endocrine responses to stressful psychological events. Psychiatric Clinics of North America 3, 251-276.

Saffran, M. & Schally, A.V. (1955). Release of corticotropin by anterior pituitary tissue *in vitro*. Can. J. Biochem. 33, 408.

Sapolsky, R.M. (1989). Hypercortisolism among socially subordinate wild baboons originates at the CNS level. Arch. Gen. Psychiatry 46, 1047-1051.

Schulte, H.M., Chrousos, G.P., & Gold, P.W. (1985). Continuous administration of synthetic ovine CRF in man. J. Clin. Invest. 75, 1781-1785.

Selye, H. (1936). Stress syndrome. A syndrome produced by diverse noxious agents . Nature 138, 32.

Smotherman, W.P. & Levine, S. (1978). ACTH effects on response supression and plasma corticosterone in the mouse. Physiol. Behavior 20, 503-507.

Sokolov, E.N. (1966). Neuronal mechanisms of the orienting reflex. Proceedings of the 18th International Congress of Psychology Symposium 5, 31-36. (Abstract.)

Southwick, S.M., Krystal, J.H., Morgan, C.A., Johnson, D., Nagy, L.M. Nicholaou, A., Henigar, G.R., & Charney, D.S. (1993). Abnormal noradrenergic function in posttraumatic stress disorder. Arch. Gen. Psychiatry 50, 267-274.

Thompson, R.C., Seasholtz, A.F., & Herbert, E. (1987). Rat corticotropin releasing hormone gene: Sequence and tissue specific expression. Mol. Endocrinol. 1, 363-370.

Trethowan, W.H. & Cobb, S. (1952). Neuropsychiatric aspects of Cushing's syndrome. Arch. Neur. Psych. 67, 283.

Ur, E. (1991). Psychological aspects of hypothalamo–pituitary–adrenal activity. Baillieres. Clin. Endocr. Metab. 5(1), 79-96.

Ur, E., Dinan, T.G., O'Keane, V., Turner, T.H., Clare, A.W., Mcloughlin, L., Rees, L.H., Grossman, A., & Besser, G.M. (1992a). The effect of metyrapone on the pituitary–adrenal axis in depression: Relation to dexamethasone suppressor status. Neuroendocrinology 56, 533-538.

Ur, E., Turner, T.H., Goodwin, T.J., Grossman, A., & Besser, G.M. (1992b). Mania in association with hydrocortisone replacement for Addison's disease. Postgrad. Med. J. 68, 41-43.

Vale, W., Speiss, J., Rivier, C., & Rivier, J. (1981). Characterization of a 41-residue ovine hypothalamic peptide that stimulates secretion of corticotropin and betaendorphin. Science 213, 1394-1397.

Vamvakopolous, N.C. & Chrousos, G.P. (1994). Hormonal regulation of human corticotropin-releasing hormone gene expression: Implications for the stress response and immune/inflammatory reaction. Endocr. Rev. 15, 409-420.

Vickers, R.R. (1988). Effectiveness of defenses: A significant predictor of cortisol excretion under stress. J. Psychosom. Res. 32, 21-29.

Weingartner, H., Gold, P., & Ballenger, J.C. (1981). Effect of vasopressin on human memory functions. Science 211, 601-603.

RECOMMENDED READINGS

Bremner, J.D. (1999). Does stress damage the brain? Biol. Psychiatry 45, 797-805.

De Bellis, M.D. Baum, A.S., Birmaker, B., Keshavan, M.S., Eccaid, C.H., Boring, A.M., Jenkins, F.J., & Ryan, N.D. (1999). Biological stress systems. Biol. Psychiatry 45, 1259-1270.

Friedman, M.J., Charney, D.S., & Deutch, A. (1995). Neurobiological and Clinical Consequences of Stress. From Normal Adaptation to Post-traumatic Stress Disorder. Lippincott-Raven, Philadelphia.

Komaroff, A.L., & Buchwald, D.S. (1998). Chronic fatigue syndrome: an update. Ann. Rev. Medicine 49, 1-13.

Magarinos, A.M. & McEwen, B.S. (1995). Stress induced atrophy of apical dendrites of hippocampal CA3c neurons: Involvement of glucocorticoid secretion & excitatory amino acid receptors. Neuroscience 69, 89-98.

Pagani, M. & Lucini, D. (1999). Chronic fatigue syndrome: A hypothesis focusing on the autonomic nervous system. Clin. Science 96, 117-125.

Simon, R.I. (1999). Chronic posttraumatic stress disorder: A review and checklist of factors influencing prognosis. Harvard Rev. Psych. 6, 304-312.

Yehuda, R. (1999). Biological factors associated with susceptibility to post-traumatic stress disorder. Can. J. Psychiatry 44, 34-39.

Chapter 7

Stress and the Immune System

CARMINE M. PARIANTE and ANDREW H. MILLER

Biological Psychiatry, pages 135-153.

ISBN: 1-55938-819-6

INTRODUCTION

Over the past several decades, a large body of data has been amassed that convincingly demonstrates the capability of stress to modulate a wide range of immune responses. Studies in the late 1950s and early 1960s on laboratory animals focused primarily on the ability of stress to influence disease expression including infections and virus-induced tumors, while more recent efforts have focused on the immunologic and neuroendocrine mechanisms of these effects and the impact of stress in humans. Although much of the interest has been on the negative impact of stress on immune function, studies have also shown that stressors can both enhance and inhibit immune responses. Many of these qualitative issues appear to be related to whether the stress is acute or chronic and which immune response in what immune compartment (blood, spleen, lymph nodes, gut, skin, lung, etc.) is being examined. In addition, stressors applied in the context of an active immune response (e.g., during an infection) may have differential effects depending on the timing of the stress. For example, there can be either an exacerbation (typically if the stressor is initiated after exposure to an immune challenge) or an attenuation (typically if the stressor is initiated and terminated prior to an immune challenge) of disease expression. Similarly, the effects of stress relate to the particular type of immune response, which is being elicited by the pathogen (e.g., humoral versus cellular response).

Most recently, data indicate that stress can elicit the release of immune mediators such as proinflammatory cytokines in the absence of a more formal immune challenge, that is, a pathogen. These cytokines have in turn been shown to induce behavioral syndromes (sickness behavior) and neuroendocrine changes that may contribute to the behavioral and neuroendocrine response to stress. Taken together, the data indicate that stressors enter into a complex equation involving interactions between the neuroendocrine and immune systems. The impact of stress therefore is highly context specific and "relative" to the many host factors "in play" at the time of stress exposure.

The following chapter is designed to provide a foundation for understanding the richness and complexity of the impact of stress on the immune and neuroendocrine systems and attempts to give a balanced perspective as to whether stress-induced changes serve to enhance or inhibit the immunologic response to challenge and the development of diseases.

STRESS AND THE IMMUNE SYSTEM IN LABORATORY ANIMALS

Effects of Stress on Viral Infections and Virus-Induced Tumors

Interest in the effects of stress on the immune response grew out of early studies in laboratory animals that examined the influence of stress on the development of diseases involving the immune system, most notably infectious diseases and tumors. For example, the studies of Rasmussen and colleagues (1957) utilized an experimental model of psychosocial stress and demonstrated that repeatedly exposing mice to a shock-avoidance procedure as well as to physical restraint and high intensity sound resulted in increased susceptibility to herpes simplex virus, poliomyelitis virus (Johnsson and Rasmussen, 1965), Coxsackie B virus (Johnsson et al., 1963), vesicular stomatitis virus (Jensen and Rasmussen, 1963), and polyoma virus infection (Rasmussen, 1969). Similar studies by Ader and Friedman (1965) demonstrated that brief daily handling and mild electric shock administration early in life modified the rate of tumor development and survival of rats injected with Walker 256 sarcoma. In order to understand the mechanisms of these effects of stress on disease development, studies logically began to focus on the direct influence of stress on the immune system.

Effects of Stress on Immune Parameters

Some of the first studies of stress and the immune system utilized electric shock as a stressor. For example, Keller and colleagues (1981) demonstrated a relationship between the nature and intensity of an acute stressor and the degree of suppression of cellular measures of the immune system. A graded series of low-level and high-level tail shock produced a progressively greater suppression of both the number of circulating lymphocytes and of the peripheral blood lymphocyte proliferative response to the mitogen, phytohemagglutinin (PHA). There was no effect of stress, however, on the response to PHA of lymphocytes isolated from the spleen. More recently, Lysle and co-workers (1987) also found that the magnitude of suppression of the mitogen response to electric shock was related to the numbers of shocks administered.

Over the years a wide variety of stressors have been examined in laboratory animals including crowding, rotation, restraint, electric footshock, noise, exposure to a predator, exercise, and so forth (Weiss and Sundar, 1992). These stressors have been shown to inhibit virtually every aspect of the immune response. Stress-induced effects range from decreasing the number of lymphocytes and monocytes in the peripheral blood to inhibiting natural killer cell activity, mitogen-induced lymphocyte proliferation, and the production of antibodies. However, in some cases, especially with very brief and/or mild stressors and repeated or prolonged exposure to a stressor (e.g., restraint, electric footshock, sound), adaptation of the immune response or enhanced immune responses have been found using a variety of immune measures including lymphocyte and splenocyte proliferative responses to mitogens or antigens, natural killer (NK) cell activity, antibody production, and cutaneous delayed-type hypersensitivity (Solomon, 1969; Lysle et al., 1990; Berkenbosch et al., 1991; Jain and Stevenson, 1991; Wood et al., 1993; Dhabhar and McEwen, 1996).

It has been proposed that stress may differentially affect various compartments of the immune system and that some immune parameters may be less stress-sensitive, possibly due to redundancy in control pathways. For example, during a viral infection, physical restraint has been shown to suppress cellular immune responses but not humoral responses in the same animal (Sheridan et al., 1991). Moreover, electric footshock stress was found to enhance antigen-specific humoral and cell-mediated immunity whereas no changes were observed in proliferative responses to polyclonal mitogens (Wood et al., 1993).

Interestingly, Keller and colleagues (1981) described stress-induced decreases in the peripheral blood lymphocyte proliferative response to the mitogen PHA but found no effect of the stress on the response to PHA of lymphocytes isolated from the spleen. In contrast, Lysle and co-workers (1987) observed a marked reduction in the response to mitogenic stimulation of both peripheral blood and splenic lymphocytes. Recently, however, Shurin and colleagues (1994) used a brief stressor (a single, 5 s electric foot shock) in order to limit the activation of multiple hormonal pathways and found that splenic lymphocyte mitogenic activity was suppressed for 1 to 60 min after foot-shock. Blood lymphocyte mitogenic function, on the other hand, was significantly enhanced after one foot-shock, returned to control level 10 min later, was significantly suppressed 30 min later, and was then back to normal an hour later. Differences in stress-induced patterns of changes between spleen and peripheral blood lymphocytes may indicate that different hormonal pathways are responsible for immune changes in the various immune compartments or may represent changes in the lymphocyte subsets in the two compartments secondary to altered cell trafficking patterns as a function of stress (see section below on mechanisms).

Effects of Stress on Proinflammatory Cytokines

It has been shown that different mild stressors like open-field exposure, electric foot-shock, and restraint induce an elevation in plasma levels of interleukin-6 (IL-6) in rodents (LeMay et al., 1990; Zhou et al., 1993). IL-6 is a proinflammatory cytokine that is important as a mediator of the acute phase response (Abbas et al., 1994). Like other proinflammatory cytokines such as IL-1β and tumor necrosis factor α, IL-6 has potent neuroendocrine effects including the capacity to stimulate the release of corticotropin releasing factor (CRF) and adrenocorticotrophic hormone (ACTH)(Besedovsky and Del Rey, 1996)(Figure 1).

Proinflammatory cytokines, like IL-6, also exhibit the capacity to induce a group of behavioral symptoms referred to as "sickness behavior" (Kent et al., 1992). Sickness behavior is a behavioral syndrome that commonly accompanies serious viral or bacterial infections and includes fatigue, loss of appetite, sleep disturbance, social withdrawal, decreased libido, depressed mood, and

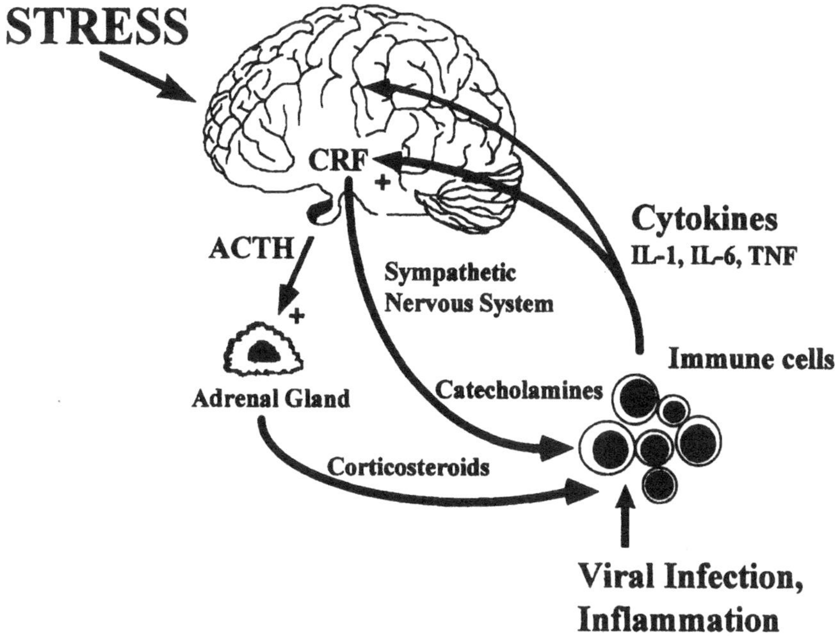

Figure 1. Pathways mediating stress effects on the immune system. Depicted are two major outflow pathways, which are involved in mediating the effects of stress on the immune system: the hypothalamic–pituitary–adrenal axis, which ultimately leads to the release of corticosteroids, and the sympathetic nervous system, which co-releases catecholamines and peptides. Corticotropin releasing factor (CRF) is a neuropeptide that serves as a pivotal regulator of these two pathways. The ability of the immune response to elicit a CRF response through the release of cytokines following a viral infection or inflammation is also indicated, and emphasizes the bi-directional nature of the communication between the brain and immune system.

general malaise. Sickness behavior also occurs in patients undergoing high dose cytokine therapies for neoplastic or viral illnesses. The overlap of symptoms in certain psychiatric disorders (especially major depression) with sickness behavior has raised the possibility that cytokines elicited during stress may contribute to the expression of behavioral alterations in stress-related disorders like major depression. Finally, since IL-6 has potent effects on the immune system, this cytokine may contribute to stress-related changes in immune function.

STRESS AND THE IMMUNE SYSTEM IN HUMANS

Stressful Life Events

Laboratory and clinical studies with humans also have indicated that stress influences the immune response. One of the first studies to suggest a link between stressful events and the immune system in humans was conducted by Bartrop and colleagues (1977) who found that bereaved individuals had lower mitogen-stimulated lymphocyte proliferative responses compared to controls. Schleifer and coworkers (1983) confirmed these findings in a prospective study of spouses of women with advanced breast carcinoma. In this study, lymphocyte proliferative responses to the mitogens PHA, concanavalin A (Con A), and pokeweed mitogen (PWM) were significantly lower during the first two months following bereavement compared to prebereavement responses, while no differences in lymphocyte subpopulations in the peripheral blood were found. In some of the individuals, the impaired proliferative responses were still present up to one year after the death of the spouse.

Kiecolt-Glaser and Glaser (1991), who have investigated the association of a range of stressful life events with the immune response, initially focused their attention on academic stress among medical students as a commonplace stressful situation. They found a decrease in natural killer (NK) cell activity during the final examination period as compared to the preexamination baseline response. The examination stress was also associated with decreases in the number of T cells, decreased mitogen responses and interferon production, increased antibody titers to latent herpes viruses (a putative marker of decreased cellular immune function), and a reduced antibody response to recombinant hepatitis B vaccine (Glaser et al., 1992).

Kiecolt-Glaser and colleagues also evaluated the effects of chronic life stressors such as caregiving for Alzheimer's patients and found alterations in lymphocyte subpopulations and increased antibody titers to herpes simplex virus (Kiecolt-Glaser et al., 1987). In a prospective study on a similar population, caregivers to Alzheimer's victims also exhibited decreased proliferative responses to mitogens and more days of illness from infectious disease compared to matched controls (Kiecolt-Glaser et al., 1991). Finally, and most noteworthy from a clinical point of view, subjects exposed to chronic stress also were found to exhibit impaired antibody responses to an influenza virus vaccine (Kiecolt-Glaser et al., 1996) and a longer latency in wound healing (Kiecolt-Glaser et al., 1995).

It is an interesting observation that advancing age increases the responsivity of the neuroendocrine and immune systems to the physiologic effects of stress and depression. For example, studies with laboratory animals suggest that older animals show both a greater sensitivity to stress-induced immune alterations (Lorens et al., 1990) and an impaired capacity to terminate the hypothalamic–

pituitary–adrenal (HPA) axis (Sapolsky et al., 1986) and the sympathetic nervous system (Lorens et al., 1990; Milakofsky et al., 1993) responses to stress.

In humans, age has been shown to be a factor along with chronic stress and social support in determining the cardiovascular reactivity to stress (Uchino et al., 1992). In the largest study to date investigating immune abnormalities in major depression (Schleifer et al., 1989), examination of 91 depressed patients and 91 age- and sex-matched controls failed to provide evidence of between-group differences but revealed significant age-related differences. For example, compared to controls, depressed patients showed decreased lymphocyte responses as a function of advancing age. Similar age-related differences between depressed subjects and controls were found for CD4+ lymphocytes.

The interaction between stress, depression, and age is important because older people also show age-related alterations in immune function (immunosenescence), a phenomenon that may leave them more vulnerable to stress-induced immune alterations and the development of immune-related diseases including infectious diseases and cancer. A case in point is the findings of Pariante and colleagues (1996), who examined female caregivers of handicapped people. Caregivers had a significantly lower percentage of T cells, a significantly higher percentage of T-suppressor/cytotoxic cells and a significantly lower T-helper:suppressor ratio. It is a matter of interest that older caregivers (>45 years, median of age) also had lower numbers of T cells and T-Helper cells and higher antibody titers for cytomegalovirus. Altogether, such results confirm the view that age plays a role in modulating the influence of stress on the immune system.

Laboratory Stressors

In addition to life stressors, brief experimental stressors like mathematical tests and puzzle solving, with or without "frustrating" conditions, have also been shown to influence immune parameters. Typically, the immune alterations include decreased proliferative responses to mitogens and increased numbers of circulating NK and T-suppressor/cytotoxic cells (Kiecolt-Glaser et al., 1992). Although it is difficult to compare these paradigms with stressful life events, laboratory studies have provided insight into the possible mechanisms of stress-related immune alterations in humans. For example, immune changes have been reported to occur within 5 minutes of the onset of the stressor and are more pronounced in those subjects who have the greatest cardiovascular reactivity (measured by systolic and diastolic blood pressure and heart rate). Such results have led to the suggestions that catecholamines play a predominant role in mediating these effects (Herbert et al., 1994). In addition, these modifications in immune parameters induced by experimental stress appear to be influenced by previous life experiences such as increased daily hassles being associated with greater stressor-induced decrease in T and NK cells in peripheral blood. These findings support the hypothesis that psychological stress, chronic psychological stress in particular, may not only have a direct effect on immune system per se, but also may modify the reaction of the immune system to subsequent acute stressful events (Brosschot et al., 1994).

MECHANISMS OF STRESS EFFECTS ON THE IMMUNE SYSTEM

It is widely recognized that the primary biological components of the stress response are the HPA axis and the autonomic nervous system (ANS), specifically the sympathetic branch. Both outflow pathways are involved in mediating stress-induced alterations in immune parameters, depending upon the type of stress and/or immune compartment (Figure 1).

Hypothalamic-Pituitary-Adrenal Axis

The HPA axis is responsive to three distinct physiological stimuli: circadian rhythm inputs from the suprachiasmatic nucleus, physical and psychological stress, and inflammatory and immune reactions. Stressful experiences of a wide range are capable of stimulating the HPA axis with the resultant release of corticosteroids (i.e., cortisol in humans and nonhuman primates, and corticosterone in mice and rats). Corticosteroids mediate their actions on target tissues through two distinct intracellular receptor subtypes referred to as the mineralocorticoid receptor (MR) or type-1 adrenal steroid receptor, and the glucocorticoid receptor (GR) or type-2 adrenal steroid receptor (Reul and DeKloet, 1985). MRs have a high affinity for endogenous corticosteroids and are believed to play a role in the regulation of circadian fluctuations in these hormones (especially the regulation of ACTH secretion during the diurnal trough in cortisol secretion). GRs, on the other hand, have a lower affinity for endogenous corticosteroids than MRs and are therefore believed to be more important in the regulation of the response to stress when endogenous levels of glucocorticoids are high.

Immune tissues exhibit a high degree of heterogeneity in the expression of adrenal steroid receptor subtypes. For example, the thymus expresses one of the highest concentrations of GR in the body (~1000 fmol/mg protein), followed by the spleen (~500 fmol/mg protein) and the peripheral blood mononuclear cells (~250 fmol/mg protein). In addition, where only GR is expressed in the thymus, both GR and MR are expressed in the spleen (Miller et al., 1990; Spencer et al., 1991).

Corticosteroids have potent effects on the immune system and are capable of both decreasing immune cell function and inducing lymphocyte subset redistribution between blood and other bodily compartments. Clinically, pharmacologic doses of corticosteroids (the adrenal hormones and their synthetic analogues) are used because of their well known antiinflammatory and immunosuppressive effects, mainly due to the inhibition of the synthesis of several cytokines and mediators of inflammation (Schleimer et al., 1989). However, under physiologic conditions these hormones have been found to be important modulators of immunity (McEwen et al., 1997). For example, peaks in corticosterone levels related to the diurnal cycle or experimental stress have been shown to induce selective changes in peripheral blood leukocyte subsets (Dhabhar et al., 1994, Dhabhar et al., 1996). Moreover, it has been hypothesized that glucocorticoids shift the balance of an ongoing immune reaction from a TH1-directed (cell-mediated) response to a TH2-directed (antibody-mediated) response, by regulating the production of specific interleukins (Mosmann and Coffman, 1989; Mason, 1991). Studies on adrenalectomized animals (Keller et al., 1988) have shown that stress-induced lymphopenia in rats occurs in association with stress-induced secretion of corticosteroids and can be prevented by adrenalectomy, while other stress-induced modifications seem unrelated to adrenal hormones. For example, in a study by Keller and colleagues (1983), adrenalectomized animals undergoing a tail shock continued to exhibit a significant decrease in PHA-induced proliferative response of peripheral blood lymphocytes, while the stress-induced lymphopenia previously noted in stressed intact animals was no longer apparent.

The role of adrenal steroid hormones in blood leukocyte distribution has been systematically evaluated in a series of studies conducted by Dhabhar and colleagues (1994; 1995; 1996) on the effects of diurnal cycle, acute stress, and corticosteroid secretion on immune cell distribution. Using a mild acute stress model (two hours of restraint), significant and selective changes in peripheral blood cell distribution were found including decreased numbers of white blood cells, diminished numbers and percentages of monocytes and lymphocytes—namely, B cells, NK cells, and, to a lesser degree, T cells—and a slight increase in neutrophil numbers and percentages. These changes were associated with a concomitant increase in plasma corticosterone and were almost completely abolished by adrenalectomy or by treatment of the animals with an inhibitor of the synthesis of corticosterone. Moreover, administration of corticosterone (MR and GR agonist) or the selective GR agonist RU28362 to adrenalectomized animals resulted in close replication of the stress-induced changes observed in intact animals. This suggests that corticosteroid activation of the GR plays a major role in stress-induced leukocyte redistribution (Miller et al., 1994). Finally, these stress-induced changes in leukocyte subpopulations were remarkably similar to those obtained in nonstressed animals at the beginning of their active period, the time when corticosteroids reach their highest peak during the diurnal cycle (Dhabhar et al., 1994).

It is noteworthy that stress-induced changes in immune cell numbers and distribution between the blood and various immune compartments may have profound effects on the effectiveness and functioning of the immune system. For example, it has been suggested that circadian variation in lymphocyte proliferation in response to polyclonal mitogens is related to diurnal changes in peripheral blood leukocyte subsets (Tavadia et al., 1975; Dhabhar et al., 1994) and that stress-induced suppression of splenic and peripheral blood NK activity is related to stress-induced migration of NK cells out of these compartments (Ghoneum et al., 1987). Lymphocyte redistribution may imply that cells are directed toward various immune compartments (i.e., lymph nodes, spleen, mucosa, skin), where they may be more likely to encounter antigens (Dhabhar et al., 1994; Dhabhar et al., 1996).

Sympathetic Nervous System

The second fundamental pathway of the stress response, the ANS, also plays a relevant role in the immune response to stress. Historically, the identification of nerve fibers derived from the ANS in immune tissues was one of the first indications that communication between the central nervous system and immune system was possible. Parasympathetic and sympathetic nerve fibers have been identified in organs that are responsible for the development, growth, and function of lymphocytes: bone marrow, thymus, spleen, and lymph nodes. Sympathetic nerve fibers typically enter these lymphoid tissues in association with the vascular supply. However, inside the organs, the nerves are associated with both smooth muscle cells of the blood vessels, as they play a role in vascular tone, and also in the parenchyma associated with lymphocytes and other immune cells. Therefore, the ANS can influence the immune system either by changing the vascular tone and the blood flow into the lymphoid organs or through a direct effect of the released neurotransmitters, especially catecholamines (norepinephrine) and peptides (neuropeptide Y, substance P, vasoactive intestinal peptide, calcitonin gene-related peptide) that in turn interact with specific receptors on nearby immune cells (Bellinger et al., 1992).

Indeed, animal studies have demonstrated that surgical or chemical sympathectomy alters immune responses in rodents as well as attenuates stress-induced immune changes (Hori et al., 1995). As might be anticipated from the pattern of nervous innervation of lymphoid tissues, abrogation of stress-induced immune changes by antagonizing the sympathetic nervous system is most apparent in solid immune tissues such as the spleen. For example, a previous study by Cunnick and colleagues (1990) showed that stress-induced suppression of splenic lymphocyte proliferation to polyclonal mitogens is not influenced by adrenalectomy but is markedly attenuated by β-adrenergic receptor antagonists.

In humans, the sympathetic nervous system might play a role in changes induced by brief experimental stressors as suggested by the rapid onset of the immune changes and the higher sensitivity of those subjects with increased cardiovascular responses (Herbert et al., 1994). However, evidence of elevated sympathetic activity has also been described in subjects experiencing a chronic, stressful situation (Uchino et al., 1992) and it is therefore conceivable that the sympathetic nervous system may play a role in some of the immune changes occurring in association with naturalistic stressors as well.

Corticotropin Releasing Factor

Interestingly, in laboratory animals, intracerebroventricular (ICV) administration of CRF, which is copiously released during the stress response and activates the HPA axis, is also able to stimulate the sympathetic nervous system, which in turn leads to suppression of splenic NK activity (Irwin et al., 1988). In fact, the effect of ICV CRF on splenic NK activity is reversed by sympathetic nervous system blockade using the sympathetic ganglionic blocker chlorisondamine.

It should also be mentioned that CRF is capable of directly modulating immune and inflammatory responses (Karalis et al., 1991). Local production of CRF has been demonstrated in inflammatory diseases such as ulcerative colitis (Kawahito et al., 1995) and arthritis (Nishioka et al., 1996) where it is proposed to act as a local proinflammatory agent. Recent evidence also suggests that CRF may act as a protective factor against inflammation-induced pain (Schafer et al., 1994) and plasma extravasation (Yoshihara et al., 1995).

HPA axis hormones and catecholamines are not the only factors involved in the modulation of the immune response following stress. In fact, studies conducted on hypophysectomized rats (Keller et al., 1988) show that the stress-induced suppression of peripheral blood lymphocyte proliferative response to the mitogen PHA is more pronounced in stressed, hypophysectomized animals than in stressed intact animals (controls). These findings suggest that a pituitary hormone may be involved in counteracting stress-induced immunosuppressive mechanisms. The specific pituitary-dependent mitigating or compensating hormones are not known, but they probably involve multiple hormones with immunoenhancing properties, for instance, growth hormone and prolactin (Kelley, 1991; Bernton et al., 1991). These findings also hint that a regulatory network of hormonal and nonhormonal systems is involved in the maintenance of immunologic capacity following exposure to stressors. The restraining influence of the pituitary on stress response may be of relevance to the understanding of homeostatic maintenance of critical body functions.

As far as the biochemical mechanisms of the effects of stress on the immune system are concerned, it is important to mention that nitric oxide (NO) has been shown to be involved in the physiological and pathological responses to stress in various organs including the immune compartments. NO is a ubiquitous molecule that is involved in very different phenomena such as blood vessel tone, gastric mucosa protection, neurotoxicity, and macrophage function, and that acts mainly by forming covalent linkages to several targets such as enzymes. For example, NO stimulates cyclic-GMP synthesis by linking and therefore activating the guanylyl cyclase enzyme. Evidence is now available that NO production in the immune system is induced in acutely stressed rats (Persoons et al., 1995). Stress-induced changes in NO production by macrophages are relevant for the stress-induced decrease in the lymphocyte proliferative response, since both depletion of macrophages and addition of the NO synthesis inhibitor L-NMMA attenuate this stress-induced phenomenon (Coussons-Read et al., 1994).

RELEVANCE OF STRESS EFFECTS ON THE IMMUNE SYSTEM TO MEDICAL DISEASE

An important area of investigation in humans is the connection between stressful events and the progression or outcome of medical illness involving the immune system, especially cancer and viral infections including acquired immune deficiency syndrome (AIDS).

Stress and Cancer

That psychosocial interventions such as group therapy could improve the survival in patients with cancer was first suggested by Grossarth-Maticek and colleagues (1984) and then elegantly confirmed by two independent series of studies conducted by Spiegel and colleagues (1981; 1989) and Fawzy and colleagues (1990 a,b; 1993). Both describe the influence of psychosocial interventions on the course of illness in patients with different kinds of cancer using a randomized study design.

Spiegel and colleagues (1989) treated women with metastatic breast cancer by means of a weekly group therapy. The intervention focused on encouraging discussion of how to cope with cancer and expression of feelings about the illness and its physical consequences and on increasing social supports by developing relationships among the members. While self-hypnosis was used for pain control, the patients did not use imagery of the immune system fighting against the tumor (as used by Achterberg et al., 1977) nor were they told that group therapy might affect the course of their illness. Following the first report about the usefulness of this approach in improving the quality of life of the patients (Spiegel et al., 1981), a survival analysis was conducted comparing a group receiving one year of the psychosocial treatment against a control group (both undergoing routine oncological care). After 10 years of follow-up, the survival time of patients in the intervention group was almost double that of the controls (36.6 mo vs. 18.9 mo, starting from the onset of the intervention). The divergence in survival was not evident during the treatment but appeared 20 mo after study entry, almost 8 mo after the end of the treatment.

Fawzy and colleagues (1990 a,b; 1993) also used a group therapy approach in the treatment of patients with malignant melanoma, but the intervention consisted of a shortterm (6 weeks) structured approach focused on health education, enhancement of problem-solving skills, stress management with relaxation techniques, and psychological support. They found that at a 6 mo follow-up, this intervention (compared to a control group obtained by a randomization design) was not only effective in reducing psychological distress of patients (1990a) but was also capable of inducing changes in immune parameters; results showed increases in NK cell percentage and NK cytotoxic activity (two assays of possible relevance in the reaction of the immune system against cancer cells) and small decreases in CD4+ (T-helper cells)(1990b). More interesting was the fact that after 5 to 6 years of follow-up (1993), patients who underwent the psychosocial intervention had a lower rate of death and cancer recurrence than controls. However, the relationship between immune parameters and outcome was not conspicuous in that only the baseline NK activity was predictive of recurrence (not of survival) and the changes over time of the immune parameters had no apparent effect on the course of the illness. While these studies testify to the positive effect of psychosocial intervention on the course of illness (and quality of life) of cancer patients, the role of putative changes in the immune system in determining this influence has yet to be firmly established.

Stress and Viral Infection

Data concerning the effects of stress on viral infections show that individuals who exhibit higher levels of perceived life stress are significantly more likely to be infected and develop a "cold" following intranasal inoculation of standardized doses of a series of respiratory viruses (Cohen et al., 1991). This approach has been extended to include an investigation of the influence of stressful life events in subjects infected with the human immunodeficiency virus (HIV), especially HIV-positive asymptomatic subjects. This issue is not trivial, because the hypothesis that psychological factors might influence the progression from the seropositive asymptomatic state to actual AIDS has led to the development of strategies for avoiding exposure to stressful events or for controlling the psychological reaction to these events (Kessler et al., 1991). However, the scientific evidence of an association between psychosocial factors and disease progression remains controversial. Conflicting findings have been reported on the effects of stress on immune parameters in this population; for example, three studies have failed to find an influence of psychosocial factors on CD4+ cells alone (Rabkin et al., 1991; Kessler et al., 1991), CD4+ and CD8+ cells (Perry et al., 1992), and markers of disease progression including the onset of fever and thrush (Kessler et al., 1991). In contrast, two recent studies have found that HIV-positive subjects who experienced severe life stress have relevant changes in immune parameters. In one study (Evans et al., 1995) on HIV-positive asymptomatic homosexual men, the presence of severe stress in the previous 6 months was associated with lower CD8+ and lower NK cell counts, whereas no such association was evident in a control group of HIV-negative controls. In another study (Kemeny et al., 1995), HIV-positive and HIV-negative homosexual men were followed prospectively to evaluate the effect of the death of their intimate partner on immune parameters. Those who experienced bereavement during the follow-up had a significant increase in the level of serum neopterin (a marker of immune activation) and a significant decrease in the proliferative response to PHA compared to their prebereavement evaluation. These changes were not present in the control groups composed of HIV-positive and HIV-negative nonbereaved men. It is worth mentioning that the latter study showed no effect of stress on lymphocyte subsets, including CD4+, CD8+, and NK cells.

Several methodological issues may account for the discrepancies in the literature such as differences in the stage of illness, the behavioral influences on the immune system, and above all, the immune measures evaluated (Goodkin et al., 1994). In fact, negative studies by Rabkin and colleagues (1991) and Perry and colleagues (1992) evaluated mainly CD4+ cells, a lymphocyte subset that is heavily damaged by HIV infection. Indeed, the studies by Evans and colleagues (1995) and Kemeny and colleagues (1995) also failed to find any influence of stress on this parameter. These negative findings concerning the CD4+ subset count are intriguing given the importance of this cell subset as a marker of progression of the illness. However, it is possible that this immune measure may simply not be appropriate or meaningful in this context because the relevant role of the HIV virulence in determining the CD4+ level might confound the influence of psychosocial factors (Stein et al., 1991).

On the other hand, studies by Evans and colleagues (1995) and Kemeny colleagues (1995) have shown that, in HIV-positive subjects, stress influences CD8+ cells, NK cells, serum neopterin, and proliferative responses to PHA. Each of these parameters bear relevance to disease progression: the CD8+, cytotoxic T lymphocytes, and NK cells are important in the immune responses against viral infections and thus may have a role in controlling HIV infection; both increased neopterin level and decreased PHA proliferation in HIV-positive subjects have been described to predict the development of AIDS. Recent work by Evans and colleagues (1997) also supports the notion that severe life stress through effects on the immune response predicts early disease progression in HIV infected individuals.

CONCLUSIONS

We have reviewed the most compelling evidence demonstrating that stress influences the immune system. The state of the field is far more complicated than it appeared to be when investigation in this area first began. Results of studies examining various aspects of the immune response have demonstrated that, based on the type and timing of the stressors, different hormonal and/or neurotransmitter pathways can be activated with distinct results in the various immune compartments. Although epidemiological studies in humans have shown that stress can influence the outcome of medical illnesses, most notably infectious diseases (Graham et al., 1986), the role of the stress-induced changes in the immune system in determining these findings remains to be clarified. Moreover, the relevance of changes in immune assays, lymphocyte subsets, or plasma soluble factors to the function of the immune system is still unclear. Therefore, more detailed knowledge about brain–neuroendocrine–immune relationships is required if the basic mechanisms of stress–immune interactions and their significance for health and illness in humans are to be elucidated.

ACKNOWLEDGMENTS

This work was supported in part by NIMH grants MH00680 and MH47674 (AHM), University of Cagliari, Italy (CMP), and Italian Consiglio Nazionale delle Ricerche Grants AI95.00290.04 and AI96.00156.04 (CMP).

REFERENCES

Abbas, A.K., Lichtman, A.H., & Pober, J.S. (Eds.) (1994). *Cellular and Molecular Immunology, 2nd ed.* Saunders, Philadelphia.

Achterberg, J., Matthews-Simonton, S., & Simonton, C. (1977). Psychology of the exceptional cancer patient: A description of patients who outlive predictive life expectancies. Psychother. Theory Res. Practice 14, 416.

Ader, R. & Friedman, S.B. (1965). Differential early experiences and susceptibility to transplanted tumors in the rat. J. Comp. Physiol. 59, 361-364.

Bartrop, R.W., Lazarus, L., Luckherst, E., & Kiloh, LH. (1977). Depressed lymphocyte function after bereavement. Lancet 1, 834-836.

Bellinger, D.L., Felten, S.Y., & Felten, D.L. (1992). Neural-immune interactions. In: *American Psychiatric Association Annual Review* (Tasman, A., Ed.). Vol. 11, pp. 127-144. American Psychiatric Press, Inc., Washington, D.C.

Berkenbosch, F., Wolvers, D.A.W., & Derijk, R. (1991). Neuroendocrine and immunological mechanisms in stress induced immunomodulation. J. Steroid. Biochem. Molec. Biol. 40, 639-647.

Bernton, E.W., Bryant, H.U., & Holaday, J.W. (1991). In: *Psychoneuroimmunology*, 2nd ed. (Ader, R., Felten, D.L., & Cohen, N., Eds.), pp. 403-428. Academic Press, New York.

Besedovsky, H.O. & Del Rey, A. (1996). Immune-Neuro-Endocrine Interaction: Facts and Hypotheses. Endocrine Rev. 17, 64.

Brosschot, J.F., Benschop, R.J., Godaert, G.L.R., Olff, M., De Smet, M., Heijnen, C.J., & Ballieux, R.E. (1994). Influence of life stress on immunological reactivity to mild psychological stress. Psychosom. Med. 56, 216-244.

Cohen, S., Tyrell D.A., & Smith, A.P. (1991). Psychological stress and susceptibility to the common cold. N. Engl. J. Med. 325, 606.

Coussons-Read, M.E., Maslonek, K.A., Fecho, K., Perez, L., & Lysle, D.T. (1994). Evidence for the involvement of macrophage-derived nitric oxide in the modulation of immune status by a conditioned aversive stimulus. J. Neuroimmunol. 50, 51-58.

Cunnick, J.E., Lysle, D.T., Kucinski, B.J., & Rabin, B.S. (1990). Evidence that shock induced immune suppression is mediated by adrenal hormones and peripheral beta-adrenergic receptors. Pharmacol. Biochem. Behav. 36, 645-651.

Dhabhar, F.S. & McEwen, B.S. (1996). Stress-induced enhancement of antigen-specific cell-mediated immunity. J. Immunol. 156, 2608-2615.

Dhabhar, F.S., Miller, A.H., McEwen, B.S., & Spencer, R.L. (1995). Effects of stress on immune cell distribution: Dynamics and hormonal mechanisms. J. Immunol. 154, 5511-5527.

Dhabhar, F.S., Miller, A.H., McEwen, B.S., & Spencer, R.L. (1996). Stress-induced changes in blood leukocyte distribution. Role of adrenal steroid hormones. J. Immunol. 157, 1638-1644.

Dhabhar, F.S., Miller, A.H., Stein, M., McEwen, B.S., & Spencer, R.L. (1994). Diurnal and acute stress-induced changes in distribution of peripheral blood leukocyte subpopulations. Brain Behav. Immun. 8, 66-79.

Evans, D.L., Leserman, J., Perkins, D.O., Stern, R.A., Murphy, C., Tamul, K., Liao, D., van der Horst, C.M., Hall, C.D., Folds, J.D., Golden, R.N., & Petitto, J.M. (1995). Stress-associated reductions of cytotoxic T lymphocytes and natural killer cells in asymptomatic HIV infection. Am J Psychiat 152, 543-550.

Longmate, J.A., Silva, S.G., van der Horst, C.M., Hall, C.D., Folds, J.D., Golden, R.N., & Petitto, J.M. (1997). Severe life stress as a predictor of early disease progression in HIV infection. Am. J. Psych. 154, 630-634.

Fawzy, F.I., Cousins, N., Fawzy, N.W., Kemeny, M.E., Elashoff, R.,& Morton, D. (1990a). A structured psychiatric intervention for cancer patients. I. Changes over time in methods of coping and affective disturbance. Arch. Gen. Psychiat. 47, 720-725.

Fawzy, F.I., Kemeny, M.E., Fawzy, N.W., Elashoff, R., Morton, D., Cousins, N.,& Fahey, J.L. (1990b). A structured psychiatric intervention for cancer patients. II. Changes over time in immunological measures. Arch. Gen. Psychiat. 47, 729-735.

Fawzy, F.I., Fawzy, N.W., Hyun, C.S., Elashoff, R., Guthrie, D., Fahey, J.L., Morton, D.L. (1993). Malignant melanoma: Effects of an early structured psychiatric intervention, coping, and affective state on recurrence and survival 6 years later. Arch. Gen. Psychiat. 50, 681-689.

Ghoneum, M., Gill, G., Assanah, P., & Stevens, W. (1987). Susceptibility of natural killer cell activity of old rats to stress. Immunology 60, 461-465.

Glaser, R., Kiecolt-Glaser, J.K., Bonneau, R.H., Malarkey, W., Kennedy, S., & Hughes, J. (1992). Stress-induced modulation of the immune response to recombinant hepatitis B vaccine. Psychosom. Med. 54, 22-29.

Goodkin, K., Mulder, C.L., Blaney, N.T., Ironson, G., Kumar, M., & Fletcher, M.A. (1994). Psychoneuroimmunology and Human Immunodeficiency Virus type 1 infection revisited. Arch. Gen. Psych. 51, 246-247.

Graham, N.M.H., Douglas, R.M., & Ryan, P. (1986). Stress and acute respiratory infection. Am. J. Epidem. 124, 389-401.

Grossarth-Maticek, R., Schmidt, P., Veter, H., & Arndt, S. (1984). Psychotherapy research in oncology. In: *Health Care and Human Behaviour.* (Steptoe A. & Mathews A., Eds.), p. 325. Academic Press, London.

Herbert, T.B., Cohen, S., Marsland, A.L., Bachen, E.A., Rabin, B.S., Muldoon, M.F., & Manuck, S.B. (1994). Cardiovascular reactivity and the course of immune response to an acute psychological stressor. Psychosom. Med. 56, 337-344.

Hori, T., Katafuchi, T., Take Shimizu, & Niijima, A. (1995). The autonomic nervous system as a communication channel between the brain and the immune system. Neuroimmunomodulation 2, 203-215.

Irwin, M., Hauger, R.L., Brown, M., & Britton, K.T. (1988). CRF activates autonomic nervous system and reduces natural killer cell cytotoxicity. Am. J. Physiol. 255, R744-R747.

Jain, S. & Stevenson, J.R. (1991). Enhancement by restraint stress of natural killer cell activity and splenocyte responsiveness to concanavalin A in Fischer 344 rats. Immunol. Invest. 20, 365-376.

Jensen, M.M. & Rasmussen, A.F. Jr. (1963). Stress and susceptibility to viral infection. Sound stress and susceptibility to vesicular stomatitis virus. J. Immunol. 90, 21-23.

Johnsson, T., Lavender, J.F., Hullin, E., & Rasmussen, A.F., Jr. (1963). The influence of avoidance-learning stress on resistance to coxsackie B virus in mice. J. Immunology 91, 569-575.

Johnsson, T. & Rasmussen, A.F., Jr. (1965). Emotional stress and susceptibility to poliomyelitis virus infection in mice. Archiv. Gesamte Virusforsch. 18, 390.

Karalis, K., Sano, H., Redwine, J., Listwak, S., Wilder, R.L., & Chrousos, G.P. (1991). Autocrine or paracrine inflammatory actions of corticotropin-releasing hormone *in vivo*. Science 254, 421-423.

Kawahito, Y., Sano, H., Mukai, S., Asai, K., Kimura, S., Yamamura, Y., Kato, H., Chrousos, G.P., Wilder, R.L., & Kondo, M. (1995). Corticotropin releasing hormone in colonic mucosa in patients with ulcerative cholitis. Gut 37, 544-551.

Keller, S.E., Schleifer, S.J., Liotta, A.S., Bond, R.N., Farhoody, N., & Stein, M. (1988). Stress-induced alterations of immunity in hypophysectomized rats. Proc. Natl. Acad. Sci. U.S.A. 85, 577-566.

Keller, S.E., Weiss, J.M., Schleifer, S.J., Miller, N.E., & Stein, M. (1981). Suppression of immunity by stress: Effects of a graded series of stressors on lymphocyte stimulation in the rat. Science 213, 1397-1400.

Keller, S.E., Weiss, J.M., Schleifer, S.J., Miller, N.E., & Stein, M. (1983). Stress-induced suppression of immunity in adrenalectomized rats. Science 221, 1301-1304.

Kelley, K.W. (1991). Growth hormone in immunobiology. In: *Psychoneuroimmunology*, 2nd ed. (Ader R., Felten D.L., & Cohen N., Eds.), pp. 377-402. Academic Press, New York.

Kemeny, M.E., Weiner, H., Duran, R., Taylor, S.E., Visscher,B., & Fahey, J.L. (1995). Immune system changes after the death of a partner in HIV-positive gay men. Psychosom. Med. 57, 547-554.

Kent, S., Bluthe, R.M., Kelley, K.W., & Dantzer, R. (1992). Sickness behavior as a new target for drug development. TIPS 13, 24-28.

Kessler, R.C., Foster, C., Joseph, J., Ostrow, D., Wortman, C., Phair, J., & Chmiel, J. (1991). Stressful life events and onset in HIV infection. Am. J. Psychiat. 148, 733-738.

Kiecolt-Glaser, J.K., Cacioppo, J.T., Malarkey, W.B., & Glaser, R. (1992). Acute psychological stressors and short-term immune changes: What, why, for whom and to what extent? Psychosom. Med. 4, 680-685.

Kiecolt-Glaser, J.K., Dura, J.R., Speicher, C.E., Trask, O.J., & Glaser, R. (1991). Spousal caregivers of dementia victims: Longitudinal changes in immunity and health. Psychosom. Med. 53, 345-362.

Kiecolt-Glaser, J.K. & Glaser, R. (1991). Stress and immune function in humans. In: *Psychoneuroimmunology*, 2nd ed. (Ader, R., Felten D.L., & Cohen N., Eds.), pp. 849-867. Academic Press, New York.

Kiecolt-Glaser, J.K., Glaser, R., Gravenstein, S., Malarkey, W.B., & Sheridan, J. (1996). Chronic stress alters the immune response to influenza virus vaccine in older adults. Proc. Natl. Acad. Sci. U.S.A. 93, 3043-3047.

Kiecolt-Glaser, J.K., Glaser, R., Shuttleworth, E.C., Dyer, C.S., Ogrocki, P., & Speicher, CE. (1987). Chronic stress and immunity in family caregivers of Alzheimer's Disease victims. Psychosom. Med. 49, 523-535.

Kiecolt-Glaser, J.K., Marucha, P.T., Malarkey, W.B., Mercado, A.M., & Glaser, R. (1995). Slowing of wound healing by psychological stress. Lancet 346, 1194-1196.

Lorens, S.A., Hata, N., Handa, R.J., Van de Kar, L.D., Guschwan, M., Goral, J., Lee, J.M., Hamilton, M.E., Bethea, C.L., & Clancy, J. Jr. (1990). Neurochemical, endocrine, and immunological responses to stress in young and old Fisher 344 male rats. Neurobiol. Aging 11, 139-150.

LeMay, L.G., Vander, A.J., & Kluger, M.J. (1990). The effects of psychological stress on plasma interleukin-6 activity in rats. Physiol. Behav. 47, 957-961.

Lysle, D.T., Cunnick, J.E., & Rabin, B.S. (1990). Stressor-induced alteration of lymphocyte proliferation in mice: Evidence for enhancement of mitogenic responsiveness. Brain Behav. Immun. 4, 269-277.

Lysle, D.T., Lyte, M., Fowler, H., & Rabin, B.S. (1987). Shock-induced modulation of lymphocyte reactivity: Suppression, habituation, and recovery. Life Sci. 41, 1805-1814.

Mason, D. (1991). Genetic variation in the stress response: Susceptibility to experimental allergic encephalomyelitis and implications for human inflammatory disease. Immunology Today 12, 57-60.

McEwen, B.S., Biron, C.A., Brunson, K.W., Bulloch, K., Chambers, W.H., Dhabhar, F.S., Goldfarb, R.H., Kitson, R.P., Miller, A.H., Spencer, R.L., & Weiss, J.M. (1997). The role of adrenocorticoids as modulators of immune function in health and disease: Neural, endocrine, and immune interactions. Brain Res. Rev. 23, 79-133.

Milakofsky, L., Harris, N., & Vogel, W.H. (1993). Effect of repeated stress on plasma catecholamines and taurine in young and old rats. Neurobiol. Aging 14, 359-366.

Miller, A.H., Spencer, R.L., Hassett, J., Kim, C., Rhee, R., Ciurea, D., Dhabhar, F., McEwen, B.S., & Stein, M. (1994). Effects of selective type I and II adrenal steroid receptor agonists on immune cell distribution. Endocrinology 135, 1934-1944.

Miller, A.H., Spencer, R.L., Stein, M., & McEwen, B.S. (1990). Adrenal steroid receptor binding in spleen and thymus after stress or dexamethasone. Am. J. Physiol. 259, E405-E412.

Mosmann, T.R. & Coffman, R.L. (1989).Th1 and Th2 cells: Different patterns of lymphokine secretion lead to different functional properties. Ann. Rev. Immunology 7, 145-173.

Nishioka, T., Kurokawa, H., Takao, T., Kumon, Y., Nishiya, K., & Hashimoto, K. (1996). Differential changes of corticotropin releasing hormone (CRH) concentrations in plasma and synovial fluids of patients with rheumatoid arthritis (RA). Endocr. J. 43, 241-247.

Pariante, C.M., Carpiniello, B., Del Giacco, G.S., Farci, A.M.G., Orrú, M.G., Piludu, G., Piras, A., Sitzia, R., & Miller, A.H. (1997). Chronic caregiving stress alters peripheral blood immune parameters: The role of age and severity of stress. Psychother. Psychosom. 66, 199-207.

Perry, S., Fishman, B., Jacobsberg, L., & Frances, A. (1992). Relationships over 1 year between lymphocyte subsets and psychosocial variables among adults with infection by Human Immunodeficiency Virus. Arch. Gen. Psychiat. 49, 396-401.

Persoons, J.H.A., Schornagel K., Breve, J., Berkenbosch, F., & Kraal, G. (1995). Acute stress affects cytokines and nitric oxide production by alveolar macrophages differently. Am. J. Resp. Crit. Care Med. 152, 619-624.

Rabkin, J.G., Williams, J.B.W., Remien, R.H., Goetz, R., Kertzner, R., & Gorman, J.M. (1991). Depression, distress, lymphocyte subsets, and Human Immunodeficiency Virus symptoms on two occasions in HIV-positive homosexual men. Arch. Gen. Psychiat. 48, 111-119.

Rasmussen, A.F. Jr. (1969). Emotions and Immunity. Ann. N.Y. Acad. Sci. 164, 458-461.

Rasmussen, A.F. Jr., Marsh, J.T., & Brill, N.Q. (1957). Increased susceptibility to herpes simplex in mice subjected to avoidance-learning stress or restraint. Proc. Soc. Exp. Biol. Med. 96, 183-189.

Reul, J.M. & de Kloet, E.R. (1985). Two receptor systems for corticosterone in rat brain: Microdistribution and differential occupation. Endocrinology 117, 2505-2511.

Sapolsky, R.M., Krey, L.C., & McEwen, B.S. (1986). The neuroendocrinology of stress and aging: The glucocorticoid cascade hypothesis. Endocr. Rev. 7, 284-301.

Schafer, M., Carter, L., & Stein, C. (1994). Interleukin 1-beta and corticotropin-releasing factor inhibit pain by releasing opioids from immune cells in inflamed tissue. Proc. Nat. Acad. Sci. U.S.A. 91, 4219-4223.

Schleifer, S.J., Keller, S.E., Bond, R.N., Cohen, J., & Stein, M. (1989). Major depressive disorder and immunity: Role of age, sex, severity, and hospitalization. Arch. Gen. Psychiatry 46, 81-87.

Schleifer, S.J., Keller, S.E., Camerino, M., Thornton, J.C., & Stein, M. (1983). Suppression of lymphocyte stimulation following bereavement. JAMA 250, 374-377.

Schleimer, R.P., Claman, H.N., & Oronsky, A. (Eds)(1989). *Anti-inflammatory Steroid Action, Basic and Clinical Aspects*. Academic Press, San Diego.

Sheridan, J.F., Feng, N., Bonneau, R.H., Allen, C.M., Huneycutt, B.S., & Glaser, R. (1991). Restraint stress differentially affects anti-viral cellular and humoral immune responses in mice. J. Neuroimmunol. 31, 245-255.

Shurin, M..R., Zhou, D., Kusnecov, A., Rassnick, S., & Rabin, B.S. (1994). Effect of one or more footshocks on spleen and blood lymphocyte proliferation in rats. Brain Behav. Immun. 8, 57-65.
Solomon, G.F. (1969). Stress and antibody response in rats. Intl. Arch. Allergy 35, 97-104.
Spencer, R.L., Miller, A.H., Stein, M., & McEwen, B.S. (1991). Corticosterone regulation of type I and type II adrenal steroid receptors in brain, pituitary and immune tissue. Brain Res. 549, 236-246.
Spiegel, D., Bloom, J.R., & Yalom, I.D. (1981). Group support for patients with metastatic cancer: A randomized outcome study. Arch. Gen. Psychiat. 38, 527-533.
Spiegel, D., Bloom, J.R., Kraemer, H.C., & Gottheil, E. (1989). Effects of psychosocial treatment on survival of patients with metastatic breast cancer. Lancet 2, 888-891.
Stein, M., Miller, A.H., & Trestman, R.L. (1991). Depression, the immune system, and health and illness. Arch. Gen. Psychiat. 48, 171-177.
Tavadia, H.B., Fleming, K.A., Hume, P.D., & Simpson, H.W. (1975). Circadian rhythmicity of human plasma cortisol and PHA-induced lymphocyte transformation. Clin. Exp. Immunol. 22, 190-193.
Uchino, B.N., Kiecolt-Glaser, J.K., & Cacioppo, J.T. (1992). Age-related changes in cardiovascular response as a function of a chronic stressor and social support. J. Pers. Soc. Psychol. 63, 839-846.
Yoshihara, S., Ricciardolo, F.L.M., Geppetti, P., Linden, A., Hara, M., Chan, B., & Nadel, J.A. (1995). Corticotropin-releasing factor inhibits antigen-induced plasma extravasation in airways. Eur. J. Pharmacol. 280, 113-118.
Weiss, J.M. & Sundar, S. (1992). Effects of stress on cellular immune responses in animals. In: *American Psychiatric Association Annual Review.* (Tasman, A., Ed.), Vol. 2, pp. 145-168. American Psychiatric Press, Inc., Washington, D.C.
Wood, P.G., Karol, M.H., Kusnecov, A.W., & Rabin, B.S. (1993). Enhancement of antigen-specific humoral and cell-mediated immunity by electric footshock stress in rats. Brain Behav. Immun. 7, 121-134.
Zhou, D., Kusnecov, A.W., Shurin, M.R., DePaoli, M., & Rabin, B.S. (1993). Exposure to physical and psychological stressors elevates plasma interleukin 6: Relationship to the activation of the hypothalamic–pituitary–adrenal axis. Endocrinology 133, 2523-2530.

RECOMMENDED READINGS

Dhabhar, F.S. (1998). Stress-induced enhancement of cell-mediated immunity. Ann. N.Y. Acad. Sci. 840, 359-372.
Kiecolt-Glaser, J.K. (1999). Stress, personal relationships, and immune function: Health implications, Brain, Beh. Immnity 13, 61-72.
Ottaviani, E. & Franceschi, C. (1996). The neuroimmunology of stress from invertebrates to men. Progr. Neurobiology 48, 421-440.
Shephard, R.J., Castellani, J.W., & Shek, P.N. (1998). Immune deficits induced by strenuous exertion under adverse environmental conditions: Manifestations and countermeasures. Crit. Rev. Immunology 18, 545-568.

Chapter 8

Fear and Anxiety Mechanisms of the Brain:
CLINICAL IMPLICATIONS

JAAK PANKSEPP

Biological Psychiatry, pages 155-177.

ISBN: 1-55938-819-6

INTRODUCTION

Earlier in this century it was not uncommon for theorists to believe that fear was simply the belief that something is dangerous. Accordingly, it was assumed that we would come to clarify the nature of fear by talking to people about what makes them scared and anxious. Although cognitive appraisals obviously can precipitate emotions, this level of analysis is no longer deemed sufficient for a scientific understanding of fear. To understand fear we must specify the brain systems that elaborate not only the cognitive and bodily features of fear but also the emotional feelings that characterize fear. At present, many theorists believe that even though the cognitions and beliefs associated with fear may vary widely among species, depending on their degree of encephalization, the bodily expressions and affective feelings of fear emerge directly from ancient emotional circuits (or "affect programs") of the mammalian brain that can be analyzed with the tools of modern neuroscience (for one comprehensive survey, see Burrows et al., 1990). Through such strategies, we are currently learning a great deal about human fear by studying the animal brain.

In line with the outdated psychological perspectives, earlier animal models were also premised on the supposition that fear simply reflected *learned anticipation* of harmful events. Now, however, it is evident that the capacity for fear is a genetically ingrained and endogenously coordinated function of the mammalian nervous system that responds to unconditional threatening stimuli in the environment, and with experience, various neutral associated stimuli (i.e., conditional stimuli) can come to evoke the integrated fear response. This should come as no surprise. An organism's ability to perceive and anticipate dangers was of such obvious importance during evolution that it was not left simply to the vagaries of individual learning. Even though learning is essential for animals to utilize their fear systems effectively in the real world, learning does not create fears by pasting together a variety of external experiences or internal bodily components. Evolution created a coherently operating neural substrate for this emotional response, as it did for many others (see Chapter 2). Thus, to understand the deep experiential nature of fear in humans and the intrinsic bodily consequences of such brain states, we must seek to fully understand the genetically dictated but developmentally refined neural processes that mediate homologous neuroaffective states in other mammals (Panksepp, 1990).

The essence of fear in humans consists of an aversive state of the nervous system, characterized by apprehensive worry, general nervousness, and tension, which automatically informs them that their safety is threatened. It is accompanied by specific forms of autonomic and behavioral arousal. The driving force for the experiential and behavioral coherence of fear appears to arise from a distinct, widely ramifying, subcortical system of the brain that prompts animals to hide (freeze) if danger is distant or escape (flee) if danger is close. When such states become conditioned by aversive events being associated with previously neutral stimuli and perceptions, animals and humans develop the ability to anticipate dangers and hence protect themselves in advance of imminent threats. When activity in such systems becomes free-floating, or disconnected from external events, individuals come to exhibit generalized anxiety disorders and posttraumatic stress disorders. Thus, in adult organisms, fear is always a mixture of specific types of instinctual and learned brain activities.

The neural trajectory of one major fear system of the brain (Figure 1) arises from the central amygdala (and perhaps other higher brain zones such as the lateral septal area) and projects downward through the anterior and medial hypothalamus to the periaqueductal gray (PAG) of the midbrain and adjacent tegmental fields. Henceforth, this system will be called the FEAR system, to distinguish it from other, less well understood, negative affective systems such as those that precipitate panic attacks and social separation anxiety. It is not yet known precisely how the FEAR system helps create the phenomenological experience of anxiety, but we can be reasonably certain that it does. Electrical stimulation along the circuits generates apparently fearful states along with many fear-related behaviors in both experimental animals and humans (Panksepp, 1985; Depaulis and Bandler, 1991).

Also, the system appears to be hierarchically arranged, with higher areas collecting perceptual/cognitive information, the middle hypothalamic zones controlling various hormonal and other autonomic responses, and the most essential lower zones putting together an integrated instinctual behavioral/bodily response. The more caudally the artifical stimulation is applied along this system, (e.g., within the PAG), the easier it is to generate a coherent fear response with the smallest amount of electrical current. During stimulation of the higher reaches such as the amygdala and lateral septal area, fear responses emerge more slowly and with less intensity and require electrical current levels higher than those needed in the PAG or anterior hypothalamus. Responses from higher areas are integrally dependent on the integrity of the lower areas but not vice versa. Pharmacological and surgical dampening of activity along this circuit makes animals and humans placid, and the consequences of activity in the FEAR system for other brain functions are substantial (for summary, see Panksepp, 1991).

THE DISPERSION OF THE FEAR SYSTEM WITHIN THE BRAIN

Modern neuroscience techniques have finally detailed the widespread extent of the FEAR system in the brain. This has been most clearly highlighted by techniques such as visualization of the expression of early genes such as *c-fos* in animals that have received predictable fear provoking stimuli including foot shock (Beck and Fibiger, 1995), exposure to nonpainful threatening stimuli such as scary environments (Silviera et al., 1993), and direct electrical stimulation of the

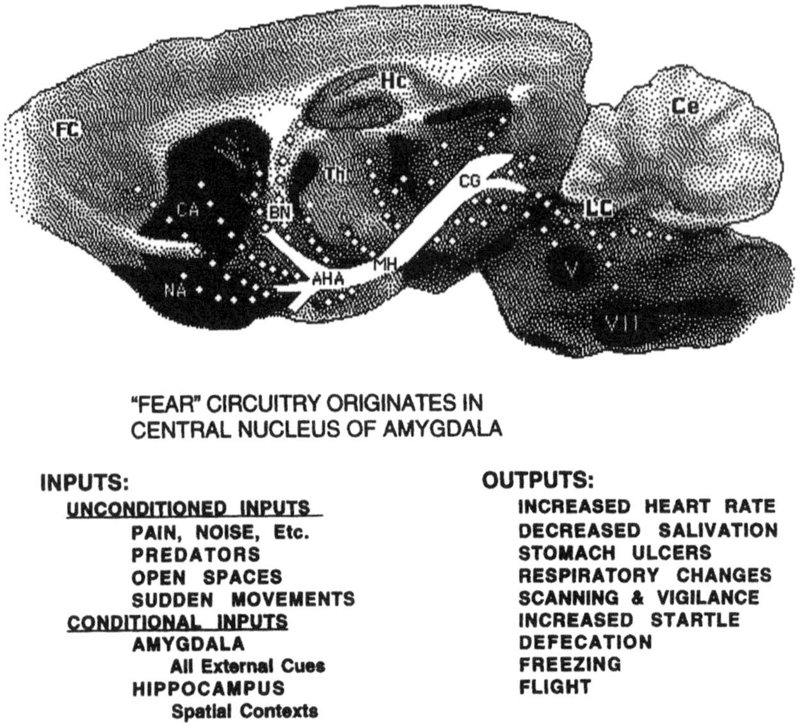

Figure 1. Schematic of the FEAR system depicted on a saggital section of the rat brain (the background of which highlights high density acetylcholine esterase staining in black). This transhypothalamic executive system for FEAR orchestrates many cognitive, affective, behavioral, hormonal, and physiological changes that characterize various fearful states. The executive circuit is a two-way avenue of communication between central regions of the amygdala (which transmits information caudally primarily by the ventral amygdalofugal pathway and the mesencephalic periaqueductal or central gray (CG). This circuit courses through the anterior and medial hypothalamic areas of the diencephalon where it is especially easy to elicit fearful behaviors (both freezing and flight) using electrical stimulation. There are multiple entry and exit points in this circuit (as depicted by the branching bubbles) that synchronize the many brain and bodily processes, which must be concurrently influenced when an animal is threatened. Anatomical designations are as follows: AHA, anterior hypothalamic area; CA, caudate nucleus; Ce, cerebellum; CG, central gray; BN, bed nucleus of the stria terminalis; FC, frontal cortex; Hc, Hippocampus; LC, locus coeruleus; MH, medial hypothalamus; NA, nucleus accumbens; Th, thalamus; V, motor nucleus of the trigeminal nerve; VII, nucleus of the facial nerve (Adapted from Panksepp, 1996).

underlying brain circuitry (Silviera et al., 1995). Similar brain patterns are seen in animals that have been defeated while fighting (Kollack-Walker et al., 1997) and in those that have been exposed to the fearful sounds of conspecifics (Beckett et al., 1997). These widespread changes must be contrasted with the rather modest brain changes that have been documented in human brain imaging studies. Functional magnetic resonance imaging (fMRI) and positron emission tomography (PET) studies conducted on anxious, individuals yield only modest regional arousal in areas such as the amygdala (Rauch et al., 1995; Irwin et al., 1996). This is partly explained by the the fact that, typically, rather weak cognitive fear stimuli must be used in human studies. Also, such technologies may not yet have the resolution to highlight many of the subcortical brain areas that are, in fact, aroused during fear. Thus, it is reasonable to suppose that functional imaging studies of the human brain have not yet provided reasonably comprehensive estimates of the neural systems that mediate anxiety. So far they have probably only highlighted some of the higher brain areas (i.e., only the canopy of neural systems) that mediate anxiety as well as some extracerebral artifacts (Drevets et al., 1992). Most preclinical investigators now tend to agree that the animal work is clarifying the fuller extent of fear systems that may also operate within the human brain. Such work is revealing neuroanatomical, neurophysiological, and neurochemical details that no human brain imaging technique can yet approximate. Totally new types of anti-anxiety drugs will eventually emerge from such animal work (Blanchard et al., 1993; Heilig et al., 1994; Johnson and Lydiard, 1995).

THE SYMPTOMS OF ANXIETY

Under anxiety disorders, the *Diagnostic and Statistical Manual of Mental Disorders* (*DSM-IV*) includes eight major types of disturbances, including panic attacks, agoraphobia, specific phobias, social phobias, obsessive–compulsive disorders, posttraumatic stress disorder, acute stress disorder, and generalized anxiety disorder. The most common clinical symptom of all these disorders is excessive anxiety and worry that causes sustained distress. Most of the focus of this chapter will be on these last dimensions of anxiety. Among the common symptoms of generalized anxiety, we often find a variety of psychological distrubances such as uncontrollable apprehensive expectations, jumpiness, and a tendency for excessive vigilance and fidgeting. The various autonomic symptoms commonly include gastrointestinal symptoms like upset stomach, diarrhea, and frequent urination, as well as other visceral symptoms like tachycardia, chronic dryness of the mouth, and increased but shallow respiration. Some are bothered more by the physical symptoms, whereas in others, psychological distress is the prevailing concern. Practically all of these changes can be evoked by artificial activation of the FEAR system.

From a clinical perspective, it is especially important to understand the neurophysiological and neurochemical mechanisms that can counteract these processes. Optimal medical control of anxiety must ultimately be based on this knowledge. Progress in this area was a matter of chance and exceptional good fortune during the early days of modern biological psychiatry. For example, the efficacy of the future minor tranquilizers such as chlordiazepoxide in dampening emotionality in animals was identified in 1960 during the final experiment just prior to the scheduled termination of a relatively unfruitful research program on benzodiazepines (BZs) at Hoffman-LaRoche labs. Over the past decade, emerging knowledge about the underlying brain systems has allowed investigators to proceed in more systematic and logical ways, and we can predict where the breakthroughs will be.

Among the new generation of anti-anxiety medicines will be ones that can reduce the arousal of the corticotrophin releasing factor (CRF) system, which seems to be a major neuropeptide component of various anxieties (Dunn and Berridge, 1990; Chalmers et al., 1996). The utility of nonpeptide CRF receptor anatgonists has already been validated in animal models (Schultz et al., 1996), and such agents will soon undergo clinical evaluation. Neuropeptide-Y and ACTH antagonists may soon follow (Heilig et al., 1994). While cholecystokinin (CCK) antagonists have already been evaluated and little benefit has been observed, the preclinical data were also often rather contradictory and confusing (Harro et al., 1995). In any event, all of these peptides and many others are concentrated along the trajectory of the FEAR system.

Before proceeding with a discussion of the amygdalo–hypothalamo–PAG FEAR circuit, let me briefly summarize the vast number of animal models of anxiety that have been developed by behaviorists during the past few decades. Most of the preclinical pharmacological work on new anti-anxiety agents is emerging from the systematic analysis of such animal models. Unfortunately, it is not yet clear how most of these models relate to the various FEAR circuit components. A great deal of basic preclinical work remains to be done and thus the existing behavioral literature still gives the impression of being chaotic and unintegrated. Let me briefly summarize in a systematic way, the diversity of models that are available.

PRECLINICAL MODELS FOR THE STUDY OF FEAR

The various preclinical laboratory models for the study of anxiety can be conveniently broken down into those that use painful stimuli to produce symptoms of anxiety (i.e., punishment procedures) and those that use no explicit punishments. In addition, each of these categories includes some models that analyze changes in learned behaviors while others rely on measures of unconditional, or instinctual behaviors (yielding the four general types of models, as

Table 1. A Taxonomy of Animal Models of Fear

	With Punishment	*No Punishment*
Learned	• Active avoidance tasks • Conditioned emotional responses • Punished behavior tasks • Passive avoidance tasks	• Partial reinforcement extinction effect
Spontaneous	• Freezing to shock • Defensive burying • Stimulation of fear circuits • responses to loud sounds (startle)	• Open-field exporation • Avoidance of bright lights • Social interaction tests • Plus-Maze test • Predatory odors

summarized in Table 1). More detailed descriptions of these models are available elsewhere (Panksepp, 1998).

Most of these models are quite sensitive to the effects of minor tranquilizers, suggesting that they share common motivational features, but a conceptual problem runs through much of the literature on this topic. Anti-anxiety effects are generally *assumed to* exist when drugs release previously punished (and hence inhibited) behaviors. For instance, animals that have received shocks upon pressing a bar for food tend to inhibit lever pressing, while under presumed anti-anxiety agents they resume bar-pressing more readily. Unfortunately, there are at least two distinct explanations for such effects: Animals may be less anxious or they may simply be more disinhibited. This second possibility was rarely considered in earlier discussions of how animal models can relate to our understanding of human anxiety. Another problem is that there are so many differences in formal procedures and drug sensitivity among the various experimental models that it is not yet possible to argue for a shared anxiety process for all of them. There are also no accepted guidelines as to which models are the best predictors for which clinical problems and why. Accordingly, the literature may actually be describing many different types of fears and/or different ways in which the brain handles the same type of fear. Although this is not the place to contrast and compare these models in any detail, the organizational scheme of Table 1 helps visualize how the various models conceptually interrelate, and there is considerable room for further development.

The large variety of models that exist for the analysis of fear may also highlight the fact that multiple processes exist for the elaboration of trepidation in the brain. Fearful states can be evoked by painful stimuli, cues that have been previously associated with aversive stimuli, various nonpainful stimuli that have indicated danger in the evolutionary history of a species, and perhaps even certain frustrating events. Since these models are often differentially sensitive to various pharmacological manipulations, it is clear that there are many distinct cognitive and motivational controls over fear processes at the neurological level. However, for present purposes, we shall assume that many of the stimuli that evoke anxiety do derive their motivational impact from a shared neural circuit, namely, the transhypothalamic FEAR system.

THE NATURE OF A MAJOR BRAIN SYSTEM THAT MEDIATES FEAR

As already emphasized, brain-stimulation studies have long suggested that a coherently operating FEAR system extends from the temporal lobe (from specific areas of the amygdala) through the anterior and medial hypothalamus to the lower brain stem (through the PAG substance of the mesencephalon) and then down to specific autonomic and behavioral output components of the lower brain stem and spinal cord, which control the physiological symptoms of fear. (These symptoms include increases in heart rate, blood pressure, the startle response, elimination, and perspiration; for summary, see Figure 1). A consensus is emerging that a fundamental form of unconditional fear is mediated by this neural system (Panksepp, 1982, 1990; Davis, 1992; 1994; LeDoux, 1993; 1995; Maren and Fanselow, 1996). When this system is activated by electrical stimulation, animals exhibit a variety of fearlike behaviors ranging from an initial freezing response at low current levels, which may reflect the appropriate response to fearful objects at a distance, to an increasingly precipitous flight response at higher current intensities, reflecting the appropriate response to fearful objects at a closer, more threatening, distance. The responses evoked artificially by brain stimulation look very similar to the behavior of an animal being pursued by a predator or receiving repeated foot shock. Even though such animals appear to be severely distressed, it seems unlikely that the brain stimulation is activating a pain pathway since animals typically do not squeal or exhibit other apparent symptoms of pain. Minor tranquilizers exert at least part of their anti-anxiety effect by reducing arousal of this brain system. This chapter will focus most on this preeminent FEAR system, but it will also touch upon several other brain systems that mediate aversive central states that have often been subsumed under the concepts of fear and anxiety (e.g., separation anxiety).

Although we cannot directly measure subjective experience, the behavioral evidence from all mammals that have been studied with sensitive avoidance measures such as "place aversions" strongly suggests that a coherent internal state of dread is elaborated by the FEAR system. Humans verbally report powerful feelings of foreboding as stimulation is applied to these brain sites during the course of neurosurgery. Animals readily learn to escape such stimulation suggesting that they experience strong negative feelings. If given the opportunity, they avoid environments in which they received such stimulation (Panksepp, 1996), and if no avenue of escape is provided, they will freeze as if in the presence of a predator (Panksepp et al., 1991).

Whether the subjective experience of fear is mediated directly by this circuit or via interactions with other interconnected brain areas will have to be resolved by further research. However, the evidence favors the idea that affective experience is an intrinsic subcortical function, for decorticate animals still exhibit escape and fear behaviors when such circuits are stimulated. Although we do not have any detailed understanding of how affective experience is actually generated by basic emotional circuits such as these or whether higher reaches in the amygdala or lower reaches in the PAG are more essential for the experiential response, it is reasonable to suppose that this whole circuit is essential for the integrated fear response and that it lies at the very heart of many human and animal anxieties. I have advocated the view that the feeling states of the various emotions are created by such emotional systems interacting with a primitive, but extended, neurosymbolic representation of the body that originates within the midbrain (Panksepp, 1998). In any event, these circuits can be sensitized by repeated activation, whether induced by direct brain stimulation (Adamec, 1993) or by life experiences (Adamec and Stark-Adamec, 1989). Once this type of "limbic permeability" is established, it is difficult to permanently reverse the hypersensitivities, but many consider such states to be an important testing ground for evaluating new therapies for free-floating anxieties and posttraumatic stress disorders (PTSDs) (van der Kolk, 1987).

ON THE VARIETIES OF ANXIETY SYSTEMS IN THE BRAIN

We must not oversimplify the complexities that face us in our attempts to obtain a definitive understanding of trepidation within the mammalian brain. Almost certainly, several distinct forms of "anxiety" may exist that we are not yet able to differentiate. Indeed, the FEAR system probably has differential inputs such as the simple cue-based systems from the amygdala and spatial information systems from the hippocampus (Phillips and Le Doux, 1992; Fanselow et al., 1994). In addition, several distinct types of anxiety-related information may be mediated in one brain zone such as the amygdala (Killcross et al., 1997).

Before proceeding with a more detailed discussion of the FEAR system, it is worth briefly highlighting another distinct neuroemotional sytem that can mediate aversive states related to anxiety. For instance, there is clearly a distinct separation distress system (as indexed by measures of separation calls in species as diverse as primates, rodents, and birds) which runs from the preoptic area and bed nucleus of the stria terminalis down through the dorsomedial thalamus to the mesencephalic PAG region where virtually all emotional systems appear to converge and interdigitate (Panksepp et al., 1988). This system has been postulated to mediate such negative feelings as loneliness, grief, and other forms of separation anxiety and to contribute substantially to the precipitous forms of acute distress known as panic attacks.

Clinically, a distinction between the brain mechanisms that control panic attacks and those that produce anticipatory anxiety first became apparent when it was found that the best available anti-anxiety agents (e.g., BZs such as chlordiazepoxide and diazepam) neither inhibited panic attacks in humans nor separation anxiety in animals. However, drugs such as tricyclic antidepressants (e.g., imipramine and chlorimipramine), which were found to exert excellent anti-panic effects in humans and reduced separation distress in animals, had little effect on anticipatory anxiety. Quite remarkably, people whose panic attacks had been effectively attenuated with tricyclics still suffered from the fear that their attacks might recur (Klein and Rabkin, 1981).

A clinical distinction can also be made between fearful anxiety and separation anxiety. The former is characterized by generalized apprehensive tension with a tendency toward various autonomic symptoms such as tachycardia, sweating, and gastrointestinal symptoms. The latter, especially in intense forms such as grief, is accompanied by feelings of weakness and depressive lassitude with autonomic symptoms of a parasympathetic nature (e.g., strong urges to cry) often accompanied by tightness in the chest and the feeling of having a lump in the throat. The former beckons one to escape situations that intensify the anxiety, whereas the latter tends to motivate thoughts about the lost object of affection and impels one to seek succor and the company of special loved ones.

A compelling case has been put forward asserting that panic attacks emerge from primitive suffocation–alarm systems of the brain that may be closely coupled with separation distress systems (Klein, 1993). Although separation distress/PANIC systems and FEAR systems can be distinguished in the brain, it is to be expected that they can also operate synergistically: Chronic anxiety can increase the incidence of panic attacks, and panic attacks can lead to chronic anxiety. The ability of some new anti-anxiety agents to reduce panic may also indicate that the two emotional systems share some common inhibitory influences. For instance, alprazolam can quell panic (Schweizer et al., 1993) perhaps because, in addition to BZ receptor effects, it can also promote serotonin transmission and reduce β-adrenoreceptor activity.

The chronic mood changes that commonly accompany PTSD, including mixtures of anger and anxiety, may also need to be distinguished from the types of anxiety already discussed. The severity of PTSD can be diminished with anti-seizure medications such as carbamazepine, an agent that is not consistently effective in the control of either panic attacks or anticipatory anxiety (Charney et al., 1993). In addition to facilitating GABA activity, carbamazepine has a spectrum of other neurochemical effects. It is noteworthy that this drug can block "kindling"—the induction of chronic seizure potentials in the brain via the periodic application of brief electrical stimulation to seizure-prone areas of the temporal lobe such as the amygdala. It is not unusual for kindled animals to exhibit chronic emotional changes (increases in irritability as well as heightened sexuality), further suggesting that similarities may exist between this type of evoked brain change and PTSD.

In addition, other psychiatric disorders are commonly accompanied by anxiety. For instance, obsessive–compulsive behaviors and rituals may often represent an attempt to ward off various encroaching anxieties. We do not know how or whether these incipient anxieties are mediated by any of the systems discussed above. It is noteworthy, however, that the serotonin uptake inhibitor chlorimipramine, which is so effective in controlling obsessive–compulsive behaviors, is also an effective anti-panic agent suggesting a shared neurochemical substrate for both (Altemus, 1995). It is also effective in reducing separation distress in animals, although its efficacy in other anxiety disorders remains to be adequately evaluated.

It currently seems evident that brain systems that mediate anticipatory and chronic generalized anxiety and those that mediate panic attacks, separation anxiety, and posttraumatic stress disorders can be distinguished in terms of brain mechanisms. However, all strong emotional states appear to share some nonspecific alarm or alerting components when threatening stimuli first appear on the psychological horizon. For instance, generalized cerebral arousal/attentional systems such as cholinergic and noradrenergic arousal circuits of the brain stem (see Chapter 2, Figure 1) would fall into such a category. Also, most anxieties are accompanied by arousal of the pituitary–adrenal stress response, which helps recruit bodily and cerebral resources to cope with various stressors (for reviews, also see Puglisi-Allegra and Oliverio, 1990).

RELATIONSHIPS BETWEEN PAIN AND FEAR

In addition to the different forms of anxiety, there are many other types of aversive internal feelings—ranging from pain to various types of hungers, thirsts, and other bodily needs. It is especially important to consider the role of pain in the genesis of anxiety, since that has been the traditional way to produce fear-conditioning in animal models. Animals readily learn to escape from and avoid places where they have been hurt. Current evidence suggests that pain and fear systems can be dissociated even though they interact strongly at various locations within the neuroaxis. One of the most compelling lines of evidence of this is that fearlike behaviors in animals and fear states in humans are not readily produced by electrical stimulation of the classical spinothalamic pain systems. It is only at midbrain levels, where the classical pain systems diverge into reticular fields, that localized brain stimulation begins to yield fear behaviors such as freezing and flight. Thus, even though pain systems do send inputs into areas of the brain that mediate fear (such as the PAG of the mesencephalon), electrical activation of the FEAR system does not appear to readily evoke the sensation of pain in humans or animals. Humans who have been stimulated in these latter brain areas typically report fear rather than pain, and animals exhibit flight and escape with no apparent pain (at least as indicated by squealing). Likewise, lesions of brain areas that contain fear circuitry do not typically affect pain thresholds in animals (for summary of evidence, see Panksepp et al., 1991).

In addition to the pathways whereby pain gains access to the FEAR system, it may well be the case that the FEAR system reduces pain sensitivity. It is well documented that animals and humans do not focus on their bodily injuries when they are scared (Fanselow, 1991), and there is now some evidence that fear-induced analgesia emerges from arousal of pain inhibition circuits such as serotonin and endogenous opioids near the PAG of the mesencephalon.

The rest of this chapter will highlight some of the neurochemistries that are implicated in the FEAR system. Although a great deal of work remains to be done, it is becoming clear that along with mediating the unconditional aspects of fear, this system also permits the construction of learned behavioral strategies to cope with fear. In addition to controlling urges to flee, FEAR circuitry also establishes brain conditions for the expression of passive fear tendencies such as freezing (and hence perhaps the up-tight feeling of nervous tension that characterize chronic anxiety neuroses).

THE NEUROCHEMISTRY AND PHARMACOLOGY OF FEAR

The most useful knowledge about fear and anxiety for medical purposes will emerge from an understanding of the neurochemical systems that elaborate and modulate fearful impulses within the brain (Leonard, 1992). It is clear that BZ receptors are concentrated along the trajectory of the FEAR circuit (Figure 1), from the central amygdala, downward via the ventral amygdalofugal pathway, through the anterior and medial hypothalamus, across the substantia nigra to the PAG and the nucleus reticularis pontis caudalis (where fear modulation of the startle reflex occurs). The discovery of the BZ- GABA receptor complex has been the single most important discovery for explaining how traditional minor tranquilizers (i.e., alcohol, barbiturates, and the benzodiazepines) inhibit anxiety (Haefely, 1990). This receptor complex has distinct GABA binding sites and BZ binding sites as well as binding sites for the older anxiolytics such as barbiturates, all of which can act synergistically to increase neuronal inhibition (by facilitating chloride ion flow into the cell). In other words, anxiety is quelled by BZs through the hyperpolarization of the neuronal elements that pass the anxiety message through the neuroaxis. While agonists for the BZ receptor, such as the many variants of BZ-type minor tranquilizers, suppress activity in the FEAR circuit, they may also modulate higher processing of associated stimuli (e.g., anxious thoughts and appraisals), perhaps via effects on the relatively abundant BZ receptors in the neocortex. Antagonists for the receptor (such as flumazenil) are usually behaviorally inactive by themselves, suggesting that endogenous anxiety signals are not tonically present at BZ receptor sites. Of course, these antagonists can block the anti- anxiety effects of exogenously administered BZs and the anxiety provoked by β-carboline "inverse agonists" (see below).

One of the key questions has been what type of endogenous molecule normally acts on the BZ receptor. Even though definitive evidence is not available, one likely candidate continues to be the endogenous neuropeptide called diazepam binding inhibitor (DBI), which appears to promote anxiety when it is released onto BZ receptors (Ferrarese et al., 1993). Thus, it is suspected that DBI exerts an "inverse agonist" effect at BZ receptors, actively decreasing chloride flow and hence increasing activity in the basic brain substrates for anxiety. This "inverse agonist" concept was first generated by the discovery of various ß-carboline drugs, which had the opposite effect to BZs, that is, they actively inhibited chloride entry into neurons via interaction with the BZ–GABA complex and thereby could promote anxiety in both humans and animals. This is surely only part of the story.

Which other neurotransmitters convey the signal of fear through the neuroaxis? There are presently several reasonable candidates. Although norepinephrine (NE) and serotonin have been touted as signals for transmitting anxiety, they are unlikely to be specific anxiety transmitters. Certain drugs that can increase certain types of NE and serotonin activity (e.g., yohimbine and *m*-chlorophenylpiperazine (MCPP), respectively) do promote the experience of anxiety in humans (Charney et al., 1987), but these effects could easily reflect indirect general-arousal effects (amplifying whatever tendencies already exist in the nervous system) rather than specific emotional responses. There are many other neuropeptide and amino-acid candidates that appear to modulate anxiety-type behaviors more powerfully and more specifically; hence they will probably figure more heavily in our future understanding of fear substrates in the brain.

At present, one compelling candidate is the simple amino-acid neurotransmitter glutamate. A remarkably powerful behavioral syndrome that resembles the arousal of fear can be evoked by administering the glutamate agonists kainic acid and *N*-methyl-*D*-asparatate (NMDA) into the lower ventricular system. Within minutes, animals begin to run around in apparent psychic anguish (often in a semicrouched posture) with rapid head scanning, persistent vocalization, and bulging eyes suggestive of profound terror. The fact that these episodes can be inhibited by appropriate glutamate receptor antagonists (those that block kainate or NMDA receptors) may suggest that new anti-anxiety agents could be created through the selective pharmacological manipulation of these systems. However, since so many fundamental brain processes are controlled by glutamate circuits (from simple sensory processes to memory), such a strategy may be impractical due to the probability of many side-effects.

In addition to DBI, there are several other anxiogenic neuropeptide transmitters that operate along anatomical pathways parallel to the trajectory of the FEAR system (compare Figures 1 and 2 of Chapter 2). Most strikingly, central administration of the neuropeptide CRF can promote symptoms of anxiety in animals. Briefly, the effects of this peptide are as follows: CRF causes agitated arousal and can reduce a variety of positively motivated behaviors including feeding, sexual activities, and other social activities (for review, see Dunn and Berridge, 1990). Animals also tend to freeze in environments in which they previously received CRF, and normal shock-induced freezing can be diminished by CRF antagonists (Candor et al., 1992). Since CRF arising from the paraventricular nucleus of the hypothalamus also controls the pituitary–adrenal stress response (which accompanies many emotions as well as many psychiatric disturbances, such as depression), it is generally believed that CRF receptor antagonists will prove to be potent anti-anxiety and anti-stress agents (Heilig et al., 1994).

Also, the neuropeptide α-melanocyte stimulating hormone (αMSH) promotes camouflage-type pigmentary changes in many fish and reptiles, and not surprisingly, a vigorous freezing/hiding pattern can be evoked by central administration of this peptide in chicks (Panksepp, 1993). Adrenocorticotrophic hormone (ACTH), which comes from the same segment of the proopiomelanocortin (POMC) gene as αMSH, has essentially identical effects. There is little comparable data on mammalian behavior patterns, even though microinjections of high doses of ACTH into the PAG can precipitate vigorous jumping and flight. The affective effects of such treatments remain to be evaluated using conditioned freezing and place-avoidance paradigms, but antagonists of these neuropeptide receptor systems may well reduce certain types of fearful behavioral inhibition in humans. Another intriguing anxiogenic peptide is cholecystokinin (CCK), which can precipitate panic attacks in humans and a broad spectrum of anxiogenic symptoms in animals with the use of a variety of models described above (Harro et al., 1993). As mentioned earlier, however, preliminary clinical trials with CCK antagonist have not been promising.

A variety of other neuropeptides appear to reduce anxiety symptoms following central administration. Centrally administered opioids (acting at μ receptor sites) as well as oxytocin and prolactin are very effective agents for reducing separation anxiety (Panksepp, 1991; 1993). Neuropeptide Y (NPY) has emerged as a major anti-anxiety system that may have receptors that are independent of those that regulate feeding behavior (Heilig, 1995). Recently neurosteroids that can modulate GABA receptors have also appeared as potential anti-anxiety agents (Paul and Purdy, 1992).

An especially important dimension of future research is to specify precisely how these various neurochemical vectors mediate aspects of processes subsumed by the broad category of anxiety. Do certain neurochemistries convey specific fears while others are indirect modulators (e.g., providing gain settings, duration controls, etc.) within the FEAR system? Future research with animal models should be able to tease apart the distinct functions of the diverse chemistries that control anxiety.

LEARNING WITHIN FEAR SYSTEMS

Learning mechanisms allow organisms to effectively channel their specific fears into appropriate responses within the environment. The FEAR system of the brain has some intrinsic sensitivities (e.g., responding unconditionally to painful stimuli and the presence of potential predators), but it also has the ability to establish new input components to inform the organism about various specific threats that require learning. Some environmental circumstances lead to rapid conditioning, presumably because certain perceptions have comparatively easy access to the FEAR system within the brain, whereas totally neutral stimuli take longer to condition. For instance, it has been found that in humans, autonomic fear responses condition more rapidly when shock is paired with angry faces than when it is paired with smiling faces (Öhman et al., 1989). Presumably, the brain is predisposed to associate the potentially threatening configuration of an angry face with fear more easily than that of a pleasant face. In other words, we can anticipate that the neural configurations (which are probably in the amygdala and surrounding inferotemporal cortex) that decode emotional expressions have "sensitized" input channels to the FEAR circuit (Adolphs et al., 1994).

There is probably a variety of such channels to the FEAR system that are different in different species. Thus, humans are presumably more apt to develop fears of dark places, high places, approaching strangers (especially those with angry faces), and snakes and spiders. Rats are especially apt to fear well-illuminated areas, open spaces, and the smell of cats and other potential predators. However, completely neutral stimuli and the imagination can also access the FEAR system of the brain. One can come to fear unseen bacteria. During the past few years, great progress has been made in unraveling the manner in which this system is capable of being conditioned by specific learning experiences.

Stimuli that have been paired with painful events can access FEAR circuits directly from subcortical sensory analyzers of the thalamus or from the more complex analyzers of the cortex. It is especially important to emphasize that FEAR circuits can be conditioned directly via thalamoamygdala connections, with no need for cortical processing of the signals that predict aversive events (LeDoux, 1993). There are both direct anatomical entry points from the thalamus into the central nucleus of the amygdala and more indirect cortical ones (LeDoux et al., 1990; Davis et al., 1995). The intraamygdaloid circuitries that mediate the neural computations that impinge on the unconditional FEAR circuits, a process that starts in the central nucleus of the amygdala, are now being detailed (Pitkanen et al., 1997). Learned fear associations may be mediated by glutamatergic synapses and β-adrenergic synapses, which are concentrated in the amygdala (Chaill et al., 1994), but it is expected that many other amines and neuropeptides localized there will also prove to be influential. Also, it is likely that conditioning can be elaborated at lower levels of the fear circuit (i.e., at hypothalamic and mesencephalic levels), but that remains to be demonstrated. Once the details of the learning mechanisms have been worked out (Davis 1992; 1994), it should be possible to facilitate the deconditioning of learned fears with new pharmacological maneuvers (Muller et al., 1997). Indeed, it has been demonstrated that glutamate antagonists can retard the extinction of conditioned fears (Falls et al., 1992), suggesting that fears need to be deconditioned by an active form of new learning, which is mediated by glutamate.

Although it is clear that the amygdala is essential for processing various cognitive stimuli associated with fear (Bechara et al., 1995; Young et al., 1996), it is not certain that the exclusive or even major area of the brain that generates the affective experience of fear are the higher reaches of the brain that process knowledge concerning fearful stimuli. Indeed, recent work indicates that autonomic conditioning of fear can still occur in humans even after higher limbic structures such as amygdala and hippocampus have been destroyed (Tranel and Damasio, 1990). It seems likely that a great deal of affective experience can still be generated by the FEAR system even though its highest reaches,which harvest cognitive inputs, are severely impaired.

THE TREATMENT OF ANXIETY IN CLINICAL PRACTICE

Until the development of benzodiazepine-type minor tranquilizers, the drugs that could successfully control human anxiety were opioids, alcohol, barbiturates, and meprobamate (Gray, 1987). The treatment of anxiety was revolutionized by the serendipitous discovery that chlordiazepoxide (CDP) could calm wild animals. The entry of this agent into pharmaceutical practice, under the trade name of Librium, was rapid because of its remarkable specificity and safety margin as compared to anything that had been used before. CDP could reduce anxiety at less than a hundredth of the lethal dose, which was a remarkable improvement over any other agent that existed. Not long thereafter, potent versions such as diazepam (Valium) became available and many others have surfaced since. The mild sedative effects commonly observed at the beginning of drug therapy exhibited rapid tolerance, whereas the anxiolytic effects were sustained, evoking little tolerance during longterm use. The efficacy of BZs in diminishing anticipatory anxiety and chronic anxiety neurosis was demonstrated in a large number of well controlled clinical studies (Nutt, 1990). These drugs produced no apparent physical dependence during modest use, even though longterm use of high doses, which became a common practice, did yield dependence and a withdrawal syndrome resembling that of alcoholism. The utility of BZs in inhibiting alcohol withdrawal symptoms further affirmed that these agents work upon common GABA-related substrates in the brain (Tallman and Gallagher, 1985).

The BZs rapidly supplanted all other anti-anxiety medications on the market. Several additional medical uses were soon discovered, including relaxation of muscular spasms, and effective control of certain types of seizures, especially those that emanate from the limbic system. BZs also found a receptive market as sleep-promoting agents, and because of their cross-tolerance, they became effective medications for the alleviation of alcohol withdrawal (Roy-Byrne and Cowley, 1991).

Although a variety of BZs came on the market, it was not until 1979, that the BZ receptor was finally identified in the brain (Young and Kuhar, 1980). A different type of receptor was later identified in the periphery. Since then, a great number of the subsequently developed BZs are now tailor-made and marketed for specific disorders even though their basic mode of neuronal action remains fundamentally unchanged. The practice of using different agents for different disorders, such as fast-acting BZs for sleep disorders and long-acting ones for alcohol withdrawal and anxiety, is not based on any fundamental difference in their mode of action but rather on differences in potency and speed of entry into and exit from the brain.

Even though BZs turned out to be remarkably safe medicinal agents, various shortcomings have been revealed during the ensuing years, including (as mentioned above) a dependence syndrome during longterm use. Other side-effects include increased appetite, disorientation and memory loss (especially in the elderly), and the release of aggressive tendencies in passive–aggressive individuals. In clinical practice, the sedation produced by BZs is usually of short duration, although it tends to persist among the elderly. Due to these drawbacks, particularly the potentiation of confusional states, these drugs should not be used in conjunction with alcohol, which operates, at least in part, through the same neural systems. Although these drawbacks of BZs are modest compared to most other psychotropic medications, care in monitoring side-effects is essential. Problems that can arise with such agents could prove troublesome in our litigious society. Indeed, the development of novel BZ receptor antagonists for the treatment of drunkenness was aborted because of the potential liabilities that might result from individuals having traffic accidents after taking such drugs. Because of such shortcomings of BZs, and the additional profits to be made, there has been a concerted effort to find additional anti-anxiety agents (Kunovac and Stahl, 1995).

Even though there are many candidates in the wings, the only major item that has reached the market is buspirone (Buspar), which has a profile of action quite distinct from the BZs (Eison and Temple, 1986). The therapeutic effect of this agent appears to be based on the ability of the serotonin system to modulate anxiety. Buspirone has the relatively selective effect of stimulating 5-HT_{1A} receptors, which are predominantly concentrated on serotonin cell bodies. At this site, buspirone reduces serotonin neuronal activity, and hence it acutely diminishes serotonin release in higher brain areas, which can lead to longterm upregulation of postsynaptic serotonin receptors. Although some investigators believe that buspirone alleviates anxiety by reducing 5-HT activity, the benefits may be due to a compensatory functional elevation of brain serotonin activity.

It should be remembered that there are also abundant postsynaptic 5-HT_{1A} receptors in the brain, and it presently remains possible that the postsynaptic effects of buspirone contribute as much to the control of anxiety as the effects of the drug on presynaptic ones. Thus, even though certain types of serotonin activity are still widely considered to be anxiogenic, the anti-anxiety effects of buspirone could well be due to a longterm facilitation of serotonin sensitivity in the brain (Stahl et al., 1992).

In any event, the therapeutic effects of buspirone tend to be milder than those obtained with BZs, but many fewer side-effects are encountered. Buspirone neither produces sedative effects nor shortterm psychic effects (hence it is not subject to abuse) nor dependence withdrawal upon discontinuation. Unfortunately, buspirone appears to exhibit very little benefit in those individuals who have previously benefited from BZs (Schweizer et al., 1986). This suggests that buspirone should be the initial treatment of choice at the onset of longterm pharmacotherapy. Unlike some of the newer BZs such as alprazolam, buspirone has no efficacy as an anti-panic agent.

For common physiological symptoms of anxiety, such as palpitations and sweating, β-noradrenergic blockers such as propranolol still appear to be the items of choice. "Beta-blockers" are generally deemed to be useful medication for the symptomatic control of anxiety that accompanies certain activities such as public presentations and performances.

Finally, it can be noted that MAO inhibitors such as phenelzine have been found highly effective for the control of social phobias and other neurotic personality disorders (Liebowitz, 1988). On the other hand, tricyclic drugs that are effective in reducing the incidence of panic attacks are also useful for reducing childhood anxiety-related disorders such as school phobias and enuresis, both of which may arise from overactive separation–distress systems of the brain.

CONCLUDING REMARKS

An understanding of the nature of fear and other emotions depends on the issue of precisely how affective experience is constructed within the brain. Since psychodynamic issues are almost impossible to address directly using empirical procedures, we still have to infer such processes from behavioral and psychological endpoints. The utilization of brain imaging techniques, including new EEG procedures, may eventually be able to monitor emotions directly, but that possibility still seems to be remote. Also, substantive progress on such issues will require that we continue to work out the neurobiological details in appropriate animal models. If other mammals also have subjective experiences such as fear, then it will be difficult to understand the underlying brain mechanisms without attempting to deal forthrightly with the problem of subjectivity in humans as well as animals. One might even argue that subjective feelings may be a key to certain aspects of brain organization and that it will be impossible to construct an accurate neurophysiological model of the mind without taking the existence of such brain functions seriously and studying them intensively, albeit indirectly, with the tools that presently exist.

Along the way, we will have to come to terms with the evils and advantages of anthropomorphism in psychobiological research. To the best of our knowledge, the subcortical systems that subserve FEAR are still shared by all mammals in a remarkably homologous fashion. Even though the cognitive components that accompany fear are more diverse, it currently seems likely that the essential components that generate the affective state of fear are more closely related to the neural components that we share with the animals.

Although the issue of subjective emotional experience in animals continues to be downplayed (LeDoux, 1996), it is quite easy to envision that a subjective state of fear could have a concrete function in the nervous system. It may be an efficient way to encode evolutionary values that enhance learning strategies. In other words, the affective states that accompany emotional behaviors may provide animals with a simple neural "mnemonic" for previous experiences. Rather than having to learn a complex behavior sequence for each new emergency, animals may be able to utilize affective states to efficiently encode new situations with common evolutionary values and to respond rapidly with a class of instinctive/intrinsic emotional states and behaviors that can be refined by subsequent learning. Unfortunately, in our modern world, these ancient solutions can often be unproductive, so it will be important not only to continue searching for new ways to dampen activity in the FEAR system, but also to find new pharmacological ways to help extinguish learned fears within the brain.

SUMMARY

Our understanding of the neurobiology of human fears has emerged largely from basic research on the brains of "lower" animals. Evidence from behavioral neuroscience strongly suggests that the unconditional (innate) capacity to experience fear, along with fear-typical patterns of autonomic and behavioral arousal, emerge from specific systems of the brain—the most prominent being a FEAR circuit, which courses between the amygdala and the central gray of the midbrain. Fear behaviors can be evoked by artificially activating this circuit, and the response is coordinated with autonomic and experiential effects. The neurochemical controls of this emotional system include a variety of neuropeptides that can modulate anxiety. Minor tranquilizers of the benzodiazepine (BZ) class work, in part, by dampening activity in this emotional system through increased GABA-mediated neural inhibition. The traditional use of minor tranquilizers in the control of clinical anxiety is also summarized.

REFERENCES

Adamec, R.E. (1993). Lasting effect of FG-7142 on anxiety, aggression and limbic physiology in the rat. J. Neurophysiol. 7, 232-248.

Adamec, R.E. & Stark-Adamec, C. (1989). Behavioral inhibition and anxiety: Dispositional, developmental, and neural aspects of the anxious personality of the domestic cat. In: *Perspective on Behavioral Inhibition.* (Stevens, J., Ed.), pp. 93-124. University of Chicago Press, Chicago.

Adolphs, R., Tranel, D., Damasio, H., & Damasio, A. (1994). Impaired recognition of emotion in facial expressions following bilateral damage to the human amygdala. Nature 372, 669-672.

Altemus, M. (1995). Neuroendocrinology of obsessive-compulsive disorder. In: *Advances in Biological Psychiatry, Vol. 1.* (Panksepp, J., Ed.), pp. 215-233, JAI Press, Greenwich, CT.

American Psychiatric Association (1994). *Diagnostic and statistical manual of mental disorders, DSM IV.* American Psychiatric Association, Washington, D.C.

Bechara, A., Tranel, D., Damasio, H., Adolphs, R., Rockland, C., & Damasio, A.R. (1995). Double dissociation of conditioning and declarative knowledge relative to the amygdala and hippocampus in humans. Science 269, 1115-1118.

Beck, C.H.M. & Fibiger, H.C. (1995). Conditioned fear-induced changes in behavior and in the expression of the immediate early gene *c-fos*: With and without diazepam pretreatment. J. Neurosci. 15, 113-121.

Beckett, S.R.G., Duxon, M.S., Aspley, S., & Marsden, C.A. (1997). Central *c-fos* expression following 20kHz/ultrasound induced defence behavior in the rat. Brain Res. Bull. 42, 421-426.

Blanchard, R.J., Yudko, E.B., Rodgers, R.J., & Blanchard, D.C. (1993). Defense system psychopharmacology: An ethological approach to the pharmacology of fear and anxiety. Behav. Brain. Res. 58, 155-156.

Burrows, G.D., Roth, M., & Noyes, R., Jr. (Eds.)(1990). *Handbook of Anxiety, Vol. 3, The Neurobiology of Anxiety.* Elsevier, Amsterdam.

Candor, M., Ahmed, S.H., Koob, G.F., Le Moal, M., & Stinus, L. (1992). Corticotropin-releasing factor induces a place aversion independent of its neuroendocrine role. Brain Res. 597, 304-309.

Chaill, L., Prins, B., Weber, & McGaugh, J.L. (1994). β-Adrenergic activation and memory for emotional events. Nature 371, 702-704.

Charney, D.S., Deutch, A.Y., Krystal, J.H., Southwick, S.M., & Davis, M. (1993). Psychobiologic mechanisms of posttraumatic stress disorder. Arch. Gen. Psychiat. 50, 294-305.

Charney, D.S., Woods, S.W., Goodman, W.K., & Heninger, G.R. (1987). Serotonin function in anxiety: II. Effects of the serotonin agonist MCPP in panic disorder patients and healthy subjects. Psychopharmacol. 92, 14-24.

Chalmers, D.T., Lovenberg, T.W., Grigoriadis, D.E., Behan, D.P., & De Souza, E.B. (1996). Corticotrophin-releasing factor receptors: From molecular biology to drug design.Trends Neurosci. 17, 166-172.

Davis, M. (1992). The role of the amygdala in fear and anxiety. Ann. Rev. Psychology 43, 353-375.

Davis, M., Campeau, S., Kim M., & Falls, W.A. (1995). Neural systems of emotion: The amygdala's role in fear and anxiety. In: *Brain and memory: Modulation and Mediation of Neuroplasticity.* (McGaugh, J.L., Weinberger, N.M., and Lynch, G., Eds.), pp. 3-40. Oxford Univ. Press, New York.

Davis, M. (1994). The role of the amygdala in emotional learning. Internat. Revs. Neurobiol. 36, 225-266.

Depaulis, A. & Bandler, R. (Eds.) (1991). *The midbrain periaqueductal gray matter: Functional anatomical and neurochemical organization.* Plenum Press, New York.

Drevets, W.C., Videen, T.O., MacLeod, A.K., Haller, J.W., & Raichle, M.E. (1992). PET images of blood flow changes during anxiety: Correction. Science 256, 1996.

Dunn, A.J. & Berridge, C. (1990). Physiological and behavioral responses to corticotropin-releasing factor administration: Is CRF a mediator of anxiety or stress responses? Brain Res.Revs 15, 71-100.

Eison, A.S. & Temple, D.L. (1986). Buspirone: Review of its pharmacology and current perspective on its mechanism of action. Am. J. Med. 80 (Suppl., 3B), 1-9.

Falls, W.A., Miserendino, M.J.D., & Davis, M. (1992). Extinction of fear-potentiated startle: Blockade by infusion of an NMDA antagonist into the amygdala. J. Neurosci. 12, 854-863.

Fanselow, M.S. (1991). Analgesia as a response to aversive Pavlovian conditional stimuli: Cognitive and emotional mediators. In: Fear, Avoidance and Phobias: A Fundamental Analysis. (Denny, M.R., Ed.), pp. 61-86. Lawrence Erlbaum Associates Pubs., Hillsdale, New Jersey.

Fanselow, M.S., Kim, J.J., Yipp, J., & De Oca, P. (1994). Differential effects of the *N*-methyl-D-aspartate antagonist D,L-2-amino-5-phosphonovalerate on acquisition of fear of auditory and contextual cues. Behav. Neurosci. 108, 235-240.

Ferrarese, C., Appollonio, I., Bianchi, G., Frigo, M., Marzorati, C., Pecora, N., Perego, M., Pierpaoli, C., & Frattola, L. (1993). Benzodiazepine receptors and diazepam binding inhibitor: A possible link between stress, anxiety and the immune system. Psychoneuroendocrinol. 18, 3-22.

Gray, J.A. (1987). The Psychology of Fear and Stress, 2nd ed. Cambridge University Press, Cambridge, England.

Haefely, W.E. (1990). The GABA-benzodiazepine receptor: Biology and pharmacology. In: Handbook of anxiety: Vol. 3. The Neurobiology of Anxiety. (Burrows, G.D., Roth, M., and Noyes, R., Jr., Eds.) Elsevier, Amsterdam.

Harro, J., Vasar, E., & Bradwejn, J. (1993) CCK in animal and human research on anxiety. TIPS 14, 244-249.

Harro, J., Vasar, E., Koszycki, D., & Bradwejn, J. (1995). Cholecystokinin in panic and anxiety disorders In: Advances in Biological Psychiatry, Vol 1. (Panksepp, J., ed.) pp. 235-262, JAI Press, Greenwich, CT.

Heilig, M. (1995). Antisense inhibition of neuropeptide Y (NPY)-Y1 receptor expression blocks the anxiolytic-like action of NPY in amygdala and paradoxically increases feeding. Regulatory Peptides, 59, 201-205.

Heilig, M., Koob, G.F., Ekman, R., & Britton, K.T. (1994). Corticotropin-releasing factor and neuropeptide Y: Role in emotional integration. TINS 17, 80-85.

Irwin, W., Davidson, R.J., Lowe, M.J., Mock, F.J., Sorenson, J.A., & Turski, P.A. (1996). Human amygdala activation detected with echo-planar functional magnetic resonance imaging. NeuroReports 7, 1765-1769.

Johnson, M.R. & Lydiard, R.B. (1995). The neurobiology of anxiety disorders. Psychiatric Clin. N. Amer. 18, 681-725.

Killcross, S., Robbins, T.W., & Everitt, B.J. (1997). Different types of fear-conditioned behaviour mediated by separate nuclei within amygdala. Nature 388, 377-380.

Klein, D.F. (1993). False suffocations alarms, spontaneous panics, and related conditions. Arch. Gen. Psych. 50, 306-317.

Klein, D.F. & Rabkin, J. (Eds.) (1981). Anxiety: New Research and Changing Concepts. Raven Press, New York.

Kollack-Walker, S., Watson, S.J., & Akil, H. (1997). Social stress in hamsters: Defeat activates specific neurocircuits within the brain. J. Neurosci. 17, 8842-8855.

Kunovac, J.L. & Stahl, S.M. (1995). Future directions in anxiolytic pharmacotherapy. Psychiatric Clin. N. Amer. 18, 895-909.

LeDoux, J. (1993) Emotional networks in the brain. In: *Handbook of Emotions.* (Lewis, M. & Haviland, J.M., Eds.), pp. 109-118. Guildford Press, New York.

LeDoux, J.E. (1995). Emotion: Clues from the brain. Ann. Rev. Psychol. 46, 209-235.

LeDoux, J.E. (1996). *The Emotional Brain.* Simon & Schuster, New York.

LeDoux, J.E., Cicchetti, P., Xagoraris, A., & Romanski, L.M. (1990). The lateral amygdaloid nucleus: Sensory interface of the amygdala in fear conditioning. *J. Neurosci.* 10, 1062-1069.

Leonard, B.E. (1992). *Fundamentals of Psychopharmacology.* Wiley, New York.

Liebowitz, M.R. (1988). Pharmacotherapy of personality disorders. In: *Emotions and Psychopathology.* (Clynes, M. & Panksepp, J., Eds.), pp. 77-94, Plenum Press, New York.

Maren, S. & Fanselow, M.S. (1996). The amygdala and fear conditioning: Has the nut been cracked? Neuron 16, 237-240.

Muller, J., Corodimas, K.P., Fridel, Z., & LeDoux, J.E. (1997). Functional inactivation of the latral and basal nuclei of the amygdala by muscimol infusion prevents fear conditioning to an explicit CS and to contextual stimuli. Behav. Neurosci. 111, 683-691.

Nutt, D.J. (1990). The pharmacology of human anxiety. Pharmac. Ther. 47, 233-266.

Öhman, A., Dimberg, U., & Esteves, F. (1989). Preattentive activation of aversive emotions. In: *Aversion, Avoidance and Anxiety.* (Archer, T. & Nilsson, L.-G., Eds.), pp. 169-193. Lawrence Erlbaum, Hillsdale, NJ.

Panksepp, J. (1982). Toward a general psychobiological theory of emotions. Behav. Brain Sci. 5, 407-467.

Panksepp, J. (1985). Mood changes. In: *Handbook of Clinical Neurology, Revised Series, Vol. 1(45), Clinical Neuropsychology.* (Vinken, P.J., Buyn, G.W., & Klawans, H.L., Eds.), pp. 271-285.

Panksepp, J. (1990). The psychoneurology of fear: Evolutionary perspectives and the role of animal models in understanding human anxiety. In: *Handbook of Anxiety, Vol. 3, The Neurobiology of Anxiety.* (Burrows, G.D., Roth, M., & Noyes, R., Jr., Eds.), pp. 3-58. Elsevier, Amsterdam.

Panksepp, J. (1991). Affective neuroscience: A conceptual framework for the neurobiological study of emotions. In: *International Review of Studies on Emotion, Vol. 1.* (Strongman, K.T., Ed.), pp. 59-99. Wiley, Chichester, England.

Panksepp, J. (1993) Neurochemical control of moods and emotions: Amino acids to neuropeptides. In: *Handbook of Emotions.* (Lewis, M. & Haviland, J.M., Eds.), pp. 87-107. Guildford Press, New York.

Panksepp, J. (1996). Modern approaches to understanding fear: From laboratory to clinical practice. In: *Advances in Biological Psychiatry, Vol. 2.* (J. Panksepp, Ed.), pp. 209-230, JAI Press, Greenwich, CT.

Panksepp, J. (1998). *Affective neuroscience: The Foundations of Human and Animal Emotions.* Oxford University Press, New York.

Panksepp, J., Normansell, L., Herman, B., Bishop, P., & Crepeau, L. (1988). Neural and neurochemical control of the separation distress call. In: *The Physiological Control of Mammalian Vocalizations.* (Newman, J.D., Ed.), pp. 263-300). Plenum Press, New York.

Panksepp, J., Sacks, D.S., Crepeau, L.J., & Abbott, B.B. (1991). The psycho- and neurobiology of fear systems in the brain. In: *Fear, Avoidance, and Phobias.* (Denny, M.R., Ed.), pp. 7-59, Lawrence Erlbaum Assocs., Hillsdale, NJ.

Paul, S.M. & Purdy, R.H. (1992). Neuroactive steroids. FASEB J. 6, 2311-2322.

Phillips, R.G. & Le Doux, J.E. (1992). Differential contribution of amygdala and hippocampus to cued and contextual fear conditioning. Behav. Neurosci. 106, 274-285.

Pitkanen, A., Savander, V., & LeDoux, J.E. (1997). Organization of intra-amygdaloid circuitries in the rat: An emerging framework for understanding functions of the amygdala. TINS 20, 517-523.

Puglisi-Allegra, S. & Oliverio, A. (Eds.) (1990). *Psychobiology of Stress.* Klower Academic Pubs., Dordrecht, The Netherlands.

Rauch, S.L., Savage, C.R., Alpert, N.M., Miguel, E.C., Baer, L., Breiter, H.C., Fischman, A.J., Manzo, P.A., Moretti, C., & Jenike, M.A. (1995). A positron emission tomographic study of simple phobic symptom provocation. Arch. Gen. Psychiat. 52, 20-28.

Roy-Byrne, P.P. & Cowley, D.S. (1991). *Benzodiazepines in clinical practice: Risks and benefits.* American Psychiatric Press, Washington, D.C.

Schultz, D.W., Mansbach, R.S., Sprouse. J., Braselton, J.P., Collins, J., Corman, M., Dunaiskis, A., Faraci, S., Schmidt, A.W., Seeger, T., Seymour, P., Tingley, F.D., III, Winston, E.N., Chen, Y.L., & Heym, J. (1996). CP-154,526: A potent and selective nonpeptide antagonist of corticotropin releasing factor receptors. Proc. Nat. Acad. Sci. U.S.A. 93, 10477-10482.
Schweizer, E., Rickels, K., & Lucki, I. (1986). Resistance to the anti-anxiety effect of buspirone in patients with a history of benzodiazepine use. N. Eng. J. Med. 314, 719-720.
Schweizer, E., Rickels, K., Weiss, S., & Zavodnick, S. (1993). Maintenance drug treatment of panic disorder. I. Results of a prospective, placebo-controlled comparison of alprazolam and imipramine. Arch. Gen. Psychiat. 50, 51-60.
Silviera, M.C.L., Sandner, G., & Graeff, F.G. (1993). Induction of Fos immunoreactivity in the brain by exposure to the elevated plus maze. Behav. Brain Res. 56, 115-118.
Silviera, M.C.L., Sandner, G., Di Scala, G., & Graeff, F.G. (1995). C-fos immunoreactivity in the brain following electrical or chemical stimulation of the medial hypothalamus of freely moving rats. Brain Res. 674, 265-274.
Soubrie, P. (1986). Reconciling the role of central serotonin neurons in human and animal behavior. Behav. Brain Sci. 9, 319-364.
Stahl, S.M., Gastpar, M., & Keppel Hesselink, J.M. (Eds.)(1992). Serotonin 1A Receptors in Depression and Anxiety. Raven Press, New York.
Tallman, J.F. & Gallagher, D.W. (1985). The GABA-ergic system: A locus of benzodiazepine action. Ann. Rev. Neurosci., 8 21-44.
Tranel, D. & Damasio, A.R. (1990). The covert learning of affective valence does not require structures in hippocampal system or amygdala. J. Cog. Neurosci. 5, 79-88.
van der Kolk, B.A. (Ed.)(1987). Psychological Trauma. American Psychiatric Press, Washington, D.C.
Young, A.W., Hellawell, D.HJ., van de Wal, C., & Johnson, M. (1996). Facial expression processing after amygdalotomy. Neuropsychologia 34, 31-40.
Young, W.S. & Kuhar, M.J. (1980). Radiohistochemical localization of benzodiazepine receptors in the rat brain. J. Pharmcol. Exp. Ther. 212, 337-346, 438-451.

RECOMMENDED READINGS

Conner, K.M., & Davidson, J.R.T. (1998). Generalized anxiety disorders: Neurobiological and phamaco-therapeutic perspectives. Biol. Psychiatry 44, 1286-1294.
Coplar, J.D. & Lydiard, R.B. (1998). Brain circuits in panic disorder. Biol. Psychiatry 44, 1264-1276.
LeDoux, J. (1998). Fear & the brain. Where have we been, and where are we going? Biol. Psychiatry 44, 1229-1238.
Stein, M.B. (1998). Neurobiological perspectives on social phobia: From affiliation to zoology. Biol. Psychiatry 44, 1277-1285.

Chapter 9

Biological Aspects of Depression

B.E. LEONARD

Biological Psychiatry, pages 179-205.

ISBN: 1-55938-819-6

INTRODUCTION

The Oxford Dictionary defines depression as a state of "low spirits or vitality". Clearly, this state has been experienced by most people at some stage during their lives. However, the psychiatrist is seldom concerned with such a mood change unless it persists for such a long time that it incapacitates the individual. Should the depressed mood be associated with feelings of guilt, suicidal tendencies, and disturbed bodily functions (such as weight loss, anorexia, loss of libido, or a disturbed sleep pattern characterized by early morning wakening) and persist for weeks or even months often with no initiatory cause, then psychiatric assistance is usually required. The various types of depression that have been identified will not be discussed herein because the drug treatment is essentially similiar irrespective of whether or not there appears to be an initiatory cause. For example, bereavement is often associated with a severe depressive episode, particulary in the elderly, and while counseling may be of considerable assistance in enabling the patient to adjust to the change in circumstances, the use of an anti-depressant is often advisable.

Many psychiatrists still divide depression into the endogenous (i.e., no apparent external cause) and reactive (i.e., an identifiable external cause) types and, while such a division may be of some value regarding ancillary treatment, there is presently no evidence to suggest that the biochemical changes that may be causally linked to the illness differ nor is their any evidence that the ways in which the patient should be assisted by drugs differ substantially. Other international classifications of depression are based on the mono- and bipolar dichotomy, a system of classificaton that separates those patients with depressive symptoms only from those that fluctuate between depression and mania (i.e., manic-depression) or have only manic symptoms. In such cases, treatment strategies differ: specific and anti-manic drugs such as lithium or the neuroleptics would be used to abort an acute attack of mania, whereas anti-depressants are the drugs of choice to treat the depressive episodes. (Readers are referred to the various classification manuals such as the *Diagnostic and Statistical Manual of Mental Diseases of the American Psychiatric Association* (DSM-IV) or the *International Classification of Disease, 10th Revision* (ICD 10), for further details.) It should be emphasized, however, that depressed patients frequently show symptoms of anxiety that may require additional treatment. In such cases, the use of a sedative anti-depressant such as amitriptyline, mianserin or mirtazepine, or one of the newer 5-hydroxytryptamine (5-HT) reuptake inhibitor anti-depressants, which also appear to have anxiolytic properties, may be of value.

EVIDENCE FOR NEUROTRANSMITTER ABNORMALITIES IN DEPRESSION

Some 25 years ago, Bunney and Davis as well as Schildkraut and Smith formulated the amine theory of depression. This theory was partly based on the earlier discovery by Woolley and Shaw that serotonin may play a role in the regulation of mood. Coincidentally, reserpine, an alkaloid obtained from the Indian snake plant *Rauwolfia serpentia* (which was being used to treat hypertension), was found to cause depressive symptoms in a small proportion of otherwise psychiatrically normal individuals. Further studies showed that, in the rat brain, reserpine depleted neurotransmitter storage granules of serotonin, noradrenaline, and dopamine. The amine theory of depression was consolidated by the finding that tricyclic anti-depressant drugs prevented the neuronal uptake of norepinephrine and/or serotonin and, as a consequence, presumably corrected the deficit in central biogenic amine neurotransmission, which was hypothesized to occur in depression.

It was soon realized that the inhibition of monoamine uptake into neurons within the brain could not explain the therapeutic effects of anti-depressants. Thus, there was no temporal relationsip between the inhibition of neurotransmitter uptake and the onset of therapeutic response to treatment with tricyclic anti-depressants. Depending on the drug concentration, tricylic anti-depressants prevented the transport of biogenic amines either at the neuronal plasma membrane or intracellularly at the storage vesicle membrane. Furthermore, the effect of tricyclic anti-depressants occurs fairly rapidly following acute administration, whereas it has been well established since the initial introduction of imipramine more than 30 years ago that several weeks of drug treatment are required before the optimal therapeutic effect is established. Clearly, there is no temporal relationship between the onset of the therapeutic response and the direct effect that anti-depressants appear to have on monoaminergic uptake processes. A further complication to the amine theory arose when it was established that some effective anti-depressants, such as iprindole, mianserin, and such recently introduced novel anti-depressants as mirtazepine, have little effect on monoamine uptake systems after either acute or chronic and administration. In contrast, drugs such as cocaine and the amphetamines inhibit amine uptake but are of little therapeutic value in the treatment of depression. Thus, by the 1970s, it was self-evident that the amine theory needed drastic revision. As such, the attention of neuropharmacologists was increasingly directed away from amine uptake mechanisms to receptor mechanisms, particularly toward the pre- and postsynaptic monoaminergic receptors, the responsiveness of which may be altered as a consequence of the changes in the concentrations of biogenic amine neurotransmitters that followed the chronic administration of anti-depressants.

One of the most seminal publications in the 1970s that helped to establish that chronic anti-depressant treatment leads to an adaptation in central neurotransmitter function was that of Vetulani and Sulser, who showed that various anti-depressant treatments and electroconvulsive shock treatment reduce the sensitivity of β-adrenergic receptor–coupled adenylate cyclase activity in the rat forebrain. Crews and Smith further showed that chronic desipramine treatment was associated with a subsensitivity of the presynaptic α-2 adrenergic autoreceptor in the rat heart; further studies suggested that similar changes could also occur in the forebrain. From these discoveries, two experimental approaches have been applied. Firstly, one approach has concentrated on studying the effects of anti-depressant treatment on the changes in the density and functional activity of secondary messenger systems associated with monoamine receptors located primarily in the limbic cortex of the rat. Secondly, psychopharmacologists have attempted to correlate changes in behavior with the chronic administration of different classes of anti-depressant drugs, and ECT. It is, of course, hoped that there will be a synthesis between those two approaches so that a conclusion may be reached regarding the temporal relationship between the changes in neurotransmitter receptor function, a specific pattern of behavior, and the delay that occurs before an anti-depressant or ECT produces a beneficial therapeutic effect.

Despite the advances that have been made in recent years in developing animal models of depression, the depressed human being, in the final analysis, is the only model requiring investigation if one is to understand the biochemical basis of the condition. There are four major approaches to understanding the biochemical etiology of depression:

a. analysis of postmortem brain of depressed suicide victims for neurotransmitters and their metabolites
b. analysis of cerebrospinal fluid from depressed patients
c. investigation of peripheral markers of neurotransmitter receptors by studying blood cells (e.g., platelets, lymphocytes)
d. endocrine disturbances which appear to be linked to the onset and duration of depression

Postmortem Studies

Direct evidence implicating the involvement of biogenic amines in the etiology of depression comes from an analysis of postmortem material. However, there are major difficulties in interpreting the data obtained from postmortem studies that arise from the following:

1. There is an effect by postmortem changes on the metabolism of neurotransmitters.
2. The presence of drugs taken coincidentally often affect the metabolism of the amines being investigated. Alcohol, together with an overdose of the anti-depressant taken to relieve the illness, is increasingly used as the means of committing suicide and now accounts for approximately 14% of the causes of all accidental and intentional deaths due to chemical agents in the UK. Such drugs have a profound effect on the concentrations and metabolism of the biogenic amines, thereby leading to an incorrect conclusion being drawn regarding the neurotransmitter status at the time of death.
3. There is difficulty in obtaining relevant controls with which to compare the brain from the suicide victim.
4. The precise diagnostic classification of the victim at the time of death is often elusive. It is by no means certain, for example, that depressive illness is the only primary cause of suicide.

In the light of these difficulties, perhaps it is not surprising to find that the evidence implicating an abnormality in the concentrations of the major neurotransmitters thought to be involved in depression, namely norepinephrine and serotonin, is inconclusive. In those studies conducted prior to 1976, there was evidence that the concentrations of 5-hydroxytryptamine (5-HT) and its major metabolite 5-hydroxyindole acetic acid (5-HIAA) were reduced, whereas results of the more recent studies are less conclusive. In the last decade, studies have tended to concentrate on changes in the densities of neurotransmitter receptors in the suicide brain, the advantage of such an approach being that alterations in receptor density may give a more dynamic representation of neurotransmitter function and are less vulnerable to postmortem change. Thus Stanley and Mann reported an increase in the density of 5-HT_2 receptors in limbic regions of the suicide brain, a finding that would be consistent with a prolonged deficiency of the neurotransmitter.

Mann and colleagues have reviewed the changes reported in the concentrations of 5-HT and 5-HIAA in the postmortem brain of suicide victims compared to controls and have concluded that the concentrations of 5-HIAA have been shown to be decreased in the frontal cortex. However, three studies have reported these indoles to be unchanged. By contrast, six out of seven studies have reported that the density of 5-HT_2–type receptor is increased in the prefrontal cortex of suicide victims. Such findings support the view that the same serotonin receptor types increase in number in limbic regions of the brain to compensate for the decrease in the availability of the neurotransmitter. It should be added, however, that a detailed study of 5-HT_2 receptor binding sites in 19 depressed suicide victims by Cheetham and colleagues failed to find any evidence of changes in the 5-HT_2 receptor density in any of the five brain regions investigated. One possible explanation for these differences could lie in the nature of the tritiated ligand used to label the 5-HT_2 receptor sites.

While the changes in the density of these receptors in the cortices of suicides are inconsistent, there is evidence that those patients who died violently have an increased number of 5-HT_2 receptors in the cortex. Such changes may be associated with aggressive and impulsive behavior that is independent of the psychiatric diagnosis. This view is supported from the findings of low CSF-5HIAA concentration in compulsive gamblers, alcoholics, and violent criminals. Thus it may be speculated that impulsive and aggressive behavior is associated with a reduced functional activity in the serotonergic system resulting in a compensatory increase in the number of $5HT_2$ receptors.

Catecholamines

The catecholamines norepinephrine (NE) and dopamine (DA) and their respective central metabolites 3-methoxy-4-hydroxyphenylethylene glycol (MHPG) and homovanillic acid (HVA) have received less attention than 5-HT and 5-HIAA. Of the five studies of postmortem brain in which NE, DA, and their major metabolites were determined, no significant differences were found in their concentrations in the brains of suicides when compared to their respective controls. Only two studies have so far detected lower concentrations of MHPG and therefore it must be concluded that the studies of the catecholamines and their metabolites have not revealed consistent abnormalities in the brains of suicide victims. There is evidence that the tyrosine hydroxylase levels are raised in suicide victims who suffered from depression and that chronic, but not acute, administration of antidepressants to rats results in a reduction in the concentration of this enzyme. The results suggest that chronic stress leads to increased firing of noradrenergic neurons, which is attenuated by chronic antidepressant treatment. Antidepressants may therefore act at the level of gene expression of tyrosine hydroxylase.

To date, little attempt appears to have been made in detecting the possible changes in the subtypes of 5-HT_1 receptor in postmortem brains. Studies of the frontal cortex of depressed suicides failed to detect any changes in total 5-HT_1 or in 5-HT_{IA} receptor densities. By contrast, a number of 5-HT_2 sites were shown to be lower in the hippocampus of brains from anti-depressant-free suicide victims.

Meyson and colleagues reported an increase in the density of central muscarinic receptors in suicide brains; changes were also found in the density of 5-HT receptors but not β-adrenoceptors. Whereas a great deal of attention has focused on the serotonergic system, little interest has been shown in the central cholinergic system in depression. Perhaps in the light of these findings, and those of Leong and Brown who suggested that the central cholinergic system may be hyperactive in depression, it is timely to consider the possible involvement of central cholinergic symptoms, particularly as a possible causal association has been found between a defect in the function of this system and Alzheimer's disease the symptoms of which often include depression.

Despite the experimental evidence to show that most anti-depressants increase the density of the $GABA_A$–type receptors in the frontal cortex of the rat, there is no evidence to show that the density of these receptors is altered in the brains of depressed suicides, nor is there any evidence that the concentrations of GABA (or the neurotransmitter amino acids taurine, glutamate, aspartate, or glycine) are altered in the limbic regions of suicide victims.

Changes in Neurotransmitter Tranporter Sites in Postmortem Brain

^{3}H-imipramine binding sites may be allosteric regulators of the serotonin uptake sites and could provide an index of the density of presynaptic serotonergic neurons. Of seven studies of the density of the imipramine binding site density in the frontal cortex of suicide victims, three studies reported a decrease, one found an increase, and three studies showed no change. Reasons for the discrepancy in these findings and those in the density of cortical 5-HT_2 receptors may be related to the nature of the suicide (whether violent or non-violent) and the precise psychopathology of the patient. Clearly, patients suffering from a number of serious psychiatric abnormalities are liable to commit suicide, depression being only one of these possible causes.

There are a number of reasons for the inconsistent findings of changes in ^{3}H imipramine binding sites in the brains of depressed patients. In addition to the factors that have already been referred to, further confounding factors include the hemispheric asymmetry in ^{3}H-imipramine binding in postmortem frontal cortex. Thus the number of ^{3}H-imipramine binding sites appear to be much greater in the right than in the left hemisphere of control brains, but this asymmetry is reversed in suicides and in homicide victims with a history of psychiatric illness. An additional factor is in the choice of radioligand used to detect the 5-HT uptake site in brain tissue. ^{3}H-imiprimane therefore labels multiple sites in the human brain only one of which is thought to be related to the 5-HT uptake site so that conflicting results may relate to the proportion of non–5-HT sites that are labeled.

More recently, the more selective 5-HT uptake inhibitor paroxetine has been developed as a radioligand for the 5-HT uptake site. Studies have shown that in human frontal cortex, the number of ^{3}H-paroxetine binding sites is much lower than the number of ^{3}H-imipramine sites. Researchers using brain tissue from suicides and controls have failed to find any asymmetry in ^{3}H-paroxetine binding;

neither have differences been found in the number of binding sites for this ligand between the suicides and their controls.

It may be concluded that postmortem studies do not provide convincing evidence of an abnormality in 5-HT uptake sites in the brains of depressed patients or suicide victims. It is also of interest to find that no change could be detected between the number of ^{3}H-desipramine binding sites, which serve as the ligand used to detect norepinephrine transport sites in suicides and their controls.

Corticotrophin Releasing Factor in Postmortem Brain

The hyperactivity of the hypothalamic–pituitary–adrenal axis in depression is thought to be partly due to a hypersecretion of corticotrophin releasing factor (CRF) from the hypothalamus. CRF, in addition to its role in the control of cortisol synthesis and release via adrenocorticotrophic hormone (ACTH) in the periphery, is also known to act as a neurotransmitter in the brain.

In the brains of depressed suicides, there is evidence to show that the number of CRF receptors is lower in the frontal cortex of suicides relative to their controls. This suggests that there is an adaptive decrease in CRF receptors in depressed patients that may result from the chronic hypersecretion of CRF. The possible function of CRF in anxiety and depression will be considered in detail later.

Cerebrospinal Fluid and Urine: Changes in Amine Metabolites in Depression

Differences in the concentration of 5-HIAA in the CSF-depressed patients compared with nondepressed controls have been replicated by several groups of investigators. Thus, Sjöström found that both 5-HIAA and the major dopamine metabolite homovanillic acid (HVA) were reduced in the CSF of depressed patients. Later, Åsberg and colleagues reported that the 5-HIAA concentrations in a group of depressed patients followed a bimodal distribution, one group having normal and the other a low concentration of the CSF metabolite. There was evidence that the latter group of patients were more likely to attempt suicide by violent means than those with relatively normal 5-HIAA concentrations.

There has been much interest in the possible relationship between aggressive behavior and reduced availability of serotonin. Experimental studies have shown that the 5-HT_{IA} and 5-HT_{IB} receptor subtypes might be involved in aggressive behavior at least in some animal species. Thus eltoprazine (a mixed 5-HT_{IA} and 5-HT_{IB} agonist) and metatrifluormethylphenyl piperazine (TFMPP), a selective 5-HT_{IB} agonist, specifically reduce aggressive behavior in several animal species. It is not without interest that chronic anti-depressant treatment results in a hypersensitivity of 5-HT_{IA} postsynaptic receptors and a hyposensitivity in presynaptic sites. It may be concluded that anti-depressants may not only improve the mood of the depressed patient but also reduce the impulsive behavior that could be responsible for the suicidal thoughts and attempts that are frequent symptoms of depression.

The introduction of probenecid to block the efflux of acid metabolites from the CSF has enabled investigators to assess the turnover of 5-HT in the consciously depressed patient. The use of this technique to calculate 5-HT turnover is based upon the assumption that the rate of synthesis is equal to the rate of clearance of 5-HIAA, a process that should occur in a kinetically single compartment system. This approach has been extensively applied by Van Praag and coworkers who have shown that the rate of accumulation of 5-HIAA is significantly lower in depressed patients than in control subjects; it would also appear that the 5-HIAA accumulation returned to normal following clinical improvement in approximately 50% of the patients. These researchers postulated that those patients who failed to show a normalization of the 5-HIAA accumulation rate on clinical recovery were more likely to show an increased tendency to further episodes of depression. Other investigators have also shown that the accumulation of 5-HIAA following probenecid was lower in depressed patients than controls.

Montgomery studied the effects of maprotiline and zimelidine (which are antidepressants assumed to show specificity in impeding the reuptake of noradrenaline and serotonin, respectively) on the CSF 5-HIAA concentrations and clinical response in groups of depressed patients who had normal or decreased metabolite levels before the commencement of treatment. Both drugs were equally effective in alleviating the depressed symptoms but no correlation could be found between the clinical response and the pretreatment 5-HIAA status. The results of such a study cast doubt on relevance of the presumed specificity of an anti-depressant in inhibiting the reuptake of norepinephrine and serotonin *in vitro* to their actions following administration to depressed patients. Potter and colleagues, in a detailed study of the effects of desipramine, zimelidine, and clorgyline on the 5-HIAA, HVA, and 3-methoxy-4-hydroxyphenylglycol (MHPG; the main noradrenaline metabolite in the CNS) concentrations of depressed patients, also failed to show that the drugs specifically altered the concentrations of these metabolites. This is in spite of the fact that it would have been anticipated that desipramine would preferentially affect MHPG, zimelidine 5-HIAA, and that clorgyline (a selective MAO-A inhibitor) would elevate both norepinephrine and 5-HT. Unfortunately, a control group was not included in this study, so it is not possible to determine whether the pretreatment values of these metabolites were decreased.

There is evidence that, under carefully controlled conditions of diet, exercise, and time of day at which the body fluid is collected, an equilibrium is established in the distribution of MHPG between the CSF, blood, and urine. Thus studies on central noradrenergic function in depression have been largely restricted to an analysis of urinary MHPG. Maas reported that some depressed patients have a subnormal MHPG excretion, and later studies concluded that such a subgroup were those diagnosed as suffering from primary affective disorder. It is noteworthy that others have failed to find any change in the urinary MHPG concentration in patients with this form of depression. Furthermore, the range of urinary MHPG values quoted by the various authors fall well within the range quoted by Hollister and colleagues for normal subjects. Unless such factors as diet, physical activity, and physical or mental stress are strictly controlled, the relevance of urinary MHPG studies to the understanding of the etiology of depression and response to anti-depressant treatment is doubtful.

Despite the limited value of the analysis of urinary metabolites in the assessment of central neurotransmitter function in depressed patients, changes in urinary metabolites of norepinephrine have been shown by some investigators to be of predictive value in response to anti-depressant treatment. Thus Maas and colleagues have shown that low CSF 5-HIAA concentrations and high urinary metanephrine values were associated with a higher incidence of response to anti-depressant treatment. However, it should be noted that Corona and colleagues, in their study of the response of depressed women to amitriptyline treatment, did not find any correlation between the urinary MHPG concentration and the response of the patients to anti-depressant treatment.

One of the most critical studies of changes in neurotransmitters in the CSF of depressed patients has been conducted by Gjerris and colleagues who concluded that the results obtained from studies of the changes in CSF amines, their metabolites and various neuropeptides after recovery from depression, are ambiguous. Only CSF epinephrine concentrations would appear to be a state marker of treatment response in depression, while CSF arginine vasopressin also has a tendency to increase following clinical recovery. The changes in CSF serotonin following recovery would appear to reflect the biochemical properties of the anti-depressant used and not necessarily the change in the mood state.

Equally serious criticisms may be made of the significance of CSF 5-HIAA studies. Although Garelis and colleagues have argued that 5-HIAA in the CSF reflects central changes in 5-HT metabolism, others have demonstrated that CSF 5-HIAA is primarily of spinal cord origin. Curzon and colleagues showed that whereas the lumbar CSF from depressed patients is lower than controls, the ventricular 5-HIAA concentrations of these patients was normal. Such studies suggest that the lowered lumbar CSF 5-HIAA concentration may reflect reduced spinal neuronal activity, which could occur as a result of the motor retardation that is a common feature of depressed patients. However, not all studies show that the concentration of 5-HIAA returns to control values following clincial recovery, so clearly such a factor cannot be the only explanation. Another difficulty arising from the CSF 5-HIAA data concerns its relevance in assessing amine function. Green and Deakin have argued that, while such changes may reflect the metabolism of 5-HT, they do not necessarily reflect the functional state of the 5-HT system. Such a view would accord with the clinical findings, which show that an improvement in mood of depressed patients was not always associated with a normalization of the CSF 5-HIAA concentrations. It may be concluded that, while studies on CSF and urinary metabolites may be indicative of a derangement in central serotonergic neurotransmission, it would be incautious to rely on such evidence alone for a verification of the amine hypothesis of depression.

Peripheral Markers for the Study of Depression

The need for researchers to obtain a peripheral model of the nerve terminal in order that changes in transmitter receptor function and transport might be investigated in patients has led to an interest in the platelet as a possible model. Structurally, both the nerve terminal and the platelet contain mitochondria and dense core vesicles that store 5-HT or other amines. MAO-activity is associated with the mitochondrial fraction of both nerve endings and platelets. The major difference is that, in the platelet, only one type of MAO, MAO-B, occurs in the platelet, whereas in the nerve terminal, both MAO-A and -B are found, the subtype of enzyme depending on the function of the terminal. The storage vesicles in the platelet contain only 5-HT but, as in the nerve terminal, the release of this amine is a calcium-dependent process. Unlike nerve terminals, platelets cannot synthesize 5-HT, but this role is taken over by the closely associated amine precursor, uptake, and decarboxylation cells (APUD cells), which are of the same embryonic origin as the nerve terminals and platelets, all of which contain the unique neuron-specific enolases. The APUD cells and platelets may therefore be considered to represent a paraneuronal system.

There are, however, a number of important differences between the platelet and the nerve terminal. Platelets store glycogen, whereas nerve terminals do not. Structurally, platelets are not directly connected with any other cells or with the nervous system, whereas neurons are directly and indirectly connected with hundreds of other nerve cells via their axonal and dendritic processes. It may therefore be concluded that, although platelets may be considered to be a useful model of the serotonergic nerve terminal and although the shape changes, which may be induced by selective α- and 5-HT_2 agonists, can give some degree of insight into the changes in receptor function that may occur in psychiatric illness, their relevance to all types of nerve terminal is necessarily limited.

Changes in Peripheral Transmitter Receptors in Depression

One method that has been used to monitor postsynaptic adenoceptor changes in the depressed patients *in vivo* involves measuring the pressor response to intravenously administered α agonists such as phenylephrine or norepinephrine, or the indirectly acting amine tyramine. An enhanced pressor response has been reported to occur in untreated depressed patients, suggesting that the postsynaptic α receptors are hyperactive. Other studies have concentrated on changes in the density and aggregating function for the α-2 receptors on the platelet membane. While there would appear to be little evidence that the functional activity of these receptors is altered, there is conflicting evidence regarding the density of these receptors—with some investigators reporting an increase, others a decrease, and yet others no change. A major problem in evaluating these studies stems from the nature of the ligand used to detect the receptors and the relative specificity of the ligand for the α receptors. Studies with the highly specific ligand ^{3}H-rauwolscine consistenly show that the density of these receptors is unchanged.

Studies on the density and function of β adrenoceptors on the lymphocyte membrane have also yielded equivocal results. Some of the changes reported are undoubtedly due to the heterogeneity of the patients studied, lack of control for age, different periods being allowed for the washout of any psychotropic medication being taken by the patients, and the specificity of the ligand for the β receptors. The most consistent studies, in which attempts have been made to control for some of these factors, show that the β receptor density is raised in the untreated patients and returns to control values in those patients responding to treatment. Thus the β receptor density would appear to be a useful state marker of depression. Regarding the β receptor function, there is some evidence that the receptor-linked cyclase activity is decreased. Thus it may be surmised that there is a deficit in the link between the β receptor and the secondary messenger system, which may be causally related to depression.

While there has been widespread interest in possible changes in adrenoceptor activity that may be linked to depression, less attention seems to have been paid to serotonin receptor function. There is evidence that 5-HT_2–induced platelet aggregation is significantly reduced in the untreated, depressed patients but normlizes on effective treatment irrespective of the nature of that treatment. This provides clinical evidence for the view that 5-HT_2 receptor function is augmented following effective treatment and that the time of onset of the clinical response coincides with improvement.

Adrenergic System

The results of experimental studies on the chronic effects of anti-depressant drugs on postsynaptic adrenoceptors have led to a greater insight into the adaptive changes that occur following drug administration and have also enriched the methods that have been developed for the selection of new putative anti-depressants. However, has such an approach led to a greater understanding of the state-dependent biochemical changes that might occur in the depressed patient? If anti-depressants decrease the functional activity of the adrenoceptor-linked cyclase system, then it may be postulated that this receptor–enzyme complex is hyperactive in the untreated patient. Furthermore, as the activity of the postsynaptic receptor system reflects changes in the concentration of neurotransmitters in the synaptic cleft, it must also be assumed that the mechanisms governing the release of the neurotransmitter are abnormal in the depressed patient. As the release of norepinephrine from the terminal is regulated, at least partially by the presynaptic α-2 adrenoceptors, it seems likely that the activity of this receptor system is also abnormal.

Although there is broad agreement that chronic anti-depressant treatment attenuates postsynaptic adrenoceptor activity in the limbic cortex of the rat brain (presumably as a consequence of a decrease in the activity of the inhibitory α-2 adrenoceptors that are located presynaptically), the results of studies on changes in monoamine receptor activities on the lymphocyte and platelet membrane are equivocal.

A possible explanation for the variation in the results obtained from studies of α-2 adrenoceptors on platelets in depressed patients could be associated with the nutritional status of the patients during the different phases of the study. It is well known that anorexia and a loss in body weight are frequent symptoms of depression; increased appetite is generally taken to be a sign of clinical improvement, and many of the conventional (tricyclic) anti-depressants increase body weight as a consequence of the changes in the intermediary metabolism of carbohydrates that they induce. There is clinical evidence that an inverse relation exists between platelet α-2 adrenoceptor density and the plasma catecholamine concentrations. Fasting has also been shown to induce a fall in the concentration of plasma norepinephrine in patients with anorexia nervosa, the plasma norepinephrine concentrations being significantly lower than in age- and sex- matched controls. The decrease in plasma norepinephrine was associated with a rise in the platelet α-2 adrenoceptor density. It may be concluded that the nutritional status of the depressed patient must be critically assessed and taken into account when attempts are made to correlate changes in the adrenoceptor density with the psychiatric status. In the studies of changes in platelet adrenoceptor activity in the depressed patient, already referred to, such factors as the loss in body weight and nutritional status of the patient before and following therapy have largely been ignored.

Serotonergic System

It is well established that many effective anti-depressants impede the reuptake of ^{3}H-serotonin into cortical synaptosomes from rat brain and into platelets of non-depressed subjects who have received a single dose of anti-depressant. Several investigators have found that the transport of ^{3}H-serotonin into the platelets of depressed patients is reduced before treatment but returns to normal following effective therapy, irrespective of the presumed mechanism of action of the anti-depressant (i.e., whether it is a specific norepinephrine or a serotonin uptake inhibitor); ECT has also been equally as effective as anti-depressants in normalizing serotonin uptake.

The sensitivity of the serotonin receptor on the platelet membane also appears to be subnormal in the untreated patient and returns to control values following effective drug treatment. There is experimetnal evidence to suggest that the serotonin receptor on the platelet membrane is of the 5-HT_{IA} type, which is widely distributed in limbic regions of the brain and the activity of which is increased following longterm anti-depressant treatment. These findings therefore suggest that the sensitivity of 5-HT receptors is decreased in the untreated patients and returns to normal values following effective treatment. Brusov and colleagues, in a study of platelet shape change velocity in a population of drug-free bipolar and unipolar depressed women, found that the 5-HT_2 receptor sensitivity was enhanced initially and normalized following effective anti-depressant treatment. These authors suggest that the upregulation of the 5-HT_2 receptors is a consequence of reduced presynaptic serotonergic activity, possibly caused by a reduction in the number of functioning nerve terminals in the untreated depressed state.

Arora and Meltzer, who used ^{3}H-lysergic acid diethylamide (LSD) as the ligand to determine the density of 5-HT_2 receptors on the platelet membranes of depressed patients, reported that the density of these receptors was increased in comparison with age- and sex-matched controls. However, the increase in 5-HT_2 receptor density was increased in female depressed patients only, and no changes were reported in the male depressives. Neither the density of the ^{3}H-imipramine-binding sites nor the rate of uptake of ^{3}H-serotonin into the platelets of either the male or female patients was changed. Clearly, there is a major discrepancy in these findings and those of many other investigators, who used the more specific 5-HT_2 receptor ligand (^{3}H-ketanserin) to label the receptor sites.

Leysen and colleagues have convincingly argued that both the ligand-binding sites and the platelet aggregation responses are mediated by 5-HT_2 type receptors. In addition to their ability to transport and store 5-HT, catecholamines are also found in platelets, but the evidence for the existence of a high affinity transport site for the uptake of these amines is lacking. Nevertheless, Omenn and Smith have suggested that dopamine is transported into the platelet by the 5-HT uptake site whereas norepinephrine appears to be taken up by a process of passive diffusion. Platelets, therefore, may be used as a model for serotonergic and possibly dopaminergic nerve terminals.

An indication of the accuracy of platelet 5-HT uptake, the process of reflection in the human brain is difficult to obtain. Even so, Paul and colleagues compared the ^{3}H-serotonin uptake and ^{3}H-imipramine binding under similar experimental conditions in a synaptosomal fraction from rat brain and in human platelets. These investigators showed that, for a series of anti-depressants, there was an excellent correlation ($r = 0.9$) between the ability of these drugs to inhibit 5-HT uptake and in their efficacies in displacing ^{3}H-imipramine from its binding site. Electrolytic lesions of the raphe nucleus paralleled the reduced ^{3}H-serotonin uptake into the hypothalamus, whereas an irreversible inhibitor of ^{3}H-imipramine binding, 2,8-dinitroimipramine, produced a dose-dependent decrease in ^{3}H-serotonin uptake without affecting the uptake of ^{3}H-norepinephrine or ^{3}H dopamine. These investigators concluded that ^{3}H-imipramine selectively labels 5-HT- uptake sites in both the brain and platelets. Therefore, it would seem reasonable to use studies of changes in platelet function in depressed patients to gain an insight into possible alteration of central neurotransmitter processes in patients before and during treatment.

Amine Uptake Sites

There is good experimental evidence to suggest that tricyclic anti-depressants bind specifically, and with high affinity, to platelet and neuronal membranes. These binding sites are associated with, but probably not identical to, the transport site for 5-HT and norepinephrine. Tricyclic anti-depressants have been shown to have a high affinity for the ^{3}H-imipramine or ^{3}H-desipramine binding sites, whereas other centrally acting drugs, such as phentolamine or chlorpromazine, have only a low affinity for these sites. The physiological relevance of these binding sites is the subject of debate. If these sites function as presynaptic receptors controlling the uptake of 5-HT or norepinephrine, is there an endogenous ligand for these sites? Alternatively, does the ^{3}H-imipramine-binding site reside on the 5-HT carrier mechanism and alter the uptake of 5-HT by allosteric conformational changes that are not physiologically relevant? Such questions, together with those suggesting that the imipramine-binding sites in the cortex are heterogenous and involve selective-binding sites for ^{3}H-cocaine and ^{3}H norzimelidine, can only be answered by more detailed biochemical and pharmacological studies.

During the years 1980 to 1987, some 36 studies have been undertaken into the density of ^{3}H-imipramine binding sites on the platelet membranes of untreated depressed patients. Twenty-three of these studies showed that there was a decrease in the density of the imipramine binding sites, varying from 10% to 54%; one study showed a 9% increase in the number of binding sites, whereas 12 of the studies reported no change. It seems probable that several experimental and clinical variables contribute to the changes reported. Examples include the method used to isolate the platelets and the preparation of the platelet membranes, and the duration of the washout period before blood was taken from those patients that had been on anti-depressant therapy. In general, it would appear that a washout period of four weeks is necessary to ensure that the changes in platelet imipramine binding are not affected by the anti-depressant that has been administered.

Paul reported that the decreased ^{3}H-imipramine binding to the platelet membrane of depressives is influenced by the hypercortisolemia often found in these patients. In many studies, a parallel decrease in ^{3}H-imipramine binding and 5-HT uptake has been observed, but it would seem that the two systems are differentially controlled, particularly as the activity of the ^{3}H-serotonin uptake system appears to normalize following effective treatment. Differences between the imipramine binding sites and the 5-HT transport sites probably reside in the differences in genetic control over such sites. Most studies in which changes in ^{3}H-imipramine binding were investigated in depressed patients suggest that this is a trait, rather than a state marker of the illness, although a few studies have shown that the imipramine binding sites return to normal on clinical recovery. Overall, however, the evidence would suggest that the ^{3}H-imipramine binding site is a trait marker of depression—a possibility that, if true, may be useful in the evaluation of the vulnerability of a person to the illness. Whether changes in ^{3}H-imipramine binding are unique for depression still remains an open question. For example, the results for changes in the density of these binding sites on platelet membranes from schizophrenics are so far inconclusive. Furthermore, only about 50% of the studies have described a significant decrease in the ^{3}H-imipramine binding site on the platelet membrane.

Summary

As man would appear to be unique among primates in suffering from depression, most of the evidence that implicates neurotransmitter abnormalities in the etiology of the condition has relied on an analysis of postmortem brain of depressed suicide victims. While this would appear to be the most direct method for studying the neurotransmitters, their metabolites, and receptors in the human brain, the results implicating specific transmitter abnormalities in the cause of depression has been limited. This is because a high proportion of those committing suicide may not primarily be suffering from depression. Furthermore, factors such as the method of committing suicide and the postmortem delay further complicate the interpretation of the results obtained. Nevertheless, there is some evidence to implicate a disorder in serotonergic function in severe depression, with other transmitters (such as norepinephrine, acetylcholine, and some of the neuropeptides) playing an ancilliary role.

The analysis of cerebrospinal fluid and urine from depressed patients has been the subject of detailed investigation over the past 20 years. Again, the results of such studies have been disappointing largely because of difficulty in correlating the changes in lumbar cerebrospinal fluid metabolites of transmitters with their brain concentrations. Some evidence suggests that the turnover of serotonin is decreased in depression. Apart from some studies that indicate changes in the main central metabolite of norepinephrine, MHPG, in the urine of depressives, the findings resulting from urine analyses have been equivocal.

The most profitable experimental approach to the study of neurotransmitter changes in depression would appear to come from the investigation of peripheral markers of transmitter receptors on blood cells, particularly platelets and lymphocytes. Such studies suggest that, in the depressed patient, there is a defect in adrenergic and serotonergic receptor function and in serotonin transport, which may be a state marker of the illness. Other studies have suggested that the ^{3}H-imipramine binding site on the platelet membrane, which may function as a modulatory site for serotonin transport, is a trait marker of the illness.

The advent of modern imaging techniques, such as single photon emission tomography (SPECT) and positron emission tomography (PET), may well enable receptor changes to be located in specific brain regions of conscious patients. In this regard, it has already been shown that the cortical transport of ^{13}C labeled tryptophan is defective in depressed patients, which serves to emphasize the defect in transport processes that could underlie the illness.

CIRCADIAN CHANGES IN DEPRESSION

The periodicities of mood and its disorders have been the subject of research for centuries. It has been well established that severe depression is associated with a diurnal variation in mood (the patient is usually most depressed in the morning), early morning awakening, gross disturbance in the sleep architecture (usually characterized by a shortened latency for the onset of the first phase of rapid eye movement sleep), and the regular patterns of recurrence of depressive episodes.

In recent years, chronobiology, the study of time-related processes in living organisms, has made an important contribution to our understanding of the circadian disturbances that occur in depression. It has been established that, in mammals, the principal circadian pacemaker is located in the suprachiasmatic nucleus of the anterior hypothalamus. Light has been identified as the most powerful external stimulus and synchronizer of this pacemaker, while melatonin, an indoleamine found in a high concentration in the pineal gland of mammals including man, may act as an internal synchronizer of circadian rhythms. There is some evidence to suggest that the secretion of melatonin is impaired in depressed patients, which might be a cause of the circadian disturbances.

Chronobiological studies have demonstrated changes in the responsivity of physiological processes in mammals to the actions of psychotropic drugs; this has become the subspeciality of chronopharmacology. It has been established that anti-depressants and mood-stabilizing drugs such as lithium can modify the phase, period, and amplitude of the circadian changes in a variety of animal species including man. Most of the anti-depressants investigated have been shown to delay the circadian phase and lengthen its period, an effect that could explain the normalization of the sleep–wake cycle and other circadian processes that occur in depressed patients following effective anti-depressant treatment. The precise locus of action of anti-depressants on the circadian rhythm is uncertain, although it may be speculated that the change in the activity of the suprachiasmatic nucleus must play a central role. It seems possible that the processing of light stimuli via the retina could also be important. The role of light therapy in the treatment of seasonal depression, particularly autumn–winter seasonal depression, has been the subject of considerable interest, but the success of such treatment is still the subject of controversy.

In conclusion, both clinical and experimental studies have shown that the numbers of transmitter receptors and amine transport processes have shown circadian changes. It is well established that depression is associated with a disruption of the circadian rhythm as shown by changes in a number of behavioral, autonomic, and neuroendocrine aspects. One of the main consequences of effective treatment is a return of the circadian rhythm to normality. For example, it has been shown that the 5-HT uptake into the platelets of depressed patients is largely unchanged between 0600 h and 1200 h, whereas the 5-HT transport in control subjects has shown a significant decrease over this period. The normal rhythm in 5-HT transport is only reestablished when the depressed patient responds to treatment. Thus it may be hypothesized that anti-depressants normalize disrupted circadian rhythms. Only when the circadian rhythm has returned to normal can full clinical recovery be established. Whether this is a primary or co-incidental effect of anti-depressants is uncertain.

NEUROENDOCRINE CORRELATES OF DEPRESSION

There is physiological evidence that the control of anterior pituitary hormone releasing factors by the hypothalamus is largely under the influence of the biogenic amine neurotransmitters. It would therefore seem a reasonable assumption to predict that any change in the activity of these transmitters that may be causally associated with depression would also be reflected in an abnormality of hormone release from the pituitary. It has long been known that depressed patients show an abnormality in the circadian fluctuations of most hormones, whereas the hypersecretion of cortisol that is not readily amenable to suppression by the glucocorticoid dexamethasone has been widely advocated as a biological marker of depression. A hypersecretion of cortisol is also found in patients with Cushing's disease and, as such, patients frequently show evidence of depression, it can be speculated that changes in the function of glucocorticoid receptors in the brain may play some role in the etiology of the illness. Largely as a result of the seminal studies of Carroll and colleagues, the dexamethasone suppression test has been shown to be of predictive value for the diagnosis of further episodes of endogeous (or melancholic) depression in up to 95% of patients. However, it is now apparent that positive tests can occur in patients with senile dementia, alcoholism, anorexia nervosa, and malnutrition, and in those subject to inadequate renal dialysis. Whether the changes in cortisol secretion in such patients are attributable to secondary symptoms of depression still awaits elucidation.

Nevertheless, despite its limitations, the dexamethasone suppression test (DST) is still the most widely used biological test for depression. The test consists of administering 1 mg of the synthetic glucocorticoid dexamethasone at 2300 h followed by a measurement of the serum cortisol concentration at 1600 the following day. Dexamethasone nonsuppression is defined as the serum cortisol concentration in excess of 5 μg/100 ml (137 nmol/l) at 1600 h. In most cases, the DST normalizes with effective anti-depressant treatment, suggesting that it is a state marker of depression. Should the patient still show DST nonsuppression following effective treatment, it is probable that such patients will relapse.

The release of adrenocorticotropic hormone (ACTH) by CRF (an action that effects regulation of the secretion of cortisol from the adrenal cortex) is also reduced in the depressed patient. This blunted response may be due to a "down-regulation" (decreased functional activity) of CRF receptors that are located in the anterior pituitary. Similar changes in glucocorticoid receptors on lymphocytes have been found in depressed patients. Such findings suggest that depressed patients have a decreased glucocorticoid receptor plasticity, which could account for the lack of feedback inhibition of CRF/ACTH release by the elevated cortisol concentration that occurs in depression. It is of interest to note that chronic anti-depressant treatment with tricyclic anti-depressants or selective serotonin reuptake inhibitors leads to an upregulation (return to normal sensitivity) of glucocorticoids in the rat brain. Clearly, this ability of anti-depressants to suppress the hypothalamic–pituitary–adrenal hyperactivity by normalizing the functional activity of central glucocorticoid receptors, may be of fundamental clinical importance not only in understanding the etiology of depression but also how anti-depressants cause their beneficial effects.

Neuroendocrine Challenge Tests for Norepinephrine Function

Neuroendocrine challenge tests of norepinephrine function in depressed patients have suggested that abnormalities in the functioning of postsynaptic α-1, α-2, and β adrenoceptors occur in such patients. Thus the secretion of growth hormone following the intravenous administration of the α-2 receptor agonists clonidine and guanfacine is blunted in the depressed patient. Similarly, the secretion of cortisol caused by D-amphetamine is blunted, an effect that has been ascribed to a subsensitivity of postsynaptic α-1 adrenoceptors. However, it should be emphasized that amphetamine enhances norepinephrine and dopamine release and blocks the reuptake of those amines so that the conclusion that a specific defect in α-1 receptor function occurs in depressed patients must be interpreted with caution. With regard to postsynaptic β adrenoceptors, the effect of nocturnal melatonin secretion, which is mediated by the direct stimulation of β adrenoceptors in the pineal gland, has been shown to be blunted in depressed patients. This suggests that these receptors are subsensitive, a finding that conflicts with the changes in the lymphocyte receptors and those occurring in some regions of the suicide brain. Regarding the presynaptic α-2 receptors, plasma MHPG concentrations have been shown to increase in response to the administration of the selective norepinephrine reuptake inhibitor desipramine and the α-2 antagonist yohimbine. Such findings suggest that the presynaptic α-2 receptors are hypersensitive in depression.

Despite the clinical evidence that the adrenoceptors are abnormal in the untreated depressed patient, there is little evidence so far to indicate the anti-depressant treatment normalizes the functional activities of these receptors. Thus, it must be concluded that the changes in both the serotonergic and adrenergic receptors that occur in response to the various neuroendocrine challenge tests are epiphenomena and not directly related to the therapeutic response of the patient to anti-depressant treatments.

So far, there is no convincing evidence that neuroendocrine abnormalities of the dopaminergic system occur in the depressed patient.

Neuroendocrine Challenge Tests for Serotonin Function

One of the most extensively studied of the neurendocrine challenge tests has been the response to the intravenous administration of the serotonin precursor l-tryptophan (TRP). The synthesis of brain 5-HT is dependent on the availability of TRP, which enters the brain through an amino acid transport mechanism that also carries large neutral amino acids. There is experimental evidence to show that large changes in dietary TRP can cause fluctuations in the concentration of brain 5-HT, at least in the rat.

It has been shown recently that a liquid diet composed of 15 amino acids but lacking in TRP, when given to depressed patients during remission on an SSRI anti-depressant causes a rapid relapse. This suggests that in both animals and man, the rate of synthesis of brain 5-HT is considerably influenced by the availability of TRP.

The intravenous administration of TRP has been used to investigate changes in the endocrine status of depressed patients. There is considerable evidence to show that TRP enhances the release of prolactin and, to a lesser extent, growth hormone and ACTH. In depressed patients, it has been widely reported that the prolactin response to an acute TRP challenge is blunted. Such changes have been ascribed to a hyposensitivity of 5-HT_1–type receptors while the possible involvement of 5-HT_2 receptors in the control of the prolactin response triggered by TRP remains controversial. Longterm treatment of depressed patients with tricyclic anti-depressants, MAO inhibitors, or lithium enhances the prolactin response to the TRP challenge, which suggests that the anti-depressant may sensitize the 5-HT_1 receptors. However, the nontricyclic anti-depressants such as buproprion, mianserin and trazodone were without effect, nor was there any apparent correlation between the clinical response to treatment and the normalization of the prolactin response.

Qualitatively similar results have been obtained following the administration of the 5-HT releasing agent D-fenfluramine. The partial 5-HT_{IA} agonist ipsapirone has been shown to lower the body temperature of depressed patients to a lesser extent than control subjects, an effect that has been interpreted as evidence for a subsensitivity of the presynaptic 5-HT_{IA} receptors.

Neuroendocrine Challenge Tests for Cholinergic Function

While the monoamine hypothesis of depression has received considerable experimental and clinical support, less attention has been paid to the role of the cholinergic system. This is somewhat surprising as there is evidence that cholinomimetic drugs such as physostigmine can have a profound effect on the mood of both normal subjects and depressed patients. Janowski and coworkers, on the basis of their studies in patients with different types of affective disorders, have hypothesized that excessive cholinergic activity was associated with depression, whereas mania resulted from an inbalance between the central noradrenergic system (which is increased) and the cholinergic system (which is decreased).

The cholinergic hypothesis of depression has been further investigated by studying the effects of cholinomimetic drugs such as physostigmine (a nonselective, reversible anti-cholinesterase) and pyridostigmine (a short-acting, reversible anti-cholinesterase). The intravenous administration of these drugs have been shown to increase the release of cortisol, ACTH, and β endorphin; the release of these hormones is enhanced in depressed patients, suggesting that the responsiveness of the muscarinic receptors (possibly the M_3 subtype) is enhanced in depression. However, such changes do stress the need for further studies on the possible role of the cholinergic system in the affective disorders.

Summary

The results of the neuroendocrine challenge tests imply that the noradrenergic and serotonergic postsynaptic receptors are functionally hypoactive in the depressed patient. There is some evidence that the responsiveness of these receptors normalizes after effective treatment. Recent studies suggest that some muscarinic (cholinergic) receptors are supersensitive in depression, findings that lend support to the hypothesis that depression is due to an inbalance in the norephinephrine–acetylcholine neuronal pathway.

CHANGES IN THE IMMUNE SYSTEM IN DEPRESSION

The term "psychoneuroimmunology" applies to the reciprocal integration of the brain with the endocrine and immune systems. The concept that mental states of patients can influence the causes of physical illness can be traced back to antiquity. For example, Galen, in the Second Century, described the increased incidence of cancer of the breast in women who had melancholic temperaments. Since then, the medical literature has contained numerous anecdotal reports of chronic physical illness and even sudden, unanticipated death following bereavement among elderly subjects.

Only with the recent knowledge of the complex interactions between mental events and the immune and endocrine systems has it been possible to understand how such dramatic life events may arise. Clearly, the role of stress, irrespective of its cause, is instrumental in causing profound changes in the immune and endocrine systems and the brain. All forms of stress result in the activation of the pituitary–adrenal axis, with a consequent rise in circulating catecholamines and glucocorticoid hormones from the adrenal gland. The secretion of adrenocorticotrophic hormone (ACTH) from the pituitary gland (which is controlled by CRF) triggers the secretion of adrenal glucocorticoids, whereas stress-induced activation of the sympathetic system is responsible for the catecholamine secretion. It is now apparent that ACTH secretion can also be increased by thymic peptides (e.g., thymopoietin), whereas interleukin-1 (IL-1), a product of macrophage activity, has been shown to regulate ACTH secretion.

Such events show how the immune, endocrine, and central nervous systems are integrated in their responses to any form of stress. It is well established that physical or psychosocial stress causes increased secretions of prolactin, growth hormone, thyroid hormone, and gonadal hormones, in addition to ACTH. Endogenous opioids are secreted under such conditions and function as immunomodulators while also elevating the pain threshold. Receptors for such hormones exist on immunocompetent cells, along with receptors for catecholamines, serotonin (5-HT), and acetylcholine.

In addition to the regulatory effects of the nervous system on the immune system, there is now convincing evidence that the immune system can influence brain function. Thus, changes in the activity of specific nuclei in the hypothalamuses of rats have been described following the formation of antibodies to specific antigen challenges. Such changes in electrical activity appear to be linked to specific decreases in noradrenaline concentrations in these nuclei. Changes in the activity of the serotonergic neurons in the hippocampus also occur shortly after the occurrence of the immune response. These findings illustrate how the immune system, presumbly via the release of immunoregulatory peptides (also called immunotransmitters) such as the interleukins from macrophages, can influence the activity of the hypothalamic–pituitary axis and also higher centers of the brain (e.g., the hippocampus), which are involved in shortterm memory processing.

Immune System in Affective Disorders

Susceptibility to bacterial and viral infections and to the establishment of tumors is reported to arise more frequently in those who are depressed than in those who are not. An analysis of the immune systems of those suffering from severe psychological stress of bereavement has shown that the activity of those immune cells that are fundamentally involved in the host defense against infections (e.g., T-lymphocytes and natural-killer cells, NKCs) is dramatically reduced. Such an effect can occur following chronic and subchronic stress.

The past 20 years have witnessed a broad interest in the role of the hypothalamic–pituitary–adrenal axis in the psychobiology of affective disorders. In depressed patients, rises in serum cortisol are frequently reported in addition to disruptions of circadian patterns of cortisol secretion and an insensitivity of cortisol secretion to suppression by glucocorticoids (e.g., dexamethasone). One consequence of hypercortisolemia is a suppression of immune function. However, it would appear the immune changes are not a direct reflection of such an event.

Kronfol and colleagues studied the T-cell response to mitogen stimulation *in vitro* and showed that the suppression of T-cell replication occurred in all depressed patients independently of their plasma cortisol levels. Other studies have shown that the number of T cells is decreased in ambulatory-depressed patients but not in those with other types of major psychiatric illness, such as schizophrenia. While the precise cause of the decreased T-cell function is unclear, it would appear that peptides (e.g., endogenous opioids, neurotensin, vasoactive intestinal peptide, substance P, somatostatin) are causally involved. It has been speculated that such peptides depress T-cell function via their action on the hypothalamic hypophysial pathway. It should be emphasized, however, that not all aspects of cellular immunity are reduced in depression. Thus, macrophage activity (with the consequent increase in the release of proinflammatory cytokines is increased irrespective of the plasma glucocorticoid concentration.

Changes in Neutrophils and Monocytes

In addition to a suppression of T-cell and NKC activity in depressed patients, changes have also been found to occur in the activity of neutrophils and monocytes. Thus, studies in this laboratory have shown that the activity of monocytes is enhanced and that of neutrophils decreased in untreated depressed patients. On recovery, the activity of the phagocytes returns to control values, suggesting the changes are state, rather than trait, markers of depression. Such changes in phagocytosis are qualitatively different from those found in patients with schizophrenia, mania, anxiety states, alcoholism, or Alzheimer's disease.

We have attempted to determine the nature of the factors responsible for the abnormal phagocytic responses of monocytes and neutrophils in depressed patients. One possible explanation lies in the elevation of prostaglandin concentrations in the plasma of depressed patients. There is convincing evidence that immunoregulation is partly mediated by the peripheral and central sympathetic systems and that prostaglandins can modulate central neurotransmitter release, possibly by modifying the passage of calcium into neurons. It is also known that drugs that are therapeutically effective in treating depression (e.g., monoamine oxidase inhibitors, tricyclic anti-depressants, and lithium) can inhibit the synthesis of prostaglandins. Monocytes are an important source of IL-1 and prostaglandins, and as the activity of these cells is enhanced in depressed patients, it may be hypothesized they play an important role in elevating peripheral and (owing to the activity of the microglia in the brain) central immunotransmitters.

Such changes could contribute to behavioral state commonly associated with depression, namely, lowering of mood, anhedonia, altered sleep patterns, anorexia, and loss of libido. Then changes are simulated by the proinflammatory cytokinin in both animals and man.

Obviously, more detailed studies must be undertaken to validate this hypothesis, but these preliminary findings link neurotransmitter changes in depressed patients to the delays in onset of action of anti-depressants and the changes in cellular immunity.

Summary

It now seems probable that specific disturbances occur in the immune system in psychiatric illness that are not artifacts of nonspecific stress factor, institutionalization, or medication. The known effects of the neuroendocrine system on the immune response, and the recent evidence that receptor sites for neurotransmitters and neurendocrine factors occur on lymphocytes and macrophages, support the hypothesis that immunologic abnormalities may assist in precipating the symptoms of anxiety and depression—common symptoms of major affective disorders. The next step in research is to define more precisely which specific immune factors are involved.

DO ANTI-DEPRESSANTS ACT BY CHANGING ENDOGENOUS ENDOCOIDS IN THE DEPRESSED PATIENT?

Several groups of investigators have produced evidence to indicate that the plasma and brain contain factors that affect the action of serotonin on the platelet membranes and the transport of serotonin into synaptosomes. As has been mentioned, Brusov and colleagues thus showed that, in a population of drug-free bipolar and unipolar depressed women, a concentration of serotonin sufficient to induce half-maximal shape change in the platelet (a measure of 5-HT receptor function) was significantly lower than that of age-matched controls. Chronic anti-depressant treatment resulted in a normalization of the 5-HT_2 receptor function. Similar changes have been reported by others. Such changes in serotonin receptor sensitivity that are associated with effective anti-depressant treatments have been attributed to a change in the concentration of an endogenous peptide(s). These have been reported present in bovine forebrain as inhibiting both ^{3}H-mianserin and ^{3}H ketanserin binding to 5-HT_2 receptors. McAdams and colleagues have shown that the aggregatory response of platelets to serotonin was reduced by a plasma factor (or factors) that occurred only in those patients who were depressed or had not responded to anti-depressant treatment. These factors would also inhibit the aggregatory response when incubated with platelets from control subjects. Once the patients recovered, the patients' plasma no longer affected the serotonin-induced aggregation. In these same patients, the plasma also partially inhibited the uptake of ^{3}H-serotonin into the platelets. Although it is uncertain whether the factors that suppress 5-HT_2 receptor function and ^{3}H-serotonin transport are the same, the results do suggest that it such factors are only present while the symptoms of depression persist. This raises the possibility that anti-depressants produce their beneficial effects by changing the endogenous concentration of some depressogenic endocoids.

Other investigators have studied changes in ^{3}H-serotonin uptake into platelets from depressed patients and found evidence that bovine brain and human platelet extracts contain substances that inhibit both serotonin uptake and ^3H imipramine binding. Some of the endogenous substances responsible for the inhibition of ^3H serotonin uptake were recognized by rabbit antibodies against imipramine, thereby suggesting that they may possess structures similar to those identified by the antibodies. An aqueous plasma fraction from patients with premenstrual syndrome was also shown to inhibit platelet serotonin uptake. Although the precise identity of the factor, or factors, responsible for those effects is unknown, there is a report that α-1 acid glycoprotein, which is present in higher concentration in the depressed patient and normalizes following effective treatment, is one possible cause of these changes in serotonin function. However, recent studies have not supported this.

It is interesting to speculate that the cause of depression may reside in the presence of endogenous endocoid(s) that initiate the neurotransmitter defects associated with the symptoms. How such factor(s) also cause changes in the sensitivity of central glucocorticoid receptors and some of the cytokines that also appear to be malfunctioning in depression is currently unknown. More than 2000 years ago, Hippocrates speculated that melancholia was due to an excess of "black bile". Is it possible that we are now in a position to identify the chemical nature of "black bile" as a possible cause of depression?

CONCLUSION

Our search for an understanding of the biochemical basis of depression and of the mechanism of action of the various treatments used to alleviate depression has been devoted almost entirely to an exploration of the noradrenergic and serotonergic systems. Furthermore, many of the concepts which form the basis of the amine hypothesis of depression are based on extrapolations from the acute effects of high doses of anti-depressants on rat brain. Despite the recent interest in the chronic effects of anti-depressants on monoamine receptor function in depressed patients, it is still uncertain whether such receptor changes noted in the platelets and lymphocytes are epiphenomena. Clearly, future studies will need to concentrate on investigating longterm adaptive changes in both aminergic and nonaminergic neurotransmitter systems. Perhaps the development of sophisticated brain-imaging devices may enable better insight into such interacting process to be attained.

Another essential development relates to the need to produce better animal models of depression rather than models for the detection of antidepressants, the latter of which we presently have. This requires a more detailed and sophisticated appraisal of behavioral changes elicited by brain lesions, or environmental manipulations that produce an abnormal animal whose behavior is selectively normalized by effective anti-depressant therapy. Such models may not only assist in our understanding of the relationsip between biochemistry and behavior, but also help to lay the basis for the development of genuinely novel anti-depressants. It is a sobering thought that, with one or two exceptions, there is little evidence that the 80 or more compounds in clinical development will add anything worthwhile to our understanding of depression or in improving the efficacy of treatment of the depressed patient. This is despite most of the new changes having fewer adverse side effects (Bloom and Kupfer, 1995).

The decade of the 1970s was largely devoted to an explanation of the mode of action of anti-depressants on the concentrations of amines and their metabolites. The 1980s was a decade in which adaptive changes in receptor function were envisaged as important in understanding the action of the anti-depressants. The 1990s has concentrated our attention on the link between the receptors and their secondary messenger systems. It is always possible that, by the year 2000, we shall begin to understand how anti-depressants work!

REFERENCES

Bloom, F.E., & Kupfer, D.J. (Eds.)(1995). Psychopharmacology: The fourth generation of Progress. Raven Press, New York.

Brunello, N., Perez, J., Tinelli, D. et al. (1990). *Biochemical and molecular changes in rat cerebral cortex after chronic antidepressant treatment:* in vitro *and* in vivo *studies.* Pharmacol. Toxicol. 66 (Suppl. 3), 112-120.

Charney, D.S., Southwick, S.M., Delgado, P.L. et al. (1990). Current status of the receptor sensitivity hypothesis of antidepressant action. In: *Pharmacotherapy of Depression.* (Amsterdam, J.D., Ed.), p. 13. Marcel Dekker, New York.

Cooper, J.R., Bloom, F.E., & Roth, R.H. (1991). *The Biochemical Basis of Neuropharmacology.* Oxford University Press, Oxford.

Cowen, P.J. (1992). Serotonin receptor subtypes: Implications for psychopharmacology. Br. J. Psychiatry 159, 7-14.

de Boer, J. A., Sitsen, J.M.A. (Eds.)(1994). *Handbook of Depression and Anxiety—A Biological Approach.* Marcel Dekker Inc., New York.

De Vane, C.L. Pharmacokinetics of the selective serotonin reuptake inhibitors. J. Clin. Psychiatry 53 (Suppl. 2), 13-20.

Feighner, J.P. & Boyer, W.F. (1991). *Diagnosis of Depression, Vol 2. Perspectives in Psychiatry.* John Wiley and Sons, Chichester, England.

Feighner, J.P. & Boyer, W.F. (1991). *Selective Serotonin Reuptake Inhibitors, Vol. 1. Perspectives in Psychiatry.* John Wiley and Sons, Chichester, England.

Frazer, A., Molinoff, Penny, & Winokur, A. (Eds.)(1994). *Biological Bases of Brain Function and Disease.* Raven Press, New York.

Gonzalez-Heydrich, J. & Peroutka, S.J. (1990). Serotonin receptor and reuptake sites: Pharmacologic significance. J. Clin. Psychiatry 51, 5-12.

Henry, J.A., Antao, C.A. Suicide and fatal antidepressant poisoning. Eur. J. Med.1, 343-348.

Horton, R. & Katona, C. (Eds.)(1991). *Biological Aspects of Affective Disorders.* Academic Press, New York.

Leonard, B.E. (1989). From animals to man: Advantages, problems, and pitfalls of animal models in psychopharmacology. In: *Human Psychpharmacology: Measures and Methods.* (Hindmarch, I., & Stonier, P.D., Eds.), pp. 23-66. John Willey and Sons, London.

Leonard, B.E. (1992). Pharmacological differences of serotonin reuptake inhibition and possible clinical relevance. Drugs 43, 3-1O.

Leonard, B.E. (1992). Effect of antidepressants on neurotransmission : A common mechanism of action? In: *Current Aspects of the Neurosciences, Vol. 4.* (Osborne N.N., Ed.), pp. 205-238. MacMillan, Basingstoke, England.

Leonard, B.E. (1997). *Fundamentals of Psychopharmacology.* 2nd ed. John Wiley and Sons, Chichester, England.

Leonard, B.E. (1993). The comparative pharmacology of new antidepressants. J. Clin. Psychiat. 54 (Suppl.), 3-15.

O'Brien, J.T., Ames, D., Schweitzer, E. (1993). HPA axis function in depression and dementia: A review. Int. J. Geriat. Psychiat. 8, 887-898.

Price, L.H., Charney, D.S., Delgado, P.L. et al. (1990). Clinical data on the role of serotonin in the mechanism(s) of action of antidepressant drugs. J. Clin. Psychiatry 51 (Suppl. 4), 44-50.

Strange, P.G. (1992). *Brain Biochemistry and Brain Disorders.* Oxford University Press, Oxford.

Tollefson, G.D. (1991). Antidepressant treatment and side effect considerations J. Clin. Psychiatry 52 (Suppl. 5), 4-13.

Van Praag, H.M., Brown, S.L., Asnis, G.M. et al. (1990). Beyond serotonin: A multiaminergic perspective of abnormal behavior. In: The Role of in Serotonin Psychiatric Disorders. (Brown, S.L. & Van Praag, H.M., Eds.), pp. 3O3-332. Brunner/Mazel, New York.

RECOMMENDED READINGS

Briley, M. & Montgomery, S. (Eds.). (1998). Antidepressant Therapy at the Dawn of the Third Millennium. Martin Dunitz, London.

Dinan, T.G. (1994). Glucocorticoids and the genesis of depressive illness: A psychobiological model. Br. J. Psychiat. 164, 365-371.

Drevets, W.C. (1998). Functional neuroimaging studies in depression: The anatomy of melancholia. Ann. Rev. Med. 49, 341-361.

Dumar, R.S., Heninger, G.R., & Nestler, E.J. (1997). A molecular & cellular theory of depression. Arch. Gen. Psych. 41, 23-32.

Frazer, A. Molinoff, P., & Winokur, A. (1994). Biological Basis of Brain Function and Disease. Raven Press, New York.

Honig, A. & Van Praag, H.M. (Eds.). (1997). Depression: Neurobiological, Psychopathological, and Therapeutic advances. John Wiley and Sons. Chichester, England.

Leonard, B.E. & Song C. (1997). Changes in the immune-endocrine interrelationships in anxiety and depression. Stress Medicine, 13, 217-227.

Leonard, B.E. (1997). Action mechanism of antidepressants. In: Depression: Neurobiological, Psychopathological and Therapeutic Advances (Hong, A., and Van Praag, H.M. (Eds.) pp. 459-470. John Wiley and Sons, Chichester, England.

Leonard, B.E. (1997). Fundamental of Psychopharmacology (2nd ed.) John Wiley & Sons, Chichester, England.

Rajkowska, G., Miguel-Hidalgo, J., J., Wei, J., Dilley, G., Pittman, S.D., Meltzer, H.Y., Overholser, J.C., Roth, B.L., & Stockmeier, C.A. (1999). Morphometric evidence for neuronal and glial prefrontal cell pathology in major depression. Biolog. Psychiatry 45, 1085-1098.

Shah, P.J., Ebmeier, K.P., Glabus, M.F., & Goodwin, G.M. (1998). Cortical grey matter reductions associated with treatment resistant chronic unipolar depression. Br. J. Psych. 172, 527-532.

Chapter 10

Psychobiology of Suicidal Behavior

KEVIN M. MALONE

Biological Psychiatry, pages 207-216.

ISBN: 1-55938-819-6

INTRODUCTION

Epidemiology of Suicidal Behavior

There are over 30,000 suicides per year in the USA. Suicide is the eighth leading cause of death for all age groups and the second leading cause of death in adolescents and young adults. Over 300,000 people a year attempt suicide in the U.S.A. (Weissman, 1974). The crude annual incidence rate for attempted suicide for the 1970s has been estimated in the order of 160 to 300 per 100,000 (Weissman, 1974) but may be as high as 730 per 100,000 when all data sources are included. Estimates of lifetime suicide attempt rates among adults range between 10% for adults (Petrovic et al., 1983; Johnson and Jennison, 1992; Meehan et al., 1992) and 14% for adolescents (Soliman et al, 1986). Attempted suicide constitutes a significant proportion of accident and emergency workload in a general hospital and accounts for more than 10% of acute medical admissions (Kessler et al., 1988) and 5% of admissions to intensive care units. It is a major public health problem, is a serious complication of Major Depressive Disorder (Guze and Robins, 1970; Klerman and Weissman, 1992), and may predate a completed suicide. (Hawton et al., 1993; Rosen, 1970; Pokorny, 1983; Morgan, 1993; Nordentoft et al., 1993a,b). the high rates of attempted suicide in at-risk groups represent substantial morbidity. Study of risk factors for suicide attempts may not only assist in prevention of future suicide but also in prevention of future attempted suicide, an even more common event.

Suicidal Acts—A Spectrum of Behavior

The spectrum of suicidal behavior includes completed suicide, failed suicide, suicidal gestures (sometimes called parasuicide), and suicidal ideation. Different risk factors are associated with completed suicide, attempted suicide, and suicidal ideation without an attempt. It is thus conceivable that the neurologic bases, biological correlates, and pharmacologic treatment for these conditions may overlap and also be diverse.

Integrating Psychopathology and Neurobiology of Suicidal Behavior

Many models of suicidal behavior have been suggested. Our research group has favored a stress-diathesis model (Malone et al., 1995a; Mann et al., 1999). In this model, state-dependent factors such as the onset of major depression or an interpersonal crisis can *trigger* suicidal behavior in those with a trait disposition to suicidal behavior and determine its timing. The trait-dependent factors determine the *threshold* for acting on suicidal thoughts when they emerge.

NEUROBIOLOGY OF SUICIDAL BEHAVIOR

Serotonin Abnormalities in Completed Suicide

Evidence now exists for reduced serotonin function independent of psychiatric diagnosis in completed suicide based on postmortem brain studies in suicide victims. (Beskow et al., 1976; Stanley and Mann, 1983; Arango et al., 1995). Over the past 20 years, research groups including ours have reported in suicide victims a decrease in brainstem serotonin, a decrease in 5-hydroxyindole acetic acid (5-HIAA), which is the major metabolite of serotonin, a decrease in brain serotonin transporter protein imipramine binding, and an increase in brain postsynaptic $5HT_{2A}$ receptor binding. Recently, we have additionally reported in suicide victims a possible localization of some of the serotonin abnormalities to the ventrolateral prefrontal cortex, such as increased $5HT_{1A}$ receptor binding and decreased serotonin transporter binding (Arango et al., 1995). This brain area is involved in mediating inhibition and in executive reasoning. Thus, abnormalities in this brain area or reduced serotonin input this ventrolateral area of prefrontal cortex may result in a failure to adequately inhibit suicidal feelings and impulses. A great deal of work doubtlessly remains to be done to localize the brain areas that modulate suicide risk.

The Role of Genetic Factors in Suicidal Behavior

Some of the most compelling evidence of the importance of genetic factors in suicidal behavior comes from studies on twins that concordance rates for suicidal behavior is higher in monosygotic twins than in dizygotic twins (Roy, 1983; Roy et al., 19991; Roy, 1993). The familial association of suicidal behaviors, along with adoption studies, have found the risk of suicidal behavior to be biologically transmitted, and both provide evidence of the influence of genetic factors (Roy. 1983). It remains to be determined how genetic factors may modulate suicide risk. One possibility is that polymorphisms associated with key elements in the serotonin system may generate or code for the synthesis of suboptimal peptides necessary for serotonin neurotransmitter or receptor synthesis. This area of research may yield highly significant results over the next few years.

Serotonin Abnormalities in Suicide Attempters

Similar to the neurobiologic findings in completed suicide, several significant abnormalities in serotonin function have been reported in suicide attempters. A reduction in CSF 5-HIAA levels have been reported in 10 out of 15 studies of suicide attempters compared to nonattempters. Some of the negative studies may have arisen due to confounding effects of combining low lethality suicide attempters (parasuicide) with high lethality suicide attempters (failed suicides). We have hypothesized that biologic (serotonin) dysfunction is most likely to be found in subjects whose suicidal acts behaviorally most closely approximate completed suicide. We reported that CSF 5-HIAA levels are significantly lower in patients with a history of high lethality suicidal acts as compared with equally depressed patients who have a history of low lethality suicidal acts only (Mann and Malone, 1997).

Neuroendocrine Serotonin Challenge Tests

Neuroendocrine responses to the serotonin-releasing agent fenfluramine represents another paradigm with which to study serotonin function in patient populations. The prolactin response (serotonin-mediated) to fenfluramine has been consistently reported as being reduced in depressed patients (Mann et al., 1996). In a recent study, we controlled for depression and found that the prolactin response was lowest in patients who had a history of a high lethality suicide attempt compared to patients who had made a low lethality suicide attempt only (Malone et al., 1996). Moreover, we also found a blunted prolactin response to fenfluramine in patients who had co-morbid borderline personality disorder. Borderline personality disorder is characterized by mood lability, impulsivity, and anger. Patients whose major depression is co-morbid with borderline personality disorder show elevated rates of suicidal behavior (Corbitt et al., 1996; Brodsky et al., 1997).

Platelet Serotonin Measures

Several platelet measures have been linked to suicidality (Pandey et al., 1995). However, platelet indices convey little information about brain function or localization of brain function, and interpretation of platelet findings is subject to confounding factors of platelet age, patient age, and sex as well as seasonality and other effects. We and others have reported that aggressive traits and medically damaging suicide attempts are associated with alterations in platelet serotonin content and with increased platelet $5HT_2$ receptors but that these receptors are subresponsive to serotonin stimulation (Mann et al., 1992).

ENVIRONMENTAL AND OTHER INFLUENCES ON SEROTONIN AND SUICIDAL BEHAVIOR

Cholesterol

Recent evidence has found that reductions in serum cholesterol levels may increase the risk of death from suicide and nonaccidental death (Muldoon et al., 1990). In animal studies, we have found that lowering dietary cholesterol in monkeys causes increased aggression and also causes a blunted prolactin response to fenfluramine. Cholesterol levels may be important to maintain the integrity of central serotonin responsivity. Further research is necessary to elucidate the significance of this association.

Substance Abuse

Chronic alcohol use causes disproportionate suicidality in depressed patients with co-morbid alcohol abuse (Murphy and Wetzel, 1990). In animal studies, chronic alcohol causes serotonin depletion. Since those studies, we have reported that, in depressed and schizophrenic patients, increased cigarette use was associated with more suicidal behavior and lower serotonin function. It this finding is replicated, it will have a significant impact on our understanding of the potential brain effects of nicotine, and how it may increase the risk of suicidal behavior in high risk patients.

Estrogen

Serotonin responsivity has been found to be higher in women than in men, an effect that declines with age, and postmenopausal women have serotonin responsivity values more similar to males (McBride et al., 1990). Estrogen levels appear to modulate serotonin responsivity across the menstrual cycle (O'Keane et al., 1991). It has not yet been established how estrogen may modulate serotonin responsivity, and how his effect may or may not influence suicidal behavior.

TREATMENT STRATEGIES FOR SUICIDAL BEHAVIOR

Primary Prevention

Efforts at reducing the rate of suicidal behavior including completed suicide in the population are difficult to design and costly to implement, and significant results will not be seen immediately because of the relatively low base rate of completed suicide. It is likely that significant progress will be made only if a well planned and cohesive program is implemented that targets primary, secondary, and tertiary prevention methods. Single strategies are likely to work for a short time but effects will not be sustained. A good example of this is that the introduction of natural gas instead of coal gas in the UK in the 1960s led to a significant drop in suicide fatalities over the following five years. This effect was not sustained as the rate of suicide deaths gradually increased from other methods. It is possible that if a concurrent and concerted effort at reducing suicide by other strategies such as those outlined below had been implemented at the time of natural gas introduction, the lowered suicide rates may have been sustained. Current primary prevention methods include efforts to reduce easy availability of high lethality methods. Evidence that this may have an effect comes from studies of firearms and adolescent suicide victims. Several studies have shown that adolescents are more like to commit suicide from a firearm if there is a firearm in the home, and the evidence to date suggests that removal of firearms from the home lessens the risk of suicide in that home (Brent et al., 1987; Brent et al., 1988; Brent et al., 1991). Other examples of primary prevention include fencing off access to high bridges and installing double doors on subway platforms. In addition to these preventive strategies having a mechanical deterrent effect, such strategies also send a message to high risk individuals that there is community concern about the problems of suicidal behavior, and potentially may deflect suicidal people toward seeking help. This raises the question of whether secondary prevention services should be implemented for higher risk or more vulnerable populations.

Secondary and Tertiary Prevention

Specific populations are at higher risk for suicidal acts than others. Male adolescents, substance abusers, individuals with chronic diseases, and those with major psychiatric disorders such as major depression or schizophrenia are some of the highest risk subgroups (Pokorny, 1983; Henrikson et al., 1993; Isometsä et al., 1994).

Psychoeducation and health education in general in high schools is currently either ignored or poorly presented and mostly focuses on the dangers of substance abuse and unsafe sex. Psychological difficulties and suicidal ideation remain taboo topics, whereas children are hearing in the community that assisted suicides may be becoming a social option. Such messages can only contribute to the vacillating ambivalence between choosing living versus dying that some adolescents experience. Preliminary studies suggest that adolescents are willing and sometimes relieved to discuss suicidal ideation. It remains to be determined whether a sufficiently sensitive screening program can be devised to identify high risk adolescents for suicidal behavior.

Health care professionals who are engaged in treating psychiatric populations require the most clinical vigilance, and studies suggest that more vigilance is required. We recently found that a considerable amount of suicidal behavior goes undetected in the average clinical psychiatric interview (Malone et al., 1995b). It is difficult for communities to identify suicidal patients if health care professionals are failing to do so. Therefore, more effort needs to be focused on standardizing clinical assessment techniques for suicidal behavior and training professionals to implement these assessments in high risk populations.

Special tertiary prevention services need to be developed for patients who have already made a suicide attempt. They are 40 times more likely to kill themselves compared to the population, and if they have been hospitalized, they are more likely to commit suicide in the year following discharge. Although programs targeting this population will engage patients who will not kill themselves, nonetheless, up to 40% of suicide victims have made one previous unsuccessful suicide attempt, commonly of high lethality. The current mainstay of tertiary intervention is antidepressant pharmacotherapy reviewed below. Additional strategies that may augment current antisuicide strategies need to be developed such as outreach programs to follow nonattenders of outpatient follow-up.

Pharmacotherapy

Studies in adults have shown that antidepressants including monoamine oxidase inhibitors, tricyclic antidepressants, and selective serotonin reuptake inhibitors (SSRIs) are effective in the treatment of major depression. We and others have found that antidepressant medications are effective in reducing suicidal ideation and that those may decrease the frequency of suicide attempts and suicide. It is important to study how SSRIs and other drugs affect the suicide threshold in addition to evaluating their efficacy in treating major depression. It is possible that antidepressants may reduce suicide rates by a direct action on the severity of depression. Alternatively, our early evidence suggests that antidepressants, particularly SSRIs, may have a separate and additional effect on the suicide threshold (Malone et al., 1994). Certain antidepressant drugs (i.e., non-SSRIs) might effect depression without altering the suicide threshold.

Psychotherapy

There are some suicidal patients for whom medication may not be suitable due to intolerable side effects as in the case of pregnant or lactating women. Until recently, researchers have been reluctant to systematically study the role of any particular psychotherapy in the suicidal patient. A specific psychotherapy aimed at reducing suicidal impulses has been described (Linehan et al., 1994). The therapy, known as dialectical behavior therapy (DBT) focuses on reducing the number of suicidal episodes through a combination of psychoeducation, support, and skills training. Preliminary results using this therapy in suicidal patients has been promising. Further work on these and other psychological treatments for the management of the suicidal patient is needed. It will also be useful to know whether such psychologic treatments may augment pharmacologic strategies both in terms of reducing symptoms and, equally importantly, in preventing relapse.

SUMMARY

Considerable evidence has accumulated for a biological component to suicide risk, and preliminary work suggest that the focus of the biological abnormality may lie within the serotonin projections to the prefrontal cortex. There are early suggestions that similar factors may operate in adolescent suicides and suicide attempters. The efficacy of antidepressant medication in preventing suicidal acts in depressed patients has not yet been clearly demonstrated, although early evidence suggests that selective serotonin reuptake inhibitors may have an antisuicide component in addition to their antidepressant effects. Much research remains to be done to understand the neurochemical basis underlying restraint from suicidal ideas (threshold) during depression and other major stressors (trigger). The development of more specific treatments targetting (a) a biological elevation in the suicide threshold and (b) an early and reduced impact of triggers for suicidal acts may significantly reduce morbidity and mortality from suicidal behavior.

REFERENCES

Arango, V., Underwood, M.D., Gubbi, A.V., & Mann, J.J. (1995). Localized alterations in pre- and postsynaptic serotonin binding sites in the ventrolateral prefrontal contex of suicide victims. Brain Res. 688, 121-133.

Avery, D., & Winokur, G. (1978). Suicide, attempted suicide, and relapse rates in depression. Arch. Gen. Psychiatry 35, 749-753.

Beskow, J., Gottfries, C.G., Roos, B.E., & Winblad, B. (1976). Determination of monoamine and monoamine metabolites in the human brain: Post mortem studies in a group of suicides and in a control group. Acta Psych. Scand. 53, 7-20.

Brent, D.A., Perper, J.A., & Allman, C.J. (1967). Alcohol, firearms, and suicide among youth. Temporal trends in Allegheny County, Pennsylvania, 1960 to 1983. J. Am. Med. Assoc. 257, 3369-3372.

Brent, D.A., Perper, J., Goldstein, C.E., Kolko, D.J., Allan, M.J., Allman, C.J., & Zelenak, J.P. (1988). Risk factors for adolescent suicide: A comparison of adolescent suicide victims with suicidal inpatients. Arch. Gen. Psychiatry 45, 581-588.

Brent, D.A., Perper, J.A., Allman, Moritz, G.M., Wartella, M.E., & Zelenak, J.P., (1991). The presence and accessibility of firearms in the homes of adolescent suicides. A case-control study. J. Am. Med. Assoc. 266, 2989-2995.

Brodsky, B.S., Malone, K.M., Ellis, S.P., Dulit, R.A., & Mann, J.J. (1997). Borderline personality disorder characteristics associated with suicidal behavior. Am. J. Psychiatry 154, 1715-1719.

Corbitt, E.M., Malone, K.M., Haas, G.L., & Mann, J.J. (1996). Suicidal behavior in patients with major depression and comorbid personality disorders. J. Affect. Disord. 39, 61-72.

Guze, S.B., & Robins, E. (1970). Suicide and primary affective disorders. Br. J. Psychiatry 117, 437-438.

Hawton, K., Fagg, J., Platt, S., & Hawkins, M. (1993). Factors associated with suicide after parasuicide in young people. Brit. Med. J. 306, 1641-1644.

Henriksson, M.M., Aro, H.M., Marttunen, M.J., Heikkinen, M.E., Isornetsä, E.T., Kuoppasalmi, K.I., & Lönnqvist, J.K. (1993). Mental disorders and comorbidity in suicide. Am. J. Psychiatry 150, 935-940.

Isometsä, E.T., Henriksson, M.M., Aro, H.M., Heikkinen, M.E., Kuoppasalmi, K.I., & Lönnqvist, J.K., (1994). Suicide in major depression. Am. J. Psychiatry 151, 530-536.

Johnson, K.A., & Jennison, K.M. (1992). The drinking-smoking syndrome and social context. Int. J. Addict. 27, 749-792.

Kessler, R.C., Downey, G., Milavsky, R.J., & Stipp, H. (1988). Clustering of teenage suicides after television news stories about suicides: A reconsideration. Am. J. Psychiatry 145(11), 1379-1383.

Klerman, G.L., & Weissman, M.M. (1992). The course, morbidity, and costs of depression. Arch. Gen. Psychiatry 49, 831-834.

Linehan, M.M., Tutek, D.A., Heard, H.L., & Armstrong, H.E. (1994). Interpersonal outcome of cognitive behavioral treatment for chronically suicidal borderline patients. Am. J. Psychiatry 151, 1771-1776.

Malone, K.M., Corbitt, E.M., Szanto, K., & Mann, J.J. (1994). Antidepressant pharmacotherapy and suicidal behavior during major depression. NCDEU 34th Annual Meeting (Abstract).

Malone, K.M., Haas, G.L., Sweeney, J.A., & Mann, J.J. (1995a). Major depression and the risk of attempted suicide. J. Affect. Disord. 34, 173-185.

Malone, K.M., Szanto, K., Corbitt, E.M., & Mann, J.J. (1995b). Clinical assessment versus research methods in the assessment of suicide behavior. Am. J. Psychiatry 152, 1601-1607.

Malone, K.M., Corbitt, E.M., Li, S., & Mann, J.J. (1996). Prolactin response to fenfluramine and suicide attempt lethality in major depression. Br. J. Psychiatry 168, 324-329.

Mann, J.J., McBride, P.A., Brown, R.P., Linnoila, M., Leon, A.C., DeMeo, M.D., Mieczkowski, T.A., Myers, J.E., & Stanley, M. (1992). Relationship between cental and peripheral serotorin indexes in depressed and suicidal psychiatric inpatients. Arch. Gen. Psychiatry 49(6), 442-446.

Mann, J.J., Malone, K.M., Sweeney, J.A., Brown, R.P., Linnoila, M., Stanley, B., & Stanley, M. (1996). Attempted suicide characteristics and cerebrospinal fluid amine metabolites in depressed inpatients. Neuropsychopharmacology 15, 576-586.

Mann, J.J., & Malone, K.M. (1997). Cerebrospinal fluid amines and higher lethality suicide attempts in depressed inpatients. Biol. Psychiatry 47, 162-171.

Mann, J.J., Waternaux, C., Haas, S.L., & Malone, K.M. (1999). Towards a clinical model of suicidal behavior in psychiatric patients. Am. J. Psychiatry 156, 181-189.

McBride, P.A., Tierney, H., DeMeo, M., Chen, J.-S., & Mann, J.J. (1990). Effects of age and gender on CNS serotonergic responsivity in normal adults. Biol. Psychiatry 27, 1143-1155.

Meehan, P.J., Lamb, J.A., Saltzman, L.E., & O'Carroll, P.W. (1992). Attempted suicide among young adults: Progress toward a meaningful estimate of prevalence. Am. J. Psychiatry 149, 41-44.

Morgan, G. (1993). Long term risks after attempted suicide. Identifying risk factors should help to reduce subsequent suicide. Brit. Med. J. 306, 1626-1627.

Muldoon, M.F., Manuck, S.B., & Matthews, K.A. (1990). Lowering cholesterol concentrations and mortality: A quantitative review of primary prevention trials. Br. Med. J. 301, 309-314.

Murphy, G.E., & Wetzel, R.D. (1990). The lifetime risk of suicide in alcoholism. Arch. Gen. Psychiatry 47, 383-392.

Nordentoft, M., Breum, L., Munck, L.K., Nordestgaard, A.G., Hunding, A., & Laursen Bjældager, P.A. (1993a). High mortality by natural and unnatural causes: A 10 year follow up study of patients admitted to a poisoning treatment center after suicide attempts. Brit. Med. J. 306, 1637-1641.

Nordentoft, M., Breum, L., Munck, L.K., Nordestgaard, A.G., Laursen Bjældager, P.A., & Hunding, A. (1993b). Attempted suicide. Are affective disorders missed? Reply. Brit. Med. J. 307, 866.

O'Keane, V., O'Hanlon, M., Webb, M., & Dinan, T. (1991). *d*-Fenfluramine/prolactin response throughout the menstrual cycle: Evidence for an oestrogen-induced alteration. Clin. Endocr. 34, 289-292.

Pandey, G.N., Pandey, S.C., Dwivedi, Y., Sharma, R.P., Janicak, P.G., & Davis J.M. (1995). Platelet serotonin-2A receptors: A potential biological marker for suicidal behavior. Am. J. Psychiatry 152, 850-855.

Petrovic, S.L., McDonald, J.K., Snyder, G.D., & McCann, S.M. (1983). Characterization of β-adrenergic receptors in rat brain and pituitary using a new high-affinity ligand, [^{125}I]iodocyanopindolol. Brain Res. 261, 249-259.

Pokomy, A.D. (1983). Prediction of suicide in psychiatric patients. Report of a prospective study. Arch. Gen. Psychiatry 40, 249-257.

Rosen, D.H. (1970). The serious suicide attempt. Epidemiological and follow-up study of 886 patients. Am. J. Psychiatry 127(6), 64-70.

Roy, A. (1983). Family history of suicide. Arch. Gen. Psychiatry 40, 971-974.

Roy, A., Segal, N.L., Centerwall, B.S., & Robinette, C.D. (1991). Suicide in twins. Arch. Gen. Psychiatry 48, 29-32.

Roy, A. (1993). Genetic and biologic risk factors for suicide in depressive disorders. Psych. Quart. 64, 345-358.

Soliman, K.F.A., Mash, D.C., & Walker, C.A. (1986). The effect of altered 5-hydroxytryptamine levels on β-endorphin. Contents in rat brain (42326). Proc. Soc. Exp. Biol. Med. 182, 187-193.

Stanley, M., & Mann, J.J. (1983). Increased serotonin-2 binding sites in frontal cortex of suicide victims. Lancet 1, 214-216.

Weissman, M.M. (1974). The epidemiology of suicide attempts, 1960 to 1971. Arch, Gen. Psychiatry 30, 737-746.

RECOMMENDED READING

Burmeister, M. (1999). Basic concepts in the study of diseases with complex genetics. Biol. Psychiatry 45, 522-532.

Chapter 11

Bipolar Disorder

JAMES C.-Y. CHOU and ROBERT CANCRO

Biological Psychiatry, pages 217-232.

ISBN: 1-55938-819-6

INTRODUCTION

Bipolar disorder, also known as manic–depressive illness, is a severe mental disorder characterized by episodes of abnormal mood and related cognitive and behavioral changes which lead to various degrees of functional impairment. While the characteristic syndromes of mania and hypomania are hallmarks of bipolar disorder, bipolar depressed patients may be, in a cross-sectional evaluation, clinically indistinguishable from patients with unipolar depression.

The lifetime prevalence of bipolar disorder is about 1% throughout the world. Patients who suffer a manic episode have a 90% chance of having another episode within 5 years. One fourth of all bipolar patients attempt suicide, and one tenth complete suicide. If unidentified or untreated, a patient who has a first episode of bipolar disorder with an onset at age 25 will suffer, on average, an estimated loss of 9 years of life, 14 years of productivity, and 12 years of good health. This same patient, with optimal diagnosis and treatment, will recover on average 6.5 years of life, 10 years of productivity, and 8.5 years of good health. Clearly, this is an illness which is managed rather than cured.

While a wide range of neurobiological and psychological abnormalities have been found in bipolar patients, a specific etiology and pathogenesis remain elusive. The majority of recent research has focused on biological approaches to this disease. Clearly, there is a genetic component in many if not all cases, and bipolar disorder tends to run in families. However, genetics alone cannot explain all cases of bipolar disorder; genetic factors probably render patients more vulnerable to the effects of certain environmental stressors which together ultimately lead to expression of the disease.

DIAGNOSIS

Arataeus, in 150 A.D., is usually credited as being the first to record observations of mania following depression in the same patient. Similar observations were made by others throughout history with various etiologies thought to underlie this set of illnesses. Kraepelin, in 1899, used the term "manic–depressive" to describe circular psychoses and mania. He emphasized the need to consider carefully both longitudinal course, as well as cross-sectional symptoms. In modern psychiatry, this concept has evolved into one of a disorder of mood with characteristic episodes of major depression, mania, and/or hypomania.

According to the *Diagnostic and Statistical Manual of Mental Disorders, Fourth Edition, (DSM-IV)* (1994) published by the American Psychiatric Association (which has been coordinated with the *ICD-10*), a diagnosis of bipolar disorder is made if the following are true: (1) either a manic episode or a mixed episode has occurred (bipolar I disorder) or (2) a hypomanic episode has occurred in a patient with a history of at least one episode of major depression (bipolar II disorder). A diagnosis of bipolar disorder should not be made if the manic, mixed, or hypomanic episodes were caused by medical conditions such as hyperthyroidism, brain tumors, or HIV encephalitis, or if the episode is secondary to drugs such as cocaine, amphetamines, antidepressants, or steroids. Psychotic symptoms, if present, must have been limited to the period of the episodes of either depression or mania and cannot have been present for more than two weeks either before or following the mood episode. If psychosis has persisted beyond such periods, then diagnoses of schizoaffective disorder or schizophrenia should be considered.

A manic episode is defined as a distinct period of elevated, expansive, or irritable mood lasting at least one week (unless treated) with three or more of the following seven associated symptoms: grandiosity, talkativeness, decreased sleep, flight of ideas or racing thoughts, distractibility, increased goal-directed activity, and increased involvement in potentially risky activities. If the mood is irritable, then four of the above symptoms are needed to make the diagnosis. The symptoms must lead to marked functional impairment, and the episode cannot be secondary to drugs or medical conditions.

A hypomanic episode is similar to a manic episode but is less severe and may be of shorter duration. This diagnosis may be made if there is a distinct period of elevated, expansive, or irritable mood lasting at least four days with at least three of the seven symptoms of mania but associated with only mild-to-moderate functional impairment. The impairment must be observable by others but cannot be severe enough to warrant a diagnosis of manic episode. Furthermore, if psychotic features are present, the episode must be diagnosed as a manic episode. Hypomania is differentiated from normal periods of good mood by the unequivocal presence of functional impairment and the requirement that the episode is noticed by observers.

While a minority of bipolar patients have only manic episodes, the majority also experience major depressive episodes. A major depressive episode is a period of at least two weeks of impaired functioning with at least five of the following symptoms: persistent depressed mood, loss of interest or pleasure, a change in weight (not due to deliberate dieting), insomnia or hypersomnia, psychomotor changes, fatigue or loss of energy, feeling worthless or excessively guilty, impaired concentration, and death thoughts. Either depressed mood (or irritable mood in children or adolescents) or loss of interest or pleasure must be present, and the episode cannot be the consequence of substances or a medical condition.

While it may seem paradoxical, more than 40% of manic patients have features of depression during a manic episode. If the depressive symptoms associated with mania are severe enough, the diagnosis should be changed to mixed episode. A mixed episode is defined as a distinct episode during which the patient has sufficient symptoms to warrant a diagnosis of manic episode, as well as sufficient depressive symptoms to warrant a diagnosis of major depression (excluding duration) nearly every day for at least seven days. The episode cannot be caused by drugs or medical conditions. There must also be marked impairment in functioning. Psychotic symptoms and agitation are common in mixed episodes. A mixed episode is the equivalent of a manic episode in terms of leading to a diagnosis of bipolar I disorder. The diagnosis of mixed episode is important because it is associated with a worse prognosis as well as a decreased likelihood of having a good response to lithium in acute and maintenance treatment.

Additional specifiers reveal bipolar disorder as having these features: a rapid cycling course, a seasonal pattern, a chronic course, a postpartum onset, poor interepisode recovery, psychotic symptoms, catatonia, melancholia, or atypical features. In particular, rapid cycling, psychosis, and chronic course are important specifiers because they are associated with a poorer prognosis and a decreased likelihood of a good response to lithium in prophylactic maintenance treatment.

Substance abuse is common in patients with mood disorders who often use alcohol or illicit drugs in an unconscious effort to medicate themselves. If the mood episodes are judged to be caused by the substances, then bipolar disorder should not be diagnosed. If symptoms of both substance dependence and bipolar disorder not related to the substances are present, then both diagnoses can be made. Anxiety is another common associated feature of mood disorders, especially depression and, for bipolar patients, an additional diagnosis of an anxiety disorder should only be made if the anxiety symptoms clearly persist beyond the mood episodes. Another area of problematic differential diagnosis is in bipolar patients who have symptoms of a personality disorder, in particular borderline personality disorder. While both bipolar disorder and a personality disorder may be diagnosed in the same patient, mood episodes must be carefully evaluated using the above diagnostic criteria in order to differentiate them from the rapid and more transient changes in mood and affect associated with a personality disorder.

EPIDEMIOLOGY

Estimates of the lifetime prevalence of bipolar disorder range from 0.6% to 1.6% (Goodwin and Jamison, 1990) and the age of onset has a unimodal distribution with a mean at about 30 years. However, some patients have manic episodes before age 20, and risk of developing the disorder increases with age through age 35. A new onset of bipolar disorder in the elderly is not common, and people who reach age 50 without experiencing a manic episode have a markedly decreased risk of developing bipolar disorder. While unipolar depression is more common in females, bipolar disorder is found at equal rates in males and females. There is no clear evidence of racial or ethnic differences in prevalence rates of bipolar disorder. Patients with bipolar disorder are more likely than the general population to be single, separated, or divorced, but this may be partially due to of the detrimental effects that having bipolar disorder can have on relationships. Studies comparing prevalence rates of bipolar disorder in urban versus rural settings have been inconclusive.

Studies that have compared prevalence rates between different countries have been plagued with many methodological problems, and consistent international differences in prevalence have not been found. Although several studies have suggested a high prevalence in Israel, others have suggested a low prevalence in Northern European countries. While degree of industrialization of a country does not seem to affect the prevalence rate of bipolar disorder, it may affect the symptoms of mood disorders, as depressed patients in nonindustrialized countries are more likely to have somatic complaints and are less likely to make suicide attempts. Bipolar disorder in the United States has been found to be more common in immigrants than in nonimmigrants.

The relationships between bipolar disorder and social class are fascinating and complex. A majority of studies have found higher rates of bipolar disorder in middle and upper socioeconomic classes, and no studies have found bipolar disorder to be more common in lower socioeconomic classes. This is strongly suggestive that bipolar disorder is associated with higher social class but the reasons for this remain uncertain. This association implies that bipolar disorder does not lead to a downward social drift as do some other serious chronic mental disorders such as schizophrenia. In spite of this, bipolar disorder is still a common diagnosis among the homeless mentally ill.

PATHOGENESIS

Genetics

Twin, family, and adoption studies have shown that genetic factors are important in the transmission of bipolar disorder (Gershon et al., 1987). The disorder is clearly heritable and does cluster in families, facts that support genetic etiologic factors. An identical twin of a patient with bipolar disorder has a 65% chance of developing the disorder while a fraternal twin has a 20% chance. Furthermore, related mood disorders such as schizoaffective disorder and cyclothymia may be transmitted genetically together with bipolar disorder. No single gene has yet been conclusively proven to be associated with transmission of bipolar disorder, although there is interest in an area of the X chromosome related to color blindness. For genetic counseling for patients and families, it is useful to describe the risks of different relatives developing bipolar disorder, but genetic testing is not yet a part of usual clinical practice.

Environmental Factors

Several theories have emerged to explain the pathogenesis of bipolar disorder (Goodwin and Jamison, 1990). One theory proposes a dysregulation of normal dampening mechanisms for one or more amine neurotransmitter systems including serotonin and possibly others (perhaps through affecting second messenger systems such as cyclic AMP or polyphosphoinositide). Other theories focus on a dysregulation of circadian rhythms or other mechanisms such as kindling. These theories provide little insight regarding the relative importance of genetic and environmental factors in the pathogenesis of disease. Environmental factors have been clearly recognized as important because not all persons genetically loaded for bipolar disorder (e.g., an identical twin of an affected patient) develop the illness. These factors will lead to the disorder when they act on genetically vulnerable individuals. Potential environmental factors include substance use, medical illness, psychological stresses, prescribed medications, and others.

Biological Factors

Alcohol and substance abuse have important effects on the course of bipolar disorder. Also, substance abuse may mask symptoms of mood disorders and confound attempts at diagnosis. Typically, the onset of substance use parallels the onset of the mood disorder, but there are strong data indicating that, for most patients with both diagnoses, the mood symptoms preceded the substance abuse. Co-morbid substance abuse worsens the course of bipolar disorder and increases the risk of suicide. In particular, alcohol, stimulants, and hallucinogens have been associated with switching from depression to mania, and with a worsened course for mixed or rapid cycling bipolar patients. Since there is a high co-morbidity between bipolar disorder and substance abuse, accurate diagnosis and treatment of substance problems in bipolar patients is critically important.

The traditional description of the course of untreated bipolar disorder is that, with age, episodes become more severe, frequent, and prolonged, and that the disorder progresses to a more difficult-to-treat, rapid cycling subtype. This observation led to the explanation of the disorder through the use of a "kindling" model, which is especially appealing because anti-convulsant drugs that suppress kindling (e.g., carbamazepine and valproic acid) have been shown to have efficacy in bipolar disorder. However, the kindling model cannot explain the entire disorder. Furthermore, recent data from methodologically improved studies reveal that the course of bipolar disorder is highly variable (Winokur et al., 1993). The use of anti-depressants during depressive episodes may induce switching into mania, a series of events associated with a worse prognosis. Finally, discontinuation of maintenance lithium treatment may lead to subsequent refractoriness and a worse prognosis.

Another biological factor that can precipitate episodes of bipolar disorder is medical illness. Abnormalities of the hypothalamic–pituitary–renal axis and the hypothalamic–pituitary–thyroid axis can cause episodes of mood disorder. In particular, thyroid disease is more prevalent in patients with mood disorders, and untreated thyroid disease is a common reversible cause of episodes of bipolar disorder or may worsen its course.

Psychological Factors

Psychological stresses have received less research attention. The many psychological stresses of adolescence are probably important in the onset of the disorder in many patients. Criticism from family members may worsen the course of bipolar disorder. Denial of the illness can reduce compliance with medications and other treatments. In particular, patients experiencing successful prophylaxis of bipolar disorder often wonder why they need to continue taking medications since they are free of symptoms. They are at risk of premature discontinuation of medications, relapse, and a worsened course.

NEUROCHEMICAL STUDIES

There are many limitations to the vast available literature relevant to neurochemical studies of bipolar disorder. Most of these studies were of depressed patients, and few of them separately reported on those patients who were bipolar depressed. Most of these studies examined patients only at a single cross-section in time, thereby looking only at the acute state rather than the progression or course of the disorder. Neurochemical studies of bipolar mania are few, and studies that have reported on the same patients during episodes of depression, mania, and between episodes are even more scarce.

The existing neurochemical studies have examined the following areas: neurotransmitters or their metabolites in blood, CSF, or urine; neurotransmitter receptors; postmortem brain concentrations of neurotransmitters or their metabolites; enzymes that synthesize or metabolize neurotransmitters; neurohormones; neuropeptides; electrolyte metabolism; and membrane transport (of ions, neurotransmitter precursors, and drugs, including lithium).

Early studies of neurotransmitters in bipolar disorder suggested that depression may be related to decreased catecholeamine activity while mania is related to excessive catecholeamine activity. However, recent studies have shown that the relationship between neurotransmitter activity and clinical states in bipolar disorder is far more complex than a simple "more versus less" model. Neurotransmitter studies in bipolar disorder have focused on norepinephrine, dopamine, serotonin, γ-aminobutyric acid (GABA), and acetylcholine, all of which show changes in bipolar disorder.

Norepinephrine (NE) is clearly important in both unipolar and bipolar depression (Schatzberg and Schildkraut, 1995). Synaptic NE levels are markedly increased by many anti-depressants that either inhibit NE reuptake (tricyclic antidepressants) or reduce its metabolism (monoamine oxidase inhibitors). Chronic treatment with such antidepressants leads to a decreased number of NE receptors through downregulation. NE plasma levels and CSF, plasma, and urine levels of its principal metabolite, 3-methoxy-4-hydroxyphenylglycol (MHPG) have been studied in bipolar disorder. CSF MHPG studies have been inconclusive. Plasma NE levels may be higher in melancholic unipolar depressed than in bipolar depressed patients while differences between bipolar depressed patients and controls are conflicting. Urinary MHPG has been shown to be decreased in both bipolar and unipolar depressed patients with bipolar depressed patients showing the lowest levels. Low plasma or urinary MHPG has been associated with a greater likelihood of a manic response to amphetamine or tricyclic anti-depressants. In mania, CSF MHPG has been found to be higher than in controls or in bipolar depressed patients. Two available studies of bipolar manic patients suggest that urinary MHPG is also increased in mania. Platelet α_2-adrenergic receptors have been studied in depressed patients. Even though these studies did not separate unipolar and bipolar depressed patients, they suggest that the number of these receptors may be increased while their sensitivity is decreased in depression.

Dopamine has also been implicated in bipolar disorder (Willner, 1995). Dopamine precursors (e.g., L-dopa) and receptor agonists (e.g., bromocriptine and piribedil) as well as drugs that increase dopamine release (e.g., cocaine and amphetamine) have been shown to precipitate mania and may reduce depressive symptoms. Furthermore, dopamine receptor antagonists (e.g., neuroleptics) are effective treatments for mania. CSF levels of homovanillic acid (HVA), a dopamine metabolite, are probably decreased in both unipolar and bipolar depression suggesting decreased dopamine turnover. CSF HVA levels have been found to be elevated in mania. Plasma HVA studies in bipolar disorder are equivocal. Thus, although far from conclusive, a variety of different approaches lead to the conclusion that activity of both NE and dopamine is decreased in bipolar depression and increased in mania.

Serotonin, a neurotransmitter with important neuromodulatory effects, has been of particular interest in bipolar disorder (Goodwin and Jamison, 1990). Serotonin is believed to reduce a wide range of effects caused by other neurotransmitters. One theory of the pathogenesis of bipolar disorder assumes that serotonin activity is decreased in both depression and mania allowing disinhibited extreme mood states to occur. This theory was advanced following early findings of decreased CSF levels of the principal serotonin metabolite, 5-hydoxyindoleacetic acid (5-HIAA) in both mania and depression, with the decrease persisting beyond the acute episode. However, numerous subsequent studies have not replicated this consistently. CSF 5-HIAA levels in bipolar depression studies have produced conflicting results with some studies reporting no difference from controls or unipolar depressed. Studies of CSF 5-HIAA in mania are similarly equivocal. Studies that compared CSF 5-HIAA between bipolar depressed and manic subjects have found levels to be similar during these two states.

Most antidepressants increase synaptic serotonin levels, although the effects on serotonin receptor concentrations are not as robust as are those that were found for NE receptors. Postmortem studies of suicide victims have found serotonin and 5-HIAA levels to be lower than controls, but these findings may not be specific for unipolar or bipolar depression.

GABA is an inhibitory neurotransmitter that inhibits both dopaminergic and NE neurons. Some drugs used to treat mania, including valproate, carbamazepine, lithium, and propranolol, increase GABA-ergic activity. Furthermore, GABA-mimetic drugs have been suggested in animal studies to have antidepressant effects, which are blocked by GABA antagonists. This suggests that reduced GABA activity might be consistent with both mania and depression. Furthermore, GABA-ergic drugs including lithium and the anti-convulsants valproate and carbamazepine may be effective in prophylaxis of both mania and depression. These effects may be related to the ability of GABA-ergic anti-convulsants to reduce kindling. CSF GABA has been found to be reduced in bipolar depressed patients when compared to controls, while no difference was found between euthymic bipolar patients and controls. CSF GABA in manic patients has not been found to be different from controls. Plasma GABA levels were higher in manic or euthymic bipolar patients when compared to controls. When the patients were depressed, however, plasma GABA levels were no different from controls. Three studies of bipolar patients in remission have had conflicting results but were confounded by the fact that lithium discontinuation causes decreased plasma GABA levels.

Acetylcholine has been associated with bipolar disorder largely through two general lines of thinking. Firstly, acetylcholine activity is known to be in a balance with both NE and dopamine. Mania may be a consequence of increased activity of either NE or dopamine relative to acetylcholine with depression as a consequence of the opposite. Secondly, acetylcholinesterase inhibiting drugs, which increase synaptic acetylcholine levels, have been shown to decrease manic symptoms and induce depressive symptoms. However, these effects are not specific to bipolar patients.

Enzyme studies have largely measured either plasma concentrations of enzymes or concentration of enzymes in platelets or erythrocytes. Platelet monoamine oxidase has been found to be reduced in bipolar depression. Catechol-o-methyltransferase has been studied in erythrocytes, but results for bipolar patients are inconclusive. Dopamine-β-hydroxylase levels may reflect the level of dopamine activity. Several studies suggest that plasma levels of dopamine-β-hydroxylase are decreased in bipolar patients when compared to normals. One study reported that patients with a family history of mood disorders had significantly lower plasma dopamine-β-hydroxylase suggesting that this was a potential genetic marker for bipolar disorder. Acetylcholinesterase in erythrocytes has been reported to be lower in both unipolar and bipolar depressed patients and continued to be lower after remission of the depressed episode; however, this finding has not been replicated. Similarly, GABA-transaminase has also been reported to be decreased in bipolar depressed patients. This finding also persisted beyond the depressive episode suggesting that low levels of these last two enzymes might be a trait of bipolar disorder.

Neuroendocrine studies examining the hypothalamic–pituitary–adrenal axis have found that bipolar depressed patients often have elevated blood cortisol levels and are more likely to have nonsuppression of cortisol following administration of dexamethasone (the dexamethasone suppression test, DST), a finding that is also present in unipolar depression. The results in mania are unclear. Response of adrenocorticotropic hormone after stimulation with corticotropin releasing factor (CRF) has been shown to be decreased in bipolar depressed patients but unchanged in bipolar manic patients. Regarding the hypothalamic–pituitary–thyroid axis, both depressed and manic bipolar patients have been shown to have a reduced elevation of thyroid stimulating hormone following administration of thyrotropin releasing hormone. Melatonin, prolactin, and the growth hormone response to clonidine challenge have been found to be decreased in bipolar depressed patients. In manic patients, melatonin may be increased, whereas the growth hormone response may be decreased. What remains unclear from neuroendocrine studies is whether any of these findings in bipolar patients are pathogenic in bipolar disorder or might instead be a result of stress following rapid and intense mood episodes or psychosis.

Neuropeptide studies have shown that somatostatin is decreased but CRF is elevated in bipolar depression and that these findings occur only during the depressed state. Vasopressin and oxytocin may also have state-dependent changes in bipolar disorder. Electrolyte studies remain inconclusive, although sodium metabolism is of considerable interest in bipolar disorder since lithium, a monovalent cation closely related to sodium, is an effective treatment. Similarly, calcium metabolism has been found to be affected in bipolar disorder, with CSF levels increased in bipolar depression, and calcium channel blockers have been suggested as potential treatments for mania. Membrane transport studies are important but so far have been inconclusive. Another area of considerable interest has been platelet imipramine binding, which may reflect neuronal ability to transport 5-HT, a serotonin precursor. Both unipolar and bipolar depressed patients have been shown to have state-dependent decreased platelet imipramine binding when compared to controls.

A limited number of studies have examined bipolar patients between episodes. However, this type of study is important because of the potential to identify trait markers of bipolar disorder. Cholinergic supersensitivity has consistently been found in euthymic bipolar patients as has an enhanced melatonin response to light. The pathophysiologic or clinical significance of these findings has yet to be determined.

CT/MRI STUDIES

Computed tomography (CT) and magnetic resonance imaging (MRI) measure brain structure. The identification of structural abnormalities that occur consistently in bipolar disorder might pinpoint which structures are important in its etiology or pathogenesis. The most consistently reported CT finding in bipolar disorder is that of an increased ventricle-to-brain ratio, suggesting neuronal cell loss (Pearlson and Schlaepfer, 1995). This finding, however, is nonspecific and is also found in schizophrenia, unipolar depression, dementia, alcohol dependence, and other disorders. Reports of clinical correlates of these findings remain inconclusive. Other reported CT findings include sulcal widening, cerebellar vermian atrophy, asymmetry, and others, but these have not been consistently reported.

MRI, with its increased resolution and ability to image deep structures, has opened new doors to studying brain structure. One of the most provocative MRI findings has been that of subcortical white matter hyperintensities, which have been found in bipolar disorder and persist beyond acute mood episodes. What remains unclear is the significance of these hyperintensities, which are also found in some other groups including the elderly depressed. Are they fundamental markers for bipolar disorder, or are they a result of drug treatments such as by lithium? Could they be a result of repeated episodes of the disorder itself?

FUNCTIONAL STUDIES

Studies of brain function in bipolar disorder include positron emission tomography (PET), regional cerebral blood flow (rCBF), single photon emission computerized tomography (SPECT), electroencephalography (EEG), and quantitative EEG (QEEG).

While methods examining cerebral blood flow (rCBF and SPECT) have been used to study bipolar patients, definitive conclusions from these studies are not yet possible and there are many conflicting results (Goodwin and Jamison, 1990).

There are suggestions that, in bipolar depression, cortical blood flow is decreased while the ratio of nondominant–dominant frontal cortical blood flow may be elevated. Both of these findings are in bipolar depressed patients as compared to manic patients and are probably state-dependent. PET studies that reflect metabolism of glucose or oxygen rather than blood flow have supported these results by showing that bipolar depressed patients have significantly decreased cortical glucose metabolism during the acute episode compared to either normal controls or manic patients. Several studies have followed the same patients longitudinally and found that this decreased cortical glucose metabolism is present only during the acute episode. PET studies have also shown that, during bipolar depression, there is a larger decrease in cortical glucose metabolism in the dominant hemisphere overall; this decrease is most pronounced in the dominant frontal lobe—a fact that is also consistent with rCBF findings.

Although there are many reports of EEG in bipolar disorder, these studies have yielded few insights regarding its pathogenesis. The majority of bipolar patients have a clinically normal EEG. The presence of EEG abnormalities has been associated with a negative family history of mood disorders as well as with an increased likelihood to have nonsuppression on the DST. EEG studies that examined laterality suggest that there may be greater EEG abnormalities in the nondominant hemisphere than in the dominant hemisphere during episodes of bipolar depression. Expanding on this, QEEG studies have sought to differentiate bipolar depressed patients from unipolar depressed patients and normal controls. Using multivariate methods to identify differences between diagnostic groups on a large number of parameters of the QEEG, discriminate functions have been used to successfully differentiate 80% to 90% of clinically diagnosed bipolar depressed patients from unipolar depressed patients and normal controls.

SOMATIC TREATMENTS

Somatic treatments for mania include drugs and electroconvulsive therapy (Chou, 1991). Drugs used for bipolar disorder are usually divided into several categories: mood stabilizers including lithium, valproate, and carbamazepine; anti-depressants; and tranquilizers, which include antipsychotics and benzodiazepines. The term "anti-manic" is used for both mood stabilizers and tranquilizers.

Treatment of Mania

The most important step in treating acute mania is identifying it. Mania may be obscured by symptoms of psychosis, substance dependence, and personality disorder. Once identified, the primary treatment of acute mania is the use of psychotropic medications and the placing of the patient, if necessary, in a controlled environment. Manic patients characteristically have poor judgment and, in a euphoric or irritable state, may be dangerous and require hospitalization. The primary medications used to treat mania are lithium, antipsychotics, anti-convulsants, and benzodiazepines. The choice of drug treatment during the acute episode depends on the target symptoms: lithium or anti-convulsants are typically used for elevated mood whereas antipsychotics or benzodiazepines are for treating psychosis or agitation. However, the acute drug treatment should logically lead to a regimen for continuation and maintenance treatment.

Although some studies have found anti-convulsants and antipsychotics to be as effective as lithium during the acute episode, lithium remains the drug of choice because of its well documented efficacy in maintenance. Lithium is usually prescribed in divided doses with blood level monitoring every four to seven days during the acute episode. The dose should be increased as tolerated to achieve a steady-state level of 1.0 to 1.5 mEq/L. However, many patients develop side effects, and up to one-third of patients are refractory. Furthermore, the onset of action of anti-manic effects of lithium may be slow, requiring up to seven to 10 days after achieving a therapeutic blood level. Often, lithium is combined with antipsychotics or benzodiazepines, which provide rapid behavioral control. However, antipsychotics should be avoided in longterm treatment because of the potential for the sometimes irreversible side effects of tardive dyskinesia to which mood disorder patients appear to be particularly prone. Benzodiazepines also should be minimized in longterm treatment because of possible dependence.

The anti-convulsants, valproate and carbamazepine, have also been shown to be effective in mania, and valproate has been approved by the U.S. Food and Drug Administration as a treatment for acute mania. Furthermore, valproate produces less side effects than lithium. Many psychiatrists are using these drugs for patients who are either refractory or only partially responsive to lithium or who cannot tolerate its side effects. Although research data supporting longterm efficacy of the anti-convulsants in maintenance of bipolar disorder is sparse, many psychiatrists consider these as alternatives to lithium and preferable to antipsychotics.

Atypical anti-psychotics including risperidone and olanzapine are currently being studied for treatment of acute mania. Since these drugs have a decreased risk of tardive dyskinesia in comparison to typical antipsychotics, they may be quite useful for bipolar patients.

For manic patients who do not respond to any of the above treatments, electroconvulsive therapy, typically over a course of up to 10 bilateral treatments is usually effective in truncating the manic episode (Mukherjee et al., 1994). Other drugs that have been used to treat mania include other anti-convulsants such as lamotrigine, calcium channel blockers, clozapine, beta blockers, clonidine, and thyroid supplementation.

Treatment of Mixed Episode

The presence of symptoms of both depression and mania in a single episode is associated with a poor response to lithium and has a worse prognosis than acute mania without mixed features. Several studies have shown that valproate is superior to lithium in treatment of a mixed episode, although both lithium and carbamazepine should also be considered. In addition to a mood stabilizer, patients with mixed episodes may require tranquilizers for control of acute agitation.

Treatment of Bipolar Depression

Both lithium and anti-depressants have been shown to be effective in treating bipolar depression (Zornberg and Pope, 1993). However, in treating depression, it is important to avoid inducing mania through a phenomenon known as switching. Every known anti-depressant has been reported to induce switching. Patients who experience switching from a depressive episode to a manic episode have a worsened prognosis and are more likely to become rapid-cycling. However, it is difficult to withhold treatment from acutely depressed patients. For such depressive episodes, lithium is the treatment of choice since it has been shown to have efficacy for bipolar depression as well as in maintenance treatment. However, many bipolar depressed patients will require additional treatment. If other anti-depressants such as tricyclic anti-depressants or selective serotonin reuptake inhibitors are used, they should be combined with an anti-manic treatment such as lithium, valproate, or carbamazepine in order to provide prophylaxis against any potentially emerging mania. Several studies suggest that selective serotonin reuptake inhibitors are less likely than tricyclic anti-depressants to induce mania. There is also some interest in lamotrigine as a promising treatment for bipolar depression.

Maintenance Treatment

Lithium is also the best studied treatment for maintenance and has clearly been shown to be superior to placebo (Prien and Koscis, 1995). Furthermore, the blood level of lithium during maintenance treatment has been shown to be related to outcome with higher blood levels associated with decreased rates of relapse (Gelenberg et al., 1989). However, since side effects are related to blood level, it is important to minimize them—especially in asymptomatic patients during maintenance treatment who are likely to question the need for the drug.

Less is known about carbamazepine and valproate in maintenance, although these drugs are also often used (Prien and Koscis, 1995). One recent maintenance trial indicates that patients who respond to valproate in an acute manic episode do well on subsequent maintenance treatment with valproate. Because of the risk of tardive dyskinesia, antipsychotics are not recommended routinely for maintenance treatment of bipolar disorder. However, their use is common because they are often initiated during an acute episode and simply continued into maintenance because of concern that stopping the neuroleptic will lead to a relapse.

Psychotherapy is important during maintenance treatment, particularly to enhance compliance with maintenance drug treatments. In addition, interpersonal, cognitive, family, and group therapies can assist in coping with the many sequelae of bipolar disorder. Psychoeducation is also needed, and both patients and their families must learn about the importance of complying with longterm treatment as well as identifying any signs of relapse as soon as possible. In an individual patient, characteristic signs (e.g., making sudden travel plans or not sleeping) may herald the onset of a manic episode. The patient and the family must know the importance of getting into treatment immediately should such signs be noticed.

Another clinical issue that always arises in maintenance treatment is that of the duration of treatment. Patients who are on effective maintenance treatment are, by definition, not having active symptoms and may wonder why they have to continue taking medication. Furthermore, the maintenance medication may have been started in the hospital during an acute episode and be associated with a variety of unpleasant experiences, including the loss of manic euphoria and possibly the onset of depression.

The decision about when to discontinue maintenance treatment must be made on an individual basis, taking many factors into consideration. How likely is the patient to have another episode? If so, how severe might it be? Is this the best time in the patient's life when a relapse would be least disruptive? Are there side effects? Would early signs of relapse be easily identified by the patient or family?

Another issue further complicating the decision about discontinuing maintenance treatment is recent data suggesting that stable patients successfully maintained on lithium if discontinued from taking lithium, may become lithium refractory and suffer a worse course and prognosis than they presumably would have had if they had stayed on lithium maintenance (Suppes et al., 1991). These provocative data suggest that longterm and perhaps lifetime lithium maintenance may be appropriate for bipolar 1 patients even after having only one manic episode.

SUMMARY

Research on bipolar disorder was considerably advanced by the recognition of this disorder as clearly different from both unipolar depression and schizophrenia. This recognition has only clearly solidified within the last several decades. Nonetheless, there is a wealth of both clinical and basic studies of this devastating disorder. While the definitive etiology and pathogenesis remain unknown, there are promising clues coming from a widely diverse range of sources. These include the following: genetic studies showing clear genetic transmission in some families; neurochemical studies implicating norepinephrine, serotonin, and dopamine; functional studies indicating state-dependent changes in laterality, and structural MRI studies showing provocative white matter hyperintensities.

Bipolar disorder is often considered one of the more treatable of the chronic and persistent mental disorders, yet one-third of bipolar patients may be considered refractory. Furthermore, the course of the disorder is unpredictable with many patients suffering a deteriorating downhill course. This is further complicated by a high co-morbidity with substance dependence and high rates of noncompliance for stable asymptomatic patients during effective maintenance treatment.

Treatment of bipolar disorder is complex, and virtually all available psychotropic drugs have been suggested for at least some phase of the disorder. Treatment of acute episodes often requires combining two or more drugs. In addition, drugs that are effective in acute episodes may be contraindicated in maintenance, and the issue of how long successful maintenance should be continued remains unclear.

Bipolar disorder continues to present significant challenges for both clinical and basic science research. Increased understanding of the disorder will come only through connecting basic science advances to clinical correlates in bipolar patients.

REFERENCES

Chou, J.-C.Y. (1991). Recent advances in treatment of acute mania. J. Clin. Psychopharmacology 11, 3-21.

Gelenberg, A.J., Kane, J.M., Keller, M.B., Lavori, P., Rosenbaum, J.F., Cole, K., & Lavelle, J. (1989). Comparison of standard and low serum levels of lithium for maintenance treatment of bipolar disorder. N. Engl. J. Med. 321, 1489-1493.

Gershon, E.S., Berrettini, W., Nurnberger, J., & Goldin, L.R. (1987). Genetics of affective illness. In: *Psychopharmacology: The Third Generation of Progress.* (Meltzer, H.Y., Ed.), pp. 481-491. Raven Press, New York.

Goodwin, F.K. & Jamison, K.R. (1990). *Manic–depressive Illness.* Oxford University Press, New York.

Licht, R.W. (1998). Drug treatment of mania: A critical review. Acta Psych. Scand. 97, 387-397.

Mukherjee, S., Sackeim, H.A., & Schnur, D.B. (1994). Electroconvulsive therapy of acute manic episodes: A review of 50 years experience. Am. J. Psych. 151, 169-176.

Pearlson, G.D. & Schlaepfer, T.E. (1995). Brain imaging in mood disorders. In: *Psychopharmacology: The Fourth Generation of Progress.* (Bloom, F.E. & Kupfer, D.J., Eds.), pp 1019-1028. Raven Press, New York.

Prien, R.F. & Koscis, J.H. (1995). Long-term treatment of mood disorders. In: *Psychopharmacology: The Fourth Generation of Progress.* (Bloom, F.E. & Kupfer, D.J., Eds.), pp 1067-1079. Raven Press, New York.

Schatzberg, A.F. & Schildkraut, J.J. (1995). Recent studies on norepinephrine systems in mood disorders. In: *Psychopharmacology: The Fourth Generation of Progress.* (Bloom, F.E. & Kupfer, D.J., Eds.), pp. 911-920. Raven Press, New York.

Suppes, T., Baldessarini, R.J., Faedda, G.L., & Tohen, M. (1991). Risk of recurrence following discontinuation of lithium treatment in bipolar disorder. Arch. Gen. Psych. 48, 1082-1088.

Willner, P. (1995). Dopaminergic mechanisms in depression and mania. In: *Psychopharmacology: The Fourth Generation of Progress.* (Bloom, F.E. & Kupfer, D.J., Eds.), pp. 921-931. Raven Press, New York.

Winokur, G., Coryell, W., Keller, M., Endicott, J., & Akiskal, H. (1993). A prospective follow-up of patients with bipolar and primary unipolar affective disorder. Archives of General Psychiatry 50, 457-465.

Zornberg, G.L. & Pope, H.G., Jr. (1993). Treatment of depression in bipolar disorder: New directions for research. J. Clin. Psychopharmacology 13, 397-408.

RECOMMENDED READINGS

Berrettini, W. (1998). Progress and pitfalls; bipolar molecular linkage studies. J. Affect. Disorders 50, 287-297.

Calabrese, J.R., & Rapport, D.J. (1999). Mood stabilizers and the evolution of maintenance study designs in bipolar I disorders. J. Clin. Psychiatry 60, (Suppl. 5), 5-13.

Chen, S.T., Altshuler, L.L. & Spar, J.E. (1998). Bipolar disorder in late life: A review. J. Geriatric. Psych. Neurol. 11, 29-35.

Ghaemi, S.N., Boiman, E.E. & Goodwin, F.K. (1999). Kindling and second messengers: An approach to the neurobiology of recurrence in bipolar disorder. Biological Psychiatry 45, 137-144.

Licht, R.W. (1998). Drug treatment of mania: A critical review. Acta Psych. Scand. 97, 387-397.

Viguera, A.C. & Cohen, L.S. (1998). The course and management of bipolar disorder during pregnancy. Psychopharm. Bulletin 34, 339-346.

Chapter 12

The Biological Basis of Schizophrenia

PHILIP WINN

Biological Psychiatry, pages 233-261.

ISBN: 1-55938-819-6

INTRODUCTION

The state now recognized as schizophrenia was first classified by Emil Kraepelin (1896). A German psychiatrist, Kraepelin (1856-1926) predicted a significant strand in twentieth century thought as he emphasized the likely physiological origins of mental illness and the importance of accurate and detailed measurements of psychological state. Kraepelin described *dementia praecox*, an early onset progressive intellectual deterioration ("*dementia*"=progressive intellectual deterioration; "*praecox*"=early onset). In this were included several definable states: paranoia (a delusional state), hebephrenia (a disorder of mood, thought, and affect), and catatonia (a psychomotor disturbance). These states had been separately described previously by others, but Kraepelin drew them together for the first time, believing that they had a common core.

The term "schizophrenia" itself was introduced by Eugen Bleuler (1857-1939), a Swiss psychiatrist. In *Dementia Praecox, or the Group of Schizophrenias*, Bleuler (1911) argued that this disorder was one in which a split had occurred between emotions and reason—that there had been a loss of "associative threads" leading to a breakdown in behavior and thought. Unlike Kraepelin, Bleuler believed that the disorder was neither progressive nor of early onset and so the term "dementia praecox" was inappropriate. Schizophrenia was coined from the Greek *schizein* (to split) and *phren* (mind) and served Bleuler's purpose in emphasizing the loss of mental associations within the mind (rather than implying the existence of two minds).

As well as placing different emphasis on the progress, onset and unitary nature of the disorder, Kraepelin and Bleuler differed in one other fundamental respect. Kraepelin separated dementia praecox from manic–depressive illness (which he regarded as the other major psychosis) and retained a narrow description of schizophrenia, albeit one that was extensively subdivided eventually into 36 major categories. In contrast, Bleuler adopted a much wider definition of schizophrenia, including manias and depressive illnesses, which Kraepelin had specifically excluded.

Since these early speculations, there have been many attempts to classify and account for schizophrenia, but the tone of the debate and its broadest issues were set by Kraepelin and Bleuler. How are the various features of schizophrenia best classified: do they have a common core etiology or are there distinct pathologies associated with what are in fact quite distinct characteristics? Is schizophrenia best classified as a *biological* disorder fundamentally linked to an abnormality in the way in which a brain is built or operates? Or is it better described as a *psychological* dysfunction, perhaps reflected in abnormal brain function, but better described as essentially a disorder of mental life? In this chapter, I will examine both neurological and psychological aspects of schizophrenia in order to comprehend better the nature of the condition.

PROBLEMS WITH THE CLASSIFICATION OF SCHIZOPHRENIA

The classification of mental illness is held to be important in that it allows one to identify correctly disorders that can be specifically treated and for which there might be some predictable outcome. It is also important when considering etiology. The relationship between a disorder and its cause is reciprocal. Kraepelin classified the disease into a number of different types: hebephrenia (a disorganization syndrome involving hallucinations and delusions), catatonia (a rare form involving immobility interspersed with periods of excitement), and paranoia (a delusional form). Problems became apparent with this classification since patients often did not fit neatly into one category or the other but showed features of different types. Several other authors have attempted to classify the signs and symptoms of schizophrenia. Here I will discuss only two recent approaches.

Possibly the most productive attempt in recent years was made by Tim Crow who classified schizophrenia into two syndromes: positive (Type I) and negative (Type II) (Table 1) (Crow 1980a,b; Andreasen, 1982). The positive signs and symptoms included such things as hallucinations, delusions, and thought disorders, while the negative syndrome included such features as poverty of speech, flattening of affect, and social withdrawal. The syndromes are further contrasted: the positive syndrome Crow thought of as acute whereas the negative was chronic; intellectual impairment was present in the negative, but not the positive, syndrome; and the positive signs and symptoms were amenable to treatment by antipsychotic drugs whereas the negative were not. More interestingly, Crow also suggested that different pathologies were present in each syndrome. He associated the positive syndrome with increased numbers of dopamine receptors. In essence, the brain was intact but neurochemically abnormal. In contrast, the negative syndrome was associated with structural abnormalities in the brain; patients with negative syndrome had enlargement of the cerebral ventricles.

Over the years, the distinction between positive and negative signs and syndromes has been accepted, but whether or not two distinct disease states can be identified is open to question. For instance, neither positive nor negative syndrome adequately embraces stereotyped thoughts and actions. First of all, stereotypies—repetitive, purposeless, invariant thoughts and actions—have been described in patients for several centuries and are unambiguously present in many schizophrenic patients (Frith and Done, 1983; 1990). Secondly, ventricular enlargement is not unique to negative syndrome schizophrenia. Thirdly, ratings of positive and negative symptoms may be obtained in the same patient. Thus, while the positive–negative dichotomy is valuable in organizing the signs and symptoms, it does not appear to describe two separate syndromes in patients.

More recent approaches to the classification of schizophrenic signs and symptoms have used factor analysis to identify clusters of symptoms rated using both

Table 1. The Principal Signs and Symptoms of Schizophrenia

POSITIVE SIGNS AND SYMPTOMS	*(Abnormal by their presence)*
Disorganized Speech and Thought Disorder	
• Incoherence	
• Loose associations	
• Derailment	
• Tangentiality	
• Distractibility	
Delusions	
• Thought insertion	thoughts placed in one's mind by an external agent
• Thought withdrawal	thoughts being stolen
• Made feelings	feelings projected into one from outside
• Made motivational acts	other agents controlling one's actions
• Made impulses	impulses to act caused by others
• Somatic passivity	unwilling recipient of sensations imposed by an external agent
• Persecution	
Hallucinations	
• Audible thoughts	the belief that the thoughts of others are audible
• Voices arguing	hearing voices
• Voices commenting	a continuous commentary on one's actions
• Mechanization	the feeling of being robotic
• Derealization of the world	
• Changes in bodily sensations	
• Size of body parts apparently changed	
• Numbness, burning, or tickling sensations	
• The sensation of having snakes inside one	
• Hypersensitivity to stimuli	
• Flat and colorless surroundings	
• Concentration deficits	
NEGATIVE SIGNS AND SYMPTOMS	*(Abnormal by their absence)*
• Avolition	lack of energy and interest
• Anhedonia	inability to experience pleasure
• Alogia	poverty of speech (amount); poverty of the content of speech (content)
• Flat affect	stimuli fail to elicit emotional responses (though inner, non-expressed, emotions may be unimpaired)
• Lack of vocal inflection	
• Increased speech latency	
• Poor eye contact	
OTHER SIGNS AND SYMPTOMS	*(Difficult to classify as positive or negative)*
• Catatonia	immobility, waxy flexibility
• Inappropriate affect	emotional responses inappropriate to the context
• Stereotypy	thought and/or action becomes repetitious, invariant and inappropriate; perseveration in thought and/or action

the Present State Examination (PSE; Wing et al., 1974) and the Comprehensive Assessment of Symptoms and History (CASH; Andreasen, 1985). Liddle (Liddle, 1987; Liddle and Barnes, 1990) has described three separate clusters of symptoms: psychomotor poverty syndromes, a disorganization syndrome, and a reality distortion syndrome. While such an approach might have some predictive value, there are limitations. Factor analytic studies can systematize observations (and clearly do, for Liddle's work has been reinforced by that of others; Arndt et al., 1991), but ultimately they are still only correlative: certain signs and symptoms tend to associate together, but the relationship between them is not described by the correlation and many patients will show signs and symptoms, simultaneously or sequentially, from more than one cluster. Until one can clearly describe an underlying mechanism to account for all the observations within a cluster, one is doing no more than systematize observations and, valuable though that is, it does not address the question of what causes schizophrenia.

The problem of classification has plagued studies of schizophrenia consistently to the point at which some authors are now suggesting that it may not be of great value to classify it at all (Bentall, 1990). For the purposes of treatment, what one wishes to do is to alleviate the signs and symptoms. Similarly, for the purpose of explanation, what one wishes to do is to explain why a particular sign or symptom is present. If the same mechanism can account for more than one sign or syndrome, that is all well and good. If it requires two mechanisms to explain, say, hallucinations and poverty of speech, then two mechanisms are what will have to be described.

THE ETIOLOGY OF SCHIZOPHRENIA

A Genetic Basis for Schizophrenia?

Studies of the incidence of schizophrenia within families have suggested that schizophrenia might have a heritable component. The closer the biological relationship between family members, the higher the incidence of schizophrenia becomes (see reviews in Gottesman et al., 1987; Plomin et al., 1997). Thus the spouses of diagnosed schizophrenics show approximately a 1% incidence of schizophrenia (equivalent to the general incidence of 0.85% in the general population) while over 40% of the monozygotic twins of diagnosed schizophrenics also develop schizophrenia. For dizygotic twins, the incidence falls to about 12% while less than 8% of the siblings of schizophrenics develop the disease. The fact that there is such an increased incidence of schizophrenia the closer a biological relative is to a patient suggests that there is a genetic component to the disease, but the fact remains that it is clearly not inherited by some straightforward mechanism. Huntington's disease, a disorder transmitted as an autosomal dominant trait showing complete penetrance, is the clearest example of a neuropsychiatric disorder the etiology of which can be described wholly in terms of a genetic mechanism. Schizophrenia, while possibly having a genetic component, cannot be described in the same way. This has not deterred molecular biologists from seeking a genetic marker for schizophrenia but, despite considerable effort and enthusiasm, no unambiguous genetic factor has as yet been isolated (see Peltonen, 1995). The best case that can be made for a genetic component is probably in the hypothesis that schizophrenia is a neurodevelopmental disorder in which there is abnormal genetic control of processes such as neuronal migration and synaptic pruning (see Bloom, 1993 for a brief review). Candidate gene analysis may in time identify genes that are important (Ross and Pearlson, 1996).

Social and Environmental Factors in the Etiology of Schizophrenia

There is a relationship between social class and the incidence of schizophrenia, though it is not a linear relationship: the incidence of schizophrenia is twice as high in the very lowest social classes as in the next highest (Hollingshead and Redlich, 1958; Kohn, 1968). This has led to controversy: does this indicate a sociogenic factor at work (is schizophrenia *caused* in some sense by low social class?) or does it show social selection operating (i.e., a tendency for schizophrenics to fall into the lowest social classes)? Attempts to examine social mobility in schizophrenia have been inconclusive: some have shown a downward social mobility consistent with a social selection hypothesis, others have not (see Kohn, 1968).

At a more microscopic level, the importance of the family in the etiology of schizophrenia has been emphasized by several authors. The concept of a schizophrenogenic mother has been proposed, most notably in the "double bind" hypothesis in which contradictory messages are presented causing psychological trauma (Bateson et al., 1956). Birth complications have been associated with an increased incidence of schizophrenia, but there does not appear to be a particular association between these and schizophrenia (Done et al., 1991). Other psychiatric disorders are also associated with obstetric complications. Indeed, there are a variety of maternal epidemiological factors associated with psychiatric disturbances in their children (Asarnow, 1988; Goodman, 1987). The idea has also been presented that the level of expressed emotion within a family can have a predisposing effect (Vaughn and Leff, 1976). There is no conclusive evidence that the behavior of parents can, in and of itself, cause schizophrenia, though there is evidence that schizophrenic out-patients relapse less frequently in home environments that are stable, caring, and structured appropriately (Brown et al., 1966).

The attempt to describe the antecedents of schizophrenia is an important exercise. If one was able to describe particular psychological or behavioral traits in children or adolescents, there would be a possibility that appropriate therapy could prevent or at least ameliorate the signs and symptoms of schizophrenia. The behavior of children preschizophrenia has been studied closely, and it has been suggested that premorbid personality abnormalities predict a poor outcome in schizophrenia compared to schizophrenics whose adolescent, premorbid personality ratings were normal (Parnas et al., 1982; Remschmidt et al., 1994; Werry et al., 1994). It is important to note however that the personality traits per se do not appear to be causal. They might predict *outcome*, but they do not predict *incidence*. Much attention has been given to indices of information processing capacities, motor functions, and psychophysiological processes as well as personality traits. So, for instance, in a recent comprehensive study of children at risk for psychiatric disturbances, the offspring of schizophrenics (but not depressive probands) were found to have intrusive anticipatory saccades, a possible indicator of attentional disturbances (Rosenberg et al., 1995). Motor development has also been found to be impaired in schizophrenics. Developmental milestones such as sitting, standing, or walking unassisted have been shown to be delayed compared to controls (Jones et al., 1994).

The issue of familial causes of schizophrenia has been controversial. There is no conclusive evidence that families or parents cause schizophrenia and, while many psychiatrists continue to probe family background in an effort to describe the antecedents of schizophrenia, others have argued that it is cruel to suggest to parents of schizophrenics, already experiencing considerable distress at the condition their children are in, that their behavior in some way is, or has been, to blame. More critical is the lingering question about developmental histories: schizophrenia, as recognized by DSM-IVR (American Psychiatric Association, 1987), typically develops in late adolescence and early adulthood. Why, if there are causal factors operating in childhood, is schizophrenia not manifest until early adulthood?

Brains and Schizophrenia

It is a central assumption of the majority of neuroscientists that brain functions can, at some level, explain psychological functions and that no psychological event takes place without there being some causal brain operation underlying it. Whatever the cause of schizophrenia—neuronal development, genetics, familial disturbance, or even viral infection (Mednick et al., 1988)—there is a general belief that psychiatric illnesses such as schizophrenia are caused by the malfunctioning of brains. To account for some or all of the signs and symptoms of schizophrenia, a number of different classes of brain disorder have been proposed: (1) those involving structural changes in the brain, (2) those accompanied by neurochemical changes in the brain, and (3) those associated with abnormal development of the brain. In the following sections, instances of all of these will be reviewed.

THE NEUROBIOLOGY OF SCHIZOPHRENIA

Structural Changes in the Brain

Ventricular Enlargement

Early studies using CAT scans and MRI analyses provided evidence of enlargement of the cerebral ventricles in schizophrenia measured in terms of the cross-sectional area of the lateral ventricles (Johnstone et al. 1976; Weinberger et al., 1979a). This discovery had tremendous impact, suggesting as it did that there were fundamental changes in the structure of schizophrenic brains. Moreover, ventricular enlargement was identified as being associated principally with the negative symptoms of schizophrenia, suggesting that the dissociation of the disorder into types I (positive symptoms) and II (negative symptoms) could be underpinned by different brain abnormalities (Johnstone et al., 1976; 1978). The finding of increased ventricular size has been replicated in many independent laboratories as has the association with negative symptoms and lack of responsiveness to pharmacotherapy (Weinberger et al., 1980; Andreasen et al., 1982).

There are, however, many difficulties associated with these findings. The increase in ventricular size is essentially a statistical finding, not an absolute marker for schizophrenia (even of negative symptoms). While the average size of the lateral ventricles is increased in schizophrenics compared to controls, the majority of patients have no enlargement but, in fact, fall within the normal range. Only some 25% of patients are thought actually to have larger ventricles than normal (Frith, 1992). Ventricular enlargement appears in other disorders in which there is organic brain damage, when ventricles expand in order to maintain brain volume as tissue is lost. It is not a process unique to schizophrenia. Moreover, although the enlargement tends to be in that portion of the lateral ventricles most closely associated with the temporal lobe, there is no compelling or consistent evidence of tissue atrophy of gliosis in schizophrenic brains, indicating that there is no gross loss of tissue that would account for ventricular enlargement. In addition, when repeated scans of schizophrenic brains have been taken, no evidence is found of progressive enlargement as time goes by (Gattaz et al., 1991). Even more problematic is the finding that, on those occasions when brain scans have been made before symptoms of schizophrenia are manifest, there is evidence of ventricular enlargement preceding the disease (O'Callaghan et al., 1988; Weinberger, 1988). The only piece of evidence that seems to support the association between ventricular enlargement and schizophrenia is a study of nonconcordant identical twins in which the schizophrenic twin was invariably found to have larger ventricles than the nonschizophrenic twin (Revelry et al., 1982; Suddath et al., 1989).

Taken together, the case for ventricular enlargement being a critical marker of brain damage in schizophrenia is poor: it is not seen in all patients (even in subsets of patients identified by their signs and symptoms); it may precede the appearance of symptoms but is not progressive; and it is not associated consistently with other markers of injury such as gliosis.

Neuronal Organization

It is apparent that there is no gross abnormality in the schizophrenic brain. Frith (1992) notes that he has known "neuropathologists to remark facetiously that it is easy to recognize the brains from schizophrenic patients because they are the ones which look normal". Since there is no obvious gross abnormality, research has been directed to determining if there are any changes in the microstructure of the brain in schizophrenic patients. Some studies have examined particular types of neurochemically identified neurons. For instance, the distribution of neurons capable of synthesizing the novel neuromodulator nitric oxide in the frontal and temporal lobes has been investigated (Akbarian et al., 1993a,b). Changes in the distribution of these in the schizophrenic brain were identified, though the results as yet have not been replicated.

Other studies have examined the numbers of neurons and glial cells in specific structures of the CNS. Typically, it is cortical tissue (especially frontal and temporal lobes) and the interconnected limbic and striatal structures that have been examined: the nucleus accumbens, hippocampus, amygdala, and mediodorsal thalamus. There is a growing amount of evidence that there are changes in the microanatomy of at least some of these structures, though a certain degree of caution still needs to be maintained in interpreting them. Many factors have to be considered in examining these studies: the ages of the patients and their medication histories in comparison to the control groups and the exact technique used for counting cells are the two most obvious problems. Recent advances in the techniques used to count cells in the CNS have made significant improvements on what has gone before. In the past, cells have been counted in two dimensions, section by section. The total number of neurons present in a given volume of tissue can then be estimated, though correction factors have to be used. This is a notoriously difficult procedure, error being introduced into the process by such things as the variability of neuron sizes within a structure and uncertainty about the exact thickness of sections. More modern stereological techniques overcome these problems using optical dissection, effectively taking morphometric measurements of single neurons *in situ* in three dimensions.

Studies over several years have indicated changes in the structure of the cerebral cortex (Weinberger et al., 1979b) and particularly of the temporal lobe and the structures associated closely with it: the hippocampus, parahippocampal gyrus, and the entorhinal cortex (Bogerts et al., 1985; Brown et al., 1986; Falkai and Bogerts, 1986; Jeste and Lohr, 1989; Heckers et al., 1990; Altschuler et al., 1990). The prefrontal cortex is also thought to show structural changes (Selemon et al., 1995), as are the nucleus accumbens (Pakkenberg, 1990) and the mediodorsal thalamic nucleus (Pakkenberg, 1990), two subcortical structures intimately associated with prefrontal function. Further down the neuraxis, the substantia nigra (Bogerts et al., 1983), pedunculopontine tegmental nucleus (Karson et al., 1991), and locus coeruleus (Lohr & Jeste, 1988) have also been shown to be different in schizophrenic brains compared to controls. Taken together, these studies appear conclusive, at least to the degree that there is something structurally different about the schizophrenic brain compared to normal. But it is worth sounding a note of caution about these studies.

A problem that recurs in these studies concerns the variability of the control data. In two studies, examination was made of cortical tissue (Selemon et al., 1995) and of subcortical tissue (Pakkenberg, 1990). Both studies used sophisticated stereological techniques to assess neuronal number and the volumes of structures. Both report statistically significant findings: there was an increase in the numbers of pyramidal and non-pyramidal neurons in the prefrontal cortex (area 9) with some variability in different layers (layers III–VI had the greatest increase) as well as an increase in neuronal density in area 17 (in the occipital cortex). In neither area was there an increase in glial cell number. Subcortically, reductions in neuronal number were found in the schizophrenic mediodorsal thalamus and nucleus accumbens but not in the ventral pallidum or basolateral nucleus of the amygdala. There are problems, however. Subcortically, although there are clear differences in the numbers of neurons in the mediodorsal thalamus and nucleus accumbens, there are also differences in the volumes of these structures. Where there is no change in the gross volume of a structure (e.g., in the amygdala or ventral pallidum) there is no change in the numbers of neurons either. The calculation of neuronal density shows statistical significance but only marginally (by the authors admission). Indeed there is considerable overlap in the variance in neuronal density between the two groups (4127 ± 1403 in the schizophrenic group compared to 5311 ± 797 cells/mm^3 in the controls [means±SD]). While this is statistically significant, it does, however, mean that there are many schizophrenics in the sample who have neuronal densities well within the normal range. Therefore, as is the case for ventricular enlargement, it is not possible to demonstrate that there is a pathology uniquely and unambiguously associated with schizophrenia or with a subset of it: the differences between schizophrenics and controls are essentially statistical in nature and thus may be of limited diagnostic or explanatory value. A similar problem occurs in the cortex where the ranges of neuronal densities again mean that many of the schizophrenic patients must have neuronal densities within the normal range. Selemon and her colleagues are quite explicit about the problem: "the wide range of neuronal densities of the normal brains is more difficult to explain...The large degree of variability is comparable to that observed for ventricular width in the normal human population." The range of neuronal densities in the mediodorsal thalamus and nucleus accumbens is similarly very large, with the largest instances of neuronal density and of cell numbers being more than twice as great as the smallest ones.

Both of these studies used the very best techniques available and have produced data that are unimpeachable. My point in highlighting them is not to draw attention to any inadequacy, but to use them as illustrations of fundamental points that must be borne in mind when comparing schizophrenic and control brains. The variability within the "normal" population is very much larger than might have been expected, and the differences between the schizophrenic and control brains are statistical rather than absolute. That is to say, while on average there may be differences in the numbers of neurons in particular structures in schizophrenic brains compared to those of normals, many of the schizophrenics will have values within the normal range. This is, in very many ways, deeply disappointing. The hope of successive generations of neuropathologists has been to define some parameter that distinguishes schizophrenic brains from normals in the way that, for instance, loss of neurons from the substantia nigra pars compacta distinguishes Parkinsonian brains. At present, such a structural marker has not been discovered. One might even begin to suggest that such a marker is not in fact there to be discovered.

Cerebral Asymmetries

In addition to suggestions that the structure of the brain is altered, there have also been many suggestions that the normal asymmetric organization of the brain is altered in schizophrenia (Crow et al., 1989). Much attention has focused on the temporal lobe. In particular, recent studies have paid attention to the planum temporale, a structure normally asymmetric in the human brain that sits in the superior portion of the temporal lobe. Psychologically, this area is of interest because it appears to be involved in the production and comprehension of language. Damage or dysfunction in the planum temporale might therefore be expected to be associated with features of schizophrenia related to the production of language, such as thought disorder. A recent high-resolution magnetic resonance imaging study (MRI) (Petty et al., 1995) showed a reversal of the normal asymmetry (left larger than right) in the planum temporale of schizophrenic patients compared to healthy control subjects matched for age, sex, handedness, race, and parental socioeconomic status. That this was not a generalized abnormality was suggested by there having been no change in the symmetry of Heschl's gyrus (in which there is normally no asymmetry), an area of primary sensory cortex contiguous with the temporal lobe. However, two other MRI studies examining asymmetry of the planum temporale (Kulynych et al., 1995; Frangou et al., 1997) showed that schizophrenic patients had normally asymmetric planum temporale. The second of these is especially interesting. Abnormalities in the asymmetries have been thought to give evidence of a genetically based neurodevelopmental disorder, but Frangou and colleagues' study (1997), which failed to find any evidence of volumetric or asymmetric abnormality in the planum temporale, involved comparison of schizophrenic patients from families frequently affected by schizophrenia with their first-degree relatives as well as matched nonrelated controls. These studies demonstrate again the difficulty in obtaining a consistent pattern of deficit in the schizophrenic brain.

The Dopamine Hypothesis of Schizophrenia

Over the last 25 years, several different neurochemical systems have been suggested to be involved in the etiology of schizophrenia. Norepinephrine, serotonin, acetylcholine, endorphins, GABA, and glutamate have been investigated (see Feldman et al., 1997, for review) but one system has dominated the literature above all others: dopamine. The association between dopamine and schizophrenia is supported by two fundamental considerations. Firstly, amphetamine abuse can lead to the development of schizophrenic symptoms—an outcome brought about by the action of amphetamine on dopamine systems (O'Connell, 1958). Moreover, amphetamine given to patients worsens positive symptoms. Secondly, the drugs used to treat schizophrenia block dopamine receptors, and the degree and specificity of their actions is positively correlated with therapeutic success. At its simplest, the dopamine hypothesis of schizophrenia proposed that the signs and symptoms of the disease were the product of dopamine overactivity. Such a hypothesis is no longer tenable. No consistent increase in dopamine levels or turnover has been found and, although there is some evidence for dopamine receptors being supersensitive in schizophrenics, the evidence for this is contradictory and the contribution of therapeutic drugs to the supersensitivity is not well understood. There is also a growing awareness of the limitations of antipsychotic medication. It is still the case that the most effective antipsychotic drugs are dopamine receptor antagonists, but there are significant problems. For instance, it takes two weeks to generate therapeutic benefit despite the drugs' ability to bind to receptors in the brain within a matter of minutes. Moreover, positive symptoms respond quite well to medication but negative symptoms do not, and controlled trials show that 40% of patients maintained on dopamine antagonists relapse within 12 months of starting treatment. Thereafter, those that did not relapse within one year have shown an average rate of relapse of 25% per year (Johnstone, 1993). The rate of relapse among nonmedicated schizophrenics is worse than for medicated, but nevertheless, it appears that the drugs are not providing complete amelioration of symptoms. Rather they appear to be lengthening the time taken to relapse.

Problems such as these have led to reformulation of the dopamine hypothesis. There are essentially three different variants of it now available. The first suggests that there is more than one site for dopamine dysfunction in the schizophrenic brain. Davis and colleagues (1991) suggest that there is hypodopaminergia in the prefrontal cortex and hyperdopaminergic in the nucleus accumbens. The two are interactive rather than independent and are associated with different signs and symptoms: frontal hypodopaminergia is associated with negative features, nucleus accumbens hyperdopaminergia with positive. The fact that dopamine activity could be changed in two different directions simultaneously helps explain both why it has been difficult to find consistent changes in the CSF concentrations of dopamine metabolites and why positive and negative symptoms can coexist in the same patients. Grace (1991) has constructed a similar argument, suggesting that longterm reduction in tonic dopamine release in the prefrontal cortex can trigger homeostatic changes that will lead to increases in phasic dopamine release. These hypotheses add an anatomical and neurophysiological sophistication to the simple dopamine hypothesis. Neither, however, appears to put dopamine at the dead center of schizophrenia. Instead, changes in the activity of the prefrontal cortex appear to be required to drive changes in dopamine. A second reformulation of the dopamine hypothesis argues that schizophrenia is essentially a dopamine *deficiency* disorder and that negative symptoms are caused by a lack of dopamine activity within limbic striatal circuitry (Early et al., 1989a,b). It is suggested that there is functional asymmetry in limbic striatal circuitry and that the dominance of the left hemisphere over the right is impaired by dopamine underactivity, leading to negative symptoms. Positive symptoms, in contrast, are caused by release of homologous systems in the right hemisphere. A third dopamine hypothesis suggests that there is dysregulation of dopaminergic activation of the cingulate cortex (Dolan et al., 1995), which acts to disrupt remote systems and in particular to disrupt cortico–cortical interactions (especially between the prefrontal and temporal cortices).

It is difficult to summarize the current state of the dopamine hypothesis. It is no longer seriously considered that simple dopamine overactivity accounts for all of the signs and symptoms of schizophrenia. Current hypotheses stress the fact that dopamine levels in different parts of the CNS can change in different directions and emphasize the fact that changing dopamine activity in different parts of the same extended circuitry can have radically different effects. Thus it is that both states of hyper- and hypodopaminergia are thought to be involved in the generation of schizophrenic signs and symptoms. More problematic has been the gradual erosion of confidence in dopamine receptor antagonists as therapeutic tools. Nothing has improved on these as medications, and they remain an important part of any treatment regime. However, the fact is that they do not treat all of the symptoms effectively (positives are much better dealt with than negatives) and they do not offer freedom from symptoms: 40% of patients on antipsychotic medication will relapse within one year. Dopamine receptor antagonists are not treating the signs and symptoms of schizophrenia: they are holding them, for a while, at bay.

Abnormal Development of the Brain

One of the most marked features of schizophrenia is that the typical time of onset is late adolescence/early adulthood. Studies of nonhuman primates have revealed that there are substantial changes in the brain during these years; this gives rise to the speculation that it is a developmental abnormality in the brain that precipitates schizophrenia. In the prefrontal cortex, for instance, the spines of pyramidal neuron dendrites, chandelier-type interneurons, local circuit neurons, and dopamine-containing afferents all change in number and density during development, each with their own temporal pattern. Moreover, it is apparent that there is considerable migration of neurons from the cortical subplate early in development. Disruption in any of these processes could lead to the presence of neurons in inappropriate locations or to the formation of too many or too few synaptic connections (see Lewis, 1997). Further down the neuraxis, a failure of neuronal pruning during development has been proposed to account for an increase in the numbers of nitric oxide synthesizing neurons in the pedunculopontine tegmental nucleus (Karson et al., 1991).

The presence of a developmental abnormality is profoundly difficult to determine, requiring, as it does, information from human brains at various stages of development and in various clinical states. Reports of abnormalities in the synaptology of the prefrontal cortex, however, have been consistent with changes in neural development. While there is as yet relatively little evidence about the development of the prefrontal cortex and other parts of the brain implicated in schizophrenia, the possibility that developmental abnormalities are at the root of schizophrenia is intriguing. Indeed, it is in the development of the brain that genetic abnormalities are most likely to be present. The lack of complete penetrance indicates that schizophrenia is not a simple genetic disorder, but it is possible that genetic malfunction in neural development is of importance. Candidate gene analysis is an approach to the neurobiology of schizophrenia currently under investigation (Ross and Pearlson, 1996). The general hypothesis is that changes in the structure or function of the brain during infancy, a period during which striking abnormalities are not produced, later lead to the appearance of schizophrenic signs and symptoms. Later maturational events alone might trigger these, building as they would on damaged foundations. Alternatively, environmental factors (e.g., stressors) could trigger the appearance of symptoms in at-risk individuals carrying a neurodevelopmental abnormality (Weinberger, 1987). The presence of a neurodevelopmental disturbance remains an hypothesis of considerable potency.

Functional Indices of Brain Changes in Schizophrenia

In parallel with studies that have examined the structure and chemistry of the schizophrenic brain, there have also been studies that have examined its activity. The electrical activity of the brain is known to be altered in schizophrenic patients. Abnormalities have been found in the auditory P300, an event-related potential generated by novel relevant stimuli, in schizophrenic patients (as it has in other groups of patients). More intriguingly, there appears to be some relationship between this and reduction in the volume of the left posterior superior temporal gyrus (McCarley et al., 1993). However, most information over the last several years about the functional state of the brain has not come from electrophysiological studies but from positron emission tomography (PET) and other functional imaging techniques. The first studies to investigate regional cerebral blood flow (rCBF) in patients were undertaken in Sweden by David Ingvar and colleagues, who showed that there was abnormally low activity in the prefrontal cortex of many, but not all, schizophrenic patients (Ingvar and Franzen, 1974; see Ingvar, 1995, for review). The term "hypofrontality" was used to describe this phenomenon, which has since been observed in many laboratories around the world. What has become more obvious with time is that (1) not all patients show hypofrontality and (2) many other groups of patients also have the condition. Of more interest in schizophrenia lately have been the use of functional imaging techniques to determine how the brain is activated during the performance of specific tasks and how its activation differs between schizophrenic patients and normal volunteers. For instance, schizophrenics fail to show the same significant activation of the dorsolateral prefrontal cortex as do controls while performing tasks (e.g., the Wisconsin card sorting task, the Tower-of-London task [Goldman-Rakic, 1987]) that challenge prefrontal ability relatively selectively (see papers by Weinberger and colleagues: Berman et al., 1986; Weinberger et al., 1986; 1988). Similarly, schizophrenic patients show impaired activation of the cingulate cortex during performance of a verbal task (Dolan et al., 1995) though, when challenged with apomorphine, a dopamine receptor agonist, the performance of patients was significantly enhanced relative to controls. Dysfunction in dopamine activity in the cingulate cortex, which these data imply, is thought to be important since the dopaminergic input to the cingulate is to layers II and III, layers principally involved in cortico–cortical signaling. Frith (1992), in particular, has hypothesized that the transfer of information between cortical areas—especially prefrontal and temporal—is of importance in schizophrenia.

Functional imaging of the CNS during the performance of certain tasks is a rapidly developing area of investigation. There are many problems associated with it, most notably the lack of temporal resolution (a problem that reciprocates that of evoked potential studies in which there is excellent temporal, but very limited anatomical, resolution). As well as the temporal control over the task, one needs to consider factors such as the state of the patients and the control over simple movement-generated activation. However, these problems are not insoluble and the functional examination of living human brains is liable to develop greatly over the next few years.

Animal Studies in the Neurobiology of Schizophrenia

Animal studies have been used in essentially three different ways in order to help understand schizophrenia: (1) to test the actions of novel therapeutic strategies (invariably drug treatments in the case of schizophrenia), (2) to investigate specific processes known to be dysfunctional in schizophrenia, and (3) to probe the functions of specific structures thought to be dysfunctional in schizophrenia. The first class of studies are of little theoretical interest in terms of understanding what schizophrenia is and how the brain might be affected by it, though clearly the development of viable, safe treatment strategies is of great importance. The other two types of study have been of rather more value in helping to understand what might be wrong in schizophrenia.

Consistent with the idea that schizophrenic patients have less control than normal over sensory flow is the fact that patients complain of being overwhelmed by stimulation; it has been shown that patients have poor prepulse inhibition of the startle reflex (Braff and Geyer, 1990; Geyer et al., 1993). This is a simple test of sensorimotor gating: if a sudden unpredicted stimulus is presented, subjects will show a startle reflex. (Put simply, subjects jump when startled by a sudden noise.) However, if a warning stimulus—the prepulse—is presented signaling an imminent startling stimulus, then the startle reflex is diminished or even abolished. Schizophrenic patients fail to show this prepulse inhibition of the startle reflex. Prepulse inhibition is a straightforward process to measure in animals, and studies on rats have identified several interconnected structures as being important in mediating it: the prefrontal cortex, nucleus accumbens, hippocampus, and, further down the neuraxis, the pedunculopontine tegmental nucleus. All of these structures have been implicated in schizophrenia by other means, mostly neuropathological. Latent inhibition is another process that has been shown to be deficient in patients (Gray et al., 1991). Latent inhibition describes the process whereby animals, having experienced a stimulus as being irrelevant in a given context, will take longer to grasp its significance when it is subsequently made salient. Schizophrenic patients show increased latent inhibition and animal studies again have identified the hippocampus and nucleus accumbens as critical sites mediating the process. Prepulse inhibition of the startle reflex and latent inhibition are two examples of processes known to be defective in schizophrenic patients; these can be systematically measured in animals in which it is then possible to make systematic experimental investigations of brain tissue with a view to understanding the neural mechanisms underlying these processes. It is particularly important to note that it is not necessary to model schizophrenia *en masse* in animals in order to conduct meaningful and relevant experiments. It is sufficient to be able to investigate particular aspects.

Other studies have used animals to investigate processes thought to be dysfunctional in schizophrenia. Patricia Goldman-Rakic and colleagues (1991) have developed the hypothesis that "*a fundamental problem in schizophrenia is the loss of working memory processes that inexorably lead to a deficit in the regulation of behavior by internalized schemata, symbolic representations and ideas*". They have thus set about understanding how the prefrontal cortex operates working memory processes at a cellular level in primates (Williams and Goldman-Rakic, 1995). For instance, they have found that neurons increase their firing when the subject has to retain information about a target location during a period of delay between the presentation of the target stimulus and the required response. Moreover, different neurons in the prefrontal cortex appear to code different targets, and failure to maintain neuronal firing is associated with behavioral error. Microiontophoretic application of dopamine appears to enhance firing. Studies such as these are designed to operate on many levels: they examine the firing of neurons in the prefrontal cortex and relate that activation to behavioral processes—the act of completing the task—and set that action within a cognitive model of working memory. Such understanding then permits speculation about what the nature of dysfunction in such a system might be. In parallel with Goldman-Rakic's elegant studies of the neural basis of a cognitive process is the work of Robbins and colleagues, who have investigated how early experiences affect the brains and behavior of rodents (Jones et al., 1990; Jones et al., 1991; Geyer et al., 1993). They have shown, for instance, that social isolation from weaning onwards produces rats with both behavioral and neurochemical changes: they are less active and show diminished dopamine activation in the nucleus accumbens in response to challenge.

Studies such as these do not start from an activity known to be dysfunctional in schizophrenia as the studies of prepulse inhibition or latent inhibition do. They instead seek to investigate at a theoretical level how cognitive and social processes operate and how they affect specific brain structures. The relationship to schizophrenia in these instances is more theoretical than the direct investigation of schizophrenic phenomena but equally valuable.

Neuropsychology

Neuropsychology is that branch of psychological science that deals with the relationships between brain and behavior. The term is generally taken to refer to studies made of normal human volunteer subjects and brain damaged patients. An important part of neuropsychology is the attempt to describe the nature of the deficits present in patients with various forms of brain disorder and to relate those deficits to theories that both explain psychological phenomena in terms of neural systems and understand neural systems in terms of the psychological functions they subserve.

Various neuropsychological theories about the nature of schizophrenia have been proposed: Goldman-Rakic (1991) has suggested that an understanding of working memory is fundamental to understanding the schizophrenic deficit, in so far as working memory "*confers the ability to guide behavior by representations of the outside world rather than by immediate stimulation and thus to base behavior in ideas and thoughts*". Related ideas have been expressed by Gray and colleagues (1991) who argue that there are difficulties in integrating stored memories with ongoing motor programs to control current behavior. A rather different line has been taken by others who have argued that a core deficit in schizophrenia is in attention. Braff and Geyer (1990) have suggested that sensorimotor gating is critical in that schizophrenic patients have difficulty screening out irrelevant stimuli from the relevant, leading to many of the signs and symptoms. Sarter (1994) has similarly explored the idea that schizophrenic patients have difficulty "disattending" to irrelevant stimuli. Other models incorporate elements of both "stored representation" and "attentional" theories. Hemsley (1994), for instance, proposes that breakdown in the relationships between stored information and current input lead to disinhibition of behavior.

All of these theories have merit but, while they satisfactorily account for some of the symptoms, none is able to account for them all. This may not be a problem: it may in fact be the case that there is no underlying unity to schizophrenia and that what in fact is required are theories that can explain aspects of the disorder rather than all of it. The problem with this argument though is that patients rarely confine themselves to only certain aspects of the disease. At different times, for instance, patients may show both negative and positive signs and symptoms as well as symptoms such as stereotyped thoughts and behaviors, which are not best classified as either positive or negative. The virtue of an all-embracing theory to account for schizophrenia *en masse* is that it might be able to account for the diversity of signs and symptoms between and within patients.

A recent theory proposed by Frith (1992) appears to be able to account for all of the signs and symptoms of schizophrenia within a single theory. Frith suggests that schizophrenia can be divided into three fundamentals: disorders of willed action, disorders of self-monitoring, and disorders in monitoring the intentions of others (see Table 2).

Frith believes these three deficits are manifestations of one core deficit in metarepresentation. Metarepresentation is a representation of a representation, a

Table 2. Frith's Tripartite Division of Schizophrenia

*Disorders of Willed Action**
- Poverty of action
- Perseveration
- Behaviour elicited by irrelevant external stimuli
- Inappropriateness

Disorders of Self-Monitoring
- A failure properly to interpret internal representations and states
- Feelings of control by alien agencies
- Hallucinations
- Thought insertion

Disorders in Monitoring the Intentions of Others
- Paranoia
- Incoherence
- Third-person hallucinations

Note: *These resemble the disorders that follow frontal lobe damage.

higher order abstraction in which mental states are themselves represented. Metarepresentation is thought to be a form of second-order representation: first-order representations are descriptions of physical states ("my desk is brown") while second-order representations incorporate additional mental states ("Dr Whiten believes my desk is brown"). Frith proposes that there are three areas of self-consciousness in which metarepresentations are critical: awareness of our own goals, awareness of our intentions and awareness of the intentions of other. Disruption of these leads to, respectively, disorders of the will, a lack of self-monitoring (and hence abnormal experience of our own actions), and feelings of persecution and delusions (since the intentions of others are no longer properly understood). Metarepresentation deficiencies have also been used to account for the symptoms of autism which, of course, are not identical to those of schizophrenia. The key difference, however, is that in one case, metarepresentation never properly develops (giving rise to autistic aloneness) while, in schizophrenia, metarepresentation is thought to develop normally but then to break down. Establishing the ability to metarepresent and then having it break down obviously should lead to different forms of disorder to the case in which metarepresentation never develops. For instance, it has been argued that third-person hallucinations (as opposed to second-person) are the product of a breakdown in the ability to understand the minds of others (Corcoran et al., 1995). The hallucinations nevertheless depend upon the awareness that others could have minds, something of which one could not have been aware had metarepresentational skills never developed at all.

Frith and colleagues have engaged in two lines of enquiry to further examine the role of metarepresentation in schizophrenia. One line investigates the extent to which patients fail in tests designed explicitly to examine metarepresentation. From these studies, there is evidence that patients do have difficulties metarepresenting, though whether or not all patients showing any of the symptoms of schizophrenia will do so remains uncertain (Corcoran et al., 1995; Frith and Corcoran, 1996). The second line of investigation intends to examine aspects of metarepresentation in patients using PET scans. For instance, in a study of patients auditory hallucinations—which Frith suggests are the product of misattributing ones own inner speech to others—evidence of significant differences between patients and controls was found (Silbersweig et al., 1995). Data from PET scan studies has been used to suggest that subcortical systems generate the hallucination while the content is determined by an appropriate neocortical region.

THE NEUROBIOLOGY OF SCHIZOPHRENIA: AN OVERVIEW AND SPECULATION

In the preceding sections I have tried to provide a brief review of studies concerned with the neurobiology of schizophrenia. Taken as a whole, it is obvious that certain structures are associated often with schizophrenia: the prefrontal cortex, the nucleus accumbens, the temporal lobe and structures embedded in it such as the hippocampus, and the mesolimbic dopamine system that innervates the accumbens and the limbic cortex. These structures are not independent of each other but form part of an integrated system. The structures of the temporal lobe are interconnected and project to the nucleus accumbens as does the prefrontal cortex. The accumbens in turn projects, via the ventral pallidum and mediodorsal nucleus of the thalamus, back to the frontal cortex. This looped circuitry is part of a more generalized system of corticostriatopallidothalamocortical loops (Alexander et al., 1986; Joel and Weiner, 1994). Tissue within this system has been associated with a variety of functions concerned with the assessment of sensory information, the use of stored information, and the processes by which environmental stimuli gain control over behavior. Such processes are fundamental to behavior and cognition and, when they are disrupted experimentally in animals, cause great difficulty in a wide variety of tasks.

Despite this repetitive association of certain structures with schizophrenia, there remain serious problems when assessing the neurobiology of this disorder. Firstly, there is no evidence of gross structural abnormality in the schizophrenic brain. The closest thing to this remains the changes in ventricular size, but the facts are that this change is not consistent across all patients (even across subsets of patients), that it is not associated with any other degenerative process, and that the enlargement may be present before symptoms appear and does not progress as does the disease. Such observations suggest that ventricular enlargement is not of major etiological importance in schizophrenia. The response of neuropathologists to the lack of gross abnormality has been to attempt to find some more subtle abnormality in the microcircuitry of the brain and to examine the brain functionally rather than structurally. Even here, though, there is as yet no consistency. It has been possible to show changes in the numbers of neurons in specific parts of the prefrontal cortex for instance, but the changes do not appear in all patients such that while there might be an average increase in neuronal number, there is no definitive change that marks the schizophrenic brain from the nonschizophrenic. Similarly, hypofrontality has not been shown in every patient although, on average, schizophrenics may display the phenomenon. The response of patients to dopamine receptor antagonists, still the most widely used therapeutic regime, is similarly uncertain: some symptoms do not respond at all to this treatment and, even when symptomatic relief is achieved, 40% of medicated patients relapse within one year.

This lack of consistency, the failure to find a reliable neurological marker of schizophrenia, is disappointing given the effort that has been expended. It is possible to claim that progress has been made in that certain structures have been implicated fairly well while others have not. The failure to find consistent evidence of damage or dysfunction in the schizophrenic brain has led many neuroscientists to look at the brain in different ways: PET scans are used to examine it functionally, and the microcircuitry is examined in ever closer detail with the rationale that if there is no gross pathology, then there must be a microscopic one. Alternatively, however, one might consider whether the lack of consistency itself has become the issue to investigate.

One substantial problem is that schizophrenia is not, at least at a symptomatic level, a unitary disease. There is a wide range of signs and symptoms that many have tried to classify in different ways. Some of the signs and symptoms are abnormal by their presence, while others are abnormal by their absence; some are typically associated together—others not so. It is possible that specific signs and symptoms are associated with particular forms of damage or dysfunction. For instance, negative symptoms are thought to be associated particularly with prefrontal deficits, while positive symptoms are more closely associated with dopaminergic abnormality than are negative symptoms, which are not responsive to antipsychotic drugs. Different authors have tackled this problem in radically different ways. Some have adopted the argument that schizophrenia has no real value as a diagnostic entity and that it would be better to concentrate on the specific signs and symptoms that are the cause of distress in schizophrenia (Bentall, 1990; Boyle, 1990). Eliminate these and the patients' problems would disappear. The use of the label "schizophrenia" has, in this argument, no value. On the other hand, Frith has described a theoretical framework that attempts to embrace and account for all of the signs and symptoms of schizophrenia (Frith, 1992). It is too soon to know whether or not this will be successful, but the attempt has undoubtedly generated more interest in recent years than any other theory. Given the existence of a testable theory, it seems pointless to abandon the concept of schizophrenia now. However, if one accepts that the concept of schizophrenia is useful, how can one reconcile the hugely devastating nature of the disorder with the apparent lack of substantial, reliable, and definitive brain injury or dysfunction? Is it the case that what needs to be understood now is why there is much less replicability in the research literature than had been hoped for?

Authors are questioning with greater frequency the philosophical basis of neurobiology (e.g., Churchland, 1986; Frith, 1992; Farah, 1994; Andreasen, 1997). The neurologists of the nineteenth century sought to define brain changes that would account not only for pathological conditions but also for mental states. Their understanding of the brain was shaped largely by the modular approach of the phrenologists, who believed that different psychological functions could be located in different parts of the brain (though there had already begun arguments as to whether psychological functions were localized or distributed—arguments that persist still; e.g., Farah, 1994). Consistent with "localizationist" assumptions are the attempts of neurologists to show that certain pathological conditions are associated with damage to specific parts of the brain. The clearest example of this is Parkinson's disease, in which it was recognized early in the twentieth century that there was loss of the pigmented cells of the substantia nigra, cells that were much later shown to contain the neurotransmitter dopamine and to project to the caudate nucleus and putamen. Here there is clear association between a particular disease and loss of particular neurons. Nonetheless, while this might be an acceptable model for certain pathological states and particular brain processes, it is not necessarily a model that is appropriate in every instance. The attempt to apply a localizationist philosophy uniformly to describe the relationships between brain activity and psychological states may have been misleading.

It is possible to identify certain sites in the brain where specific neurons engage in specific processes. The magnocellular neurons of the hypothalamic paraventricular and supraoptic nuclei, for instance, are the only ones to synthesize and secrete vasopressin in respose to dehydration. The neurons of areas V1 to V5 in the striate cortex have very specific functions in vision: the registration of color, movement, edge, and so on. These are functions that are relatively closely localized. If damage occurs in these areas, a specific function is lost. Without area V5, for instance, the perception of motion would be seriously impaired. In the rodent brain, the vibrissae (important sensory organs) map onto specific barrel-shaped columns containing about 2500 neurons in the somatosensory cortex (Woolsey and van der Loos, 1970). Each barrel processes tactile information from a single whisker. Loss of the vibrissae produces barrel field atrophy, and loss of the barrel fields causes loss of sensation. Other functions of the mammalian brain are not localized in the same way and do not have the same status. Hunger and thirst, for instance, are labels that we learn to apply to bodily states that we experience (Bruch, 1974). Clearly the state of hunger is one that is represented by and within the brain, but it is evident that it is not uniquely represented at a specific site. One does not feel hungry with a dedicated set of neurons in the same way that one detects movement with the neurons of area V5. It is a process that is best described vectorially. There are a variety of sites at which the composition of body tissues and fluids are measured and a variety of mechanisms for making adjustment, both physiologically and behaviorally. Hunger is probably not experienced as the action of individual neurons (or even tightly localized groups analogous to the barrel fields of the somatosensory cortex) but as the combined activity of all or part of this distributed neural network. The essential point here is that some functions can be ascribed directly to the firing of specific localized sets of neurons whereas other functions are *descriptions* of the activity of widely distributed networks of neurons. These are two very different things. The first is purely functional: this neuron secretes vasopressin, that neuron is tuned to detect certain sorts of movement. These are functions that can be described at a single unit level. The second involves *labeling* of certain states with terms that have no precise relationship to brain activity. They are descriptive terms that, at best, correlate with activity in particular networks of neurons. There is a third type of process as well. It is now generally accepted that the brain is made up of distributed networks of neurons. A principal virtue of neural networks is that they have properties that are not predictable from the actions of their individual constituent units (neurons in this case). Metarepresentation is in all probability such a phenomenon—an emergent property that is *dependent on* (rather than *descriptive of*) the action of neural networks but which is neither predicted from nor dependent on any given individual node (neuron) within the net.

It is important to appreciate that different psychological constructs have different relationships with brain tissue. Some psychological functions can be described at a single unit level; some psychological functions are best understood as descriptions (not properties) that we have learned to correlate with the activity of networks of neurons; and some psychological functions are emergent properties of networks, functions that are actually the properties of networks of neurons. These different forms of coding are discussed by neurocomputing theorists as local representations (in which single units code functions), dense distributed representations (in which more than half the neurons in a network are required to represent a given state; there is a high activity ratio in the network), and sparse coding (in which relatively few units in a network are required to represent a given state—a low activity ratio; Földiák and Young, 1995). For the most part, the attempt to understand the neurobiology of schizophrenia has been an attempt to describe structural changes in particular parts of the brain in order to find changes that cause schizophrenia in the same way that loss of substantia nigra neurons causes Parkinsonism. It is an approach based on the premise that all psychological functions are of the first type, tightly localized functions that a phrenologist would recognize. It is this belief that has driven the attempt to locate schizophrenia to particular CNS tissue.

However, if schizophrenia actually is a disorder of metarepresentation as Frith claims, then several things might follow. Frith is quite explicit in his belief that metarepresentation depends on the concerted action of tissue at many sites and that it involves extensive cortico–cortical interconnection as well as connections with subcortical sites (Frith, 1992). If metarepresentation is an emergent property of a widely distributed and massively interconnected network of neurons, it is possible that disruption of the network at a variety of sites might create metarepresentational difficulties and, consequently, while one might be able always to describe structural changes in the brains of patients, *they need not always be the same*. Physical disruption of a network—either the numbers of neurons or the connections between them—at many different points could have similar outcomes. Is this the explanation for the lack of a consistent change in the brains of schizophrenics? Secondly, there is the possibility that there need be no structural damage. Networks can be made dysfunctional by damaging the nodes or the connections physically or by training them inappropriately. In this latter case, the abnormality need not be structural, but only operational. Functional imaging might reveal those operational abnormalities, but the abnormality need not necessarily be the same in every schizophrenic patient. Moreover, would a disorder pitched at this operational rather than structural level be best considered as neurological or psychological?

Neurobiology is a relatively new discipline, and the emphasis over the past century has been to understand the structure of the brain. Until recently, less thought had been given to understanding how brains give rise to psychological states or to how psychological states affect brains (which they apparently do). A better appreciation of these things—a more lucidly formulated neurophilosophy—may be a necessary precondition for understanding the neurobiology of schizophrenia and for developing an effective therapy.

ACKNOWLEDGMENTS

I wish to acknowledge the assistance of Lynsey Gemmell and Cornelia Brenninkmeyer in preparing this chapter.

REFERENCES

Akbarian, S., Bunney, W.E., Jr., Potkin, S.G., Wigal, S.B., Hagman, J.O., Sandman, C.A., & Jones, E.G. (1993a). Altered distribution of nicotinamide-adenine dinucleotide phosphate-diaphorase neurons in frontal lobe of schizophrenics implies disturbances of cortical development. Arch. Gen. Psychiat. 50, 227-230.

Akbarian, S., Viñuela, A., Kim, J.J., Potkin, S.G., Bunney, W.E., Jr., & Jones, E.G. (1993b). Distorted distribution of nicotinamide-adenine dinucleotide phosphate-diaphorase neurons in temporal lobe of schizophrenics implies anomalous cortical development. Arch. Gen. Psychiat. 50, 179-187.

Alexander, G.E., DeLong, M.R., & Strick, P.L. (1986). Parallel organization of functionally segregated circuits linking the basal ganglia and cortex. Ann. Rev. Neurosci. 9, 357-381.

Altschuler, L.L., Casanova, M.F., Goldberg, T.E., & Kleinman, J.E. (1990). The hippocampus and parahippocampus in schizophrenic, suicide and control brains. Arch. Gen . Psychiat. 47, 1029-1034.

American Psychiatric Association (1987). *Diagnostic and Statistical Manual of Mental Disorders, 4th ed.* American Psychiatric Association, Washington, D.C.

Andreasen, N.C. (1982). Negative v. positive schizophrenia. Definition and validation. Arch. Gen. Psychiat. 39, 789-794.

Andreasen, N.C. (1985). *Comprehensive Assessment of Symptoms and History (CASH).* University of Iowa Press, Iowa City, IO.

Andreasen, N.C. (1997). Linking mind and brain in the study of mental illness: A project for a scientific psychopathology. Science 275, 1586-1593.

Andreasen, N.C., Olsen, S.A., Dennert, J.W., & Smith, M.R. (1982). Ventricular enlargement in schizophrenia: Relationship to positive and negative symptoms. Am. J. Psychiat. 139, 297-302.

Arndt, S., Alliger, R.J., & Andreasen, N.C. (1991). The distinction of positive and negative symptoms: The failure of a two-dimensional model. Brit. J. Psychiat. 158, 317-322.

Asarnow, J.R. (1988). Children at risk for schizophrenia: Converging lines of evidence. Schiz. Bull. 14, 613-631.

Bateson, G., Jackson, D.D., Haley, J., & Weakland, J. (1956). Toward a theory of schizophrenia. Behav. Sci. 1, 251-264.

Bentall, R.P. (1990). The syndromes and symptoms of psychosis. In: *Reconstructing Schizophrenia.* (Bentall, R.P., Ed.) pp. 23-60. Routledge, London.

Berman, K.F., Zec, R.F., & Weinberger D.R. (1986). Physiologic dysfunction of dorsolateral prefrontal cortex in schizophrenia. II. Role of neuroleptic treatment, attention and mental effort. Arch. Gen. Psychiat. 43, 126-135.

Bleuler, E. (1911/1950). Dementia praecox or the group of schizophrenias. (Trans. Joseph Zinkin), International Universities Press, New York.

Bloom, F.E. (1993). Advancing a neurodevelopmental origin for schizophrenia. Arch. Gen. Psychiat. 50, 224-227.

Bogerts, B., Hantsch, H., & Herzer, M. (1983). A morphometric study of the dopmaine-containing cell groups in the mesencephalon of normals, Parkinson patients, and schizophrenics. Biol. Psychiat. 18, 951-971.

Bogerts, B., Meertz, E., & Schonfeld-Bausch, R. (1985). Basal ganglia and limbic system pathology in schizophrenia. Arch. Gen. Psychiat. 42, 784-791.

Boyle, M. (1990). *Schizophrenia: A Scientific Delusion?* Routledge, London.

Braff, D.L. & Geyer, M.A. (1990). Sensorimotor gating and schizophrenia. Arch. Gen. Psychiat. 47, 181-188.

Brown, G.W., Bone, M., Dalison, B., & Wing, J.K. (1966). *Schizophrenia and Social Care.* Oxford University Press, London.

Brown, R., Colter, N., Corsellis, J.A.N., Crow, T.J., Frith, C.D., Jagoe, R., Johnstone, E.C., & Marsh, L. (1986). Postmortem evidence of structural brain changes in schizophrenia: Differences in brain weight, temporal horn area, and parahippocampal gyrus compared with affective disorder. Arch. Gen. Psychiat. 43, 36-42.

Bruch, H. (1974). *Eating Disorders.* Routledge & Kegan Paul, London.

Churchland, P.S. (1986). *Neurophilosophy.* MIT Press, Boston, MA.

Corcoran, R., Mercer, G., & Frith, C.D. (1995). Schizophrenia, symptomatology, and social inference: Investigating "theory of mind" in people with schizophrenia. Schiz. Res. 17, 5-13.

Crow, T.J. (1980a). Molecular pathology of schizophrenia: More than one disease process? Brit. Med. J. 280, 66-68.

Crow, T.J. (1980b). Positive and negative schizophrenic symptoms and the role of dopamine. Brit. J. Psychiat. 137, 383-386.

Crow, T.J., Ball, J., Bloom, S.R., Brown, R., Bruton, C.J., Colter, N., Frith, C.D., Johnstone, E.C., Owens, D.G.C., & Roberts G.W. (1989). Schizophrenia as an anomaly of development of cerebral asymmetry: A post-mortem study and a proposal concerning the genetic basis of the disease. Arch. Gen. Psychiat. 46, 1145-1150.

Davis, K.L., Kahn, R.S., Ko, G., & Davidson, M. (1991). Dopamine in schizophrenia: A review and reconceptualization. Am. J. Psychiat. 148, 1474-1486.

Dolan, R.J., Fletcher, P., Frith, C.D., Friston, K.J., Frackowiak, R.S.J., & Grasby, P.M. (1995). Dopaminergic modulation of impaired cognitive activation in the anterior cingulate cortex in schizophrenia. Nature 378, 180-182.

Done, D.J., Johnstone, E.C., Frith, C.D., Golding, J., Shepherd, P.M., & Crow, T.J. (1991). Complications of pregnancy and delivery in relation to psychosis in adult life: Data from the British perinatal mortality survey sample. Brit. Med. J. 302, 1576-1580.

Early, T.S., Posner, M.I., Reiman, E.M., & Raichle, M.E. (1989a). Hyperactivity of the left striato-pallidal projection. Part I: lower level theory. Psychiat. Develop. 2, 85-108.

Early, T.S., Posner, M.I., Reiman, E.M., & Raichle, M.E. (1989b). Left striato-pallidal hyperactivity in schizophrenia. Part II: Phenomenology and thought disorder. Psychiat. Develop. 2, 109-121.

Falkai, P. & Bogerts, B. (1986). Cell loss in the hippocampus of schizophrenics. Eur. Arch. Psychiat. Neurol. Sci. 236, 154-161.

Farah, M.J. (1994). Neuropsychological inference with an interactive brain: A critique of the "locality" assumption. Behav. Brain. Sci. 17, 43-104.

Feldman, R.S., Meyer, J.S., & Quenzer, L.F. (1997). *Principles of Neuropsychopharmacology.* Sinauer Associates, Inc., Sunderland, MA.

Földiák, P. & Young, M.P. (1995). Sparse coding in the primate cortex. In: *The Handbook of Brain Theory and Neural Networks.* (Arbib, M., Ed), pp. 395-398. Massachusetts Institute of Technology, Bradford, MA.

Frangou, S., Sharma, T., Sigmundsson, T., Barta, P., Pearlson, G., & Murray, R.M. (1997). The Maudsley family study. 4. Normal planum temporale asymmetry in familial schizophrenia—a volumetric MRI study. Brit. J. Psychiat. 170, 328-333.

Frith, C.D. (1992). *The Cognitive Neuropsychology of Schizophrenia.* Lawrence Erlbaum Associates, Hove, U.K..

Frith, C.D., & Corcoran, R. (1996). Exploring "theory of mind" in people with schizophrenia. Psychol. Med. 26, 521-530.

Frith, C.D., & Done, D.J. (1983). Stereotyped responding by schizophrenic patients on a two-choice guessing task. Psychol. Med. 13, 779-786.

Frith, C.D., & Done, D.J. (1990). Stereotyped behaviour in madness and in health. In: *Neurobiology of Stereotyped Behaviour.* (Cooper, S.J. & Dourish, C.T., Eds.), pp. 232-259. Clarendon Press, Oxford, England.

Gattaz, W.F., Kohlmeyer, K., & Gasser, T. (1991). Computer tomographic studies in schizophrenia. In: *Search for the Causes of Schizophrenia.* (Hafner, H. & Gattaz, W.F., Eds), pp. 242-256. Springer, Berlin.

Geyer, M.A., Wilkinson, L.S., Humby, T, & Robbins, T.W. (1993). Isolation rearing of rats produces a deficit in prepulse inhibition of acoustic startle similar to that in schizophrenia. Biol. Psychiat. 34, 361-372.

Goldman-Rakic, P.S. (1987). Circuitry of the prefrontal cortex and the regulation of behavior by representational memory. In: *Handbook of Physiology, Part 1.* (Plum, F. & Mountcastle, V., Eds), Vol. 5, pp. 373-417. American Physiology Society, Bethesda, MD.

Goldman-Rakic, P.S. (1991). Prefrontal cortical dysfunction in schizophrenia: The relevance of working memory. In: *Psychopathology and the Brain.* (Carroll, B.J. & Barrett, J.E., Eds), pp. 1-23. Raven Press, New York.

Goodman, S.H. (1987). Emory University project on children of disturbed parents. Schiz. Bull. 13, 411-423.

Gottesman, I.I., McGuffin, P., & Farmer, A.E. (1987). Clinical genetics as clues to the "real" genetics of schizophrenia. Schiz. Bull. 13, 23-47.

Grace, A.A. (1991). Phasic versus tonic dopamine release and the modulation of dopamine system responsivity: A hypothesis for the aetiology of schizophrenia. Neuroscience 41, 1-24.

Gray, J.A., Feldon, J., Rawlins, J.N.P., Hemsley, D.R., & Smith, A.D. (1991). The neuropsychology of schizophrenia. Behav, Brain. Sci. 14, 1-84.

Heckers, S., Heinsen, H., Heinsen, Y.C., & Beckmann, H. (1990). Limbic structures and lateral ventricle in schizophrenia. Arch. Gen. Psychiat. 47, 1016-1022.

Hemsley, D.R. (1994). Cognitive disturbance as the link between schizophrenic symptoms and their biological bases. Neurol. Psychiat. & Brain Res. 2, 163-170.

Hollingshead, A.B. & Redlich, F.C. (1958). *Social Class and Mental Illness: A Community Study.* John Wiley and Sons, New York.

Ingvar, D.H. (1995). Prefrontal dysfunction and conscious awareness in schizophrenia. In: *Schizophrenia.* (Alfred Benzon Symposium, 38) (Fog, R., Gerlach, J., & Hemmingsen, R., Eds), pp. 181-187. Munksgaard, Copenhagen, Denmark.

Ingvar, D.H., & Franzen, G, (1974). Abnormalities of regional cerebral blood flow distribution in patients with chronic schizophrenia. Acta Psychiat. Scand. 50, 425-462.

Jeste, D.V., & Lohr, J.B. (1989). Hippocampal pathologic findings in schizophrenia: a morphometric study. Arch. Gen. Psychiat. 46, 1019-1024.

Joel, D. & Weiner, I. (1994). The organization of the basal ganglia–thalamocortical circuits: Open interconnected rather than closed segregated. Neuroscience 63, 363-379.

Johnstone, E.C. (1993). Schizophrenia: Problems in clinical practice. Lancet 341, 536-538.

Johnstone, E.C., Crow, T.J., Frith, C.D., Husband, J. & Kreel, L. (1976). Cerebral ventricle size and cognitive impairment in chronic schizophrenia. Lancet ii, 924-926.

Johnstone, E.C., Crow, T.J., Frith, C.D., Stevens, M., Kreel, L., & Husband, J. (1978). The dementia of dementia praecox. Acta Psychiat. Scand. 57, 305-324.

Jones, G.H., Marsden, C.A., & Robbins, T.W. (1990). Increased sensitivity to amphetamine and reward-related stimuli following social isolation in rats: Possible disruption of dopamine-dependent mechanisms of the nucleus accumbens. Psychopharmacol. 102, 364-372.

Jones, G.H., Marsden, C.A., & Robbins, T.W. (1991). Behavioural rigidity and rule-learning deficits following social isolation in rats: Neurochemical correlates. Behav. Brain. Res. 43, 35-50.

Jones, P., Rodgers, B., Murray, R., & Marmot, M. (1994). Child development risk factors for adult schizophrenia in the British 1946 birth cohort. Lancet 344, 1398-1402.

Karson, C.N., Garcia-Rill, E., Biedermann, J., Mrak, R.E., Husain, M.M., & Skinner, R.D. (1991). The brain stem reticular formation in schizophrenia. Psychiat. Res.: Neuroimaging 40, 31-48.

Kohn, M.L. (1968). Social class and schizophrenia: A critical review. In: *The Transmission of Schizophrenia.* (Rosenthal, D. & Kety, S.S., Eds.) pp. 155-173. Pergamon, New York.

Kraepelin, E. (1896). Psychiatrie. 5th ed. Barth, Leipzig.

Kulynych, J.J., Vladar, K., Fantie, B.D., Jones, D.W., & Weinberger, D.R. (1995). Normal asymmetry of the planum temporale in patients with schizophrenia—3 dimensional cortical morphometry with MRI. Brit. J. Psychiat. 166, 742-749.

Lewis, D.A. (1997). Development of the frontal cortex during adolescence: Insights into vulnerable neural circuits in schizophrenia. Neuropsychopharmacol. 16, 385-398.

Liddle, P.F. (1987). The symptoms of chronic schizophrenia. A re-examination of the positive–negative dichotomy. Brit. J. Psychiat. 151, 145-151.

Liddle, P.F. & Barnes, T.R.E. (1990). Syndromes of chronic schizophrenia. Brit. J. Psychiat. 157, 558-561.

Lohr, J.B., & Jeste, D.V. (1988). Locus coeruleus morphometry in ageing and schizophrenia. Acta Psychiat. Scand. 77, 689-697.

McCarley, R.W., Shenton, M.E., O'Donnell, B.F., Faux, S.F., Kikinis, R., Nestor, P.G., & Jolesz, F.A. (1993). Auditory P300 abnormalities and left posterior superior temporal gyrus volume reduction in schizophrenia. Arch. Gen. Psychiat. 50, 191-197.

Mednick, S.A., Machon, R.A., Huttunen, M.O., & Bonett, D. (1988). Adult schizophrenia following prenatal exposure to an influenza epidemic. Arch. Gen. Psychiat. 45, 189-192.

O'Callaghan, E., Larkin, C., Redmond, O., Stack, J., Ennis, J.T., & Waddington, J. (1988). "Early onset schizophrenia" after teenage head injury. Brit. J. Psychiat. 153, 394-396.

O'Connell, P.H. (1958). *Amphetamine Psychosis.* Oxford University Press, London.

Pakkenberg, B. (1990). Pronounced reduction of total neuron number in mediodorsal thalamic nucleus and nucleus accumbens in schizophrenics. Arch. Gen. Psychiat. 47, 1023-1028.

Parnas, J., Schulsinger, F., Schulsinger, H., Mednick, S.A., & Teasdale, T.W. (1982). Behavioural precursors of schizophrenia: A prospective study. Arch. Gen. Psychiat. 39, 658-664.

Peltonen, L. (1995). All out for chromosome 6. Nature 378, 665-666.

Petty, R.G., Barta, P.E., Pearlson, G.D., McGilchrist, I.K., Lewis, R.W., Tien, A.Y., Pulver, A., Vaughn, D.D., Casanova, M.F., & Powers, R.E. (1995). Reversal of asymmetry of the planum temporale in schizophrenia. Am. J. Psychiat. 152, 715-722.

Plomin, R., DeFries, J.C., McClearn, G.E., & Rutter, M. (1997). *Behavioral Genetics, 3rd ed.* W.H. Freeman & Co., New York.

Remschmidt, H.E., Schulz, E., Martin, M., Warnke, A., & Trott, G.E. (1994). Childhood onset schizophrenia: History of the concept and recent studies. Schiz. Bull. 20, 727-745.

Revelry, A.M., Revelry, M.A., Clifford, C.A., & Murray, R.M. (1982). Cerebral ventricular size in twins discordant for schizophrenia. Lancet i, 540-541.

Rosenberg, D., Sweeney, J., Cornblatt, B., Squires-Wheeler, B., O'Hearn, K., & Erelmeyer-Kimling, L. (1995). Eye tracking dysfunction in offspring from the New York high-risk project. Biol. Psychiat. 37, 624-625.

Ross, C.A. & Pearlson, G.D. (1996). Schizophrenia, the heteromodal association neocortex and development: Potential for a neurogenetic approach. Trends Neurosci. 19, 171-176.

Sarter, M. (1994). Neuronal mechanisms of the attentional dysfunctions in senile dementia and schizophrenia: Two sides of the same coin? Psychopharmacol. 114, 539-550.

Selemon, L.D., Rajkowska, G., & Goldman-Rakic, P.S. (1995). Abnormally high neuronal density in the schizophrenic cortex: A morphometric analysis of prefrontal area 9 and occipital area 17. Arch. Gen. Psychiat. 52, 805-818.

Silbersweig, D.A., Stern, E., Frith, C., Cahill, C., Holmes, A., Grootoonk, S., Seaward, J., McKenna, P., Chua, S.E., Schnorr, L., Jones, T., & Frackowiak, R.S.J. (1995). A functional neuroanatomy of hallucinations in schizophrenia. Nature 378, 176-179.

Suddath, R.L., Christison, G.W., Torrey, E.F., Casanova, M.F., & Weinberger, D.R. (1989). Anatomical abnormalities in the brains of monozygotic twins discordant for schizophrenia. N. Eng. J. Med. 322, 464-472.

Vaughn, C.E. & Leff, J.P. (1976). The influence of family and social factors on the course of psychiatric illness. A comparison of schizophrenic and depressed neurotic patients. Brit. J. Psychiat. 129, 125-137.

Weinberger, D.R. (1987). Implications of normal brain development for the pathogenesis of schizophrenia. Arch. Gen. Psychiat. 44, 660-669.

Weinberger, D.R. (1988). Premorbid neuropathology in schizophrenia. Lancet 2, 959-960.

Weinberger, D.R., Berman, K.F., & Illowsky, B.P. (1988). Physiological dysfunction of dorsolateral prefrontal cortex in schizophrenia, III. A new cohort and evidence for a monoaminergic mechanism. Arch. Gen. Psychiat. 45, 609-615.

Weinberger, D.R., Berman, K.F., & Zec, R.F. (1986). Physiologic dysfunction of dorsolateral prefrontal cortex in schizophrenia. I. Regional cerebral blood flow evidence. Arch. Gen. Psychiat. 43, 114-124.

Weinberger, D.R., Bigelow, L.L.B., Kleinman, J.E., Rosenblatt, J.E., & Wyatt, R.J. (1980). Cerebral ventricular enlargement in chronic schizophrenia: An association with poor response to treatment. Arch. Gen. Psychiat. 37, 11-13.

Weinberger, D.R., Torrey, E.E., Neophytides, A.N., & Wyatt, R.J. (1979a). Lateral cerebral ventricular enlargement in chronic schizophrenia. Arch. Gen. Psychiat. 36, 735-739.

Weinberger, D.R., Torrey, E.F., Neophytides, A, & Wyatt, R.J. (1979b). Structural abnormalities of the cerebral cortex in chronic schizophrenia. Arch. Gen. Psychiat. 36, 935-939.

Werry, J.S., McClelland, J.M., Andrews, L.K., & Hann, M. (1994). Clinical features and outcome of child and adolescent schizophrenia. Schiz. Bull. 20, 619-630.

Williams, G.V. & Goldman-Rakic, P.S. (1995). Modulation of memory fields by dopamine D1 receptors in prefrontal cortex. Nature 376, 572-575.

Wing, J.K., Cooper, J.E., & Sartorius, N. (1974). The Measurement and Classification of Schizophrenic Symptoms. Cambridge University Press, Cambridge.

Woolsey, T.A. & van der Loos, H. (1970). The structural organization of layer IV in the somatosensory region (S1) of mouse cerebral cortex. The description of a cortical field composed of discrete cytoarchitectonic units. Brain Res. 17, 205-242.

RECOMMENDED READINGS

Benes, F.M. (1999). Evidence for altered trisynaptic circuitry in schizophrenic hippocampus. Biol. Psychiatry 46, 589-599.

Goldman-Rakic, P.S. (1999). The physiological approach: Functional arch. lecture of working memory and disordered cognition in schizophrenia. Biol. Psychiatry 46, 650-661.

Soares, J.C. and Janis, R.B. (1999). Neurochemical brain imaging investigations of schizophrenia Biol. Psychiatry 46, 600-615.

Pearlson, G.D. & Marsh, L. (1999). Structural brain imaging in schizophrenia: A selective review. Biol. Psych. 46, 627-649.

Weinberger, D.R. (1999). Cell biology of the hippocampal formation in schizophrenia. Biol. Psychiatry 45, 395-402.

Chapter 13

Heterogeneity of Schizophrenia

Diagnostic and Etiological Approaches

MING T. TSUANG, MICHAEL J. LYONS, and
ROSEMARY TOOMEY

INTRODUCTION

In the 17th century, Thomas Sydenham made a crucial breakthrough in the nosology of medical disorders—the idea of grouping co-occurring signs and

Biological Psychiatry, pages 263-277.

ISBN: 1-55938-819-6

symptoms into syndromes. The identification of syndromes promoted important advances in understanding the etiology and pathophysiology of various illnesses. With more homogenous diagnostic categories, we increase the likelihood of detecting and identifying characteristics of a disorder that may reflect etiology and pathophysiology. Kraepelin (1971/1919) is often given credit for ushering in the current era of careful attention to diagnostic categories in psychiatry. However, more than eight decades since the publication of his influential textbook, there is still considerable disagreement about the best way to define most, if not all, psychiatric disorders. Although a great deal of attention has been paid to perfecting our specification of psychiatric clinical phenotypes, we do not know specific etiological mechanisms that are necessary and/or sufficient for the development of the large majority of psychiatric disorders that includes schizophrenia. We also lack knowledge about the exact nature of the pathophysiological mechanisms involved. The limitations of our current knowledge reflect the fact that psychiatric disorders encompass a very heterogenous collection of phenomena. In this chapter, we will review historical and current thinking on both clinical and etiological heterogeneity in schizophrenia. Ideally, in the future, we will be able to define and subdivide schizophrenia on the basis of etiology and pathophysiology.

CLINICAL HETEROGENEITY IN SCHIZOPHRENIA

The diagnosis of schizophrenia is made solely on the basis of signs and symptoms; there are no laboratory tests or imaging techniques that can be used to establish the presence or absence of the disorder. Therefore, the signs and symptoms used to define schizophrenia play a critical role in clinical practice and research. Considerable variability has been observed in virtually any domain in which schizophrenia has been examined, including age of onset, clinical characteristics, treatment response, and course. This observed heterogeneity has suggested a number of approaches to subdividing the clinical syndrome. However, the numerous psychological, physiological, and anatomic characteristics associated with schizophrenia reflect differences in group means rather than qualifying as characteristics with sufficient sensitivity and specificity to be useful in classification. Carpenter and colleagues (1993) pointed out that there are no pathognomonic signs for schizophrenia and that the diagnosis is based on a combination of signs and symptoms, none of which is unique to schizophrenia.

Historical Background

The traditional approaches to specifying subtypes of schizophrenia were based on clinical features and the course of illness (Goldstein and Tsuang, 1988). Kraepelin (1971/1919) termed the illness we now call schizophrenia "dementia praecox" to signify the deteriorating, chronic course of the illness. His approach was empirical, based on clinical signs and symptoms. He viewed the disease as a unitary process reflecting a common underlying pathophysiological process. Within this unitary process, he identified the three subtypes of hebephrenia, catatonia, and paranoia. Bleuler (1911), another seminal figure in the history of schizophrenia and a contemporary of Kraepelin, did not agree with this unitary process. Bleuler coined the term "schizophrenia" and embodied his belief in the heterogeneity of schizophrenia in his term the "group of schizophrenias". Bleuler used the three subtypes described by Kraepelin but also included simple schizophrenia, a subtype that did not require psychotic symptoms. Bleuler did not believe that schizophrenia necessarily had a chronic, deteriorating course.

Goldstein and Tsuang (1988) suggested that the differences in the approaches articulated by Kraepelin and Bleuler illustrate important considerations in the general issue of subtyping schizophrenia. Bleuler emphasized what he believed to be the mental phenomena underlying the observed characteristics of schizophrenia: a disturbance in thought association, ambivalence, autism, and a disruption in the integration of intellect and affect (the four A's). Bleuler's approach yielded a broad definition of schizophrenia that included many individuals who would not receive a diagnosis of schizophrenia using the narrower, more restrictive approach of Kraepelin. Like Bleuler, Meyer (1906) viewed schizophrenia from a psychological viewpoint and believed it could exist in less severe forms. For most of the twentieth century, the United States followed a Bleulerian approach but shifted to a Kraepelinian model in 1980 with the advent of the third edition of the *Diagnostic and Statistical Manual of Mental Disorders* (American Psychiatric Association, 1980). While less severe forms of the disorder could have conceivably been labeled schizophrenia under a Bleulerian heading, this became less frequent with the shift toward a narrower diagnostic approach.

Langfeldt (1937) subdivided individuals with schizophrenia into process and reactive categories. Process (chronic) schizophrenia referred to individuals with a slow insidious onset and an unremitting course whereas reactive (acute) schizophrenia referred to individuals with rapid onset and periods of remission. It was hypothesized that individuals with process schizophrenia had an insidious "schizoid" premorbid personality that eventually deteriorated, resulting in a less favorable prognosis than individuals with reactive schizophrenia. There was some empirical support for this view, but there was not a clear-cut bimodal distribution of outcomes related to the process-reactive distinction. In a similar vein, Gittelman-Klein and Klein (1969) proposed dividing schizophrenia into poor premorbid and good premorbid subtypes. Evidence was adduced to demonstrate an association between these subtypes and symptomatology, outcome, and some biological variables (Gittelman-Klein and Klein, 1969; Harrow et al., 1969; Klorman et al., 1977; Strauss et al., 1977; Bland et al., 1978).

Various researchers and theorists described less severe forms of schizophrenia-like disorders. A number of terms were used to describe this group, including "ambulatory schizophrenics" (Zilboorg, 1941), "preschizophrenics" (Rapaport et al., 1946), "latent schizophrenia" (Federn, 1947), "latent psychosis" (Bychowski, 1953), and "pseudoneurotic schizophrenia" (Hoch, 1949). Rado (1950) first used the term "schizotypal" as a shorter term for "schizophrenic phenotype", reflecting his belief that the disorder was an overt manifestation of some inherited predisposition or genotype. Meehl (1962) expanded on the concept of schizotypy, describing the four main traits: cognitive slippage, interpersonal aversiveness, anhedonia, and ambivalence. Meehl hypothesized that a range of disorders resulted from the interaction between the environment and an inherited, integrative neural deficit he called "schizotaxia". It has often been suggested that there is a spectrum of disorders associated with schizophrenia that include phenotypes that are less severe than schizophrenia but that reflect the same genetic vulnerability. In current nomenclature, schizotypal personality disorder (SPD) exemplifies this concept of a variant of schizophrenia that is clinically less severe.

Current Conceptions

With the advent of the *Diagnostic and Statistical Manual, 3rd ed.* (DSM-III) in 1980, the United States adopted Kraepelin's more restrictive model for defining schizophrenia. The DSM provides the most widely used classification system for psychiatric disorders in the United States and is now in its fourth edition (DSM-IV; American Psychiatric Association, 1994). DSM-IV requires the presence of certain characteristic symptoms for one month, a total duration of illness (including prodromal and residual symptoms) of at least six months, and functional impairment.

The characteristic symptoms that define the active episode of schizophrenia as stated in DSM-IV include the positive symptoms of delusions or hallucinations, disorganized speech or behavior, and the negative symptoms of affective flattening, alogia, or avolition. DSM-IV is an expansion of DSM-III-R in its consideration of alogia and avolition as more central, characteristic symptoms. In addition to the more traditional categorical subtypes, DSM-IV also includes an alternative dimensional approach for subtyping schizophrenia in its "Criteria Sets and Axes Provided for Further Study" section. This approach includes a positive symptom (psychotic) dimension, a disorganized dimension, and a negative symptom (deficit) dimension corresponding to results from factor analytic studies. The original positive-versus-negative symptom distinction was posited by Andreasen and Olsen (1982) and Crow (1980). Positive symptoms are those representing a behavioral excess generally considered psychotic (e.g., hallucinations, delusions, bizarre behavior), while negative symptoms are those representing a deficiency from normal behavior (e.g., a lack of normal social responsiveness, flat affect). However, numerous studies examining the global scores of the Scale for the Assessment of Negative Symptoms (SANS) and the Scale for the Assessment of Positive Symptoms (SAPS; Andreasen and Olsen 1982) in subjects with schizophrenia have yielded three factors: a negative symptom factor, a positive symptom factor, and a disorganization factor (Klimidis et al., 1993). Delusions and hallucinations compose the psychotic dimensions, inappropriate affect and disorganized speech and behavior compose the disorganized dimension, and the negative or deficit dimension comprises the negative symptoms of schizophrenia. It is suggested that each dimension may reflect a different pathophysiological process and may respond differently to treatment.

DSM-IV describes a number of subtypes of schizophrenia. The assignment of subtype is based on the predominant symptomatology at the time the diagnosis is made. In general, DSM-IV corresponds to the traditional subtypes of schizophrenia: paranoid, catatonic, and disorganized (hebephrenic). DSM-IV also includes undifferentiated and residual subtypes.

Catatonic Type

This subtype is characterized primarily by psychomotor disturbances. These disturbances may take the form of extremes at either end of a continuum ranging from excessive motor activity to virtually complete inactivity. Excessive activity associated with catatonia is apparently purposeless and unrelated to environmental stimulation. Motor inactivity may be characterized by waxy flexibility (catalepsy) or stupor. It may also take the form of extreme negativism in which the individual maintains a rigid posture or bizarre stance, resisting efforts to alter his or her position. There may be pronounced unusual movements such as grimacing, stereotypies, mannerisms, mimicry, or automatic obedience. Other symptoms may include echolalia (senseless repetition of a word or phrase spoken by another) or echopraxia (repetition of the movements of another).

Disorganized (Hebephrenic) Type

This subtype is characterized primarily by disorganized speech and behavior and by affect that is inappropriate or flat. The individual may demonstrate inappropriate laughter or "silliness". Behavior is disorganized and the individual may be unable to carry out the normal activities of living such as self-care. If the patient has hallucinations or delusions, they are likely to be fragmented, rather than relating to a coherent theme. The disorganized subtype may also be associated with odd behavior such as grimacing and mannerisms.

Paranoid Type

This subtype is characterized primarily by prominent delusions and auditory hallucinations that occur without substantial cognitive and affective deterioration. The delusions tend to be coherent and are typically oriented around persecutory, grandiose, religious, or jealous themes. The content of hallucinations often relates to the delusional themes. The individual may be anxious, angry, and argumentative.

Undifferentiated Type

This subtype is used for individuals who meet the general diagnostic criteria for schizophrenia but who do not meet the criteria for the catatonic, disorganized, or paranoid subtypes.

Residual Type

This category is used for individuals who do not meet current criteria for any of the other DSM-IV categories but who have had at least one past episode that met criteria.

There is a hierarchy among the subtypes with the diagnosis of catatonic subtype taking priority whenever symptoms warrant this diagnosis. If symptoms of the disorganized subtype (traditionally known as hebephrenia) are present in the absence of catatonic symptoms, the disorganized subtype is diagnosed. The diagnosis of paranoid subtype is made for patients with the appropriate symptoms in the absence of catatonic or disorganized features. DSM-IV provides the additional category of undifferentiated type for individuals who meet the diagnostic criteria for schizophrenia but who do not meet the criteria for the catatonic, disorganized, or paranoid subtypes. Residual subtype is used to classify individuals who, following an episode, no longer meet the full criteria but still manifest a continuing disturbance. DSM-IV indicates that a given individual may have the clinical features of more than one subtype, but the hierarchy is applied in the order of catatonic, disorganized, paranoid, and undifferentiated subtypes.

These DSM-IV subtypes may have implications for course and outcome, although, over time, it is common for individuals with schizophrenia to meet diagnoses for more than one subtype. The disorganized subtype is often associated with poor premorbid adjustment, an onset that is slow and insidious, and a stable course without meaningful remission. The paranoid subtype tends to have a later onset and probably a more favorable prognosis than other subtypes.

Spectrum Disorders

Premorbid personality patterns that may precede schizophrenic episodes or that may represent schizophrenia spectrum disorders are classified in the DSM under the "Axis II Personality Disorders" section. Personality disorders are enduring patterns of inner experience and behavior that deviate from the norm, that are inflexible and pervasive, and that lead to distress or impairment. The personality disorders are grouped into three clusters based on symptom descriptions. The first cluster contains disorders that are odd or eccentric, and these disorders are hypothesized to be in the schizophrenia spectrum. These disorders include paranoid, schizoid, and schizotypal personality disorders, although the latter has the most evidence to support its relationship to schizophrenia. The DSM-III criteria for schizotypal personality disorder (SPD) were drawn from the Danish Adoption study classification of borderline schizophrenia, and the criteria are similar in DSM-IV. Schizotypal personality disorder is a pattern of both interpersonal deficits and cognitive, perceptual, or behavioral distortions or eccentricities. To be diagnosed, a person has to meet most of the following criteria: ideas of reference; odd beliefs, or magical thinking; unusual perceptual experience; odd thinking and speech; suspicious or paranoid ideation; odd or eccentric behavior or appearance; lack of close friends; and excessive social anxiety that does not diminish with familiarity and tends to be associated with paranoid fears.

ETIOLOGICAL HETEROGENEITY

The clinical heterogeneity of schizophrenia discussed in the previous section obviously leads to questions regarding etiology. There is substantial evidence for genetic influence on schizophrenia. Most of this evidence comes from the traditional methods of genetic epidemiology—family, twin, and adoption studies (Lyons et al., 1991a). Numerous family studies have demonstrated a statistically significantly greater risk for schizophrenia and related disorders among the relatives of individuals with schizophrenia. However, family studies alone cannot distinguish between the family environment and genes as the reasons for family resemblance for schizophrenia. Adoption studies can disentangle environmental and genetic sources of risk and have demonstrated that genes make a major contribution in the etiology of schizophrenia. Twin studies have also demonstrated that genetic factors play an important role in the etiology of schizophrenia, with an estimation of the broad heritability to be about 67% (McGuffin et al., 1984). The fact that no single genetic model provides a good fit to the observed patterns in families has also led to the suggestion that schizophrenia is etiologically heterogenous. Slater (1947) proposed that: "In view of the fact that schizophrenia is a good deal more common than any single genetically determined disorder is otherwise known to be, heterogeneity is inherently probable." The clinical phenomena that we observe in schizophrenia probably do not correspond in a one-to-one fashion with a specific genotype that is necessary and sufficient; schizophrenia may represent a common final pathway. There are precedents in other areas of medicine for this type of phenomenon. For example, Vogel and Motulsky (1986) described 10 different forms of glycogen storage disease with similar clinical manifestations. Each form of the disease is associated with a different enzyme defect reflecting 10 different genetic loci. This indicates the potential for similar clinical phenotypes to have very distinct genetic etiologies. Alternatively, pleiotropic effects may exist in which a single etiology results in various phenotypes.

If etiologic heterogeneity exists such as in the form of individuals with different etiologies in the same sample, it may be difficult to distinguish the specific genetic etiologies.

Efforts have been made to determine whether different etiologies underlie different subtypes by examining the concordance of subtypes within affected relatives. The research shows a tendency toward similar subtypes within twins but not to a conclusive degree (McGuffin et al., 1987). Kendler and colleagues (1994) investigated the validation of subtyping by clinical features by examining the relationship between clinical features and familial risk. Specifically, the presence of certain symptoms, course, global outcome, negative symptoms, and level of functioning, were compared to risk of schizophrenia or schizophrenia spectrum disorders among first-degree relatives. No relationship was observed. The authors concluded that familial factors contribute importantly to an individual's risk of schizophrenia but that the clinical features of the illness were not associated with familial risk.

Another approach to etiologic heterogeneity is the suggestion that schizophrenia may result from either primarily genetic or environmental causes. Cases presumed to be primarily or exclusively genetic in origin are called familial cases, and cases with primarily or exclusively environmental origins are called sporadic cases. One way in which the familial/sporadic approach has tried to identify "more genetic" versus "more environmental" cases has been to divide individuals with schizophrenia into those with one or more relatives with the disorder (positive family history) and those without any relatives with schizophrenia (negative family history). It is assumed that those with a positive family history are more likely to have a genetic form of the illness, and those with a negative family history are more likely to have an environmental form of the illness. The familial/sporadic approach also examines twins. Identical twin pairs in which both members of the pair have schizophrenia may be more likely to have a genetic form of the disorder, whereas identical twin pairs that are discordant (i.e., they share the same genes but only one twin has schizophrenia) may be more likely to have a sporadic form of the illness.

Lyons and colleagues (1989) reviewed the sizable body of empirical research that has utilized the familial/sporadic distinction and this research is summarized here. A number of studies have indicated that the familial schizophrenics tend to do more poorly on measures of sustained attention (Orzack and Kornetsky, 1971; Walker and Shaye, 1982), while abnormalities on the electroencephalogram are more common among sporadic schizophrenics (Hays, 1977; Kendler and Hays, 1982). Lewis and colleagues (1987) reviewed 10 computed tomography studies and concluded that sporadic cases were more likely to have structural brain abnormalities. Perinatal complications have also been found to be more common in sporadic cases.

A preliminary report of neuropsychological deficits that characterize familial schizophrenia found that familial schizophrenics showed significant variability in problem solving, abstraction, and motor control when compared with nonfamilial schizophrenics. Cluster analysis of these neuropsychological data found that, while nonfamilial schizophrenics fell into one relatively homogeneous cluster, familial schizophrenics could be divided into three distinct clusters. The patients in these three clusters differed with respect to their morbid risk for schizophrenia-spectrum disorder: patients in cluster A showed deficits across all categories, including abstraction, problem-solving, and motor control; patients in cluster B showed severe decrements in abstraction and problem solving; and patients in cluster C had severe deficits in motor control. The first- and second-degree relatives of patients in cluster A showed a morbid risk of schizophrenia of 7.0%, relatives of patients in cluster B showed a morbid risk of 14.9%, and relatives in cluster C showed a morbid risk of 19.7% (Sautter et al., 1995). These results replicate a number of family studies indicating that deficits in attention, perceptual-motor speed, and problem-solving are associated with an increased familial liability for schizophrenia (Kremen et al., 1994).

Several authors have found insignificant or relatively weak associations between age of onset of schizophrenia in patients with or without a family history of the illness (Crow, 1980; Roy and Crowe, 1994; Roy et al., 1994; Alda et al., 1996). However, many of these studies are difficult to compare because of reduced statistical power of the familial-sporadic analysis. Alda and colleagues (1996) analyzed the variance of the combined results of three large samples of schizophrenics and found that patients with no family history of schizophrenia had a consistently higher average age of onset, although no interaction between sex and family history was found. There are some indications that females tend to fall ill with schizophrenia three to four years later than males (Hafner, 1993). However, several studies have found a lack of sex differences in age at onset of patients with familial as opposed to sporadic schizophrenia (Albus and Maier, 1995; Gorwood et al., 1995; Alda et al., 1996).

In general, findings with regard to discordant identical twin pairs have not supported the view that the schizophrenic proband is likely to have a sporadic form of the illness. However, due to the relative scarcity of appropriate subjects, the studies that have been carried out do not have great statistical power. Methodological difficulties in the familial-sporadic approach have been identified (Eaves et al., 1986; Kendler, 1987), but the number of positive findings that have been reported suggest that the approach may have some merit.

Reflecting advances in the field of psychiatric genetics, molecular genetic approaches have been brought to bear on schizophrenia. In 1988, a positive linkage between schizophrenia and a marker on chromosome 5 was described (Sherrington et al., 1988). Chromosome 5 had been implicated as a possibly relevant site based on a case description of a chromosomal abnormality associated with schizophrenia (Bassett et al., 1988). However, there has been a substantial number of failures to replicate this finding (e.g., Kennedy et al., 1988; St. Clair et al., 1989; Detera-Wadleigh et al., 1989; Diehl, 1989; McGuffin et al., 1990).

A recent development has been a linkage on chromosome 6 reported by several groups (Straub et al., 1995; Moises et al., 1995; Schwab et al., 1995; Antonarakis et al., 1995). However, several groups failed to replicate the linkage to this region (Mowry et al., 1995; Gurling et al., 1995). Consistent with the theme of heterogeneity is the suggestion from these recent linkage findings that the region identified on chromosome 6 is associated with risk of schizophrenia in 15% to 30% of families segregating for schizophrenia (Straub et al., 1995). Pulver and colleagues (1995) and Kendler and colleagues (1996) have published evidence of a vulnerability locus on chromosome 8p. This locus appeared to affect only a modest proportion of families and predispose to a spectrum of schizophrenia and non-schizophrenia disorders.

Pulver and colleagues (1994b) suggested the possibility of linkage to the long arm of chromosome 22, and consistent findings were reported by other investigators (Coon et al., 1994; Polymeropoulos et al., 1994; Vallada et al., 1995). However, a collaborative study reported by Pulver and colleagues (1994a) excluded linkage to chromosome 22 loci and another study could not find evidence of linkage (Kalsi et al., 1995). Gill and colleagues (1996) reported a combined analysis of eleven schizophrenia linkage data sets. As a group, these data provided statistically significant evidence for linkage. However, their analyses suggested that the putative schizophrenia gene accounted for about 2% of the variability in the liability to develop the disorder.

A different way in which the heterogeneity of schizophrenia may manifest itself is in disorders that are not schizophrenia per se but that may reflect the same underlying etiological and pathophysiological processes (i.e., the schizophrenia-spectrum disorders). The spectrum disorders hypothetically reflect the same genetic vulnerability as schizophrenia itself. Schizotypal personality disorder (SPD) exemplifies this concept, and as mentioned previously, the criteria for SPD were drawn from the Danish Adoption study, a genetically oriented study that found an excess of schizotypal symptoms in biological relatives of schizophrenic persons. However, once the disorder was embodied in DSM-III, clinicians began applying the diagnosis to patients. Since the original notion of SPD is predicated on its genetic association with schizophrenia and on the finding of excess risk in the relatives of schizophrenic probands, one would predict that there would be an excess risk of schizophrenia among the relatives of probands with SPD. However, the evidence for excess risk of schizophrenia among the relatives of SPD probands is at best inconclusive. Thaker and colleagues (1993) reviewed the studies that assessed the risk of schizophrenia among relatives of schizotypal probands. Six studies failed to find elevated rates of schizophrenia among the relatives of SPD probands (Soloff and Millward, 1983; Torgersen, 1984; Baron et al., 1985; Schulz et al., 1986; Siever et al., 1990). However, Lenzenweger and Loranger (1989), Battaglia and colleagues (1991), and Thaker and colleagues (1993) found higher rates of schizophrenia among the relatives of schizotypals than in the comparison groups they studied. This leaves unresolved the issue of whether schizotypals identified clinically should be considered to be in a genetic spectrum with schizophrenia.

Another putative "type" of schizotypal are those that are identified from population samples using psychometric screening, often with instruments such as the Psychosis Proneness scales of Chapman and Chapman (1977). There have also been a number of studies that have recruited symptomatic volunteers with schizotypal features through newspaper advertisements (e.g., Lyons et al., 1991b; Thaker et al., 1993). It remains unclear what proportion of schizotypals identified through these various methods have a disorder that is genetically related to schizophrenia. A number of authors have suggested that the negative symptoms of schizotypy such as inadequate rapport and social anxiety may best define the type of SPD most likely to reflect a schizophrenia genotype.

Another review by Battaglia and Torgersen (1996) found that the majority of available data support schizotypal personality disorder as a phenotype liable for the schizophrenia spectrum. More specifically, evidence from twin studies has linked the affect-constricted and eccentric aspects of schizotypal personality disorder that share important genetic influences with schizophrenia (Battaglia and Torgersen, 1996). There are also data that suggest that the more "negative" symptoms of schizophrenia are significantly more concentrated among schizotypal relatives of schizophrenic probands whereas the more "positive" symptoms may be more frequent among schizotypal relatives of probands with mood disorders (Battaglia and Torgersen, 1996).

Family studies suggest that negative symptoms may be an expression of the schizophrenia genotype (Tsuang et al., 1991; Faraone et al., 1995). Lyons and colleagues (1994) found that schizotypal subjects ascertained through a relationship with a schizophrenic proband scored significantly higher on a factor defined primarily by negative symptoms of SPD in comparison to schizotypal subjects ascertained through a relationship with probands with affective disorder. The groups did not differ on a factor reflecting positive symptoms of schizotypy. Consistent with this were the findings of Lyons and colleagues (1994), who found that schizotypal subjects related to schizophrenia probands displayed more inadequate rapport than schizotypal subjects related to affective disorder probands.

CONCLUSIONS

At the clinical or symptom level, there is no doubt that schizophrenia is a heterogenous disorder. Patients with quite different and distinct patterns of symptoms can meet the diagnostic criteria for schizophrenia. However, there is still no definitive answer as to whether patients with different clinical presentations and courses of illness reflect different etiological and pathophysiological mechanisms or whether there is a single disease process that underlies the multifarious clinical characteristic symptoms. As we learn more, we will be better able to answer questions about the extent to which schizophrenia may reflect heterogenous etiological and pathophysiological phenomena.

REFERENCES

Albus, M. & Maier, W. (1995). Lack of gender differences in age at onset of familial schizophrenia. Schiz. Res. 18, 51-57.

Alda, M., Ahrens, B., Lit, W., Dvorakova, M., Labelle, A., Zvolsky, P., & Jones, B. (1996). Age of onset in familial and sporadic schizophrenia. Acta Psychiatr. Scand. 93, 447-450.

American Psychiatric Association (1980). *Diagnostic and Statistical Manual of Mental Disorders, 3rd ed.* American Psychiatric Association Press, Washington, D.C.

American Psychiatric Association (1994). *Diagnostic and Statistical Manual of Mental Disorders, 4th ed.* American Psychiatric Association Press, Washington, D.C.

Andreasen, N.C. & Olsen, S.A. (1982). Negative v positive schizophrenia. Definition and validation. Arch. Gen. Psych. 39, 789-794.

Andreasen, N.C., Olsen, S.A., Dennert, J.W. & Smith, M.R. (1982). Ventricular enlargement in schizophrenia: Relationship to positive and negative symptoms. Amer. J. Psych. 139, 297-302.

Baron, M., Gruen, R., Rainer, J.D., Kane, J., Asnis, L., & Lord, S. (1985). A family study of schizophrenic and normal control probands: Implications for the spectrum concept of schizophrenia. Biol. Psych. 20, 94-119.

Battaglia, M., Gasperini, M., Sciuto, G., Scherillo, P., & Diaferia, G. (1991). Psychiatric disorders in the families of schizotypal subjects. Schiz. Bull. 17 (4), 659-668.

Battaglia, M. & Torgersen, S. (1996). Schizotypal disorder: At the crossroads of genetics and nosology. Acta. Psychiatr. Scand. 94, 303-310.

Bland, R.C., Parker, J.H., & Orn, H. (1978). Prognosis in schizophrenia. Prognostic predictors and outcome. Arch. Gen. Psychiatry 35 (1), 72-77.

Bleuler, E. (1950). *Dementia Praecox or the Group of Schizophrenias.* (Zinkin, J., Trans.). International Universities Press, New York. (Originally published in 1911.)

Bychowski, G. (1953). The problem of latent psychosis. J. Amer. Psychoan. Assoc. 4, 484-503.

Carpenter, W.T., Jr., Buchanan, R.W., Kirkpatrick, B., Tamminga, C., & Wood, F. (1993). Strong inference theory testing, and the neuroanatomy of schizophrenia. Arch. Gen. Psych. 50 (10), 825-831.

Chapman, L. & Chapman, J. (1977). Selection of subjects in studies of cognition. J. Abnorm. Psych. 86, 10-15.

Coon, H., Holik, J., Hoff, M., Reimherr, F., Wender, P., Freedman, R., & Byerley, W. (1994). Analysis of chromosome 22 markers in 9 schizophrenia pedigrees. Am. J. Med. Genet. (Neuropsychiatric Genetics) 54, 72-79.

Crow, T.J. (1980). Positive and negative schizophrenic symptoms and the role of dopamine. Brit. J. Psych. 139, 379-386.

Diehl, S.R. (1989). Schizophrenic syndromes in epilepsies. Schiz. Bull. 15 (3), 403-419.

Eaves, L.J., Kendler, K.S., & Schulz, S.C. (1986). The familial sporadic classification: Its power for the resolution of genetic and environmental etiological factors. J. Psych. Res. 20, 115-130.

Faraone, S.V., Chen, W.J., Goldstein, J.M., & Tsuang, M.T. (1994). Gender differences in the age of onset of schizophrenia. British Journal of Psychiatry, 164, 625-629.

Faraone, S.V., Kremen, W.S., Lyons, M.J., Pepple, J.R., Seidman, L.J., & Tsuang, M.T. (1995). Diagnostic accuracy and linkage analysis: How useful are schizophrenia spectrum phenotypes? Am. J. Psych.. 152, 1286-1290.

Federn, P. (1947). Principles of psychotherapy in latent schizophrenia. Am. J. Psychotherapy 1, 129-139.

Gittelman-Klein R. & Klein, D.F. (1969). Premorbid asocial adjustment and prognosis in schizophrenia. J. Psych. Res. 7, 35-53.

Goldstein, J.M. & Tsuang, M.T. (1988). The process of subtyping schizophrenia: Strategies in the search for homogeneity. In: *Handbook of Schizophrenia: Nosology, Epidemiology & Genetics*. Vol. 3, pp. 63-83. Elsevier, Amsterdam.

Gorwood, P., Leboyer, M., Jay, M., Payan, C., & Feingold, J. (1995). Gender and age at onset of schizophrenia: Impact of family history. Am. J. Psych.. 152, 208-212.

Harrow, M., Tucker, G.J., & Bromet, E. (1969). Shortterm prognosis of schizophrenic patients. Arch. Gen. Psych. 21 (2), 195-202.

Hays, P. (1977). Electroencephalographic variants and genetic predisposition to schizophrenia. J. Neurol. Neurosurg. Psychiatry. 40, 753-755.

Hoch, P.H. (1949). Pseudoneurotic forms of schizophrenia. Psychiatr. Quarterly 23, 248.

Kalsi, G., Brynjolfsson, J., Butler, R., Sherrington, R., Curtis, D., Sigmundssonn, T., Read, T., Murphy, T., Sharma, T., Petursson, H., & Gurling, H.M.D. (1995). Linkage analysis of chromosome 22q12-13 in a United Kingdom/Icelandic sample of 23 multiplex schizophrenia families. Am. J. Med. Genet. (Neuropsychiatric Genetics) 60, 298-301.

Kendler, K.S. (1987). Sporadic vs. Familial classification given etiologic heterogeneity: I. Sensitivity, specificity, and positive and negative predictive power. Genetic Epidemiology 4, 313-330.

Kendler, K.S. & Hays, P. (1982). Familial and sporadic schizophrenia: A symptomatic, prognostic, and EEG comparison. Am. J. Psych.. 139, 1557-1562.

Kendler, K.S., McGuire, M., Gruenberg, A.M., & Walsh, D. (1994). Clinical heterogeneity in schizophrenia and the pattern of psychopathology in relatives: Results from an epidemiologically based family study. Acta. Psychiatrica Scand. 89, 294-300.

Kendler, K.S., MacLean C.J., O'Neill, F.A., Burke, J., Murphy, B., Duke, F., Shinkwin, R., Easter, S.M., Webb, B.T., Zhang, J., Walsh, D., & Staub, R.E. (1996). Evidence for a Schizophrenia Vulnerability Locus on Chromosome 8p in the Irish Study of High-Density Schizophrenia Families. Am. J. Psych.. 153, 1534-1540.

Kennedy, J.L., Giuffra, L.A., Moises, H.W., Cavalli-Sforza, L.L., Pakstis, A.J., Kidd, J.R., Castiglione, C.M., Sjorgen, B., Wetterberg, L., & Kidd, K.K. (1988). Evidence against linkage of schizophrenia to markers on chromosome 5 in a northern Swedish pedigree. Nature 336, 167-169.

Klimidis, S., Stuart, G.W., Minas, I.H., Copolov, D.L., & Singh, B.S. (1993). Positive and negative symptoms in the psychoses. Re-analysis of published SAPS and SANS global ratings. Schiz. Res. 9, 11-18.

Klorman, R., Strauss, J.S., & Kokes, R.F. (1977). Premorbid adjustment in schizophrenia. Part IV. Some biological approaches to research on premorbid functioning in schizophrenia. Schiz. Res. 3 (2), 226-239.

Kraepelin, E. (1971). *Dementia Praecox and Paraphrenia.* (Barclay, R.M., Trans.) Krieger, Huntington, N.Y. (Originally published in 1919).

Langfeldt, G. (1937). *The Prognosis in Schizophrenia and the Factors Influencing the Course of the Disorder*. Munksgaard, Copenhagen.

Lenzenweger, M.F. & Loranger, A.W. (1989). Detection of familial schizophrenia using a psychometric measure of schizotypy. Arch. Gen. Psych. 46 (10), 902-907.

Lewis, S., Reveley, A., Reveley, M., et al. (1987). The familial/sporadic distinction as a strategy in schizophrenia research. Brit. J. Psych. 151, 306-313.

Lyons, M.J., Faraone, S.V., Kremen, W.S., & Tsuang, M.T. (1989). Familial and sporadic schizophrenia: A simulation study of statistical power. Schiz. Res. 2, 345-353.

Lyons, M.J., Kendler, K.S., Provet, A.G., & Tsuang, M.T. (1991a). The genetics of schizophrenia. In: *Genetic Issues in Psychosocial Epidemiology*. pp. 119-152. Rutgers University Press, New Brunswick, NJ.

Lyons, M.J., Merla, M.E., Young, L., & Kremen, W.S. (1991b). Impaired neuropsychological functioning in symptomatic volunteers with schizotypy: Preliminary findings. Biol. Psych. 30, 424-426.

Lyons, M.J., Toomey, R., Faraone, S.V., & Tsuang, M.T. (1994). Comparison of schizotypal relatives of schizophrenic versus affective probands. Am. J. Med. Genetics (Neuropsychiatric Genetics) 54, 279-285.

McGuffin, P., Farmer, A.E., Gottesman I.I., Murray R.M., & Reveley A.M. (1984). Twin concordance for operationally defined schizophrenia. Archives of General Psychiatry 41, 541-545.

McGuffin, P., Farmer, A.E., & Gottesman I.I. (1987). Is there really a split in schizophrenia? The genetic evidence. Brit. J. Psych. 150, 581-592.

McGuffin, P., Sargeant, M., Hetti, G., Tidmarsh, S., Whatley, S., & Marchbanks, R.M. (1990). Exclusion of a schizophrenia susceptibility gene from the chromosome 5q11-q13 region: New data and a reanalysis of previous reports. Am. J. Hum. Genet. 47 (3), 524-535.

Meehl, P.E. (1962). Schizotaxia, schizotypy, schizophrenia. American Psychologist 17, 827.

Orzack, M.H. & Kornetsky, C. (1971). Environmental and familial predictors of attention behavior in chronic schizophrenics. J. Psychiatr. Res. 9, 21-29.

Polymeropoulos, M.H., Coon, H., Byerley, W., Gershon, E.S., Crow, T.J., Rubenstein, J., Hoff, M., Holik, J., Smith, A.M., Shields, G., Bass, N.J., Poulter, M., Lofthouse, R., Vita, A., Morganti, C., Merril, C.R., & DeLisi, L.E. (1994). Search for a schizophrenia susceptibility locus on human chromosome 22. Am. J. Med. Genet., (Neuropsychiatric Genetics) 54, 93-99.

Pulver, A.E., Karayiorgou, M., Lasseter, V.K., Wolyneic, P., Kasch, L., Antonarakis, S., Housman, D., Kazazian, H.H., Meyers, D., Nestadt, G., Ott, J., Liang, K.-Y., Lamacz, M., Thomas, M., Childs, B., Diehl, S.R., Wang, S., Murphy, B., Sun, C., O'Neill, F.A., Nie, L., Sham, P., Burke, J., Duke, B.W., Duke, F., Kipps, B.R., Bray, J., Hunt, W., Shinkwin, R., Nuallain, M.N., Su, Y., MacLean, C.J., Walsh, D., Kendler, K.S., Gill, M., Vallada, H., Mant, R., Asherson, P., Collier, D., Parfitt, E., Roberts, E., Nanko, S., Walsh, C., Daniels, J., Murray, R., McGuffin, P., Owen, M., Laurent, C., Dumas, J.-B., d'Amato, T., Jay, M., Martinez, M., Campion, D., & Mallet, J. (1994a). Follow-up of a report of a potential linkage for schizophrenia on chromosome 22q12-q13.1: Part 2. Am. J. Med. Genet. (Neuropsychiatric Genetics) 54, 44-50.

Pulver, A.E., Karayiorgou, M., Wolyneic, P., Lasseter, V.K., Kasch, L., Nestadt, G., Antonarakis, S., Housman, D., Kazazian, H.H., Meyers, D., Ott, J., Lamacz, M., Liang, K.-Y., Hanfelt, J., Ullrich, G., DeMarchi, N., Ramu, E., McHugh, P.R., Adler, L., Thomas, M., Carpenter, W.T., Manschreck, T., Gordon, C.T., Kimberland, M., Babb, R., Puck, J., & Childs, B. (1994b). Sequential strategy to identify a susceptibility gene for schizophrenia on chromosome 22q12-q13.1: Part 1. Am. J. Med. Genet. (Neuropsychiatric Genetics) 54, 36-43.

Pulver, A.E., Lasseter, V.K. Kasch, L., Wolyniec, P., Nestadt, G., Blouin, J-L., Kimberland, M., Babb, R., Vourlis, S., Chen, H., Laliot, M., Morris, M.A., Karayiogou, M., Ott, J., Meyers, D., Antonarakis, S.E., Housman, D., & Kazazian, H.H. (1995). Schizophrenia: A genome scan targets chromosomes 3p and 8p as potential sites of susceptibility genes. Amer. J. Med. Genet. Neuropsychiatr Genet. 60(3), 252-260.

Rado, S. (1953). Dynamics and classification of disorders behavior. Am. J. Psych. 110, 406.

Rapaport, D., Gill, M., & Scafer, R. (1946). *Diagnostic Psychological Testing*. Yearbook Publishers, Chicago.

Roy, M-A. & Crowe, R.R. (1994). Validity of the familial and sporadic subtypes of schizophrenia. Am. J. Psych. 151, 805-814.

Roy, M-A., Flaum, M.A., Gupta, S., Jaramillo, L., & Andreasen, N.C. (1994). Epidemiological and clinical correlates of familial and sporadic schizophrenia. Acta Psychiatr. Scand. 89, 324-328.

Sautter, F.J., McDermott, B.E., Cornwell, J., Johnson, J., Borges, A., Wilson, A.F., Vasterling, J.J., & Foundas, A.L. (1995). A preliminary study of the neuropsychological heterogeneity of familial schizophrenia. Schiz. Res. 18, 1-7.

Schulz, P.M., Schulz, S.C., Goldberg, S.C., Ettigi, P., Resnick, R.J., & Friedel, R.O. (1986). Diagnoses of the relatives of schizotypal outpatients. J. Nerv. & Mental Dis. 174 (8), 457-463.

Sherrington, R., Brynjolesson, J., Petursson, H., Potter, M., Dudleston, K., Barraclough, B., Wasmuth, J., Dobbs, J., & Gurling, H. (1988). Localization of the susceptibility locus for schizophrenia on chromosome 5. Nature 336, 164-167.

Siever, L.J., Silverman, J.M., Horvath, T.B., Klar, H., Coccaro, E., Keefe, R.S., Pinkham, L., Rinaldi, P., Mohs, R.C., & Davis, K.L. (1990). Increased morbid risk for schizophrenia-related disorders in relatives of schizotypal personality disordered patients. Arch. Gen. Psych. 47 (7), 634-640.

Slater, E. (1947). Genetical causes of schizophrenic symptoms. Monatsschrift für Psychiatrie und Neurologie 113, 50-58. (Reprinted along with other papers on psychiatry and genetics by same author in *Man, Mind, and Heredity*. [1971]. [Shields, J. & Gottesman, I.I, Eds.]). Johns Hopkins Press, Baltimore.

Soloff, P.H. & Milward, J.W. (1983). Psychiatric disorders in the families of borderline patients. Arch. Gen. Psych. 40 (1), 37-44.

St. Clair, D., Blackwood, D., Muir, W., Baillie, D., Hubbard, A., Wright, A., & Evans, H.J. (1989). No linkage of chromosome 5q11-q13 markers to schizophrenia in Scottish families. Nature 339 (6222), 305-309.

Strauss, J., Kokes, R., Klorman, R., & Sacksteder, J. (1977). Premorbid adjustment in schizophrenia: Concepts, measures, and implications. Part I. The concept of premorbid adjustment. Schiz. Bull. 3 (2), 182.

Thaker, G., Adami, H., Moran, M., Lahti, A., & Cassady, S. (1993). Psychiatric illnesses in families of subjects with schizophrenia-spectrum personality disorders: High morbidity risks for unspecified functional psychoses and schizophrenia. Am. J. Psych. 150 (1), 66-71.

Torgerson, S. (1984). Genetic and nosological aspects of schizotypal and borderline personality disorders. Arch. Gen. Psych. 41, 546.

Tsuang, M.T., Gilbertson, M.W., & Faraone, S.V. (1991). Genetic transmission of negative and positive symptoms in the biological relatives of schizophrenics. In: *Negative vs. Positive Schizophrenia*. pp. 265-291. Springer-Verlag, Berlin.

Tsuang, M.T., Lyons, M.J., & Faraone, S.V. (1990). Heterogeneity of Schizophrenia: Conceptual models and analytic strategies. Brit. J. Psych. 156, 17-26.

Vallada, H.P., Collier, D., Sham, P., Lim, L., Parfitt, E., Nanko, S., Taylor, C., Asherson, P., Murray, R., McGuffin, P., Owen, M., & Gill, M. (1995). Linkage studies on chromosome 22 in familial schizophrenia. Am. J. Med. Genet. (Neuropsychiatric Genetics) 60, 139-146.

Vogel, F. & Motulsky, A.G. (1986). *Human Genetics: Problems and Approaches, 2nd ed.* Springer-Verlag, Berlin.

Walker, E. & Shaye, J. (1982). Familial schizophrenia: A predictor of neuromotor and attentional abnormalities in schizophrenia. Arch. Gen. Psychiatry 39, 1153-1156.

Zilboorg, G. (1941). Ambulatory schizophrenias. Psychiatry 4, 149.

RECOMMENDED READING

Burmeister, M. (1999). Basic concepts in the study of diseases with complex genetics. Biol. Psychiatry 45, 522-532.

Chapter 14

Child and Adolescent Psychiatry:

General Principles

CAROL E. KAPLAN and ANNE THOMPSON

Biological Psychiatry, pages 279-310.

ISBN: 1-55938-819-6

INTRODUCTION: WHAT IS CHILD PSYCHIATRY?

Child psychiatry is the medical speciality that assesses and treats mental health disorders of children and adolescents. As well as the assessment and treatment of mental illness in children, child psychiatry also encompasses delays or deviations in children's development, emotional state, and behavior. These may be the result of a variety of things such as life stresses. Approximately one in five children and adolescents suffer from a psychiatric disorder. Certain groups of children have higher rates of disorder, including children from economically deprived families and physically ill children (especially those whose illnesses involve the central nervous system). In prepubertal children, boys have higher rates of psychiatric disorder than girls. In adolescents, this gender ratio is reversed. In general, girls tend to show more internalizing (i.e., emotional) problems and boys tend to show more externalizing (i.e., behavioral) difficulties. Children sometimes suffer from combinations of psychiatric disorders. Well known associations include depression and anxiety, attention-deficit/hyperactivity disorder and conduct disorder, and conduct disorder and specific learning difficulties. Comorbidity has implications for management and prognosis.

The medical model of disease is often inadequate to explain the etiology of child psychiatric disorders. There are often multiple predisposing factors, precipitants, and perpetuating factors of mental health problems in children. A child's difficulties may be understood from a variety of theoretical standpoints, for example, analytical theories, behavioral principles, or attachment theory.

Newborn babies are completely dependent on adult carers. As children grow older, they remain dependent on parents and family members at least to some extent for their physical, emotional, and social needs. Children are intricately linked to their families. A psychiatric assessment of the child as an isolated individual is inadequate: it must be placed in the context of the quality of parenting that the child receives, parental emotional well-being, the way in which the family functions as a system, cultural values, and environmental influences. The quality of parenting available to the child is the issue at stake in child protection work.

Diagnoses in child psychiatry are made in the same way as in other medical disciplines—by using operational criteria. Medical practitioners working in the field of child psychiatry need to have skills of assessment, diagnosis, and treatment just the same as those working in other medical specialities. Children are seldom in the position of choosing to see a doctor and, under these circumstances, the practitioner needs to remember the importance of communicating directly with their patient. Children are more commonly complained about than in a position to complain themselves.

Child mental health problems are complex and have multifactorial etiology and so child psychiatrists often work together with professionals with backgrounds in education, psychology, and social sciences to assess and treat children.

Child psychiatrists have unique skills in the assessment and management of disturbed and distressed children and adolescents. They tend to be involved with children whose psychiatric disorders are severe, complex, and persistent. Although child psychiatry is a young speciality, there is an expanding body of research available to clinicians to guide practice. Child psychiatrists are in a strong position to contribute to the understanding of the natural history of psychiatric disorders throughout life.

CHILD DEVELOPMENT

In order to make clinical judgements about the significance of symptoms in child psychiatry, an understanding of child development is needed. Child development comprises a number of dimensions including an increasing capacity for social interaction, cognitive development, moral development, physical growth, the acquisition of language, emotional maturation, and psychosexual development.

Theoretical models of normal development have been developed for each of these elements. Classic examples include Erikson's (1963) theory of psychosocial development, Piaget's (1952) theory of cognitive development, Kohlberg's (1969) theories of moral development, Gesell's descriptions of the sequences in which motor and language skills are acquired (Gesell and Amatruda, 1964), Anna Freud's (1965) writings about the psychodynamic development of children, and descriptions of psychosexual development (Green, 1974).

Table 1. A Summary of Normal Child Development

	Infancy	*Preschool Years*	*Mid-Childhood*	*Adolescence*
Psychological Development (Erikson, 1963)	Trust v.s. mistrust (involvement with mother)	Autonomy v.s. doubt, Initiative v.s. guilt (involvement with parents and basic family)	Industry v.s. inferiority (involvement with neighborhood and school)	Identity v.s. confusion (involvement with peer group)
Cognitive Development (Piaget, 1952)	Sensorimotor stage	Preoperational stage	Concrete operational stage	Formal operational stage
Moral Development (Kohlberg, 1964)		Preconventional morality	- - - - - - - - - - - →	Conventional morality, + / - Postconventional morality
Physical and Language Development Gesell and Amatruda, 1964	Major advances in motor development, Rapid growth velocity	Acquisition of grammar and syntax, Bladder and bowel control	Complex motor sequencing, Laterality, Widening vocabulary	Pubertal changes, Cessation of growth
Psychodynamic Development (Freud, 1965)	Primary identification, Object relations	Oedipal fantasies, Superego formation	Latency phase (capacities for internal controls increase)	Reworking of oedipal struggles in love relationships
Psychosexual Development (Green, 1974)		Gender identity apparent	Gender role established	Sexual orientation emerges

Each indivudual is unique in terms of their generic endowment, constitutional make-up, and life experience. Each aspect of their development is governed by genetic and environmental influences. Once habitual ways of reacting to various life experiences have been established, these become identifiable as aspects of an individual's personality. The development of personality characteristics is an ongoing process throughout a person's lifetime. If an individual shows seriously maladaptive characteristics, then this may justify a diagnosis of personality disorder. In view of children and adolescent's incomplete development, this diagnosis should be made rarely.

A summary of the main dimensions of normal child development appears below and in Table 1.

Development in Infancy

In the first twelve months of life, infants make dramatic progress in all aspects of development. Healthy babies gain weight and rapidly grow in length. They relinquish the primitive reflexes present shortly after birth and develop the capacity for increasingly complex movements, enabling babies to sit up at about six months of age and walk at around one year. Infants begin to babble around the fourth month of life, and, in the presence of adequate hearing and stimulation, this babble becomes increasingly elaborate and conversational.

Babies' cognitive development has been described by Piaget as being at the "sensorimotor" stage. They make sense of the world by gazing, hearing, touching, and tasting. In the first year of life, infants come to understand that they are separate from their environment, and they learn that they can influence the world around them by their own movements and sounds. Cognitive development is not sufficiently sophisticated for infants to have the capacity for moral reasoning.

Infants recognize the smell, sound, and sight of their carers early in life, and when these carers repeatedly provide care and nurture, infants establish basic trust in their primary carers. After six months of age, babies begin to display clear signs of attachment behavior indicating a special relationship or bond with their primary carers. Psychodynamically, the infant begins to differentiate between themselves and the environment and primary identifications are made. Babies develop an inner sense or representation of other people and learn to tolerate their experience of separateness and the failure of parents to satisfy every need.

Development in Preschool Years

Preschool children continue to grow rapidly. As their nervous systems mature, these children gain capacity for bladder and bowel control and an increasing repertoire of fine and gross motor movements. Language development increases rapidly so that, around the age of five, many children will have acquired an adult range of grammar and syntax.

According to Piaget's theory of cognitive development, by the time children are two years of age, they have achieved object permanence (i.e., an appreciation that people and objects continue to exist even when they cannot be seen). After the age of two, children's cognitive development enters the preoperational stage. The explosion in language development allows the child to begin to represent people and objects in images and words, although these descriptions are crude at first. A child of this age thinks egocentrically (i.e., everything applying to themselves: e.g., "The sun gets up when I open my curtains") and animistically (i.e., all objects behave like people and have feelings: e.g., "My pillow's sad 'cos I threw it out of bed"). Children of this age do not have an appreciation of conservation of number, volume, or mass; for example, when a preschool child is shown equal numbers of biscuits, firstly close together on a plate and then spread out, he or she will say there are more biscuits on the second plate. Children at this stage of cognitive development also believe that every happening has a preceding cause. This style of thinking facilitates the earliest form of moral reasoning in young children. Preschool children learn from approval and also punishment from their parents. They generalize from this to reach a simple form of moral reasoning such as the idea that bad actions are punished and good ones are not.

Once children have the motor skills to walk and run and sufficient trust in their parents to leave their side, they begin to explore their environment. However, children of this age are still very much orientated toward their immediate family and interested in their reactions. As family members repeatedly recognize and praise the child's achievements while at the same time providing consistent limits, the child internalizes a sense of autonomy and initiative.

Psychoanalytic theory describes how, at around two years of age, children become envious of the parent of the same sex in so called oedipal fantasies. This phase resolves as the child identifies with this parent and so gains some further sense of personal and gender identity.

Around the age of two or three, children have developed their gender identity (i.e., a basic sense of themselves as either male or female). Once this is established, children seem to find it odd to be questioned about whether they are a boy or a girl.

Development in Mid-Childhood

Children of this age master tasks that demand complex motor sequencing (such as riding a bicycle), and they begin to show a preference for leading with either their right or left hand and foot in motor activities (laterality). They display a widening vocabulary in everyday conversation.

Around the age of seven, children's developing cognitive skills become those of Piaget's "concrete operational stage". Children begin to think more logically and they acquire an appreciation of conservation of number, volume, and mass. Children over the age of ten generally show "conventional morality" in which the rules of their immediate family group are seen as the prime source of authority.

Social development of children of this age centers around involvement with neighborhood and school, where play and scholastic achievement allows them to develop a sense of industry. Psychodynamically, children of this age are in the "latency stage", in which they develop increased capacities for internal control. School age children consolidate a repertoire of behavior which is culturally defined as either male or female. This gender role includes the child's style of dress, choice of play activities, and gender of close friends.

Development in Adolescence

Adolescence usually coincides with the time during which children develop secondary sexual characteristics in the sequence of hormonal and physical changes of puberty. However a minority of children, especially girls, pass through puberty earlier in childhood. Children attain their adult height during puberty.

Cognitive development reaches the formal operational stage in which adolescents achieve the capacity for abstract and hypothetical thinking. Adolescents may be preoccupied by existentialist ideas or issues of global importance. Continued moral development results in adolescents becoming increasingly influenced by the values of their wider society.

Some, but not all, older adolescents will develop a sense of postconventional morality in which they make distinctions between the relative values of different guiding principles (e.g., although it is wrong to steal, it may be acceptable to take a chocolate bar to prevent a diabetic friend from having a hypoglycemic episode).

Adolescents' social development is orientated around activities with peer groups, among whom they develop their sense of identity. Issues of identity are reworked psychodynamically by the playing out of oedipal struggles in love relationships.

By the end of adolescence, most young people will be aware that they become sexually aroused in response to members of one or the other sex. This emergence of sexual orientation may occur after a period of ambiguous sexual preference.

METHODS OF ASSESSMENT IN CHILD PSYCHIATRY

Taking a History

History taking in child psychiatry involves gathering information not only from the child but also from parents or guardians, other family members, school staff, and other professionals involved with the child. Information from these sources may be conflicting. For example, children themselves may say little about the presenting problems whereas adults are usually able to describe their view of current problems more fully. However, adults often underestimate a child's level of emotional distress (Barrett et al, 1991).

An outline of the information required from a child psychiatry history appears in the sections below.

History of Presenting Complaint

This orientates clinicians to problems as viewed by the family and helps them to understand the phenomenology and meaning of symptoms. Family members should be encouraged to give an account in their own words, before a clinician asks additional questions. Important details include the duration of symptoms, how symptoms have affected the child and others in the family, what appeared to precipitate the symptoms, and what has made them better or worse.

General Health

This enables the clinician to understand present difficulties in relation to past and present organic and psychiatric disease. Inquiry should be made about any disturbance of sleep pattern, appetite, or level of activity, any delay in acquiring expected developmental skills, or the loss of previously acquired skills. Details should be determined about current and past illnesses, accidents, and other traumatic experiences, and about treatment including any hospital admissions.

Drug and Alcohol Use

This may play a part in the etiology of child psychiatric disorder, may represent an attempt at self-medication because of underlying psychosocial difficulties, or may be part of risk taking behavior in adolescence. Questions should be asked about the use of prescribed and nonprescribed medication, alcohol, and street drugs. Information gathered should include quantities taken, the frequency of use, and the circumstances surrounding drug and alcohol use together with a description of wanted and unwanted effects.

Temperament

Temperament is the style in which children behave and react to the world around them. Chess and Thomas (1986) describe nine aspects of behavior that constitute an individual's temperament; examples are activity level, rhythmicity and adaptability to a change in environment. A common error is to equate temperament with *what* a child does rather than *how* the child does things. As an illustration, consider a room full of kindergarten children riding tricycles.

Every child is doing the same thing; however, not every child's temperament will be the same. One child's way of riding a bicycle might be to speed and crash, whereas another will pedal slowly and carefully; some will be shrieking with excitment and others will be upset by the noise. The *style* of behavior gives a clue to the child's temperament. Descriptions of temperament enable the clinician to understand the style in which a child is likely to react to developmental challenges, life events, and stress. Indications of a child's temperament may be obtained by parents' descriptions of how a child has reacted to past events. Asking about similarities to other family members is often helpful.

Developmental History

This places present difficulties in context of the developmental progress of the child. Information should be gathered about prenatal life (i.e., mother's health, drug and alcohol use in pregnancy), perinatal events (i.e., delivery, perinatal health, postnatal complications), infancy (i.e., sleeping and feeding routines, mother's and other family members reactions toward the baby), developmental milestones (e.g., ages at which a child smiled, sat up, crawled, walked, talked, and was toilet trained), separations from parents (e.g., hospitalization, shared child care, kindergarten), and relationships with other children (including siblings).

Educational History

This gives information about the cognitive, emotional, and social development of children. Areas of inquiry should include adaption to starting school, relationships with children and staff, academic progress, participation in all school activities, and behavior in school. It should also include information about different schools attended and why changes of school were made.

Social Relationships

Information about these indicates the extent of a child's social development and their capacity to form satisfying relationships with others. Details should be sought regarding past and present friendships together with any experiences of bullying—either as victim or perpetrator—and how this was dealt with.

Forensic History

This is an indicator of past socially unacceptable behavior, which has come to the attention of the authorities. Details should be obtained concerning antisocial behavior, drug and alcohol use, involvement with the police, police cautions, convictions, and judicial sentences.

Family History

This indicates the capacities of parents to meet the needs of the child, the qualities of relationships within the child's family, and the genetic and environmental influences to which the child is subject. Information about parents should include ages, occupations, general and mental health, childhood experiences, drug and alcohol use, duration and quality of current partnership, social support, and quality of relationships with other adults. Inquiry should be made about physical and mental illness in members of the extended family.

Social Conditions

This gives an indication of the material provisions of the child's family and the cultural values operating in the home and neighborhood. Differing cultural contexts must be recognized and respected. Questions should be asked about family and country of origin, religious beliefs, housing conditions, and financial circumstances.

The Psychiatric Examination of Children

General Approaches

Child psychiatric interviews should take place in quiet settings furnished in a "child-friendly" style. Children may have difficulty understanding what sort of a doctor they are seeing and they may expect to be physically examined or given physical treatments. Clinicians should take time to explain that, in their sort of clinic, the most important things are communications from children and families.

There are both advantages and disadvantages to seeing children together with their parents in assessment sessions. Joint sessions allow children to hear the reasons they have been brought to the clinic directly from their parents and younger children are generally less anxious when seen with their parents. However, children, and adolescents in particular, may be reluctant to express their own points of view in front of their parents. The authors recommend that children are seen alone at some point during the assessment.

It is important that the child perceives that the child psychiatrist is listening in an accepting and nonjudgmental manner. The psychiatrist needs to be clear about the limits of confidentiality, which apply to information given by the child.

Immediate and direct questioning of the child about the presenting problems is rarely productive. Instead, several techniques may be used to help the child relax and feel prepared to talk openly. Talking about neutral topics such as TV or sport may be a useful opening activity. The psychiatrist may ask about potentially emotionally loaded topics by interspersing these questions with comments about neutral activities.

Permission to admit to different feelings can be given by asking questions such as "Everyone feels sad sometimes. Have you felt sad sometimes too?" Encouragement to avoid polite, socially acceptable but perhaps inaccurate answers can be given by giving the child alternatives, for example "Some children like school, some think its just okay, and others tell me they don't like school. How about you?"

Understanding Children's Communications

Being able to understand children's communications is the core skill of child psychiatry. Children communicate in nonverbal behavior as well as spoken language. Nonverbal communication such as play and artwork is an important means of gathering information about younger children's inner lives. From mid childhood onwards children begin to be more able to explain their thoughts and feelings in words.

Some influences on children's communications are outlined below.

Developmental level. Cognitive, language, emotional, and social development all influence the child's ability to communicate with the psychiatrist. Knowledge of a child's developmental level helps psychiatrists choose the words they use so that the child is neither mystified nor patronized. Similarly, this will influence the kinds of activities offered to the child.

Situational anxiety. Children may be anxious in the psychiatry clinic because they do not understand why they have been brought there, because they are meeting a stranger, or because they fear the doctor will tell them off. Younger children may tend to give socially acceptable answers to questions when anxious in an effort to please the doctor. Older children may sit in embarrassed silence. The way the child feels about the consultation can be greatly influenced by what the parents have said about it.

Parental expectations. Children may have been told by their parents what to say to the clinician, or be aware that what they want to say will lead to parental disapproval or reprimand. As a result, the unwary clinician may not obtain the information needed to do a proper assessment.

Psychological defense mechanisms. Adults use a wide variety of psychological defense mechanisms. Children are in the process of developing a repertoire of them. All of them may come into play in the psychiatric interview. Children whose behavior or responses have gotten them into trouble may deny difficulties when asked about them. They may attribute or project feelings of anger, guilt, or fear onto another family member, perhaps telling the doctor that their brother or sister feels that way. Young children who are anxious or fearful may play and talk energetically, as a defense to avoid painful emotions. Traumatized children may use imaginary play in which they take powerful, bossy, or cruel roles: this may be a form of identifying with the aggressor or expressing their reaction to traumatic experiences.

The Mental State Examination

Psychiatric examination of the child provides information about a child's mental state. This is usually described under the same headings as used in the psychiatric examination of adults.

Appearance. This includes a description of a child's general demeanor, state of dress, and cleanliness, height, weight, and any dysmorphic features.

Behavior. A child's behavior may be very varied and may be described as frozen, wary, restless, uncooperative, oppositional, or disinhibited. A comment should be made about the extent and appropriateness of eye contact made by the child. Note should also be made of the child's ability to separate from their parent.

Mood. Clinicians make judgements about the appropriateness and reactivity of a child's mood and comment whether the child is euthymic, depressed, or elated.

Speech and Language. This is assessed in terms of being appropriate to the child's chronological age. Any difficulties with expression or comprehension are noted.

Thought. This is described in terms of content and process.

Abnormal perceptions. These include illusions, hallucinations, and delusions.

Developmental level. The child's activity, behavior and understanding in the interview allows the clinician to comment on motor development, social interactions, play and apparent cognitive level. All of these are viewed with respect to the child's age.

The Physical Examination

A physical examination should always be part of the child psychiatric assessment. Physical examination aims to detect signs of organic pathology that may be contributing to psychiatric disorder and to provide information about a child's growth and development.

There are three parts to the physical examination:

a. *Inspection.* Observations of the child in the clinic room provide preliminary information about a child's state of nutrition, growth, and level of development. Physical signs such as abnormal gait, involuntary movements, or poor coordination suggest pathology requiring further investigation.

b. Clinical measurement. Children's height, weight, and head circumference should be measured by experienced practitioners using accurate equipment. Data should be recorded in the case notes and plotted on appropriate centile charts.

c. Physical examination. This should take place in warm, private surroundings in the presence of a chaperone and often a parent. Particular attention should be given to detecting focal or soft neurological signs.

Laboratory Tests

In the authors' view, these should be used discriminately according to the child's presenting symptoms and the findings of the physical examination.

a. Blood and urine analysis. There are many helpful blood and urine tests which should be used as clinically indicated in individual children. A common example of psychiatric presentations that require blood tests include children presenting with clinical signs of anorexia nervosa (who require hematological investigations to exclude anemia and blood biochemistry analysis to detect electrolyte imbalance) or with developmental delay and dysmorphic features (for whom chromosome analysis should be considered). Adolescents presenting with confusional states or depressed mood or anxiety of sudden onset should be asked to provide a urine sample as soon as possible for a toxicology screen.

b. Electroencephalogram (EEG). EEG examinations are indicated in children whose symptoms suggest ictal activity such as sudden changes of behavior or momentary absences. Recordings during photic stimulation, sleep, or prolonged telemetry may sometimes be required in order to demonstrate epileptiform activity.

c. Neuroradiological investigations. Computerized tomography (CT) head scanning or magnetic resonance imaging (MRI) is indicated if physical examination or EEG investigation suggests a focal lesion in the brain. These imaging techniques are also used in the investigation of suspected perinatal brain injury and the diagnosis of degenerative disease. Positron emission tomography (PET scanning) and single photon emission computed tomography (SPECT) scanning are used more as research tools than routine clinical assessment at present.

d. Psychological tests. Clinical psychologists with training and experience in the psychological examination of children can contribute greatly to the understanding of psychiatric disorder in children. Psychological testing may provide information about a child's developmental level, cognitive abilities, personality, and inner world. The results of psychological tests need to be interpreted by experienced psychologists: numerical values mean little unless they are considered in respect to the limitations of the test itself and the broad understanding of the child.

The Formulation

Child psychiatry assessments may generate a bewildering amount of information about a child and his or her parents, family, and culture. In order to be useful to the clinician, this information needs to be assimilated in an ordered way. The authors suggest that this should be done in two ways—producing both a statement of presenting problems and an etiological formulation.

a. Statement of presenting problems. This can be arranged along five axes in keeping with the World Health Organization classification:

- Axis I~ Psychiatric Disorder
- Axis II~ Specific developmental delay
- Axis III~ Cognitive level
- Axis IV~ Medical conditions
- Axis V ~ Abnormal psychosocial conditions

It should be noted that a psychiatric diagnosis in isolation is of relatively little use when planning treatment and management. For example, a child with a major depressive illness may be more usefully described under the WHO axial system as having major depression in the context of normal development, above average cognitive abilities, mild asthma, and a family who are all grieving the loss of father from an early cardiac death. The additional information gives possible etiological factors and issues to be explored in treatment.

b. Etiological formulation. Many psychiatric disorders in children and adolescents have a multifactorial origin. The emphasis placed on different etiological factors in a formulation will reflect to some extent the theoretical standpoint of the clinician. For example, three different child psychiatrists examined a seven-year-old child who was aggressive, disobedient, and restless. One psychiatrist viewed the child as having an intrinsic deficit in impulse control as a result of birth trauma. The second described the child's behavior as functioning to unite parents in their troubled relationship and the third understood the child's behavior as a reaction to cultural expectations and deprivation.

Irrespective of theoretical orientation, etiological formulations may usefully include a description of predisposing factors, precipitating factors, and perpetuating factors. Returning to the example above, the history of birth trauma and cultural issues may be viewed as predisposing factors while the positive impact of the child's symptoms on his parents' relationship can be seen as a perpetuating factor.

Formulations of any sort serve to guide the clinician. As such, they must be considered as working hypotheses, not statements of absolute truth. In clinical practice, formulations are often revised in the light of new information.

CLASSIFICATION

Classification is essential in clinical practice as, without it, communication about different disorders would be impossible. It is also basic to scientific research, by which our knowledge of the nature, antecedents, and prognosis of disorders is expanded and developed. However, it is important that classification should confer benefit, and to this end it must be emphasized that diagnoses apply to disorders not individuals; a diagnostic label is only one aspect of the whole picture; diagnoses only provide a general guide as to the types of intervention that may be used, and the type of intervention each individual may need should be devised for them specifically.

Over the last 20 years, there has been a great evolution in the classification systems in child and adolescent psychiatry. Theoretical frameworks that lacked empirical justification held sway in the early years and the result was confusion and disagreement about which school of thought should predominate. A more operationally based approach is now generally accepted and greater clarity has been achieved.

Three general approaches to classification are now used. The first is to apply a category or "box" into which various symptoms fit as a result of previous observation that certain symptoms cluster together. The second is to regard dysfunction as having a place on a spectrum of behavior which varies from the normal to the pathological. The third is to use a psychoanalytical perspective. In child psychiatry, the categorical approach is the one most frequently applied.

The two major categorical classification systems are the *Diagnostic and Statistical Manual of Mental disorders IV (DSM IV)* which is widely used in the United States, and the *International Classification of Diseases, 10th edition (ICD-10)*, which is more often used in Europe. DSM IV has a multiaxial framework that recognizes developmental parameters and codes psychosocial stressors. It provides specific diagnostic criteria for each disorder. The ICD-10 system uses fewer axes and uses a glossary rather than specific diagnostic criteria.

Despite these differences, reliability studies of DSM IIIR and ICD-9 (the previous versions), show that they are comparable. A number of standardized structured interviews have been developed for the use of making diagnoses on both of these systems.

For some childhood mental disorders, the diagnostic criteria are the same as for adult disease. ICD-10 and DSM IV both have a section of mental disorders particular to children.

TREATMENT OF CHILD AND ADOLESCENT PSYCHIATRIC DISORDERS

Introduction

Child psychiatric disorders are complex conditions that usually have multiple causes. The result is that treatment is correspondingly complex and may often involve combinations of interventions such as individual psychotherapy for the young person as well as family therapy. This chapter does not provide an extensive description of different treatment modalities but aims rather to provide an overview, which should enable the reader to understand the principles of treatment.

The basis of any therapeutic intervention is the formulation made after the full assessment of the child and his or her family and different areas of functioning (e.g., school). It should be remembered that the formulation will need revision on a regular basis as more information becomes available and as therapy progresses. It is also worthwhile to consider not only the factors that may facilitate or hinder change, but also the consequences of change for the whole family.

It is helpful to set the goals of treatment before therapy starts. The cooperation of patients and their families is needed before this can begin. In some patients, it has proven necessary to give treatment against the patient's and/or their family's wishes. However, this is a matter that would involve the use of mental health or child care legislation and is beyond the scope of this chapter.

Working with Parents

It is seldom sufficient to work with the child or young person alone. Parents may be helped to understand the problem, how it began, and what is perpetuating it, without engendering a sense of guilt in the parents. Sometimes marital or personal issues for parents become the main focus of therapy.

Individual Psychotherapy

There are many different forms of psychotherapy with children and adolescents, but common features may be described. All these therapies use the relationship between therapist and child to help to improve the child's psychological situation. The therapist must develop a working relationship with the patient that is based on an acceptance of the individual, although this does not imply automatic approval of what they may do or feel. The therapist should be able to achieve an understanding of the patient's feelings and point of view so that the patient sees the therapist as being concerned about them. It may take a long time for a child, often one who did not ask to come and who has been threatened with various future consequences of the attendance, to trust the therapist. Many long hours of work may be needed before such a relationship can be established.

The different forms of psychotherapy include supportive psychotherapy, cognitive behavior therapy, psychoanalysis, music therapy, art therapy, and play therapy. Supportive psychotherapy helps the individual cope with stressful circumstances by the therapist listening to and accepting their feelings. Psychoanalysis explores the patient's unconscious life and deals with problems within the context of the transference relationship between patient and therapist. This can be seen as a reworking of past relationships and fantasies. Music and art therapy use creative media to facilitate expression and integration of emotions. Play therapy is most often used in very young children, where play is used as a medium of communication and therapy (Landreth, 1982).

Family Therapy

In family therapy, the problems of the individual are regarded as functions of the interactions of all the family members, or family system. The family as a whole is the focus of therapy, and the individual's difficulties are expected to remit when the functioning of the family system improves.

Once the problems have been identified, the therapist explores and helps the members to find alterations to interactions within the family. There are various different strategies that can be used according to the particular school of family therapy practiced by the therapist. As in other forms of therapy, training and supervision are essential.

Group Therapy

Group therapy uses the ability of group members to help each other. It may be used for children or parents and again a variety of different approaches have been described. The role of the therapist is to facilitate useful interactions and also to provide structure and sometimes some guidance.

The aims of group therapy may be diverse and will include things such as improving self esteem, encouraging better peer relationships, and controlling behavior. Groupwork activities may include play, cooperative activity, drama, and discussion.

Behavior Therapy

This therapy is based on the idea that behavior is learned and that maladaptive behavior can be changed by a process of learning new behavior. A detailed assessment includes a behavioral analysis. This will help the therapist understand what precedes the behavior, what the behavior consists of, and what follows the behavior. This is sometimes described as the "ABC" of behavior (i.e., A = antecedents, B = behavior, C = consequences). Desired behavior is then reinforced and unwanted behavior is left unrewarded. Parents are often used as therapists in the treatment program.

Cognitive Behavioral Therapy

This form of therapy is based on the theory that thinking affects behavior and mood and that if the form and content of thought can be altered, then a change in mood and behavior can result. This therapy is better researched in adults than children, but results to date are encouraging in both age groups.

Cognitive behavioral therapy has been used to manage a variety of different disorders in childhood and adolescence: anger and aggression, attention-deficit/hyperactivity disorder, anxiety, depression, and eating disorders. Strategies that involve parents and even school members may be devised, and this level of flexibility is one of the strengths of this form of therapy. Another advantage is that cognitive behavioral therapy is relatively time limited.

Pharmacotherapy

Medication is seldom used as a first line treatment in child psychiatry by comparison with other branches of medicine. It is more common for drugs to be used as part of a complex treatment strategy. The literature shows a marked difference in practice between child psychiatrists in the United Kingdom and the United States. British psychiatrists are more conservative than their transatlantic colleagues.

The type of medication used most frequently are major tranquillizers, psychostimulants, anxiolytic agents, and antidepressants. These and others will be discussed within the sections on different disorders.

Residential Treatment

Children may be treated away from their own homes in units that are part of a hospital. These units aim to provide physical care, meet personal and social needs, and also provide appropriate education. The therapeutic approach is geared toward returning the child to home and school and should be regarded as one step in the whole process of a treatment plan. There should be a close working relationship with community services, and the purpose of the residential unit is to provide a specialized assessment and short-term treatment facility within the context of other resources available to children and their families.

A successful inpatient unit will define the problems, specify treatment goals, describe the patient's skills and areas of difficulty; specify the treatment methods, and objectively assess progress within the context of a therapeutic milieu. This is defined as "a structured environment that provides a variety of human relationships, satisfactory emotional interactions, opportunities for new learning and experiences, mastery of new situations and the development of personal and social competence." (Hersov, 1994).

PREVENTION

It is generally accepted that three levels of preventative work can be done (Henderson, 1988). Primary prevention reduces the frequency with which a disorder occurs, secondary prevention treats the disorder, and tertiary prevention reduces disability that is associated with the disorder. Primary prevention only will be considered in this section, as secondary and tertiary prevention is dealt with under each disorder.

Graham (1994) has described the essential factors of primary prevention as being risk, protective, and vulnerability factors. In considering risk factors, not only should they be identified but their casual significance in etiological pathways and the level of risk must also be known. Protective factors are those that reduce the likelihood of a disorder occurring. The way that such a factor operates may be very variable, (e.g., interacting with a risk or perhaps operating at all times). Vulnerability is usually used to describe those children who have an increased chance of developing disorders even when the risk factors are low.

In an ideal situation, prevention programs should be based on longitudinal research, which has identified the above factors and also assessed how amenable to intervention they are. However, in the field of child psychiatry, this is often very difficult to do because over knowledge is not, at this stage, great enough to identify factors amenable to modification.

Graham (1994) has described a classification for preventative programs (1)as those directed toward the living conditions of a population as a whole, (2)those directed toward a population of children of specific ages and for their parents, and (3)programs directed toward individuals at risk for specific mental health problems. Within the first group, programs may be directed at improving housing, employment, nutrition, and economic status. Examples of the second group include programs directed at pre- and perinatal care, preschool education, educational achievement in middle childhood, and adolescence (Cromer, 1985). Within the third group, there has been a focus on children with acute and chronic illness, children with conduct disorder (Hawkins et al., 1991), children who have attempted suicide (Shaffer et al., 1988), children of mentally ill parents, children of parents with disrupted relationships, children who have been placed with substitute carers, and children faced with acute stress such as major accidents or natural disasters.

LIAISON AND CONSULTATION CHILD PSYCHIATRY

Consultation work provides advice and guidance to many professionals working with children in situations where it is not possible or appropriate for the psychiatrist to work directly with a child or the family. Although a direct assessment of mental functioning and advice about care and treatment can be given, the work is usually done by members of other agencies.

Liaison work focuses on agencies working with children. It aims to deal with the needs of the agency, not the individual child or family. This type of work commonly takes place with pediatric teams involved with children presenting with ill physical health. The review of care programs and the enhancement and enablement of the skills of staff are important areas. Dealing with stress and psychological distress in professionals can be addressed in this type of work. It is important to recognize and support the severe level of stress that may be expected by staff, including junior staff working, for example, in intensive care units. Liaison work is difficult and involves sharing of interests and understanding and exploring ways of working together, sometimes in the face of suspicion and hostility.

Indications for pediatric liaison work include the following: children who have disturbances of behavior or emotions, children with physical symptoms for which no obvious organic explanation can be found, children with poor adjustment to recurrence of chronic ill health, the presence of significant parental or family pathology, or situations of suspected or known abuse (Lask, 1985).

In order to understand issues that may arise in pediatric liaison work, the child psychiatrist should have an understanding of the psychiatric aspects of physical illness in children. These are considered in detail below.

General Considerations

Children's Adjustment to Ill Health

All physical illnesses have emotional accompaniments. Illness commonly leads to the loss of some characteristic or capability such as the loss of an unblemished face after a car crash or the loss of the ability to play vigorous sports after a bad injury.

An ill child faces the psychological task of adjusting to these losses as well as coping with the direct effects of illness. Short-lived illnesses only demand that the child tolerates a few day's discomfort, malaise, and incapacity. Chronic illnesses have more profound influences on children's lives, potentially changing their family relationships, their friendships, their school lives, and their futures. The process of adjusting to the loss of previous abilities and expectations can be viewed as a process similar to grief (i.e, involving stages of numbness and disbelief, anger, mourning, and eventual adjustment). Consider the following example:

> For the first two weeks after being diagnosed as suffering from insulin-dependent diabetes, a 10-year-old girl kept forgetting she could not eat anything at any time. Then she became more compliant with her diet but irritable and resentful of having to have injections. Two months later, at the family's Christmas meal, she broke down in tears, sobbing that her life was ruined by having to eat so differently from others. By the next Christmas, she helped plan and prepare a special sugar-free menu for herself.

Children may take weeks, months, or years to come to terms with a chronic illness. Satisfactory adjustment may be compromised by delays in diagnosis, insensitive communication by doctors, difficult temperamental features, and illnesses that are severe or disfiguring. Poorly adjusted children are often perceived as difficult patients by doctors: they may be sullen, never satisfied, poorly compliant, or excessively anxious.

Children's Understanding of Illness

Children's emotional reactions to ill health are influenced by their capacity to understand the cause and likely effects of the illness. This understanding changes as a child's cognitive development advances. (For a review, see Burbach and Peterson, 1986).

Preschool children whose cognitive functioning is in the preoperational stage believe that every event is caused by some external agent or person. Children of this age tend to blame someone when someone gets ill. The phrase "You gave me your cold" is not infrequently said by these children. The following is an example:

> A five-year-old girl was often in trouble for teasing her two-year-old brother. After her brother was admitted to hospital with asthma, she was quiet, clingy, and tearful for two weeks. Eventually she told her mother that she had made her brother sick by being bad to him.

In mid-childhood, illness is more readily attributed to external agents, but children have little sense of their own anatomy and physiology and may be confused or frightened by explanations which are too sophisticated as the following illustrates:

> A surgeon showed a nine-year-old boy with acute appendicitis where he was going to make a cut on the child's abdomen and he drew a picture of an appendix for the child. The boy later told his visitors a snake had got into his tummy and he was going to have it taken out.

By mid adolescence, most young people can appreciate the causes and effects of illness in adult fashion. They are, however, often particularly distressed by the social implications of their condition, as is shown by the following example:

A fifteen-year-old girl pleaded with her pediatrician to be allowed to miss her lunchtime medication because she didn't want to appear different from her friends.

Children with chronic illness may renegotiate their adjustment to ill health as their understanding of the condition and their view of themselves in the world increases. The following is a possible scenario:

A thirteen-year-old boy with epilepsy, who had previously coped cheerfully with his daily treatment and occasional seizures, became withdrawn, tearful, and irritable. During psychiatric assessment, the boy explained that his friends had begun to talk a lot about the future, and he had become preoccupied by the realization that his epilepsy would place restrictions on his driving and choice of career.

Influence of Ill Health on Normal Development

Chronic illness in children may affect various aspects of their development. Growth may be restricted and puberty delayed by prolonged ill physical health or side effects of treatment. Cognitive development may be impaired either by a direct effect of disease processes or by the treatment, such as the neuropsychological side effects of medication, cytotoxic drugs, or cranial irradiation. Time missed from school may lead to academic failure.

Emotional development may be compromised in a number of ways. Children between the age of six months and two years are particularly distressed by repeated admissions to the hospital. Later, their mastery of autonomy, initiative, and industry may be restricted if there are high levels of parental supervision or overindulgence. Children's self concept may be influenced by a recognition that they are different from, or not as able as, other children. In adolescence, young people may fear and resent potential difficulties with independent living, relationships, and employment. Disfiguring conditions are particularly distressing.

Children whose illness or treatment regime restricts their leisure activities may miss opportunities to make friends and be part of a peer group. In the absence of informed and sensitive handling of a child's illness in school, sufferers may be stigmatized, isolated, or bullied by other children.

Family Influences

When children become ill, emotional aspects of the illness affect not only the child but also their parents, siblings, and extended family. Family members usually work their way through the phases of numbness, anger, mourning, and eventual adaption at different rates. Adjustment may be more difficult when family members blame themselves for the child's illness, for injuring the child themselves, for delaying seeking medical treatment, or for passing on an inherited condition. Preexisting difficult relationships with the child can also make adjustment more difficult. Family members' emotional reactions may in turn have a negative impact on the child.

A chronically ill child in the family may place considerable practical demands on parents. Time and money are spent adapting family life to the needs and abilities of the child, administering or supervising treatment, and attending hospital appointments. Other children in the family may feel neglected or may join their parents in the caring role. Siblings may feel guilty that they are well and able to enjoy things while their brother or sister suffers. Marital tension and rates of behavior disorders in siblings are both increased in families where there is a chronically ill child.

Associations with Psychiatric Disorder

The Isle of Wight epidemiological study showed that physically ill children have twice the rate of psychiatric disorder found in healthy children, and children with a physical condition involving the brain have four times the rate of disorder. However, the distribution of types of psychiatric condition was the same as that in health children (Rutter et al, 1976). Poor adjustment to ill health among children and their families, parental strain, and impaired development may all act as predisposing, precipitating, or perpetuating factors for psychiatric disorder.

Specific Conditions

Details of the psychiatric aspects of epilepsy, diabetes, asthma, and HIV/AIDS are given below.

Epilepsy

a. *Influences on normal development.* Epilepsy does not interfere with physical growth itself, but in children with associated learning disabilities, there may be short stature and dysmorphic features. Prolonged treatment with phenytoin can produce unsightly gingival hyperplasia and hirsuitism.

Children with epilepsy not associated with mental handicap have been shown to have slightly lower cognitive abilities than age-matched controls. This is likely to be due to the effect of adverse environmental influences rather than a direct effect of the epileptic disorder (Ellenberg et al., 1986). Prolonged and multiple use of anticonvulsant medication may interfere with a child's capacity to learn. However, uncontrolled seizures may have an even more profound effect on educational progress.

Children who suffer daytime fits, especially grand mal seizures, often have some restrictions placed on their leisure activities such as not being allowed to swim alone. Adolescents have to adapt to being unable to pursue certain careers and having legal restrictions placed on their driving. Adolescents and young adults with temporal lobe epilepsy have been described as being less interested in personal and sexual relationships (Taylor, 1969).

b. Family influences. Epilepsy was historically associated with possession and witchcraft. Negative perceptions and social stigma are still evident today. Witnessing seizures in a child is also a very distressing experience: many parents at first believe their child is going to die. Families of epileptic children need to adapt to all these challenges.

c. Associations with psychiatric disorder. Epilepsy has a particularly high rate of associated psychiatric disorder. Epilepsy, beginning early in life, seems to be a particular risk factor for psychiatric disorder. Some anticonvulsants produce behavioral side effects. For example, phenobarbitone may lead to hyperactivity in children and vigobactrin can produce aggressive behavior.

Particular associations between epileptic syndromes and psychiatric disorder are the schizophrenialike psychoses described in some adolescents with temporal lobe epilepsy and the association between infantile spasms and autism.

Asthma

a. Influences on normal development. Severe asthma can restrict growth as a result of the chronic debilitating disease itself and systemic corticosteroid treatment. Asthma induced by exercise and common allergens can lead to a child's social activities being restricted.

b. Family influences. The inherited nature of asthma is often very evident in families. Parents may feel guilty for passing on the disease to their child, and this may influence their decision to have further children as well as having an important influence on how they relate to their sick child.

c. Associations with psychiatric disorder. Mild asthma is not associated with an increased risk of psychiatric disorder in children. However, children with severe disease do have greater rates of psychiatric morbidity and there is a particular association with depression (Miller, 1987). In a study of asthma-related deaths, Strunk and colleagues (1985) found that both emotional disturbance in the child and family dysfunction were associated with a greater risk of death.

Diabetes Mellitus

a. Influences on normal development. Promptly diagnosed and adequately treated diabetes should not restrict a child in reaching their full growth potential. Severe disease in childhood is associated with reading and learning difficulties. Some children find the dietary restrictions and repeated injections difficult to adjust to at first, and previously well adjusted children entering adolescence may struggle to maintain good diabetic control in the face of changing metabolic needs, altering lifestyle, and strong peer group influences. The prospect of longterm complications of diabetes may trouble a child for the first time as their understanding of the illness and their future increases in adolescence.

b. Family influences. Parents who supervise a reluctant child's treatment may face repeated arguments. Parental anxiety may inadvertently allow a powerful diabetic child to demand many favors in return for grudging compliance with diet and insulin injections. Family dysfunction may contribute to poor glycemic control in a child.

c. Associations with psychiatric disorder. While there is no evidence that emotional stress can lead to the onset of diabetes in a child, high levels of stress in children and adolescents do correlate with poor glycemic control. The risk of psychiatric disorder increases with duration of illness. Diabetic children with major depression cause concern because of their access to insulin as a means of committing suicide.

HIV/AIDS

a. Influences on normal development. Children who are infected with the HIV virus may show failure to thrive, developmental delay, and regression, microcephaly, and seizures. Infants who have acquired their infection congenitally from drug-abusing mothers may also have been exposed to teratogenic drugs *in utero* and suffer from drug withdrawal after birth. Children with HIV may have repeated infections with common childhood organisms and, as immunosuppression increases, they become prone to opportunistic infections and malignancies. All these physical insults may limit development and growth.

Preliminary examinations of the cognitive functioning of children infected with HIV suggest that expressive behavior and expressive language functioning may be selectively impaired. The loss of speech or sight may compromise further cognitive development. Children who have HIV infection or AIDS face social stigma, recurrent ill health and illness, and loss of other family members. These are all severe challenges to a child's emotional and social development.

b. Family influences. Childhood HIV infection and AIDS may be associated with socioeconomic deprivation. Parents of ill children may themselves be ill, or be regularly abusing drugs, or be parenting their children in a style that reflects their own childhood deprivation. Children may be neglected as a result of one or all of these factors. Families in which one or more of the members has HIV infection or AIDS face considerable social rejection and isolation, guilt, and uncertainty about the sufferer's prognosis. Families may be decimated by the loss of both children and parents.

c. Associations with psychiatric disorder. There are no psychiatric disorders specific to HIV infection and AIDS. These conditions should be part of the differential diagnosis of infants presenting with failure to thrive and of older children showing developmental and intellectual decline.

CHILD PROTECTION

Introduction

Child psychiatrists are one of the professional groups who become involved with children who have been ill treated by adults, older children, or adolescents. Child abuse and neglect may have detrimental effects on psychological growth and development. Psychiatrists are able to assess how a child has been affected by ill treatment, and they may offer therapy to children who show psychiatric symptoms associated with abuse or neglect. The psychiatrist will also offer an opinion about the relationship between the child and the alleged abuser. Therefore assessment of adults forms an important part of child protection work.

Most westernized societies have a system of state intervention to protect children who are found to be suffering from significant ill-treatment within their homes or elsewhere. Child psychiatrists may provide expert advice in legal proceedings of this kind.

Child Abuse Syndromes

Child abuse is often categorized as physical abuse, sexual abuse, neglect, emotional abuse, and factitious illness by proxy. A child may suffer more than one type of abuse. In particular, emotional abuse is generally considered to accompany all other forms of child maltreatment. It should be remembered that children may be abused within their own family by either or both parents, by strangers, and by professionals working with children. The consequences of all forms of abuse are described in the next section.

Sexual abuse will not be considered further in this chapter.

Physical Abuse.

What is physical abuse? Physical abuse occurs when a child is subject to excessive physical violence from others. It may result from overly forceful attempts to discipline a child, from an adult reacting violently to anger or frustration, or from a child becoming involved in physical violence between adults.

Signs of physical abuse. Physical abuse was brought to the attention of the medical profession by Henry Kempe's landmark description of "the battered baby" (Kempe et al., 1962). When children have sustained injuries as a result of physical abuse, the pattern of injuries may be suggestive of nonaccidental injuries such as gripping, shaking, burning, or whipping. A child may have a number of injuries of different ages and carers may give unlikely explanations of how the injuries were sustained.

The child may appear afraid of or unduly submissive towards the abuser. Physically abused children are often wary of other adults, displaying so called "frozen watchfulness". They may also show signs of low self esteem and educational failure.

Neglect

What is neglect? Children of different ages require different types of care from their parents. The definition of neglect therefore varies according to the age of the child. Neglectful parents may fail to meet their child's physical or psychological needs or both.

Infants are heavily dependent on parents for their basic needs and are most vulnerable to neglect. Parents may be neglectful by failing to feed or clothe their child adequately, by failing to talk to, play with, and stimulate the child, or by failing to seek medical attention for the child when he or she is ill. They may also fail to protect the child from dangerous situations.

Older children who are subject to neglect may for example be left alone for long periods without supervision or be expected to care for younger siblings. They may be kept out of school to meet the needs of a parent or denied opportunities to mix with other children.

Signs of neglect. Signs of neglect in infants include dirty or inappropriate clothing, a baby who is mournful or unresponsive to adults, poor weight gain and developmental delay. Characteristically, these infants thrive and gain weight rapidly when provided with more attentive, nurturing care.

Neglected older children may have short stature as a result of sustained emotional deprivation. This pattern of restricted growth is called "psychosocial dwarfism" (Skuse, 1989). These children are often over-friendly to any adult who makes a kindly or interested approach to them. They may have a serious and pseudo-mature manner, which contrasts with their unmet need for age appropriate attention and nurture. They may present with behavioral disturbances such as stealing, telling fantastic stories, encopresis, overeating and risk-taking behavior. These symptoms reflect the child's unconscious aggressive drives and their poor self esteem. Their behavior may serve to draw attention to themselves or to court favor from others.

Emotional Abuse

What is emotional abuse? Emotional abuse can be defined as a sustained pattern of interaction whereby acts of omission and commission by parents lead to damage to a child's physical, social, emotional or cognitive development (Thompson and Kaplan, 1996). Acts of omission and commission include rejecting, terrorising, exploiting, and isolating the child.

Signs of emotional abuse. Emotionally abused children may show psychosocial dwarfism, aggressive or abnormally passive reactions to other children, immature emotional development, psychiatric disorder, and educational failure. However, there are no pathognomonic signs of emotional abuse.

Factitious Illness by Proxy

What is factitious illness by proxy? Factitious illness by proxy is a condition in which a parent (almost always a mother) makes a child appear to be ill by reporting or fabricating illness. An affected child may present with unexplained, prolonged, extraordinary illnesses with odd symptoms or signs that may be present only when the mother is with the child and that prove to be resistant to treatment. Clinicians should be aware that fatalities occur in this condition. (For a fuller account, see Schreirer and Libow, 1993.)

Signs of factitious illness by proxy. Children abused in this way may or may not show a range of disturbances of behavior and emotion such as immaturity, separation problems, irritability, or aggressiveness. Older children may participate in the deception passively or actively, accept the role of invalid, display antisocial behavior, or indulge in factitious illness behavior themselves.

Abusing mothers may be seen as an "ideal parent" when their child is in the hospital, alerting doctors instantly when there is a problem with the child's physical state and being unusually calm for the severity of their child's illness. Mothers may have detailed medical knowledge and they may appear to enjoy being in hospital and form unusually close relationships with staff. History taking and information from sources other than the abusing mother may reveal multiple illnesses (sometime bizarre) in other family members, or sudden deaths of siblings. Commonly there is a poor intellectual and emotional relationship between mothers and their partners, with fathers seeming peripheral to the child's care.

Formal psychiatric examination of abusing mothers often fails to detect an abnormality, but projective testing and psychotherapy may reveal underlying narcissism and extreme abilities to deny or dissociate from aggression and other feelings.

The Consequences of Child Abuse

Children do not all react in the same fashion to the same form of child abuse. The impact of an abusive act on an individual child depends on the balance of vulnerability and protective factors as well as the nature of the abusive experience itself. Factors that may make a child especially vulnerable to abuse include developmental immaturity, physical disability or deformity, having an abnormally hard to handle or passive temperament, and not being the natural child of their carer. Protective factors may include developmental maturity and having the opportunity to make positive relationships with nonabusing adults. The nature of a child's relationship with both the perpetrator and the nonabusing parent may be important mediating factors.

Physical Consequences

Children may show signs of psychosocial dwarfism. Physical sequelae of abuse include fractures, burns, head injuries, and sexually transmitted disease. A significant number of infants die as a result of abuse and neglect each year.

Psychological Consequences

Child abuse may delay or distort a child's psychological development. Typical psychological difficulties include emotional immaturity, low self esteem, perception of self as a victim, and difficulty forming trusting interpersonal relationships. The quality of a child's attachments to their parents may be severely damaged.

Psychiatric Consequences

Children who have been abused do not necessarily present with symptoms that constitute a psychiatric diagnosis. However, some associations have been found between abusive experiences and other psychiatric disorders such as posttraumatic stress disorder, depression, and eating disorders. Many children without frank psychiatric disorders show milder and transient disturbance and distress.

Consequences in Adult Life

Sustained childhood abuse may lead to serious difficulties forming satisfying relationships in adult life. The quality of adult relationships may mirror childhood experiences (e.g., interpersonal violence). Memories of childhood abuse may be evoked by life events such as marriage and childbirth, and recall can be associated with a range of intensely distressing emotions. Some abused children may abuse their own children when they become parents. However, many abused children grow into responsible and caring parents.

Assessment

A child psychiatry assessment of an abused child will be based on a thorough history and mental state examination of the child. The child psychiatrist will identify vulnerability, protective factors operating for the child, and the risk to which the child may be exposed. They will also assess the impact the abuse has had on the child's physical and psychological development as well as the quality of the parent–child relationship. Any psychiatric disorder will be noted. It may also be appropriate for the child psychiatrist to assess parents' strengths and weaknesses, their own needs for psychiatric treatment, and their likely capacity to change.

Treatment

Not all children who have suffered maltreatment require psychiatric intervention. Treatment should be explicitly focused around delays or deviations of psychological development or psychiatric disorder revealed during the psychiatric assessment. The aims of psychiatric treatment for abused children include the promotion of normal psychosocial development, restoring the child's ability to make trusting relationships with nonabusing adults, enabling the child to relinquish inappropriate emotions and cognitions about the abuse, and the treatment of specific psychiatric syndromes. Young children whose behavior is disturbed as a result of abuse may be offered play therapy to enable them to make sense of their feelings. Older children who have intrusive memories of abusive events may be helped to recount their experiences and tolerate the accompanying distress during individual psychotherapy. Cognitive-behavioral therapy may be used to address low self esteem, negative self image, and poor social skills. Group therapy may provide support and reduce isolation in survivors. It is important that therapy that is focusing particularly on past experience of abuse should only be offered to those children and adolescents where there is a very substantial reason to believe that they have had such an experience.

Some child psychiatrists work jointly with children and parents, attempting to improve the functioning of individual family members and to enhance the quality of interactions in the family as a whole. Family therapy with abusing families should only be undertaken by skilled therapists.

SUMMARY

Child psychiatry is the medical speciality that provides assessment and treatment for children and adolescents who have mental health problems. These problems include major mental illness beginning early in life as well as delays and deviations in children's development, emotional state, and behavior. The prevalence of mental health disorders in children and adolescents is approximately 20%.

This chapter has provided an overview of many of the general principles that inform practice in child psychiatry. In order to understand the etiology and possible solutions to children's mental health problems, child psychiatrists use not only the medical model of disease but also other theoretical framework such as analytical theories, behavioral principles, attachment theory, and systemic thinking. Decision making is informed by a detailed understanding of normal and abnormal child development from infancy through to adolescence.

A thorough assessment in child psychiatry requires that detailed information must be gathered from the child, family members, and school staff. Children should be given the opportunity to talk by themselves at some point during the assessment. Talking and interacting with children of all ages in order to facilitate free communication is a core skill in child psychiatry. Theoretical understanding of child development, social behavior, and psychological defense mechanisms assists the practitioner in this. Physical examination, medical investigations, and psychometric testing all augment the clinical interviews. The whole assessment allows the clinician to develop a differential diagnosis and an etiological formulation, both of which will guide intervention.

Child psychiatrists use the full range of behavioral, psychological, and pharmacological interventions available in adult psychiatry. Research is beginning to inform evidence-based practice, and there is a growing interest in preventative work. Child psychiatric interventions can be helpful in the management of children who have chronically ill physical health. All child psychiatrists need to be aware of how child abuse may manifest in developmental, emotional, or behavioral problems of childhood, and they must practice within local child protection guidelines. Focused therapeutic work may help some children who have been victims of child abuse.

REFERENCES

Barrett, L., Berney, T., Bhate, S., Famuyiwa, O., Fundudis, T., Kolvin, I., and Tyrer, S. (1991). Diagnosing childhood depression. Who should be interviewed—parent or child? The Newcastle child depression project. Brit. J. Psych. 159 (Suppl. 11), 22-27.

Burbach, D. & Peterson, L. (1986). Children's concepts of physical illness: A review and critique of cognitive-developmental literature. Health Psychology 5, 307-325.

Chess S. & Thomas A. (1986). Temperament in Clinical Practice. Guilford Press, New York.

Cromer J.P. (1985). The Yale—New Haven primary prevention project: A follow-up study. Amer. Acad. Child Adol. Psych. 24, 154-160.

Ellenberg, J.H., Hirtz, D.G., & Nelson, D.B. (1986). Do seizures in children cause intellectual deterioration? N. Eng. J. Med. 314, 1085-1088.

Erikson, E. (1963). Childhood and Society, 2nd ed. Norton, New York.

Freud, A. (1965). Normality and Pathology in Childhood. International Universities Press, New York.

Gesell, A. & Amatruda, C. (1964). Developmental Diagnosis. (Knobloch, H., Pasamanick, B., Eds.) Harper and Row, Hagerstown, MD.
Graham, P. (1994). Prevention. In: Child and Adolescent Psychiatry: Modern Approaches. (Rutter, M., Taylor, E., & Hersov L., Eds.) Blackwell, Oxford, England.
Green, R. (1974). Sexual identity conflict in children and adults. Basic Books, New York.
Hawkins, J.D., Von Cleve, E., & Catalano, R.F. (1991). Reducing early childhood aggression: results of a primary prevention program. J. Amer. Acad. Child & Adol. Psych. 30, 208-217.
Henderson, A.S. (1988). An Introduction to Social Psychiatry. Oxford Medical Publications, Oxford.
Hersov, L. (1994). In-patient and day-hospital units. In: Child and Adolescent Psychiatry: Modern Approaches. (Rutter, M., Taylor, E., & Hersov L. Eds.) Blackwell Science, Oxford, England.
Kempe, C., Siverman, F., Steele, B., Droegmuller, W., & Silver, H. (1962). The battered-child syndrome. J. Amer. Med. Assoc. 181, 4-11.
Kolhberg, L. (1969). Stage and sequence: The cognitive developmental approach to socialization. In: Handbook of Socialization Theory and Research. (Goslin, D., Ed.), Rand McNally, Chicago.
Landreth, G.L. (1982). Play Therapy. Charles C. Thomas, Springfield, Illinois.
Lask, B. (1985). Children's problems: A parent's guide to understanding and tackling them. Dunitz, London.
Miller, B. (1987). Depression and asthma: A potentially lethal mixture. J. Allerg. Clin. Immunol. 80, 48.
Piaget, J. (1952). The Origins of Intelligence in Children. International Universities Press, New York.
Rutter, M., Tizard. J., Yule, W., Graham, P., & Whitmore, K. (1976). Research report: Isle of White studies, 1964–1974. Psych. Med. 6, 313-332.
Schreirer, A. & Libow, J.A. (1993). Hurting for Love. Guilford Press, New York.
Shaffer, D., Garland, A., Gould, M., Fisher, P., & Trautman, P. (1988). Preventing teenage suicide: A critical review. J. Amer. Acad. Child & Adol. Psych. 27, 675-687.
Skuse D. (1989). Emotional abuse and delay in growth. Brit. Med. J. 181, 17-24.
Strunk, R., Mrazek, D., Wolfson, G., & LaBrecque, J. (1985). Physiological and psychological characteristics associated with deaths from asthma in childhood: A case-controlled study. J. Amer. Med. Ass. 254, 1193-1198.
Taylor D.C. (1969). Sexual behavior and temporal lobe epilepsy. Arch. Neurology 21, 510-576.
Thompson A.E. & Kaplan, C.A. (1996). Childhood Emotional Abuse. Brit. J. Psych. 168, 143-148.

Chapter 15

Major Disorders in Child Psychiatry

CAROL E. KAPLAN and ANNE THOMPSON

Biological Psychiatry, pages 311-355.

ISBN: 1-55938-819-6

INTRODUCTION

This chapter provides an overview of common psychiatric conditions and developmental disorders of childhood. The usual age of presentation of some major disorders is displayed in Figure 1.

DISRUPTIVE BEHAVIOR DISORDERS

These disorders commonly present in middle childhood—particularly in boys. Although many healthy children may at times be restless, oppositional, and destructive, children with these clinical syndromes are severely disabled in terms of eliciting disapproval and rejection from adults and children alike. They are at risk of educational failure and poor adaptation to adult life.

Conduct Disorder

A 10-year-old boy was brought for psychiatric assessment after being excluded from school for hitting a teacher. His parents reported that he often stole from shops, bullied other children, and tormented animals. His few friends admired him but were

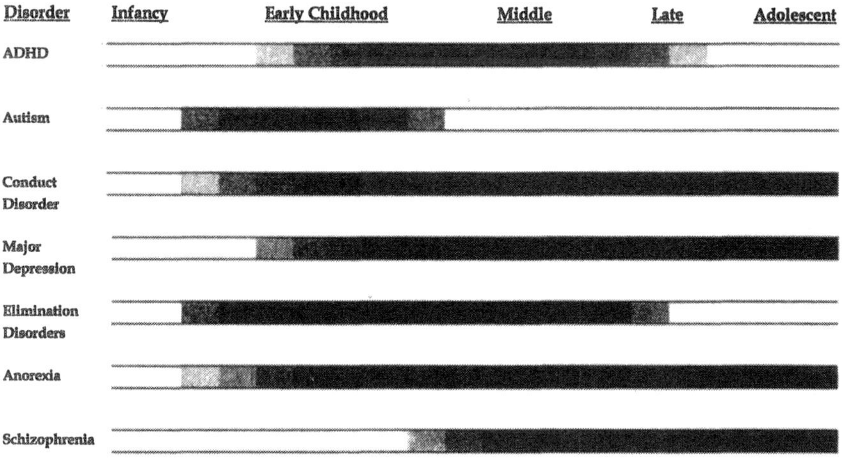

Figure 1. Usual age of presentation of disorders.

wary of him. He often truanted from school with a group of boys and, although of average ability, he achieved little when he was in the classroom.
His father and older brother had criminal records. Both parents said their son had been right to stand up for himself in the argument with his teacher.

Syndrome Characteristics

Children with conduct disorders show a range of persistently aggressive, destructive, and defiant behaviors that at times may be dangerous or criminal. Their behavior is antagonistic toward others and they show a lack of regard for others' feelings. Children with conduct disorder may be part of a peer group involved in antisocial activities or they may be isolated from and unpopular with other children. This latter group of children with so called "unsocialized" conduct disorder are sometimes regarded as being more disturbed.

Children with conduct disorder may also have other disorders including substance misuse disorders and depression.

Epidemiology

Conduct disorder is a relatively common and serious psychiatric condition in childhood with a prevalence of approximately 5%. Rates in urban areas are generally higher than in rural areas: this is probably a reflection of greater poverty in inner city areas. Conduct disorder is four times more common in males than in females.

Etiology

Early etiological theories suggested that conduct disorder was a reflection of deficient moral and social development in children. Later it was suggested that children with conduct disorder were reacting to adverse personal and environmental experiences. Modern psychiatric formulations of conduct disorder often combine elements of both these standpoints. An affected child may have individual vulnerabilities (e.g., genetic predisposition, abnormal temperament, impaired neuropsychological functioning, or past cerebral insult) together with adverse parental, family, and social factors.

Parental factors that may contribute to conduct disorder in children include antisocial behavior, substance misuse, and mental illness together with a style of parenting in which children's behavior is controlled by harsh and inconsistent techniques. Predisposing family factors include a large sibship, marital disruption, poverty, and cultural tolerance of antisocial behavior.

Management

Conduct disorder is a serious and disabling condition, which requires intervention to avert current and future morbidity. Clinicians commonly use combinations of therapeutic approaches. Individual therapy with the child may use cognitive behavioral techniques to examine negative consequences of antisocial behavior and explore alternative ways of behaving. Drug therapy may be used to control symptoms in institutional settings but is not often used for children in the community. Work with parents encourages them to adopt clear, consistent, nonpunitive behavioral approaches and to reward acceptable behavior, ignore tolerable unwanted behavior, and apply age-appropriate sanctions after the child behaves unacceptably. Family work aims to identify and alter maladaptive styles of functioning such as scapegoating. Encouraging children to participate in age-appropriate activities in the community with peers under skilled adult supervision aims to promote prosocial behavior, increase self esteem, and divert the child away from antisocial activities.

At present there is limited evidence for the effectiveness of these interventions (Offord and Bennett, 1994). Because of this, programs of primary prevention are being evaluated. These usually include educational enrichment for young children and parent training (Earls, 1994). The longterm impact of a number of these schemes is awaited with interest.

Prognosis

Children with conduct disorder are at increased risk of psychiatric disorder in adult life. A significant minority of affected children will develop an antisocial personality disorder as they mature. There is also an increased risk of dying prematurely, often by sudden and violent means. Children who have a particularly early onset of a conduct disorder have an especially poor outlook.

Oppositional Defiant Disorder

A 9-year-old boy was doing reasonably well at school, where he was part of a gang of friends who could sometimes be disruptive. At home alone with his mother he often swore at her and broke things. He would not obey her and often humiliated her in public. He was better behaved when with his father and grandparents.

Syndrome Characteristics

Oppositional defiant disorder first appeared in the DSM classification system in 1990. It has come to be regarded as a variant of conduct disorder (Schachar and Wachsmuth, 1990). In common with many children with conduct disorder, children with oppositional-defiant disorder are hostile, uncooperative, provocative, and disruptive. However, unlike conduct disordered children, they do not display seriously aggressively or antisocial behavior. Negative behavior is often displayed selectively to people the child knows well. Oppositional defiant disorder becomes evident in children before the age of 10. Affected children may also have other disorders including attention-deficit/hyperactivity disorder and learning difficulties.

Epidemiology

Children with oppositional defiant disorder are commonly represented in child psychiatry clinics. The epidemiological characteristics are similar to those described in conduct disorder.

Etiology

The etiology of oppositional defiant disorder is thought to be similar to the biopsychosocial model operating in conduct disorder. It is not yet understood why children develop one disorder rather than the other.

Management

Management takes place along the same lines as for conduct disorder. Attention to problematic family relationships is particularly important.

Prognosis

The outlook for children with this cluster of behavioral difficulties is not yet known. In the absence of seriously antisocial behavior, prognosis may be predicted to be better than for children with conduct disorder.

Attention-Deficit/Hyperactivity Disorder (ADHD)

An 8-year-old boy had a reputation at school for being disruptive, restless, and lazy. At home his parents said that he rarely settled to anything other than a computer game or TV show for more than five minutes. He woke early each morning and tried to get out of the house, oblivious to the dangers of being outside alone. His parents were exhausted and frustrated. On starting to take methylphenidate, his restlessness and inattention improved and his parents found him more responsive to a program of behavior modification.

Syndrome Characteristics

The core symptoms of ADHD are inattention, restlessness, and impulsive behavior. These behaviors are usually first evident in the preschool child. Although they may vary in severity depending on the child's environment, these symptoms are severe enough to lead to impaired social, educational, and emotional functioning. Approximately 40% of children with ADHD have another behavior disorder such as conduct disorder or oppositional defiant disorder. Children with ADHD may also have difficulties with specific aspects of learning such as reading or mathematics.

Children with ADHD commonly exhibit annoying, intrusive behavior in the home potentially causing considerable family tension. These children are frequently in trouble at home and at school and they often perform below their intellectual level in school. This pervasive failure frequently leads to the child suffering from low self esteem and also having poor peer relations.

Epidemiology

ADHD occurs throughout the world. Previous discrepancies in prevalence rates from different countries were an artifact of differences in diagnostic criteria. A methodologically sound study using internationally agreed upon diagnostic criteria found a peak prevalence of 8% in 6- to 9-year-old children with lower rates in groups of younger children and adolescents (Szatmari et al., 1989). ADHD is more common in boys and in children who live in cities.

Etiology

In most children, the origins of ADHD are multifactorial. Genetic influences, temperamental factors, parental unavailability and inconsistency, deprivation, and environmental toxins such as lead may all play a part. These adversities probably lead to an as yet undetermined abnormality of central nervous system functioning. (For a review, see Zametkin and Rapport, 1987.)

Management

A variety of treatment modalities have been suggested. Descriptions of clinical practice suggest that combinations of treatment are more effective than single therapies; however, there are no controlled trials. Stimulant medications such as methylphenidate or dextroamphetamine reduce the severity of the core symptoms, but the effectiveness of longterm treatment is not known (Weiss, 1991). Major tranquillizers and antidepressants have weaker effects and more troublesome side effects. Behavior modification, social skills training, and cognitive behavior therapy have all been advocated for treatment of the child. The major drawback of these therapies is the poor generalization from the clinic room to the home and school, and there is little improvement in academic achievement or social skills. Parent training may help stressed families feel more in control but this has not been vigorously evaluated. Remedial education is important for these children.

A number of exclusion diets have been advocated as treatments for ADHD. Although some families report that excluding food additives, food colorings, sugar, or yeast from their child's diet leads to an improvement in their behavior, scientific evidence does not support these claims (Egger et al., 1985), although some results seem to merit further research.

Prognosis

Many children continue to have attentional problems into early adult life, although some will grow out of these. However the majority will have education achievement problems and a smaller subgroup will have antisocial difficulties (Gillberg and Hellgren, 1996).

ANXIETY AND RELATED DISORDERS

Anxiety Disorders

Anxiety is a normal experience at all ages. Mild and transient episodes of anxiety precipitated by life events are a common feature of childhood and adolescence. Anxiety disorders present with the occurrence of excessive symptoms of arousal, which are distressing and/or incapacitating.

There is a developmental sequence to the experience of anxiety at different ages. Infants are afraid of loud noises and other unexpected sensory experiences. As object constancy develops, fear of loss of attachment objects is a source of anxiety. In mid-childhood, anxiety about performance is normal. In adolescence, social anxiety is common. The development of pathological anxiety also follows this age pattern.

There are many theories as to the origins of anxiety disorders, including the different theories of the mind devised by Freud and other psychoanalysts, as well as the work of the Darwinians and other ethologists. Neurophysiological theories of anxiety regulation and the influence of temperament have also been explored in some detail. The genetic influence is argued on the basis of an increased risk of anxiety disorders among relatives of a sufferer. This has been reported for panic disorders, simple phobias, and generalized anxiety.

Co-morbidity occurs more often than by chance with depression, conduct disorder, and attention deficit/hyperactivity disorder.

Multiple anxiety disorders appear to be common. Children and adolescents often present with mixtures of different symptoms, which may make diagnostic precision in terms of ICD-10 or DSM IV difficult.

The exact subdivision of anxiety disorders remains a matter of debate. ICD-10 has six categories of behavior–emotional disorders with onset specific to childhood. Separation anxiety disorder only is included in the DSM IV section on disorders of infancy, childhood, and adolescence. The classification systems have similar disorders in their sections on the anxiety disorders of adult life that are not very different from those seen in younger age groups, provided developmental differences are taken into account.

Epidemiology. Emotional disorders were found in 2.5% of all 10- and 11-year-olds studied in the Isle of White Study (Rutter et al., 1976). The Ontario study (Szatmari et al., 1989) found similar prevalence rates in boys and girls of 10.2 to 10.7 at ages 4 to 11, with differences between boys (4.9) and girls (13.6) from ages 12 to 16 years. The differences between the two studies may be due to different age ranges studied, different research methods, and also differences in the populations.

Management. Treatment usually consists of psychotherapy, which involves both child and parents and may focus on the family system. Drug treatment is occasionally used but generally should be considered as a short-term intervention. These disorders tend to be less fixed in children than adults and response to treatment is generally better.

Prognosis. The anxiety disorders in childhood have better prognoses than in adulthood. This may be because they are less entrenched and of shorter duration.

Separation Anxiety Disorder

The mother of a 4-year-old girl had never yet been able to leave her with a baby-sitter, grandparent, or nursery school teacher. Attempts to separate led to the child crying, screaming, and breath-holding with such apparent distress that the mother inevitably gave up the attempt. Other children were beginning to tease the child, calling her a "cry-baby". The child's parents' relationship was under increasing strain, and they eventually sought a referral to child psychiatry. After a careful introduction over several sessions, a play therapist managed to engage the child in play a short distance away from her parents. This separation was gradually extended in the clinic and eventually parents felt confident enough to practice short separations at home and support each other in surviving their daughter's protests.

Syndrome characteristics. The key symptom of this disorder is the reluctance of preschool children to separate from parents or care givers to such an extent that this interferes with everyday life. Children who suffer from separation anxiety are often shy and timid; they may be very clingy to their parent on whom they are overly dependent, and have difficulty in mixing with other children who often find them "babyish". They may have difficulty in getting to sleep as well as separating from parents during normal daily activities such as play groups. They may fear the loss of a family member or even abandonment.

These children may be troubled during sleep by bad dreams and frequent waking. During the day they may suffer from "free floating anxiety" (i.e., anxiety in any situation) and they may be distractible in a way that interferes with normal interactions. It should be noted that separation anxiety is distinguished from other anxiety disorders by the context of separation from a caregiver rather than by difference in symptomatology.

Over-Anxious and Generalized Anxiety Disorders

A 10-year-old girl was investigated by a pediatrician because of loss of appetite, vomiting, palpitations, and dizziness. No physical cause could be found. She was a shy girl who would rather stay at home baking with her mother than going to youth club. Her mother was very concerned that her daughter might have a serious physical illness following the recent death of her own mother.

Syndrome characteristics. The key features of both these disorders are unrealistic or excessive anxiety or worry and somatic complaints, which may arise from motor or autonomic activity.

The symptoms may cause considerable distress to both the child and the family. Children with this disorder may have loss of appetite, feelings of nausea, diarrhea, vomiting, headaches, increased frequency of micturition, palpitations, dizziness, sweating, and clammy hands. They feel tense and restless and may become very distressed. Their sleep may be disturbed by initial insomnia, nightmares, and frequent waking. These children are often shy and retiring and may be clingy to and overly dependent on their parents. They may be emotionally immature and poor at mixing with their peers. Panic attacks may occur in the course of a generalized anxiety state.

Panic Disorder

A 15-year-old girl had her first panic attack while out at a shopping mall with friends. They thought she was having a fit and called an ambulance. Physical examination by hospital staff was unremarkable and she was sent home. Over the next month the girl had 5 more episodes of severe panic symptoms and she asked to be referred for help. Her child psychiatrist offered a program of education, relaxation, and cognitive-behavioral techniques, which helped the girl to feel more in control during subsequent panic attacks. The possibility of a course of antidepressants was discussed.

Syndrome characteristics. The core feature of panic disorder is a pattern of recurring panic attacks in which the sufferer experiences a range of somatic symptoms of anxiety accompanied by frightening thoughts of impending illness or death. Symptoms during a panic attack may include shortness of breath, dizziness, tachycardia, sweating, nausea, sensation of temperature change, and other symptoms that characterize a feeling of acute fear. No organic basis for the symptoms can be found. Panic attacks characteristically arise spontaneously, although their onset is sometimes associated with specific traumas. They are often accompanied by anticipatory anxiety and the avoidance of situations associated with panic.

Epidemiology. Panic disorder is rare in prepubescent children but has been described in young children in association with other anxiety disorders and depression. Many adult sufferers first experience panic attacks in adolescence.

Etiology. Panic attacks appear to be a manifestation of the "flight or fight" response innate in mammals. Various theories have been advanced as to why this response is activated inappropriately in individuals with panic disorder. Life stresses or a low biological threshold for the triggering of the anxiety response may predispose to the condition, traumatic events may precipitate the disorder, and inaccurate cognitions about the meaning of the onset of the symptoms may perpetuate the anxiety response.

Management. The first line of management is usually anxiety education and management using basic information about physiology, relaxation exercises, and cognitive behavioral techniques to enhance the patient's understanding and sense of self-control prior to and during panic attacks. Behavioral approaches may be used to desensitize individuals to anxiety-provoking situations associated with panic. Anxiolytic and antidepressant medication may also be used.

Prognosis. Once a young person has experienced a panic attack, they are at greater risk of developing panic disorder than control subjects who have never had a panic attack. Panic disorder is a chronic condition in which symptoms wax and wane. Associated psychiatric disorders include agoraphobia, generalized anxiety disorder, and depression.

Phobic Disorders

An 11-year-old boy was brought to a child psychiatry clinic because he was terrified of dogs. Recently, he had almost been hit by a car while running into a road to avoid a dog. His mother was also frightened by dogs, having once been bitten by a so-called friendly pet. A treatment program was devised with the aim of enabling the boy to pass a dog in the street without panicking. The child and his mother both learned relaxation techniques and some basic animal psychology. They devised a routine for ignoring dogs in the street and practiced this together on daily walks. They planned a visit to see and touch a friend's tiny puppy. By the end of the program, the child still disliked dogs but no longer ran away from them.

Syndrome characteristics. A phobia is an anxiety reaction produced by a specific stimulus the magnitude of which is out of proportion to the risk posed by the stimulus. Common feared objects in childhood include dogs and spiders. Children with phobias become extremely distressed in the presence of or even in anticipation of facing their feared stimulus. They may suffer somatic symptoms of anxiety and panic reactions. Their efforts to avoid contact with the feared object or situation may interfere with the child's social or educational development and may even pose a risk such as in a child who will run into a road in order to avoid a dog. In social phobia, usually evident in older children, the feared situations are social settings in which the child has to speak or perform in front of other people. Children with phobias may have an anxious underlying temperament or a parent with the same phobia.

Epidemiology. Recent epidemiological studies suggest that phobic disorders are a common problem in childhood. However, only a proportion of sufferers are brought to child psychiatry clinics. Before puberty, the sex incidence is equal. After puberty, phobias are more common in girls.

Etiology. A child may develop a phobia as a result of an underlying anxiety-prone temperament, a traumatic encounter with the feared stimulus, or learned behavior from other family members. Once the stimulus has become a source of distress or fear, the child avoids further contact with it. This avoidance reinforces the negative reaction associated with the stimulus, and a pattern of fear and avoidance behavior begins.

Intervention. The treatment of a child with a phobic disorder centers around a behavioral program using the technique of graded desensitization. This aims to support the child while they learn to tolerate their anxiety during controlled exposure to the feared stimulus. Relaxation training and exploration of unhelpful self statements using a cognitive approach may assist in this. If possible, parents should be involved as co-therapists. Realistic aims for treatment are needed; for example, it is adaptive for children to be cautious about approaching strange dogs and realistic to aim to help a child with a dog phobia to pass a dog in the street without panicking rather than aiming for them to believe that all dogs are their friends!

Prognosis. The treatment of the simple phobias of childhood usually produces some reduction in the child's level of anxiety toward the feared stimulus and some improvement in the child's level of functioning. However progress may be limited if the phobia is a manifestation of a child's neurotic temperament or shared family anxieties. Studies show considerable continuity between phobic disorders in childhood and in adult life.

Obsessive–Compulsive Disorder (OCD)

For the last 6 months, a 13-year-old girl had been showering and changing her clothes up to six times each day. Before getting into bed she spent approximately 90 minutes going through an elaborate sequence of counting and sorting rituals. Her mother joined in with these in an attempt to help her daughter get to bed faster. If she withdrew, her daughter screamed for her to return. Psychiatric assesment suggested that this girl's obsessive-compulsive disorder had been precipitated by bullying at school and excessive worries about sexual development in a child with a premorbidly anxious temperament. Her mother's attempts to help her were seen as inadvertently reinforcing the ritualistic behavior and preventing the child from learning to tolerate obsessional anxiety without discharging it through ritual.

Syndrome characteristics. The mean age of onset is 10 years. Males seem to have an earlier onset than females. The core symptoms include obsessional thoughts (i.e., intrusive, repetitive thoughts about impending threats) and compulsive rituals (i.e., repetitive behaviors driven by anxiety). In childhood, rituals are more frequently the presenting complaint than obsessions. Washing rituals occur in 85% of patients, and in 90% of cases, the symptoms change with time (Goodman et al., 1989). Obsessional thoughts include concern with dirt or toxins, the fear of something terrible happening, a need for order, sexual thoughts, or the fear of harm occurring to self or others. For example a child may feel the need to touch a picture on the wall 12 times in order to prevent his mother crashing her car.

Compulsions include excessive washing or grooming, repeating activities (e.g., going up or down stairs, checking, touching, counting, and collecting). All of these can be seen by clinicians as measures necessary to prevent harm.

About one-third of patients report triggering stimuli for their rituals, and secrecy is often an accompaniment to these behaviors. Parents often become involved in a child's rituals; for example, helping a child with their cleaning rituals in order to get them to bed at night.

Epidemiology. Until recently, obsessive–compulsive disorder was unfamiliar to many child psychiatrists. However, work in the last decade has substantially altered this position. The lifetime prevalence rate based on adult interview is reported as 1.2 to 2.4%, with the usual age of onset from 20 to 24 years, 50% report having had some symptoms in childhood or adolescence. A study of adolescents (Whitaker et al., 1990) found a prevalence of 1.9%.

Etiology. Biological factors have been widely explored in this disorder. There has been particular interest in basal ganglia dysfunction and the striking association of OCD with disorders such as Sydenham's chorea, Huntington's chorea, Wilson's disease, and vascular disease affecting the putamen. The role of neurotransmitters such as serotonin and dopamine has also been studied. An ethological perspective has also been suggested, linking grooming behavior and its accompanying hormonal changes with the repetitive activity seen in OCD. (Swedo, 1989).

Management. In adults, behavioral treatment is well established. However, this treatment has not been studied systematically in children as yet. Response prevention is the best studied and a positive outcome has been reported in over 50% (Bolton et al., 1983).

Pharmacological treatment using serotonin re-uptake inhibitors including fluoxetine has also been tried in children and adolescents with good responses reported (Riddle et al., 1989). It should be noted that treatment may be needed for a considerable period of time as relapse when medication is discontinued is common.

An important factor recently has been the introduction of support groups for parents and their families. These are particularly effective in dealing with the avoidant and secretive behavior that often accompanies this disorder.

Outcome. The clinical course may be chronic or episodic. Spontaneous remission occurs in about one-third of patients even after years of illness. About 10% have a continuous deteriorating course. The follow up of children who received drug and behavioral treatment (Leonard et al., 1993) showed that only 33% were symptom free at 2- to 7-year follow up.

Conversion Disorder

Syndrome characteristic. Conversion disorders in children present in a similar way to the condition in adult life: the child has one or a number of symptoms or deficits that cannot be explained in terms of a neurological or general medical disease. Common symptoms include disturbances of gait, pain, and limb paralysis. The symptoms of disabilities are not produced consciously (in contrast to malingering) and they are judged to be associated with psychological factors. Predisposing factors may include conscientious, high achieving, premorbid functioning in the child and a family history of similar symptoms or deficits. The precipitant of the conversion disorder is usually stress or conflict such as academic pressure, adolescent worries, or child abuse. The disorder is usually maintained by secondary gain evident to the child in the sick role, by family members' interest in the illness, or by extensive medical investigations.

Epidemiology. Conversion disorder is said to require a certain psychological sophistication for its development and maintenance, so the condition is not seen in very young children but most commonly presents in young adolescents. It occurs approximately three times more commonly in girls.

Etiology. Conversion disorder was first explained in psychoanalytical terms as the conversion of unconscious sexually driven conflict into symbolic physical symptoms. Later, conversion disorder was proposed to represent an exaggeration of illness behavior created consciously or unconsciously in order to accrue the advantages of the sick role. In either case, precipitating stress and ongoing anxiety is assumed to play a central role.

Management. Key principles are outlined by Dubowitz and Hersov (1976). Taking a careful history and performing a vigilant physical examination may alert the clinician to symptoms or signs that do not conform to patterns of organic disease. Medical investigations should be chosen judiciously as repeated diagnostic tests often maintain the symptoms of conversion disorder. Atypical symptoms and signs accompanied by negative physical investigations should raise the possibility of conversion disorder.

Although precipitating stress may be inferred from the history, families frequently deny emotional difficulties at initial consultations. Once the diagnosis of conversion disorder is made, parents and the child should be given a clear and sensitive explanation of the disorder, stressing the good outlook for full recovery. A program of graded rehabilitation should be devised for the child to allow the child to recover without losing face. Careful coordination of the clinical team is required. The role of the child psychiatrist may include providing support for the family and team, helping the child face stress in a more adaptive way, and facilitating exploration of emotional issues within the family.

Prognosis. The majority of children presenting with conversion disorder recover within three months (Leslie, 1988). A small proportion continue to be disabled with intractable symptoms that have adverse consequences on social, emotional, and educational development.

MOOD DISORDERS

Mild, transient episodes of distress or elation are common features of normal childhood and adolescence. Some adolescents seem to find this period of life difficult to negotiate and may appear gloomy, withdrawn, and sensitive. In contrast to these variations in normal affective state, disorders of mood lead to serious psychosocial dysfunction. Previously, there was debate as to whether depression could exist in children because of their relative lack of psychological sophistication. However these ideas are refuted by research, and mood disorders are accepted as a serious and worrying group of disorders in children and adolescents.

Major Depressive Disorder

A 14-year-old girl had become increasingly withdrawn, tearful, and irritable over a four-month period. She was reluctant to attend school and see her friends, had lost weight, and was sleeping poorly. Her parents were shocked to learn that when interviewed alone, their daughter reported feeling hopeless and suicidal. Her mother recalled that a maternal aunt had spent many years in psychiatric hospitals suffering from depression. In view of her recent unexpected poor performance in a school examination, the child psychiatrist was concerned about a significant risk of suicide.

Syndrome characteristics. Depressed mood is the core symptom of major depression, accompanied by a cluster of symptoms such as loss of interest, change in appetite, sleep disturbance, fatigue, agitation or retardation, sense of worthlessness or guilt, poor concentration, and recurrent thoughts of death or suicide.

Depression in childhood can occur in association with dysthymia, anxiety disorder, conduct disorder, ADHD, anorexia nervosa, Tourettes syndrome and school refusal.

Epidemiology. The estimates of prevalence of major depression currently available among preadolescents is from 0.5 to 2.5% (Fleming et al., 1989).

There is an increase in rates of depressive disorder with age and a higher prevalence rate in adolescents ranging from 2.0% to 8.0% (Cooper and Goodyer, 1993). In addition, depression-related behaviors such as suicide and parasuicide also show a great increase in adolescence (Sellar et al, 1990). In depressed children, the sex ratio is equal, but in adolescence there is a female preponderance.

Etiology. The basis of depression in children and adolescence may be environmental and biological. Two examples of environmental effects are complicated bereavement and early parental loss. Neurochemical, neuroendocrine, sleep and genetic data have all contributed to information about possible etiological factors.

Management. Depressed children have multiple problems and, while one of the goals of treatment must be the reduction of depression, other problems such as family difficulties and impaired peer relationships, which may maintain the disorder, must not be neglected.

Multiple forms of treatment are used and include psychological approaches such as cognitive behavior therapy (Reinecke et al., 1998), interpersonal therapy (Mufson et al., 1993), and family work. Recent research suggests that individually orientated therapies may be more beneficial than family therapies (Harrington et al., 1998). Psychological therapies may be combined with the use of antidepressants. Until recently, research evidence of the efficacy of antidepressants in childhood depression was lacking. However, one of the selective serotonin reuptake inhibitors (SSRI) has now been shown to be superior to placebo in the treatment of depression in children and adolescents (Emslie et al., 1997) and interest in the use of this class of drug is growing. The SSRIs have additional advantages over the tricyclic antidepressants because their side-effect profile may be more acceptable to young people and they are safer in overdose. Very rarely ECT is used to treat potentially life threatening depressive states. There is considerable concern about the unwanted effects of this treatment in children.

Prognosis. The lifetime course for children who have suffered from a major depressive disorder is still a matter of debate, but there does seem to be some evidence that depression in children may be predictive of further episodes of depression in adulthood. There is also a suggestion that there may be an increased risk of bipolar disorder. It must also be remembered that children who have had a serious depressive disorder may suffer profound emotional and social consequences.

Bipolar Disorder

A 16-year-old boy was persuaded to come for psychiatric assessment after breaking into a neighbor's car, saying he was going to drive around the world. He had no complaints but appeared dishevelled, irritable, and expansive at interview. His parents reported that he has dropped out of college some months before and had become preoccupied with various ideas of making money, none of which were ever completed. The boy did not agree that he was ill, and he refused treatment. Later, he deteriorated into a psychotic state and was detained in an adolescent in-patient unit against his wishes in accordance with mental health law to begin treatment with major tranquillizers and lithium.

Syndrome characteristics. In several respects, mania presents with similar symptoms in childhood and adult life and the criteria for the diagnosis are the same. Grandiosity and paranoia are more common in older children, whereas irritability and emotional liability are more common in younger children. Hyperactivity, pressure of speech, and distractibility are found in both groups.

Epidemiology. Studies of adults with bipolar disorder have shown that one in five had an onset before 19 years of age (Winokur et al., 1969). The actual prevalence in childhood and adolescence is difficult to define, but is generally accepted as very low (Anthony and Scott, 1960).

Etiology. A strong genetic component is suggested, but the mode of inheritance is unclear. Similarly, the effect of environmental mediation remains unknown.

Management. Information is mostly derived from adult studies and is largely dependent on the use of neuroleptics. Lithium, sodium valproate, and carbamazepine are also used, usually augmented by major tranquillizers.

Prognosis. The outcome of bipolar disorder arising in adolescence is better than schizophrenia arising at a similar age (Werry et al., 1991). Compliance with medication is poor in one-third of cases, and the best predictor of future functioning is premorbid adjustment.

EATING DISORDERS

Concerns about eating may arise at all stages of childhood and adolescence. Parents may be concerned at their infant's slowness or reluctance to feed or about children who have fads about particular foods and may eat a very restricted diet. Very shy children may have difficulty eating in public, some children overeat and others become preoccupied by dieting. Eating and feeding is sometimes a source of conflict between adults and children; this may relate temperamental factors and also to difficulties in achieving separation and individuation. Sometimes eating behaviors may become so extreme as to be diagnosed as discrete clinical syndromes such as anorexia nervosa, bulimia nervosa, or obesity.

Anorexia Nervosa

In spite of extensive investigations, pediatricians had been unable to find a physical cause for a prepubescent 10-year-old girl's loss of weight over the last year. Psychiatric assessment showed that she was determined to lose weight and thought her body was ugly. During her next admission for investigation, uneaten food and some of her mother's diuretic tablets were found in her hospital locker. These were hidden amongst the school books she had brought with her, to ensure that she maintained her position near the top of her class.

Syndrome characteristics. The core features of anorexia nervosa in children and adolescents are the same as the adult disease: failure to maintain an expected weight, fear of gaining weight, body image disturbance, and primary or secondary amenorrhea. Children and adolescents may hide food, vomit, exercise excessively, or use purgatives or diuretics to control their weight. Diagnosis may be difficult in prepubertal children who can present with failure of weight gain rather than weight loss and who may not express typical fear of fatness and body image distortion (Treasure and Thompson, 1988). Young children may have had extensive investigations to determine an organic cause of growth failure before the diagnosis of anorexia nervosa is made. Anorexia nervosa typically arises in academically able adolescents who have conscientious, conforming premorbid temperaments.

Epidemiology. Anorexia nervosa presenting in adolescence is ten times more common in girls than boys. It presents more commonly in affluent socioeconomic groups. Prepubertal anorexia nervosa is atypical in having a more equal sex incidence and less of a skewed social class distribution.

Etiology. Differing theories of causation suggest that anorexia nervosa results from a phobic avoidance of emotional and sexual maturity, a maladaptive response to excessively intrusive parents, a primary hypothalmic dysfunction, or an extreme reaction to cultural pressures on Western females to be slim. There may be a family history of obesity or other eating disorders. Dysfunctional styles of family functioning may be evident such as failure to acknowledge and resolve conflict or enmeshed parent–child relationships.

Management. Management aims to arrest weight loss and promote healthy eating adequate for normal growth, to address maladaptive thinking about weight, body shape, and self image, and to reorientate family life away from the patient's preoccupation with food. Treatment programs are often multimodal, and treatment may need to be lengthy. Treatment of younger children centers around educating and empowering parents to take charge of their child's nutrition, and family therapy is superior to individual therapy for these children (Russell et al., 1987). The treatment of older adolescents is more like the treatment of adults but with a greater focus on the individual accepting responsibility for their own health.

Prognosis. Accumulated evidence from outcome studies suggests that 50% of patients with anorexia nervosa recover, 30% improve over time, and 20% develop chronic symptoms. Among this last group, there is a mortality rate of less than 5% (Steinhausen et al., 1991). Some children with prepubertal disease have a rather better prognosis, while others whose diagnosis was made late and who have adverse family and social factors have a worse outcome. Prolonged malnutrition in prepubertal children may have longterm consequences including failure of breast development, primary amenorrhea, and osteoporosis.

Bulimia nervosa

After a secretive period of dietary restriction and weight loss, a 17-year-old girl began to binge and vomit every day. Her parents noticed that she no longer went out with friends and they often heard her sobbing in her room. The girl learned to vomit at will and, while continuing her abnormal eating habits, still maintained a normal weight throughout her first year away at college.

Syndrome characteristics. Adolescents with bulimia nervosa eat excessive quantities of food in short periods of time (binge eating) and at other times attempt to limit weight gain by vomiting, fasting, exercising, or using medication such as laxatives or diuretics. Like sufferers of anorexia nervosa, the behavior of bulimic individuals is driven by their preoccupation with body weight and shape, although patients with bulimia may have a normal body weight. They suffer from affective instability and low self esteem and find themselves in cycles of maladaptive behavior in which their abnormal eating habits inadvertently maintain their dysphoric state.

Epidemiology. Bulimia nervosa occurs in an older population than that found to have anorexia nervosa, with a peak age of incidence of 19 years of age. Like anorexia nervosa, bulimia nervosa is much more common in women than in men. The prevalence rate among adolescents and young adult women is 1% (Fairburn and Beglia, 1990).

Etiology. Bulimic symptoms may appear during the course of anorexia nervosa and a patient may develop firstly anorexia nervosa and later present with symptoms of bulimia nervosa. As in anorexia nervosa, individual, family, and environmental factors are likely to play a part in the etiology. In comparison with anorexic individuals, bulimia sufferers tend to have greater difficulties with impulse control, depression, guilt, and tolerance of frustration (Johnson et al., 1984). The families of bulimia sufferers generally are more disorganized, more conflicted, and more orientated toward achievement than controls. Symptom profiles and the rates of depression in family members have led some authors to suggest that bulimia nervosa should be considered as a variant of the affective disorders.

Management. A variety of treatment modalities exist for this disorder. Cognitive behavioral therapy appears to be the most effective intervention. This style of work enhances the patient's sense of self control over their eating behavior by monitoring eating patterns, identifying and restricting the stimuli to binge, and challenging dysfunctional thinking. Group therapy has also shown to be effective, especially by encouraging patients to confront maladaptive beliefs and behaviors, in a supportive setting. Antidepressant medication and interpersonal psychotherapy are sometimes offered as secondary elements in treatment packages.

Prognosis. Bulimia nervosa runs a relapsing and remitting course. In his classic description, Russell (1979) described a poor outlook for this condition. However, his group of bulimia patients had developed the condition secondary to anorexia nervosa and this has come to be regarded as an indicator of a poor prognosis. Other indicators of poor prognosis are alcohol abuse, suicide attempts, and depression. Recent follow-up studies of bulimia nervosa suggest that common residual features include abnormal eating patterns and depression.

Obesity

By the age of 13, an obese girl had endured years of bullying at school. She had little self confidence and low self worth. She attempted to make friends by supplying other children with sweets, but her size and inability to take part in games because of her breathlessness resulted in rejection. Attempts to diet were repeatedly abandoned by her parents who were themselves overweight.

Adult obesity has proved to be a remarkably intractable disorder to treat. As a result there has been greater interest in identifying and treating obesity in childhood when it is hoped that intervention in the developing child will result in a better response.

Syndrome characteristics. The definition of an ideal body weight is strongly debated and criticized. Usually this is defined as the statistical average based on population measures of height and weight. Obesity is then defined as 20% or more above that level (Epstein et al., 1988). It is recommended that both percentage above ideal body weight and triceps skinfold thickness should be used to give an accurate clinical picture.

There is overwhelming evidence to show that obesity that starts in childhood persists. Once an obese child becomes adolescent, the likelihood of spontaneous remission is very low. It is also likely that the severity of obesity in childhood will correlate with the degree of adult obesity.

Physical complications of obesity may include hypertension, altered lipoprotein ratios, diabetes mellitus, respiratory problems, and an increased vulnerability to accidents, for example burns. In addition an obese child is likely to be discriminated against, have a reduced self esteem, and be stereotyped psychosocially by peers. In spite of the undoubted unhappiness of some obese children, no differences in the prevalence of psychiatric disorders have been found between obese and nonobese child populations.

Epidemiology. Estimates of prevalence of childhood obesity vary greatly. It has been suggested that it is in the 10% to 24% range, and Dietz (1989) suggests that this rate is increasing. Demographic and family variables seem to alter the prevalence of obesity. The best single predictor of childhood obesity is having two obese biological parents. Families where eating is encouraged are more likely to have an obese child.

Etiology. Childhood obesity results from an excess calorie intake relative to energy expenditure. Medical syndromes such as Cushing's syndrome, Fröhlich's syndrome, Prader-Labhart-Willi syndrome, myelodysplasia, hypothyroidism, and a number of other syndromes account for less than 5% of cases of childhood obesity.

Treatment. Cognitive behavior treatments have been found superior to other forms of intervention, and this is maintained at five-year follow-up. The role of dietary change, physical activity, parental and family change, self and parental monitoring, behavioral contracting, and social and cognitive skill development are all factors which seem to make a contribution, but await detailed evaluation.

Prognosis. It has been found that three-quarters of children who were obese at the age of four were still obese at age seven. Obesity at three to four years of age is a better predictor of adult obesity than obesity at a younger age. If an overweight child has not attained normal weight by late adolescence, he or she is highly unlikely to do so as an adult.

Pica

A three-year-old boy spent long periods alone in his family's house while his mother was out working and his unemployed father, who was responsible for the boy's daily care, was often drunk. The child regularly picked the paint off his bedroom walls and ate it. This habit proved difficult to alter until the family received an intensive package of social support, which provided the child with more appropriate care and activities.

Syndrome characteristics. Pica is the name given to the persistent eating of non-nutrient substances such as paper, crayons, paint or feces.

Etiology. There are two commonly proposed etiologies of pica. One is that nutritional deficiencies drive the sufferer to eat certain substances which may contain the lacking element such as iron or zinc. The other theory is that pica is the result of under stimulation and poor supervision of children.

Epidemiology. Normal infants explore their world by mouthing and tasting objects and so the phenomenon of abnormal behavior known as pica does not occur until after eighteen months of age. Many healthy preschool children also occasionally eat nonfood substances. Pica is more common in socioeconomically deprived children and those suffering from severe mental handicap and psychosis.

Management. The most important intervention for pica sufferers is the provision of adequate stimulation and nurture in a safe environment that denies access to nonfood substances, which are likely to be eaten. Treatment of complications such as lead poisoning, iron deficiency anemia, and intestinal parasitic infestation may be necessary. (For a review, see Bicknall, 1975.)

ELIMINATION DISORDERS

The achievement of continence is a major step toward independence and socialization. Most children achieve bowel and bladder control in the first five years of life. In general, daytime control precedes that of nighttime continence. The persistence of incontinence beyond the age at which peers are clean and dry can lead to significant social and emotional difficulties, within both the peer group and the family.

Elimination disorders may be primary (with no previous period of continence) or secondary (in which control is lost following achievement of continence). A distinction may also be made in terms of whether there is a nocturnal or diurnal pattern.

Encopresis

A 10-year-old boy had been adequately toilet trained at the age of three but had begun to defecate into his clothes at home and school in the months following his parent's divorce. Assessment revealed an unhappy and angry boy who was being scapegoated for his parents' separation and teased for being smelly at school. His family were reacting punitively to his encopresis and needed considerable persuasion and support to carry out the child psychiatric clinic's toileting program.

Syndrome characteristics. Encopresis is the passage of formed stool in inappropriate places by a child who has adequate bowel control. Children generally gain bowel control in their fourth year of life and the disorder cannot be diagnosed before then. Children may conceal or smear feces. They often have other symptoms suggestive of emotional and behavioral disturbance. Families react to encopresis with distress and disgust. Affected children may appear largely oblivious to the problem in spite of the fact that they may face considerable teasing from other children.

Epidemiology. The prevalence of encopresis is at its highest in younger children (15 per 1000) and decreases steadily so that very few adolescents have encopresis after the age of 16. The disorder is more common in boys.

Etiology. Unlike other forms of fecal soiling which may be caused by immaturity of or a deficiency in bowel functioning, encopresis is primarily an emotional disorder. There may however be a history of developmental delay or constipation, which act as predisposing factors. It may be found in children who have never been adequately taught to achieve continence (primary encopresis). Secondary encopresis is when soiling occurs after the child has achieved continence for a significant period of his life. It may be a symptom of prolonged unsatisfactory family relationships and represent an expression of aggression and regression by the child. Transient encopresis may be a reaction to a stressful life event such as the birth of a sibling, parental separation, or ill physical health. There is a high frequency of marital problems and family disharmony in the families of children who present with encopresis.

Management. Intervention starts with examination of the child to detect and treat underlying constipation. The main mode of management is behavior therapy to establish and reward a routine of appropriate defecation. Children may also be helped by play therapy. Parents need to be supported to avoid punitive management of soiling and to improve the quality of their relationship with the child.

Prognosis. Research indicates that encopresis improves and generally resolves as children grow up. However, some children with multiple psychiatric disorders from dysfunctional families may need intensive multimodal treatment to make progress.

Enuresis

A 9-year-old boy had never had a dry bed. This became a problem when he wanted to go on a skiing trip with the school. His parents had previously not been unduly concerned about their son's nighttime wetting, as the boys father had wet his bed up until the age of 14 years. A pad and bell device was used to good effect by this boy who was well motivated to be like other children on the skiing trip.

Symptom characteristics. Enuresis occurs when children repeatedly void urine into their clothes or their beds, either involuntarily or voluntarily. The diagnosis cannot be made in preschool children who have not yet achieved the developmental maturity to maintain urinary continence and should not be made if incontinence can be attributed to a medical condition. Prolonged enuresis is associated with poor social functioning of the child, parental distress and family tensions.

Epidemiology. The prevalence of enuresis varies with different population groups. Rates of wetting decline with increasing age. Rates of enuresis are higher in economically deprived communities. Daytime wetting is rarer at all ages, is more common in girls, and is more commonly associated with urinary tract abnormalities and other psychiatric disorders. Nocturnal enuresis on the other hand affects more boys than girls. There is often a family history of the condition.

Etiology. As the acquisition of urinary continence is a developmental task of the preschool years, enuresis can be viewed either as the failure to acquire this ability (primary enuresis), for example, because of inadequate or interrupted toilet training or as the loss of appropriate urination habits (secondary enuresis), for example, as a result of stresses such as the birth of a sibling or parental separation.

Management. Prior to attending the child psychiatry clinic, parents of children with nocturnal enuresis have often tried techniques such as fluid restriction prior to bedtime and "lifting" the child onto the toilet to urinate in the early part of the night. These may prevent subsequent bedwetting in some children but only effect a temporary or incomplete change in others. The most effective interventions to offer in the clinic are behavioral techniques. These include the use of "star charts" with which dry periods are monitored and rewarded, or the use of a "pad-and-bell" device (Forsythe and Butler, 1989). This inexpensive monitor is fitted onto the child"s mattress. Voided urine completes a low voltage electric circuit and causes a bell or buzzer to sound. This wakes the child during or immediately after micturition, encouraging awareness of urination and appropriate toileting behavior. These techniques need careful explanation and supervision but can produce lasting continence in approximately 80% of children. Antidepressant and synthetic antidiuretic hormone preparations can give shortterm continence for distressed families or in order to facilitate a child spending a night away from home, but medication does not assist children in achieving permanent continence. Psychotherapy is not the first choice of treatment for enuresis.

Prognosis. Good prognostic indicators for the eventual achievement of continence in children include intermittent rather than constant wetting, primary rather than secondary enuresis, being female, and living in affluent socioeconomic conditions. A small percentage of enuretic children continue to wet the bed in adult life.

DEVELOPMENTAL DISORDERS

These disorders may be divided into those which are pervasive and affect the majority of areas of functioning, and those which are specific and affect only one or two aspects of functioning. They are usually identified when a child fails to acquire appropriate skills for their age or when they lose previously acquired abilities. Sometimes a developmental delay may be hidden behind complaints of behavior difficulties, academic failure, or reluctance to attend school.

Pervasive Developmental Disorders

Autism

A four-year-old girl had developed very differently from her older sister. Although bright and happy until the age of 17 months, she had gradually become silent and disinterested in family activities. When excited or distressed she flapped her hands vigorously. Any change of daily routine was met with great distress and loud screams. At play group, she had no difficulty separating from her mother. Indeed she treated her and other members of her family, as well as staff, as objects rather than people. She repeatedly opened and closed a particular drawer and screamed if other children approached her. After a diagnosis of autism was made, her parents found her a place in a school specializing in the education of autistic children.

Epidemiology. The prevalence of autism is estimated at two to five per 10,000 and the condition occurs across all socioeconomic groups. There is a male-to-female ratio of 3:1.

Syndrome characteristics. The syndrome of autism was first described by Kanner in 1943. Three areas of dysfunction are identified: social development, communication, and restricted and repetitive behaviors and interests. Some recognition of abnormality must have occurred before the age of 36 months. The difficulties of social development are recognized as being the most handicapping feature of the disorder. This is because these children have difficulty in reciprocal social interaction and thus an impaired ability to form social relationships. As babies, they do not like physical contact or alternatively may be excessively clingy in an indiscriminate way. They do not show behaviors that indicate much interest in other people or the world about them. As older children, they are not interested in other children with whom they make unusual eye contact, and although they may be distressed at separation from their parents, they do not use them as a secure base from which to explore their environment or as a source of comfort, but rather as machines that can be used to help them devoid of feeling and responsiveness.

Dysfunctional communication is apparent in delayed or deviant language acquisition. Emotional expression and social quality are restricted. Speech is often repetitious and there may be abnormalities of tone, stress, and rhythm.

Children with autism show restricted and repetitive use of toys (e.g., spinning the wheels of a car, repeatedly feeling the material of which a toy is made) and spontaneous imaginative play is rare. Stereotypies, or repetitive movements of the hands within the visual field, are common. When older, many autistic children develop specific preoccupations (e.g., with lampposts or timetables). The need for a predictable environment is often dramatically demonstrated by catastrophic reaction to a minor change such as new curtains in a bedroom.

Many children with autism also suffer from mental handicap. Epilepsy is also found in a significant number of sufferers.

Etiology. The identification of the pathophysiology of autism is a complex and expanding field. The use of autopsy and neuroimaging techniques have not demonstrated any specific lesions in those with autism, although some studies have shown functional deficits of the prefrontal cortex in autistic subjects.

Autism has been reported over the years as associated with a variety of genetic disorders including phenylketonuria, neurofibromatis, tuberose sclerosis, and fragile X syndrome. However, these associations have been accepted as of lesser significance more recently.

Twin studies point to a strong genetic factor with large differences between concordance rates for autism in monozygotic and dizygotic twins.

Management. Intervention in autism aims to enhance socialization, communication, and learning as well as decreasing maladaptive behaviors. Autistic children require a multimodal package of management that includes specialized education, communication and social skills training, behavior modification programs, and pharmacological treatment for maladaptive behaviors. Practical help and emotional support for families is also needed. Skilled management can improve an autistic child's quality of personal interaction, but no treatment removes the basic deficits of the disorder.

Prognosis. Autism is a lifelong disorder. Only a few sufferers achieve fully independent living in adult life. Many require longterm support with daily living, employment, and relationships. Indicators of relatively good prognosis include normal cognitive abilities, preserved language skills, reasonable adaptive functioning, and the absence of behavioral disturbance.

Asperger's Syndrome

A 15-year-old boy had always been an outsider at school. Other boys found him odd: he often butted into conversations, said something inappropriate to the topic under discussion, or bored people with detailed information about trains. Although apparently oblivious to other's perceptions of him when younger, this boy presented to child psychiatry having made a superficial cut to his wrist. He had some symptoms of depression and said that he wanted a friend like other people had. He made slow progress in a social skills group tailored to the needs of young people with Asperger's Syndrome.

This is a disorder that appears for the first time in DSM IV and ICD-10. It is listed as one of the pervasive developmental disorders as it is proposed that it may be a variant of autism.

Syndrome Characteristics. The presenting symptoms include severely impaired reciprocal social interaction; an absorbing interest in a subject that is taken to extremes; attempts to impose fixed routine or the parents' special interest on all aspects of life; delayed language development and speech that is often described as "formal and pedantic"; nonverbal communication problems; motor clumsiness; obstinacy; aggressiveness; sensitivity, sometimes with paranoid ideas; rigidity; and learning problems despite above-average or average assessed intelligence. In some children, bizarre antisocial behavior may be found.

Epidemiology. Asperger's syndrome has a male-to-female ratio of 9:1, and referral is often the result of concerns about school refusal and oppositional behavior.

Etiology. The etiology of the condition is unknown.

Treatment. There is only symptomatic treatment available. The aims are to facilitate social and communicative learning, decrease problematic behavior, and support the family. Medication may be helpful but most treatment programs rely most heavily on behavioral and educative approaches.

Prognosis. Children with Asperger's syndrome have a lifelong propensity to impoverished social relationships, and limited adaptation to the demands of adult life. Independent living may be possible albeit in a restricted fashion. Insight into the sufferer's social inadequacies can lead to distress and depression.

Rett's Syndrome

> *A family with a 4-year-old child recently diagnosed as having Rett's syndrome were referred for counselling in view of their daughter's grave prognosis. Careful support was offered to deal with issues such as denial, anger, guilt, and sadness. Work continued over the period through the child's death from pneumonia at the age of 7. Eventually the family reached some acceptance of their daughter's death.*

This disorder is also a new addition to the classification systems.

Syndrome characteristics. The disorder is characterized by normal early development followed by a gradual loss of speech and language, manual dexterity, and social functioning by between seven and 24 months. Characteristic features include screaming attacks, stereotypic hand movements, and acquired microcephaly.

Epidemiology. Rett's Syndrome occurs almost exclusively in girls.

Prognosis. The prognosis is invariably degenerative, leading to severe mental handicap and motor problems.

Specific Developmental Disorders

Children who have a specific delay in the development of a cognitive or motor skill operate in that domain at a level at least two standard deviations (approximately 20%) below the level expected for their chronological age. Their general cognitive abilities are either normal or not so profoundly delayed. Specific developmental delays have a number of common features:

- They are deficits of normally acquired skills
- They are thought to be caused by central nervous system dysfunction or immaturity.
- Environmental understimulation worsens the primary organically mediated deficit.
- They are four times more common in boys.
- Children often have more than one developmental delay.

- There is often a family history.
- Improvement takes place according to normal developmental sequences.

Specific Speech and Language Disorders

Syndrome characteristics. There are a number of subtypes of specific speech and language disorders representing particular delays in the acquisition of expressive or receptive language skills. Children with these disorders generally present in the preschool years because of parental concern about poor speech or apparent difficulties with comprehension. These children have higher than average rates of disruptive behavior disorders, and the underlying language problem of the children may not be recognized at first.

Epidemiology. Prevalence figures varying from 2% to 25% have been found. Factors that may contribute to environmental understimulation include poverty, large family size, multiple births, and recurrent otitis media.

Management. Specialist assessment of speech and language, nonverbal communication, social development, play, and intelligence should be undertaken. A physical examination and assessment of hearing is required. Intervention may include increasing a child's general level of stimulation with attendance at nursery school together with the teaching of target language skills in a naturalistic setting. Children with severe language impairment may be helped by learning sign language.

Prognosis. Many very young children with speech and language delay catch up without intervention. However, children with persistent difficulties are at risk of emotional and behavioral problems, poor language use in adult life, and impaired psychosocial adjustment.

Specific Reading Delay

A 9-year-old boy seemed generally bright but was disruptive at school and often complained of headaches on school mornings. Psychological testing showed that he had the reading abilities equivalent to those of many 6-year-olds although his scores of general intelligence were in the normal range. Alerting school staff to this diagnosis prompted them to plan a program of remedial teaching and to be more understanding about his attempts to avoid reading tasks by disrupting others.

Syndrome characteristics. Specific reading delay presents in the school years as one cause of classroom learning difficulties. These children are at risk of suffering bullying and low self esteem as a result of failure in school. A proportion of them develop conduct problems in mid-childhood. Children with specific reading difficulties often have specific spelling difficulties too.

Management. Many remedial reading schemes have been advocated and all produce shortterm improvement in children's reading abilities. However, the impact on longer term reading performance has been disappointing. More encouraging outcome has been demonstrated for an intervention based on parents reading regularly with their child (Hewison, 1988). Child psychiatrists play a role in identifying the disorder, assessing and treating any secondary emotional or social handicap, and advocating adequate educational provision for the child.

Prognosis. Specific reading delay is often a relatively enduring condition that persists into adult life. In the absence of educational provision, which offers remedial help and encourages motivation and success in nonreading activities, children are at risk of educational failure, failure in employment, and difficulty forming satisfactory relationships in adult life.

Specific spelling delay. This is most commonly found in association with specific reading delay. It is also thought to occur as an isolated condition, but its prevalence and prognosis are not known.

Specific motor delay. Children with specific motor delay typically have a history or delayed motor milestones and exhibit clumsiness and poor fine motor coordination. Their experiences of failure and being less competent than other children predispose them to developing low self esteem and behavioral problems.

Specific arithmetical delay. Comparatively little is known about children with specific difficulties in arithmetic and mathematics. These skills appear to be relatively independent from language function. Deficits present as academic failure or secondary emotional problems: anxiety seems to be a particularly common form of associated psychopathology.

SCHIZOPHRENIA

A 14-year-old boy first presented to child psychiatry with a history of having become disruptive at school and irritable and withdrawn at home. Over the next 6 months he gradually revealed his preoccupation with the idea that his local water supply had been poisoned. This belief could not be shaken by pointing out that he and others had drunk the water with no ill effects. The following year, he became disturbed when watching television as he believed people were commenting on his appearance. In the following six months, further signs of symptoms diagnostic of schizophrenia emerged.

Schizophrenia in children and adolescents presents with the same core psychopathology as in adults: fundamental and characteristic distortions of thinking and inappropriate or blunted perceptions and affect. However, developmental immaturity leads to some variations in symptoms. For example, passivity phenomena are infrequent in children, formal thought disorder is not apparent in young children, and delusions may be ill formed. Hallucinations do not occur exclusively in psychotic illness in children.

Syndrome characteristics. Schizophrenia usually begins with a prodromal phase during which children show a deterioration in adaptive functioning. Psychotic symptoms then emerge and may last many months after which time a child may continue to function poorly because of residual apathy or depression. Only a minority of children will make a full recovery and return to their premorbid level of functioning.

Children and adolescents with schizophrenia are at risk of suicide, educational failure, and inadequate social development. (For a review, see Kolvin and Berney, 1990.)

Epidemiology. Schizophrenia has a lifetime prevalence of just less than 1% and an incidence of about 0.1% per year. The peak incidence of the disease in males is between the ages of 15 and 25. Females tend to develop the illness a decade later. The prevalence of early onset schizophrenia is unknown. Very early onset disease is particularly rare.

Etiology. Schizophrenia probably has a primary biological determinate with secondary environmental etiological factors. There is a substantial genetic predisposition to the disease. Factors associated with the development of schizophrenia, such as obstetric complications, winter birthdate, and immigration, do not necessarily have a direct etiological connection. The final common pathway in the genesis of the disease appears to be overactivity in the mesolimbic dopaminergic system.

Management. A multimodal approach is required (for a review, see Clark and Lewis, 1998). Specific treatment with major tranquillizers aims to reduce psychotic symptoms. The new atypical antipsychotic drugs are likely to become the first-line drug of choice because their side effect profiles are less toxic than conventional antipsychotic medication. This is likely to improve compliance and may lessen complications of longterm neuroleptic drug use. Of equal importance is the provision of a safe, supportive environment and an appropriate educational provision, and the maintenance of some appropriate social contacts. Parents need support and realistic information. Family work aims to lessen high expressed emotion which may impede recovery.

Prognosis. The course of schizophrenia is variable but generally it runs an episodic course with some degree of chronic deterioration. Affected children and adolescents are likely to become seriously disabled by impaired psychosocial functioning, and they will need ongoing social and psychiatric support. Outcome may be influenced by the child's level of premorbid functioning and the extent of recovery from the first episode (Werry and McClellan, 1992).

MISCELLANEOUS

This section describes a number of separate and distinct disorders, which the average clinician is likely to encounter in his work with children and adolescents.

Adjustment Disorders

A 7-year-old girl presented with refusal to go to school, a reluctance to leave her mother's side, and repeated demands to be given a baby's bottle of milk at bedtime. This was precipitated by a move of house and a change in her father's job, which took him away from home frequently. With consistency and support from her mother and father her difficulties lessened and disappeared after five to six weeks.

Syndrome characteristics. Adjustment disorders are short-lived emotional and behavioral reactions to a stressful event in a child's life. Symptoms are not sufficiently severe or prolonged for them to constitute another psychiatric diagnosis. Children with adjustment disorders may show a variety of symptoms such as depressive features, anxiety, and behavior disturbance. Young children tend to show a nonspecific reaction to stress characterized by regressed behavior, oppositional behavior, and increased separation anxiety. In older children, memories of and emotions surrounding the particular precipitating stressor may be evident in a child's play, dreams, drawings, conversation, or worries.

Epidemiology. Adjustment disorder is a relatively new diagnostic category and there is still debate as to whether it is a discreet and valid entity. Not all clinicians use the term in their practice, so there is little epidemiological data. In one study in Puerto Rico, however, it was found that adjustment disorders made up one-quarter of the psychiatric disorders evident in the general population (Bird et al, 1989).

Etiology. Adjustment disorders always have an obvious precipitant, which has been a significant source of stress to the child. Common precipitants include bereavements, parental separation, ill physical health, and road traffic accidents. A child may be predisposed to developing an adjustment reaction by factors such as having experienced other traumas in the past, having an anxiety-prone temperament, or having insecure attachment to caregivers. Factors that may maintain symptoms of an adjustment reaction include high levels of parental distress and secondary gain from symptoms.

Management. Most child psychiatrists treat children with adjustment reactions using combinations of individual and family therapies, aiming to contain anxiety, lessen distress, and aid adaptation to the precipitating stress. There is no research on the effectiveness of current treatments offered to children with adjustment disorders.

Prognosis. Adjustment reactions are, by definition, short-lived. It seems likely that children's prognosis varies widely after suffering an adjustment disorder. Some children who have successfully negotiated adaption to a stressful event may have learned coping mechanisms, which will increase their ability to cope with future stresses. Other children may become more vulnerable to future stresses as a result of experiencing the index trauma.

Posttraumatic Stress Disorder (PTSD)

Three months after a road traffic accident in which a 6-year-old boy sustained a broken arm and his mother received severe facial injuries, the boy became acutely fearful of small unexpected noises, refused to go to bed alone as he had nightmares about the accident, and repeatedly said that he had seen the car that hit them. He refused to travel in anything apart from a bus and his school performance deteriorated. After a course of play therapy and supportive therapy for his mother, normal routines were reestablished.

Syndrome characteristics. Following horrific experiences, some children suffer symptoms akin to posttraumatic stress disorder (PTSD) in adults. Symptoms may not be present immediately after a child experiences psychological trauma but may emerge in the subsequent six months. Parents and teachers often underestimate the psychological impact of accidents and disasters on children and may not immediately attribute a child's persistent symptoms to past trauma.

The child may reexperience the traumatic event in dreams or flashbacks; avoid places, activities, or sounds associated with the trauma; and show persistent symptoms of increased arousal such as insomnia, restlessness, and an exaggerated startle response.

Symptoms are modified to some extent by the child's age and developmental level. For example, preschool children lack the cognitive sophistication to express the full range of symptoms. They more typically show a generalized pattern of regressive behavior including sleep disturbance, separation anxiety, temper tantrums, nocturnal enuresis, and fear of the dark. School age children may relive the trauma in repetitive play or drawing. They may only talk about a tragedy if they feel parents or other adults can bear this.

Traumatized adolescents may show a decline in school performance and some experience a disabling sense of pessimism about the future. (For an overview, see Yule, 1999.)

Epidemiology. Posttraumatic reactions are likely to develop in 30% to 40% of children exposed to traumatic events, and some of these will meet the diagnostic criteria for post-traumatic stress disorder. Children as young as two and a half years of age can be affected.

Etiology. The severity of the psychological stressor seems to be the most important determinant of the severity of post-traumatic stress disorder in children. Other contributory factors include the failure of caring adults to contain a child's distress at the time of experiencing the trauma and a child's premorbid development and temperament.

Management. Professionals involved in managing the aftermath of road traffic accidents and man-made and natural disasters should provide information to parents and children about common psychological symptoms following trauma and encourage those whose symptoms persist to seek professional help.

Treatment for psychologically traumatized children uses similar interventions to those offered to adults: debriefing, group therapy, and systematic desensitization in children have all been described. Many parents will tend to make special allowances for a distressed child who has been traumatized. However, while it is important for the parent to be available and supportive, a return to normal routines and expectations are also important. Other members of the child's family may have been traumatized too and may need psychological treatment in their own right.

Prognosis. In some children, symptoms gradually resolve spontaneously. In others, perpetuating factors such as unresolved parental PTSD or pending litigation may maintain symptoms for many months.

Parasomnias

Syndrome characteristics. Parasomnias are disorders that occur in or get worse in sleep. We consider three common examples in this chapter.

Nightmares. Nightmares are frightening dreams, which may or may not wake children from sleep. Children who wake up after a nightmare are fully alert and frightened; they cry to be comforted. They are able to report the content of vivid and unpleasant dreams and may be reluctant to fall asleep again for fear of reexperiencing the nightmare. Nightmares occur in REM sleep, in the later part of the night. They are frequently a normal phenomenon, although they may become more troublesome when children are under stress. Nightmares centered around the theme of past trauma are a symptom of post-traumatic stress disorder.

Night Terrors. Children experiencing a night terror apparently wake from sleep in a frightened and aroused state, usually crying out. Children remain aroused, they may appear to be hallucinating and they may walk about or appear to be fighting off an attacker. Parents who try to comfort the child find that the child is unresponsive to this and does not seem fully awake. After 10 to 30 minutes, the child settles back to sleep and cannot recall the disturbance in the morning. Night terrors occur in stage 4 slow-wave sleep, early in the night. They are a normal event of sleep in children two years to six years of age and do not signify psychopathology, although they are very alarming for parents to witness.

Sleepwalking. Sleepwalking children get up from their bed and walk around the house or engage in mundane activity while asleep. Parents report that the child looks blank and is not fully responsive. It is difficult to wake a sleepwalking child, but they may be guided back to bed where they settle into quiet sleep. Sleep walking is another benign phenomenon of sleep. There is often a family history of this disorder.

Epidemiology. 20% of children between the ages of six and 12 years old report nightmares (Vela-Bueno et al., 1985). Night terrors occur in 3% of healthy children (Klackenberg, 1987). Up to 30% of children have had one episode of sleepwalking.

Etiology. Parasomnias are thought to result from disorders of the arousal mechanisms in the brain (Broughton, 1968). They appear to be features of the immature brain that diminish with age.

Management. The main intervention for these disorders is the education and reassurance of parents, who otherwise may inadvertently cause further distress to a child by their own reactions. Nightmares are best managed by comforting and reassuring the child as necessary. In severe cases, a short course of hypnotics may interrupt a repeating pattern of nightmares, but rebound nightmares can occur when the drugs are withdrawn. Parents can best deal with night terrors by removing any objects on which the child may hurt themselves and then sitting on the bottom of the bed until the child drifts back to quiet sleep. Prolonged phases of night terrors can be interrupted by a program of planned awakening early each night (Lask, 1988). Children who sleepwalk are best guided back to bed without trying to wake them. Doors and windows should be secured to avoid injury.

Prognosis. Nightmares, night terrors, and sleepwalking all become less frequent and troublesome with age.

School Refusal

A 12-year-old girl who was described as shy and timid developed abdominal pain several weeks after her mother's brief admission to hospital for a routine procedure. She began vomiting and having diarrhea each morning. Her parents who were concerned that she had a physical illness kept her out of school until finally, after a school holiday, she totally refused to return. Despite careful screening, her mother remained convinced that her daughter had a physical illness and was unable to support her daughter in her return to school. Day patient admission, family therapy, and supported return to school resulted in a full time return. However she remained mildly symptomatic at the beginning of each school term.

The diagnosis of school refusal is made more commonly in Europe than in the USA. DSM IV categorizes this type of difficulty with separation anxiety. The main feature is a reluctance to attend school associated with anxiety and depressive features. The term school phobia has been used to describe the apparent fearfulness about going to school.

Syndrome characteristics. The usual features are separation anxiety and a fear of leaving home, sometimes accompanied by depression and sleep disturbance. The child may also feel nauseous, have pains in the stomach or elsewhere, and may vomit and have diarrhea. These symptoms are maximal in the morning, at the time of preparation for school, and subside once the time for going to school has passed. The final complete refusal to attend school may be precipitated by an incident of bullying at school. The symptoms are aggravated by any pressure to return to school.

Children with school refusal are often described as immature, vulnerable to stress, and having difficulties with peer relationships. They can become panic-stricken and aggressive when their wishes are frustrated. The family is characterized by an overanxious and overprotective mother and a weak, ineffectual, or absent father.

Epidemiology. Weiner (1982) has described a prevalence of 1% to 2% of the school population. There seem to be peaks at the age of first entry to school, when there is a change of school, and in early adolescence.

Management. A multimodal approach to treatment is preferable. This may include individual or family psychotherapy, behavior therapy, and rarely the use of small doses of anxiolytics for a short period of time.

If the child is still attending school for part of the time, then attempts should be made to reduce the stress and increase support in the school environment. With more longstanding difficulties, day patient and residential treatment may be helpful. When the return to school is arranged, this should be as a result of careful planning involving not only the patient, but also their family, the school, and other professionals concerned. Firm handling is required in the context of an agreed plan. An "enforced return to school" is recommended by Blagg and Yule (1984), but this is not universally accepted.

Prognosis. The prognosis for this disorder is generally good but, in some children with a severe degree of disability, there may be ongoing and persistent neurotic symptoms into adulthood.

Abuse of Substances

Syndrome characteristics. Many adolescents drink alcohol, smoke tobacco, or experiment with nonprescribed psychoactive substances at some time. Because this behavior is so common and is in keeping with the drives of normal adolescent development, there is debate about whether alcohol and drug use in adolescence should be regarded as deviant behavior. However a minority of children and young people suffer medical, emotional, social, or criminal consequences of substance use. These young people may have other psychiatric disorders such as conduct disorder or depression.

Medical adverse consequences of substance abuse include sudden death from inhaling volatile substances, death or injury from trauma while intoxicated, physical dependence syndromes, and infectious diseases transmitted by intravenous drug use. Emotional problems include organic mental syndromes and psychological dependency. Abuse of substances can lead to impaired psychosocial development in adolescents who abandon school and prosocial activities in favor of alcohol and drug-using behavior. Behavior while intoxicated or the need to finance drug habits can lead adolescents to break the law.

Epidemiology. The extent and patterns of substance use vary over time and between cultures. Rates of alcohol and drug use increase throughout adolescence. Studies in the United States suggest that although fewer adolescents are beginning to use illicit drugs, there are localized, socioeconomically deprived areas in which rates of drug use among teenagers are high.

Etiology. Many interacting factors are usually evident to explain an adolescent's substance abusing behavior. Individual vulnerabilities include a tendency to risk-taking behavior, a propensity to be influenced by a deviant peer group, or underlying emotional distress, which is relieved by alcohol or drugs. Family influences include modeling of substance-using behavior and unsatisfactory homelife, which leads adolescents to seek acceptance and identity among their peers. Cultural factors such as the availability of alcohol and drugs are also influential.

Management. Intervention aims to tackle underlying vulnerabilities and equip the young person with alternative, nondeviant coping strategies in parallel with treatment for the medical and emotional consequences of substance abuse. Informed and safe use of alcohol and drugs may be an acceptable aim of treatment rather than complete abstinence. Intensive residential programs may be required to meet the educational, social, and emotional needs of vulnerable adolescents with entrenched patterns of substance abuse.

Prognosis. Heavy drug use in adolescence is associated with a number of adverse consequences in early adult life including unemployment, delinquency, divorce, and abortions (Kandel et al., 1986). It is not yet known whether current treatment programs improve the outcome of substance abusing adolescents (Wheeler and Malmquist, 1987).

Tic Disorders

Syndrome characteristics. Tics are quick, sudden, and frequently repeated movements of circumscribed groups of muscles and serve no apparent purpose. Tics may be motor or vocal, simple (e.g., blinking, coughing) or complex (e.g., squatting, repeating phrases). Tics are worsened by stress and disappear in sleep. Sufferers may have some degree of voluntary control over their tics.

There are three types of tic disorders in childhood. Children with *transient tic disorder* have frequent simple or complex vocal or motor tics, which disappear in less than 12 months. Children with *chronic motor or vocal tic disorder* have either motor *or* vocal tics, which persist for more than a year but do not impair other aspects of child development. Children with *Tourette's disorder* suffer from unremitting multiple motor and vocal tics, some of which appear as other established tics disappear. This condition is associated with impaired psychosocial and educational development.

Epidemiology. As many as 5% of children may suffer from a transient tic disorder at some time. Chronic motor or vocal tic disorder and Tourette's disorder are much less common. All tic disorders are more common in boys.

Etiology. Early psychoanalytical theories of tic disorders have largely been superseded by evidence that tic disorders occur as a result of neuroanatomical or neurochemical abnormalities. Other family members of children with Tourette's disorder may have had tics or may suffer from obsessive–compulsive disorder.

Management. Children presenting with a single tic do not require specific treatment but should be monitored in the expectation that this will resolve spontaneously. Treatment approaches for children with changing patterns of motor and/or vocal tics include behavior therapy (Azrin and Peterson, 1990) and pharmacotherapy with either haloperidol or pimozide. Neuroleptics are said to produce some improvement in 70% of Tourette's disorder sufferers (Schapiro et al., 1989). However there may be serious side effects when children are maintained on neuroleptic drugs in the longterm. Treatment should also attempt to minimalize social and educational handicaps and support families.

Prognosis. Transient tic disorders resolve fully and do not indicate vulnerability to any other psychiatric disorder. A proportion of chronic tic disorders resolve during adolescence, though some children have a lifelong condition with a fluctuating course and considerable secondary morbidity.

Elective Mutism

Syndrome characteristics. Children with elective mutism have adequate language skills but choose to stay silent in some social situations (e.g., when visitors come to the home or when in the classroom). The disorder becomes apparent as preschool children begin to expand their repertoire of social contacts. Transient reluctance to speak on first starting school is often seen in healthy children. Only sustained failure to engage in spoken communication meets the diagnostic criteria for elective mutism. The child continues to speak normally to a small number of close attachment figures in private and their reports confirm that the child has adequate hearing, comprehension, expressive language, and social development.

Epidemiology. This condition is rare with a rate of 0.7 per 1000 (Kolvin and Fundudis, 1981). It occurs more commonly in girls than in boys.

Etiology. This condition is thought to represent a disorder of social and emotional functioning rather than a deficit in the child's language skills. The child may, however, have a history of developmental language delay or speech articulation problems. Temperamental factors may contribute to the child's dogged refusal to speak, and secondary gain in the form of extra attention from other children and adults may maintain the symptom. Maternal overprotectiveness may be evident in some families. Children with intractable elective mutism often come from families in which there is some form of major psychopathology.

Management. Children with prolonged elective mutism are difficult to treat. There is some suggestion that behavior therapy reinforcing any attempts at spoken communication (Sluckin et al, 1991) may be more effective than nondirective play therapy. Parents and teachers need support to understand this condition.

Prognosis. Some children continue to speak selectively in social situations for many years and may remain socially isolated as adults. Other children eventually speak freely in all settings.

Gender Identity Disorder

Syndrome Characteristics. Children suffering from gender identity disorder identify strongly with the opposite sex. They are uncomfortable in their bodies, especially as secondary sexual characteristics develop and they choose to dress, play, and socialize like members of the other sex. They may suffer considerable personal distress and social ostracism. (For a review, see Zucker and Green, 1992.)

Epidemiology. There are no prevalence studies of this disorder, but it appears to be a rare condition. Boys present to clinics more commonly than girls. This probably reflects a lower social tolerance of this condition in boys, as well as possibly a greater biological vulnerability in males.

Etiology. Debate continues about the origins of gender identity disorder. Possible casual factors include biological influences (e.g., a genetic predisposition or an abnormal composition of the prenatal sex hormone milieu), psychological factors (e.g., close identification with the parent of the opposite sex), and social conditions (e.g., social reinforcement into opposite gender role).

Management. Children with this condition need to be engaged in a supportive and understanding therapeutic relationship that will allow the child to express and explore issues surrounding their gender identity while the clinician monitors their psychosexual development. Families require considerable support to come to terms with their child's condition. Behavior therapy aiming to decrease cross-gender behavior has been described. No outcome research is available.

Prognosis. Some gender identity problems resolve as children grow older. Persistent gender identity disorder is associated with higher rates of homosexual orientation, transexualism, and transvestism in adult life.

Suicide and Parasuicide

Suicide is a major concern, not only in those children who have a major depressive disorder. The rate of suicidal behavior in children and adolescents is rising and it has been suggested that its extent is particularly underestimated in young children (Garfinkel, 1986).

Syndrome characteristics. A number of features of suicidal behavior in children and adolescents include the identification of a plan to kill themselves that is age- and intelligence-dependent, a familial clustering of suicide behavior, an ability to use abstract reasoning, higher socioeconomic status, marital breakdown, and psychiatric disorder. Carlson (1990) noted that suicidal ideation increased in males as a function of age.

Epidemiology. The incidence of death from suicide in the USA in 1986 was 0.8 per 100,000 in five- to 14-year-old children and 12.8 per 100,000 in 15- to 24-year-olds. Suicide accounts for 16% of deaths in older teenagers (Thompson, 1987). It is more difficult to estimate the incidence of attempted suicide although it is often stated that this behavior is found more often in girls than boys (Barker, 1994). Many children and adolescents may harm themselves by taking overdoses of various drugs or cutting their wrists without seriously wishing to kill themselves. These acts are usually a response to some form of stress and may be a reflection of serious problems in the individual and/or the family: they should always be treated seriously.

Depression has been diagnosed in 30% of children who had been admitted to hospitals with a wish to kill themselves (Myers et al., 1985). The families of these children had more mothers reporting abuse, abusive fathers, and suicidal attempts by other family members. They had also experienced stressful life events more often than other children admitted to an in patient unit.

Etiology. Suicidal behavior should always be taken as an indication that something is seriously wrong in a child or adolescent's life. The behavior may be found in association with a major depressive disorder, behavioral disorder, psychosis, substance abuse, or adjustment disorder. The behavior may have been an imitation of another member of the family or peer group who has harmed themselves. Some suicide attempts will be carefully planned, others are more impulsive. The ready availability of lethal means of suicide such as drugs or guns may result in death.

Management. A careful assessment of the individual child or adolescent and their family should always be done. There should be an assessment of future suicide risk and the child should be kept in a hospital or sent home depending on this assessment. Follow-up is always a good idea, as a depressive mood may lift with all the solicitous attention given while the patient is in the hospital but return when the individual returns home.

Prognosis. If the underlying difficulties persist and there is no change in, for example, the family functioning or in amelioration of the depressive disorder, repeated attempts may be made. Planned or unplanned fatality becomes more likely in this situation.

SUMMARY

This chapter provides an overview of the major disorders found in child psychiatry. Each section uses the same approach—first considering the characteristics of the syndrome, next the epidemiology, then the etiology, then the management, and finally the prognosis.

The age of onset of the different disorders does vary, but there is considerable overlap for many. Gender is also important, with more boys than girls presenting as patients. Vignettes are used to illustrate the different presentations.

Some of the disorders have a characteristic presentation in childhood and are not comparable to psychiatric disorders in adulthood. These include the pervasive developmental disorders and some of the elimination disorders. However, in children and adolescents, some of the diagnoses are based on symptoms and complaints that are also found in adults presenting for the first time. These include the mood disorders, schizophrenia, and eating disorders.

Many disorders in child psychiatry will require the expertise of more than one professional discipline. This is an area of medicine where a team approach is particularly helpful.

The expertise of the psychologist, teacher, social worker, and others are all welded together to create a better understanding of the child, their family, and the problem.

The treatment options vary from medication to various forms of talking therapy based on different theoretical models and with different levels of proven efficiency. The research in this area is still in the early stages, and clinical judgement and experience are particularly important.

All of the serious and more chronic disorders have major implications for adult adjustment. Thus, the progress for early onset eating or psychiatric disorders is considered by many to be particularly poor the earlier the onset, and an early onset of a psychotic illness may be related to a higher rate of recurrence in later life.

REFERENCES

Anthony J. and Scott P. (1960). Manic-depressive psychosis in childhood. J. Child Psychol. 1, 53-72.

Azrin, N. & Peterson, A. (1990). Treatment of Tourette's syndrome by habit reversal: A waiting-list control group comparison. Behav. Therapy 21, 305-318.

Barker, P. (1994). Basic Child Psychiatry. 6th ed. Blackwell Science, Oxford, England.

Bicknall, J. (1975). Pica: A Childhood Symptom. Butterworths, London.

Bird, H., Gould, M., Yager, T., Staghezza, M., & Canino, G. (1989). Risk factors for adjustment in Puerto Rican children. Amer. Acad. Child Adole. Psych. 28, 847-850.

Blagg, N.R. & Yule, W. (1984). The behavioral treatment of school refusal: A comparative study. Behav. Res. Therapy 22, 119-127.

Bolton, D., Collins, S., & Steinberg, D. (1983). The treatment of obsessive–compulsive disorder in adolescence: A report of 15 cases. Brit. J. Psych. 142, 456-464.

Broughton, R. (1968). Sleep Disorders: Disorders of Arousal? Science 159, 1071-1078.

Carlson, G.A. (1990). Suicidal Behavior and Psychopathology in children and adolescents. Curr. Opin. Psychiatry 3, 449-452.

Clark, A.F. & Lewis, S.W. (1998). Practitioner Review: Treatment of schizophrenia in childhood and adolescence. J. Child Psychol. Psychiat. 39, 1071-1081.

Cooper, P.J. & Goodyer, I. (1993). A community study of depression in adolescent girls. I: Estimates of symptoms and syndrome prevalence. Brit. J. Psych. 163, 369-374.

Dietz, W.H. (1987). Childhood obesity. Ann. N.Y. Acad. Sci. 499, 47-54.

Dubowitz, V. & Hersov, L. (1976). Management of children with nonorganic (hysterical disorders of motor function). Dev. Med. Child Neurol. 18, 358-368.

Earls, F. (1994). Oppositional defiant and conduct disorders. In: Child and Adolescent Psychiatry: Modern Approaches. (Rutter, M., Taylor, E., & Hersov, L, Eds.), pp. 322-323. Blackwell Scientific Publications, Oxford, England.

Egger, J., Carter, C.M.N., Graham, P.J., Gumley, D., & Soothill, J.F. (1985). Controlled trial of oligoauligenic diet in the hyperkenetic syndrome. Lancet 1, 540-545.

Epstein, L.H., Wing, R.R., Valoski, A., & DeVos, D. (1988). Long term relationships between weight and aerobic fitness change. Health Psychol. 7, 47-53.

Emslie, G.J., Rush, A.J., Weinberg, W.A., Kowatch, R.A., Hughes, C.W., Carmody, T. & Rintelman, J. (1997). A double blind randomized, placebo-controlled trial of fluoxetine in children and adolescents with depression. Arch. Gen. Psych. 54, 1031-1037.

Fairburn, C. & Beglia, S. (1990). Studies in the epidemiology of bulimia nervosa. Amer. J. Psych. 147, 401-408.

Fleming, J.E., Offord, D.R., & Boyle, M.H. (1989). Prevalence of childhood and adolescent depression in the community. Brit. J. Psych. 155, 647-654.

Forsythe W. & Butler R. (1989). Fifty years of enuretic alarms. Arch. Dis. Childhood 64, 879-885.

Gillberg, C. & Hellgren, L. (1996). Outcome. In: Hyperactivity Disorders of Childhood. Cambridge Monographs in Child & Adolescent Psychiatry. (Sandberg, S., Ed.), pp. 477-503. Cambridge University Press, Cambridge.

Goodman W.K., Price, L.H., Rasmussen, S.A., Mazure, C., Fleischmann R.L., Hill, C.L., Heninger, G.R., & Charney D.S. (1989). The Yale-Brown obsessive–compulsive scale I. Development use and reliability. Arch. Gen. Psych. 46, 1006-1011.

Harrington, R., Whittaker, J., & Shoebridge, P. (1998). Psychological treatments of depression in children and adolescents. Brit. J. Psych. 173, 297-298.

Hewison, J. (1988). The long term effectiveness of parental involvement in reading: A follow-up to the Haringey Reading Project. Brit. J. Educ. Psych. 58, 184-190.

Johnson, C., Lewis, C., & Hayman, J. (1984). The syndrome of bulimia: Review and synthesis. Psych. Clin. North Amer. 7, 247-257.

Kandel, D., Davies, M., Karus, D., & Yamaguchi, K. (1986). The consequences in young adulthood of adolescent drug involvement. Arch. Gen. Psych. 43, 746-754.

Klackenberg, G. (1987). Incidence of parasomnias in children in a general population. In: Sleep and its Disorders in Children. (Guilleminault, C., Ed.), pp. 99-113. Raven Press, New York.

Kolvin, I. & Fundudis, T. (1981). Elective mute children: Psychological development and background factors. J. Child Psych. Psych. 22, 219-232.

Kolvin, I. & Berney, T P. (1990). Childhood schizophrenia. In: Handbook of Studies in Child Psychiatry. (Tonge, B, Burrows, D G, & Werry, J.S., Eds.), pp. 123-130. Elsevier, Amsterdam.

Lask, B. (1988). Novel and non-toxic treatment for night terrors. Brit. Med. J. 297, 592.

Leslie S. (1988). Diagnosis and treatment of hysterical conversion reactions. Arch. Dis. Child. 63, 506-511.

Leonard, H.L., Swedo, S., Lenane M., Rettew, D.C., Hamburger, M.S., Bartko J.J., & Rapoport, J.L. (1993). A two to seven year follow up study of 54 obsessive–compulsive children and adolescents. Arch. Gen. Psych. 50, 429-438.

Mufson, L., Moreau, D., Weissman, M.M., & Klerman, G.L. (1993). Interpersonal Psychotherapy for Depressed Adolescents. Guilford Press, New York.

Myers, K.M., Burke P., & McCauley, E. (1995). Suicidal behavior by hospitalized preadolescent children on a psychiatric unit. J. Amer. Acad. Child. Psych. 24, 474-480.

Offord, D.R. & Bennet, K.J. (1994). Conduct disorder: Long term outcomes and intervention effectiveness. J. Amer. Acad. Child Adol. Psych. 33(8), 1069-1078.

Reinecke, M.M., Ryan, N.E., & Dubois, D.L. (1998). Cognitive-behavioral therapy of depression and depressive symptoms during adolescence: A review and meta-analysis. J. Amer. Acad. Child. Adol. Psych. 37, 26-34.

Riddle, M., Hardin, M., & King, R. (1989). Fluoxetine treatment of children and adolescents with Tourette's and obsessive–compulsive disorders. Paper presented at Annual October meeting of the American Academy of Child & Adolescent Psychiatry. New York.

Russell, G. (1979). Bulimia nervosa: An ominous variant of anorexia nervosa. Psych. Med. 9, 429-448.

Russell, G., Szmukler, G., Dare, C., & Eisler, I. (1989). An evaluation of family therapy in anorexia nervosa and bulimia nervosa. Arch. Gen. Psych. 44, 1047-1056.

Rutter, M., Tizard, J., Yule, W., Graham, P., & Whitmore, K. (1976). Research report: Isle of Wight studies, 1964-1974. Psych. Med. 6, 313-332.

Schapiro, E., Schapiro, A., Fulop, G., Hubbard, M., Mandeli, J., Nordlie, J., & Phillips, R.A. (1989). Controlled study of haloperidol, pimozide and placebo for the treatment of Gilles de la Tourette's syndrome. Arch. Gen. Psych. 46, 722-730.

Schachar, R. & Wachsmuth, R. (1990). Oppositional disorder in children: A validation study comparing conduct disorder, oppositional disorder and normal control children. J. Child Psych. Psych. 31(7), 1089-1102.

Sellar C., Hawton, K., & Goldacre, M.J. (1990). Self poisoning in adolescents. Hospital admissions and deaths in the Oxford Region, 1980-1985. Brit. J. Psych. 156, 966-870.

Sluckin, A., Foreman, N., & Herbert, M. (1991). Behavioral treatment programmes and selectivity of speaking at follow-up in a sample of 25 selective mutes. Australian Psych. 26, 132-137.

Steinhausen, H., Rauss-Mason C., & Seidel R. (1991). Follow-up studies of anorexia nervosa: A review of four decades of outcome research. Psych. Med. 21, 447-451.

Swedo, S., Rapoport, J.L., Leonard, H.L., Lenane M., & Cheslow, D. (1989). Obsessive–compulsive disorder in children and adolescents: Clinical phenomenology of 70 consecutive cases. Arch. Gen. Psych. 46, 335-341.

Szatmari, P., Offord, D., & Boyle, M. (1989). Ontario child health study: Prevalence of attention deficit disorder with hyperactivity. J. Child Psych. and Psych. 30, 219-230.

Thompson, T.R. (1987). Childhood and adolescent suicide in Manitoba: A demographic study. Can. J. Psych. 32, 264-269.

Treasure, J. & Thompson, P. (1988). Anorexia in childhood. Brit. J. Hosp. Med. 40, 362-369.

Vela-Bueno, A., Bixler, E., Dobladez-Blanco, B., Rubo, M., Marrison, R., & Kales, A. (1985). Prevalence of night terrors and nightmares in elementary school children: A pilot study. Res. Comm. Psych. Psych. Behav. 10(3), 177-188.

Weiner, I.B. (1982). Child and Adolescent Psychopathology. Wiley, New York.

Weiss, G. (1991). Attention deficit hyperactive disorder. In: Child and Adolescent Psychiatry: A Comprehensive Textbook. (Lewis, M., Ed.), pp. 552-554. Williams and Wilkins, Baltimore, Maryland.

Werry, J. & McClellan, J.P. (1992). Predicting outcome in child and adolescent schizophrenia and bipolar disorder. J. Amer. Acad. Child Adol. Psych. 31, 147-150.

Werry, J.S., McClellan, J.M., & Chard, L. (1991). Childhood and adolescent schizophrenic, bipolar and schizophrenic disorders: A clinical and outcome study. J. Amer. Acad. Child Adol. Psych. 30, 457-465.

Wheeler, K. & Malmquist, J. (1987). Treatment approaches in adolescent chemical dependency. Pediatr. Clin. N. Amer. 34, 437-447.

Whitaker, A., Johnson, J., Shaffer, D., Rapoport, J.L., Kalikow, K., Walsh, B.T., Davies, M., Braiamon, S., & Dolinsky A. (1990). Uncommon troubles in young people. Arch. Gen. Psych. 47, 487-496.

Winokur, G., Clayton, P.J., & Reich, T. (1969). Manic Depressive Illness. Mosby, St. Louis, MO.

Yule, W. (1999). Post Traumatic Stress Disorder. Arch. Dis. Child 80, 107-109.

Zucker, K. & Green, R. (1992). Psychosexual disorders in children and adolescence. J. Child Psych. Psych. 33(1), 107-151.

Zametkin, A. & Rapoport, J. (1987). Neurobiology of attention deficit disorder with hyperactivity: Where have we come in 50 years? J. Amer. Acad. Child Adol. Psych. 26, 676-686.

Chapter 16

Child Sexual Abuse

DANYA GLASER and MALCOLM WISEMAN

Biological Psychiatry, pages 357-378.

ISBN: 1-55938-819-6

INTRODUCTION

Child sexual abuse is not a new phenomenon, although the widespread recognition and acknowledgment of its common occurrence is more recent. The hallmarks of a child sexual abuse are its secret nature and the very frequent denial of the abuse by the abuser once it is alleged to have happened. These two factors play a central part in the process of the abuse and in its aftermath. Sexual abuse occurs both within the family and outside it, but in either circumstance, the abuser is usually already known to the child. Indeed this acquaintance may be based on a deliberate befriending or "grooming" of the child by the abuser for the purposes of preplanned abuse. Involving sexuality, this form of abuse is particularly emotive and professionals do not generally work in this field in isolation from colleagues.

DEFINITIONS

The plural in the heading attests to the existence of different definitions, which, as will be shown, affect the epidemiology of the phenomenon. The most commonly accepted definition is by Schechter and Roberge:

> *Sexual abuse is defined as the involvement of dependent, developmentally immature children and adolescents in sexual activities they do not truly comprehend to which they are unable to give informed consent, or that violate the social taboos of family roles.*

The utility of this definition is in the description of the imposition of the sexual activity on the less mature, and therefore relatively powerless, child. The term "involvement" could include observation as well as physical contact. The enforced watching of pornographic videos, exhibitionism/exposure, or explicit adult sexual activity is undesirable. However, a child sexual abuse that is likely to leave the child with enduring effects usually involves physical contact between the genitalia of at least one out of abuser and child, and some part of the body of the other. It includes touching and masturbation, and oral, vaginal, and anal penetration, sometimes accompanied by threats, and more rarely involving the use of force. Since sexual abusive behavior is akin to an addiction and the wish, therefore, is to repeat this forbidden activity, the abuser tends to avoid injuring the child so that he may return and re-abuse. Many different ways of silencing the child are employed, including threats about the consequences of disclosure such as parental divorce, imprisonment of the abuser, and separation of the child from her family. Abusers often convince the children that they will not be believed if they tell (which is, not uncommonly, an accurate prediction). Some abusers describe the abuse to the child as "education", "love", a duty, or part of normal parent/child interaction.

EPIDEMIOLOGY

Child sexual abuse would appear to be a not uncommon phenomenon. However, surveys reporting the frequency of sexual abuse show different rates. This depends upon the definition of abuse that is used in the survey, the population studied, and how the information is obtained. Since sexual abuse is not a unitary phenomenon, definition is clearly important, and studies tend to define abuse in terms of the age and power differential between the perpetrator and victim, and by the type of abuse—in particular whether or not the abuse involved physical contact. The populations studied have included clinical samples such as women attending an eating disorders clinic, in-patients of psychiatric hospitals, and college students, along with surveys of particular or general populations.

In a randomly selected sample of 930 San Francisco women, Russell found that 39% of the sample reported intrafamilial or extrafamilial sexual abuse before reaching 18 years of age. 12% of these women had been abused by a relative, and 4.5% had been abused by a father or stepfather. These figures are considered to be high, and it is possible that the populations studied may be unrepresentative of the general population. An interesting aspect of the methodology of this study concerns the interview. The interviewers were specially trained and were all female. The interview schedule included several, separate questions and probes concerning sexual abuse. It is thought likely that this style of interviewing may well have increased the recall of the sample.

Baker and Duncan studied a nationally representative, and randomly selected, sample of 2,109 men and women in Britain. Using a broad definition of sexual abuse, 10% said that they had been abused before the age of 16. Mullen studied a nationally representative, and randomly selected sample from New Zealand and came up with similar findings. 10% of their sample had been sexually abused and analysis according to the age of the victim enabled them to conclude that a similar proportion had been abused in each decade band and therefore that the incidence of sexual abuse had probably not varied over the last 50 years.

Although the majority of victims are female, there is an increasing awareness of males as victims of sexual abuse. Estimates from surveys of men in the general population indicate that between 2.5% and 8.7% have been sexually abused as children. In the past, it had been believed that child sexual abuse was mainly a problem that affected girls, with the ratio of girl victims to boy victims being 9:1. More recent studies suggest that the ratio has narrowed, particularly in community studies. It is possible, therefore, that boys have particular difficulty in reporting their abuse and entering treatment.

WHO ABUSES

Perhaps the first thing to note is the marked gender difference among perpetrators, with the vast majority being male. Surveys suggest that between 95% and 98% of perpetrators that come to professional attention are men.

THEORIES OF CAUSATION

Early theories about abuse tended to focus on one aspect of the perpetrator, for example, a past history of having experienced abuse themselves. Such thoughts were supported by studies of small and highly selected groups of perpetrators—particularly those who had been convicted or those in treatment. Theories derived from these samples tend to emphasize abnormalities in the adult's development and show a marked psychopathological bias. As knowledge about child sexual abuse grew, especially that which concerned its variety and incidence, it became clear that those perpetrators who were convicted or who came into treatment were a nonrepresentative sample, and the theories were seen as inadequate to explain the totality of sexual abuse.

Feminist theorists broadened the debate, bringing in social and cultural factors through their emphasis on power relationships, with society supporting male power over dependent women and children. Male socialization, with its suppression of emotion, and its tendency to sexualize intimacy, was seen as fertile breeding ground for the male abuse of power, and specifically for sexual abuse.

Such thought has led to a debate over whether child sexual abuse represents a sexually motivated act or a desexualized problem of power relationships. Conventional sexual behavior includes behaviors that are nonsexually motivated, including a desire for affection, affiliation, and allegiance, as well as behaviors that contain an erotic component. Sexual abuse of children consists of complex and varied behaviors, and includes a wide variety of activities such as a man who on one occasion fondles his granddaughter, a man who only sexually abuses boys, a man who sexually abuses his child following stressful episodes, or a man who participates in a sex ring involving several children and several adult perpetrators. Any theory would therefore need to help us understand a large range of behaviors.

Finkelhor reviewed the various theories proposed to explain child sexual abuse and from these produced an integrated, comprehensive model. He described four preconditions of sexual abuse that will be described further in the next section. In this section we shall address various factors that relate specifically to the perpetrators.

These factors are emotional congruence, sexual arousal to children, blockage, and disinhibition. Emotional congruence describes the situation in which there is a similarity, or mutuality, between the adult's emotional needs and a child. This does not relate specifically to sexual arousal, but includes a notion of perpetrators having an arrested psychological development so that they may experience themselves as children and relate more comfortably to other children. The lowered self-esteem and inadequacy, which are often described in perpetrators in their relationship with adults, can be lessened in their relationships with children. Emotional congruence could arise in the adult as a result of childhood psychological trauma or indeed through socialization that encourages relationships governed by dominance and submission.

The second factor is sexual arousal to children. This is independent of emotional congruence and relates to the physiological sexual arousal patterns evoked by a child. There is debate about whether such responses may be present as part of normal behavior and may persist along a continuum from low arousal to high arousal or whether such responses are found only in an abnormal population. Where it occurs, however, sexual arousal to children is likely to be the result of conditioning that, in certain circumstances, persists and promotes a later sexual interest in children. Such modeling of sexual interest in children such as growing up in a family where sexual abuse occurs, even if that individual has not been sexually abused, may predispose to sexual arousal toward children. It would seem likely that intense experiences may be developed in fantasy form that, through repetition and reinforcement through masturbation, become an organizing activity. It is possible that child pornography and the sexual portrayal of children in the media (e.g., in advertising) may exert an influence through this mechanism.

The third factor is blockage, which reflects the inability of perpetrators to meet their sexual needs in adult relationships. This may be the result of early aversive sexual experiences, repressive sexual views or guilt, or marital difficulties.

The fourth factor is disinhibition. This reflects the capacity of perpetrators to overcome normal taboos regarding sexual behavior with children. Poor impulse control, such as may be found in certain personality disorders or as a result of a drug or alcohol abuse, as well as depressive periods following personal stresses such as bereavement or marital breakdown, acts to disinhibit someone predisposed to sexual arousal toward children. Similarly, social factors that may be seen to lower society's disapproval of sexual abuse of children may also act as a disinhibiting mechanism. These might include weak legal actions or sexualized portrayals of children in the media.

This model has been particularly useful in drawing attention to the several interacting factors that may be necessary before an adult sexually abuses a child. These factors include both individual and sociocultural components and suggest that an explanation of child sexual abuse requires an understanding at all four levels simultaneously.

FOUR PRECONDITIONS

The previous section outlined an approach to understanding why a perpetrator might be motivated to sexually abuse. Finkelhor, who extended this to incorporate a four-preconditions model of sexual abuse, combined psychological and social factors and also included specific environmental features that may act to increase the risk of sexual abuse occurring.

The first two preconditions, which are *motivations to sexually abuse* and *the overcoming of internal inhibitors*, have been discussed above. The third precondition, *overcoming external inhibitors*, broadens the perspective and focuses on the environment of the child, especially the presence or absence of a protective adult. The nonabusive parent, usually the mother, has a key role in protecting the child from abuse. An absent mother, whether physically or emotionally, leads to an increased vulnerability to abuse. The relationship between the nonabusing parent and the abuser is therefore important in understanding the abuse, and careful assessment of this relationship, as we shall see, is necessary in planning treatment.

The fourth precondition, *overcoming the resistance of the child*, concerns factors in the child that may increase vulnerability. An isolated, deprived child with few social supports and no trusting relationships, may be more vulnerable to the approaches of a perpetrator. Violence and threats toward the child may also reduce the child's capacity to resist, as may a close relationship between the child and the perpetrator.

This model, by incorporating a variety of other theories and ideas about sexual abuse, proposes a stepwise progression from an individual with a propensity to abuse, through that individual's capacity to overcome his own inhibitions and external protective features, in order to identify or manipulate a vulnerable child for the purposes of abuse. It has the added advantage of providing a model for assessment and for treatment.

EFFECTS OF CHILD SEXUAL ABUSE

The consequences of child sexual abuse can be divided into the shortterm effects in childhood and the longterm effects found in adults. While it is clear that child sexual abuse can have detrimental and long-lasting effects, there are few controlled studies and even fewer that separate the direct effects of the abuse from preexisting pathology—either in the child or the family. Samples tend to be drawn from those that have come to professional notice through child protection agencies or child mental health services and so are likely to be nonrepresentative. A further difficulty is that childhood sexual abuse represents a wide range of behaviors, which occur in a wide range of situations, so that the effects of abuse are likely to be very diverse.

As a group, sexually abused children are likely to be more disturbed than a nonreferred, nonclinical group of children. However, compared to children referred to child mental health services for reasons other than sexual abuse, the sexually abused group are likely to be less disturbed overall. However, the range and pattern of symptoms found in the abused group and in the clinic nonabused group tend to be similar, and both show a wide variety of symptoms. There had been a hope that reviewing studies on the effect of sexual abuse on children would lead to the description of a post–sexual abuse syndrome as a result of clustering of symptoms. However, this has not proved possible as, apart from in one or possibly two areas, there do not appear to be any specific types of disturbance found after sexual abuse. There are two areas that do appear helpful in differentiating sexually abused children from nonsexually abused children. The first is the presence of marked sexualized behavior, which is found much more commonly in the sexually abused group, although not all sexually abused children and adolescents show sexualized behavior, and nonabused groups may also exhibit sexualized behavior. Secondly, some studies have suggested that post-traumatic stress disorder is more commonly found among sexually abused children than in nonabused clinical populations. Post-traumatic phenomena include (1) hyperarousal and anxiety states, (2) intrusive memories such as nightmares and flashbacks (vivid, often visual images related to the traumatic, abusive experience), and (3) psychic numbness and dissociated states or active avoidance of potential reminders.

Shortterm Effects

Although many symptoms are likely to be shared across the age groups, it is helpful to consider the common effects as they occur in the different age groups. Studies of preschool children who have been sexually abused show a range of disturbed behavior, and they are overall more disturbed than nonabused controls. The common finding is abnormal sexualized behavior, which often has a driven quality and may comprise excessive masturbation, requesting sexual stimulation, precocious sexual knowledge, and sexual play with other children and with toys and dolls. Some studies have suggested that the preschool children who have been sexually abused tend to display more withdrawn behavior compared to aggressive behavior, although this may reflect a group composition of predominantly females, who tend to internalize rather than externalize their distress.

Among school-age children, behavioral and academic problems are commonly reported. Again, both boys and girls are likely to show sexually inappropriate behavior, masturbation, sexual preoccupation, and sexual aggression. Depressive mood is also a common feature, especially among girls.

Adolescents who have been sexually abused may present with a variety of psychiatric illnesses, but they commonly present with a constellation of depressed mood or even frank depressive illness, lowered self esteem, and suicidal ideation and deliberate self-harm. Acting out behavior such as running away from home and promiscuity are also presenting features at this stage. One can detect, therefore, a developmental pattern to the symptoms.

Factors Determining Effect

From the literature, it is difficult to identify what abuse-specific variables mediate the effects of the abuse. This is because many of the variables such as age of onset of abuse and duration of abuse are often intercorrelated and the published studies do not allow separation of these effects. However, the literature does allow some tentative conclusions to be drawn, and these can be supplemented by clinical experience. The following is a list of variables that influence the effects of the abuse:

1. *Age of the child.* Most studies support the impression that, the older the children are, at the time of presentation the more symptomatic they are likely to be. However, this is likely to be confounded by other factors such as frequency of abuse and duration of abuse.
2. *Age at onset of the abuse.* Studies of age of onset are inconclusive, and again, this is likely to be related to other confounding variables.
3. *Gender of the child.* Gender, when studied, does not reveal any consistent differences between the magnitude of effects on boys and on girls.
4. *Time elapsed since the last episode of abuse.* This variable has also not shown consistent findings, with some individuals remaining disturbed long after the abuse has stopped while others show recovery.

The next five factors relate to the abuse itself.

5. *Identity of the perpetrator of the abuse.* The perpetrator's identity does influence the effect of the abuse, with close relatives such as fathers and stepfathers having a greater deleterious effect.

6. *Number of perpetrators.* Interestingly, studies of the number of perpetrators do not reveal any clear or consistent results. Clinically, however, the impression would be that the greater the number of perpetrators, the more damaging the effect on the child.
7. *Frequency of the abuse.* Frequency influences the outcome with increased frequency causing more severe effects.
8. *Duration of the abuse.* As with greater frequency, effects are more severe with longer duration of abuse.
9. *Nature of the abuse.* Studies of this important variable reveal that penetrative abuse and the use of force lead to increased symptoms in the victim.

The final two variables are related to the child and the family circumstances.

10. *Victim's coping style or premorbid personality.* These variables mediate the effect of the abuse, with those victims with a negative outlook having a worse outcome.
11. *Lack of maternal support.* The absence of the mother's support is associated with a worse outcome.

It is also likely that the initial effects of sexual abuse are mediated by these abuse-related factors, as discussed above, interacting with family dysfunctional factors. Over time, the influence of the abuse-related factors seems to diminish, while the family factors have an increased influence in maintaining the symptoms.

Longterm Effects

Childhood sexual abuse can exert its effects acutely in childhood or chronically, lasting through to adulthood, or in a delayed fashion with apparent periods of adequate functioning interspersed with periods of disturbance.

Adults who have been sexually abused as children are more vulnerable to anxiety and phobic and depressive states, and they are overrepresented in psychiatric populations. There is thought to be a link with bulimia nervosa in adulthood as well as with hypochondriacal/psychosomatic complaints. There has been a recent, interesting literature on the link between childhood abuse and multiple personality disorder. The link is not clear, although it may be mediated through the child's experience of compound abuse, that is the experience of both physical and sexual abuse.

Suicidality is a common finding among populations of adults who have been sexually abused as children. This appears to be related to the number of perpetrators of abuse and the experience of compound abuse.

Many studies link sexual disturbance in adulthood with childhood sexual abuse. Such disturbances included reports of frigidity, confusion regarding sexual orientation, promiscuity, general dissatisfaction with sexual relationships, and sexual offending. There is also a weak link, for both men and women, with homosexuality in adulthood.

The sexual difficulties probably link with, but may be independent of, relationship difficulties and parenting difficulties that are commonly a feature of adults who have been sexually abused. Adults who have been sexually abused as children are also more likely to experience physical and sexual assaults including rape, possibly mediated through lowered self esteem, and a vulnerability to exploitative relationships.

Retrospective studies suggest that those children who have better outcomes are more likely to have had confiding relationships that have been supportive, to have been shown maternal warmth, and to have come from families that are seen as generally functioning well.

PRESENTATION

The secret nature of sexual abuse means that rarely are there witnesses to the activity, other than the abuser and the child. Occasionally, terrified or helpless siblings are made to watch or, in pedophilic rings, several children are involved. Sexual abuse is therefore only discovered if a child describes the abuse or through the very occasional instance of an abuser confessing spontaneously or if suspicions are pursued.

A few children are abused on a single occasion or only very occasionally. Their abuse may lie undiscovered for many years. Some children learn to adapt to abuse, remaining asymptomatic. Other children develop behavioral or emotional symptoms, which are, however, nonspecific presentations associated with sexual abuse sufficiently frequently to merit a high level of suspicion. Different presentations are outlined in Table 1.

Physical signs resulting from sexual abuse are found in less than 50% of children who are believed to have been sexually abused. This is so because much abuse leaves no physical trace or because rapid healing of mucosal tissue may occur before a physical examination is carried out. Most genital physical signs such as bruising, hymenal tears, enlarged hymenal opening, or reflex anal dilatation are at best suggestive or compatible with a history of sexual abuse. Conclusive, forensic evidence such as the finding of semen is only rarely present. Thus, the most useful indicator and, indeed, evidence of sexual abuse is the child's description of the abuse.

Table 1. Modes of Presentation

Disclosure		Suspicion	
Spontaneous	**Elicited**	**Suggestive**	**Possibly Indicative**
1. Intentional 2. Unintentional	In response to questions	1. *Behavioral* • Sexualized behavior 2. *Physical* • Sexually transmitted diseases • Anogenital trauma • Underage pregnancy when father 'unknown' 3. *Communication* • Mention of worrying secret 4. *Relationships* • Close contact with known sexual abuser	Nonspecific emotional and behavioral difficulties (see "effects" section of text)

MANAGEMENT OF SUSPICION AND DISCLOSURE

When sexual abuse is suspected or when a child discloses sexual abuse, the primary aim of professional action and response is to protect the child from further abuse. Protection can be achieved by one of three ways: (1) by ensuring that the abuser or his or her circumstances have changed sufficiently to no longer pose a risk to the child, (2) by effectively supervising all contact between the child and the abuser, and (3) by separating the child from the abuser or interrupting their contact. The first means of protection is not achievable in the shortterm and is therefore not appropriate as an initial response. In the context of the secret nature of sexual abuse, effective monitoring of the abuser–child relationship is not possible over longer periods and is therefore not applicable if the abuser and the child continue to live in the same household. Furthermore, for any supervision or monitoring to be effective, the person supervising must fully believe in the abuse. The frequent denial of the allegations by the alleged abuser makes believing more difficult, especially if the potentially protective person has a close relationship with the abuser. Separation of the child from the abuser is the safest and most effective protection. If the two are not living together, this is usually possible. Otherwise, it is vastly preferable for the abuser to leave the home, at least temporarily while further investigations and arrangements are set in motion.

For any of the above to be implemented, the child has to be heard and believed. The "capturing" of the child's words is therefore of vital importance. Some older children intentionally tell someone who they trust about the abuse, with the aim of bringing the abuse to a halt. Younger children are more likely to describe abuse because they are troubled by the incomprehensible nature of the experience. Having spoken about it once, children not infrequently then retract their description, particularly if they sense that they are not being believed. Furthermore, even if initial protection is achievable, it is likely that, for some children, longterm safety requires legal intervention. Sexual abuse is also a crime and some abusers are prosecuted in criminal trials. For legal purposes, the child's verbal evidence is a prerequisite. It is essential that these longterm considerations are fully incorporated into the initial professional response.

It is undesirable to interview children repeatedly in the effort to gain their account of abuse they have experienced. Moreover, the probative value of an account diminishes with the number of times it is given. It is therefore good practice to record, preferably on video, the child's account as early as possible. However, many children are initially only suspected of having been abused rather than having given an actual account of abuse. For these children, there is a need to verify or dispel the suspicion. Here, the aim is again to find a means of enabling the child to describe abuse if it has indeed occurred. The difficulty is to do so without suggesting to the child that she has been abused, if she has not. There is empirical evidence indicating that young children, below the age of 6 years, may be induced into giving false accounts in which they come to believe tenaciously. The circumstances include repeated questioning and suggestive statements.

Great skill is required in exploring with the child the underlying reasons for the suspicion. For instance, if a child is found to have physical signs suggestive of sexual abuse prior to the child having spoken about it, it is possible for the pediatrician to seek an age-appropriate explanation for the signs, from the child. Alternatively, if the child shows age-inappropriate sexualized behavior, the person who has noticed the behavior (e.g., a kindergarten teacher) can inquire of the child where he or she might have seen or learned about this activity. These are open questions to which there is no preordained answer. This is in contrast to closed questions, which require a yes/no answer, or leading questions, which have a particular assumption inherent to them. The two latter question forms are less appropriate in the first instance.

Mandatory reporting of suspected abuse may lead to a wish to video-record an interview with the child as soon as possible. However, caution is required. Children younger than eight years of age are unlikely to describe abuse for the first time in a formal setting if they have not previously talked about it to a trusted person informally. There is a further pitfall in proceeding with unplanned haste to establish whether a child has been abused. Family members who have difficulty and reluctance in believing that a child has been sexually abused may dissuade a child from persisting with her allegation or put actual pressure on her to retract. There is therefore a need to plan the approach to a child in such a way that she will not be silenced before her account is recorded. On the other hand, it is usually necessary to obtain parental permission before a child can be interviewed by child protective and law enforcement agencies. For these reasons, a specific investigation of suspected sexual abuse or the verification of a child's account must be planned by a multiprofessional group before proceeding, even if, as is usually the case, procedures have already been agreed upon. The professionals include the person reporting the suspicion, law enforcement or police, child protection agency, medical staff, and often child mental health staff and education staff. Special caution is required if there is the possibility of an organized sex ring involving more that one child or abuser. It is particularly important that there is careful written documentation of all conversations and discussions.

Beyond concerns for immediate protection and the obtaining of usable evidence in the most sensitive and child-centered manner, there are some other important considerations. Some children are frightened to talk, and not all perceive the professional response as the rescue mission, which it is intended to be. Many children continue to feel concern, loyalty, or fondness towards their abuser and merely wish for the abuse to cease. The greatest needs that the children have are to be believed, not to be blamed for disclosing the abuse, and to continue to be supported by their mother or other primary care giver(s) and family from whom children do not wish to be separated. Initial planning therefore needs to include a consideration of ways to engage the child's mother without alienating her. The mother is often forced to make a choice between the child and the abuser of whom she may be either fond or frightened and sometimes both.

ASSESSMENT OF THE LIKELIHOOD OF SEXUAL ABUSE

Although in the main diagnosis of sexual abuse depends on a disclosure by the victim, the disclosure remains part of an overall jigsaw of phenomena that must be pieced together to form a composite picture. The disclosure may be partial, where the child "tests the water" in order to assess the response.

In the assessment of the likelihood of abuse having occurred, it is helpful to consider the overall picture under four main headings: (1) the statement made by the child and/or adults, (2) the physical findings, (3) the behavioral and emotional status of the child, and (4) background or risk factors.

Statement by the Child and/or Adults

It is rarely the case that the perpetrator confesses separately from any disclosure by the victim or that the abuse is witnessed by another adult or child. The child's disclosure is often to the mother but may also be to a trusted adult such as a teacher or to a friend. The first statements made by the child, usually informally, are often the most revealing, and may contain idiosyncratic details such as memories of a particular toy or object in the room.

Physical Findings

The presence and nature of any physical findings will of course depend on the type of abuse, the time that has elapsed between the last incident of abuse and the physical examination, the expertise of the examiner, and the technique of examination used. Due to the nature of much sexual abuse including fondling, masturbation, and oral–genital contact, many children who have been sexually abused show no abnormal physical findings. Conversely, nonspecific signs that may occur in abused children may also be seen in nonabused children. Only a few signs are diagnostic of abuse in the absence of a reasonable alternative explanation. These signs include a laceration or scar of the hymen, attenuation of the hymen with loss of hymeneal tissue, and a laceration or scar of the anal mucosa extending beyond the anal verge into the perianal skin. Normal physical findings do not rule out sexual abuse.

Behavioral and Emotional Status

As we have seen, there are no symptoms that are pathognomonic of sexual abuse. However, marked sexualized behavior and posttraumatic stress phenomena appear to be more commonly found among sexually abused children than nonabused children and so should raise the level of suspicion regarding the likelihood of abuse.

Background/Risk Factors

Features described in the earlier section outlining Finkelhor's precondition model are among the risk factors for sexual abuse. Thus, they would include those features that may predispose an adult to abuse, reduce internal or external inhibitions, or make the child more vulnerable. It is often difficult to make a clear diagnosis of sexual abuse and a decision may need to be made on the balance of probabilities, following an assessment of these four areas.

False Allegations

The area of false allegations is a particularly difficult area, and there are some suggestions that their frequency may be increasing. The assessment of the likelihood of sexual abuse undoubtedly requires a full multidisciplinary assessment of all the factors mentioned above, taking into account the possibility that the statement made by the child and/or adult may be fabricated. Overall, proven false allegations are rare in younger children but are reportedly more common with older children and adolescents.

There are certain situations where false allegations tend to occur more commonly: those with a history of proven, previous sexual abuse; where there is mental illness in the parents; and, most often, where there is a custody dispute. Following a history of sexual abuse, the false allegation may sometimes arise out of a misunderstanding of normal affectionate contact between the adult and the vulnerable child. Where there is mental illness in the adult, the false allegation may become incorporated into the delusional state of the adult and be imposed on the child. Despite custody disputes representing the most common situation in which false allegations may arise, even here, the majority of allegations turn out not to be false.

The assessment of an allegation's truthfulness is extremely difficult, particularly when the only statements derive from an adult and the child makes no independent disclosures.

The quality of the disclosure tends to be important, with false allegations having a rehearsed quality and being in language more appropriate to adults and with a lack of appropriate affect. True allegations tend to contain idiosyncratic details. Usually the central experience for the child is strongly held, while peripheral details such as when or where the abuse occurred are less clearly described. A further difficulty is the possibility that, following repeated interviews or statements by the child, the child's account may acquire a rehearsed quality and the child may talk about the abuse without the same affective quality. This is one of the reasons why the first statements made by the child are actually the most helpful.

TREATMENT

Therapy for sexual abuse is no straightforward matter. There are issues concerning purpose, recipients, content, treatment modality, and process, including duration.

Purpose

Before considering the aims of therapy, it is important to recall the overall needs of the child, which form the context and direction for therapy. The child's needs are mainly fivefold:

1. to be believed
2. not to be blamed for the abuse, for not disclosing earlier, and for disclosing now
3. to be protected from further abuse
4. to be informed about developments in the post-disclosure process, including any legal proceedings
5. to live in a permanent setting, either in the original or an alternative family.

Treatment is essentially psychological, apart from rare occasions where serious injury occurs as part of the abuse or when a child contracts a sexually transmitted disease or falls pregnant. Psychological treatment is orientated both retrospectively, dealing with the aftermath and effects of the abuse, prospectively with the aim of ensuring the best possible future adjustment for the child.

Recipients and Content

Apart from the child, there are other important members of the abuse and family network who need to be considered in planning a systemic treatment approach or program. They include the mother, nonabusing parents, or primary care giver(s). They are as likely to be affected, although differently than the child, by learning of the abuse. The child's siblings, who may have been silent witnesses, unknown victims, or merely "forgotten", have their own needs. Finally, the abuser will require treatment if he is able to admit to the abuse. At different times in the process, these members of the child's family may be seen together, singly, or in different combinations, dependent on need.

The Child

The minimum requirement of postabuse help to a child is to offer the opportunity for the child to talk about the abuse and the *feelings* associated with the experience, including the inevitable *guilt* and *confusion*. There is also always a need to address issues of *sexuality* about which children who have been sexually abused hold many misconceptions. There are many reasons for children feeling to blame, despite the fact that the children are, by definition, not guilty. The reasons include the child's erroneous belief that she should have been able to stop the abuse or to talk about it earlier. Other children feel guilty for disclosing the abuse, particularly if the consequences are distressing for the mother or are perceived to have *caused* a subsequent family breakup. Some children come to feel guilty for the subsequent imprisonment of the abuser.

Beyond these irreducible minima, the child may have many more needs to be addressed in therapy. A significant proportion of children who have been sexually abused suffer from posttraumatic phenomena or actual posttraumatic stress disorder. In this condition, (i) unwanted vivid recollections of the traumatic event(s) intrude into the child's present, either during the waking state by way of flashbacks triggered by reminders or as nightmares during sleep; (ii) the child fearfully avoids possible reminders of the abuse and its circumstances; and (iii) the child may be generally more anxious.

Other issues include circumstances belonging to the child's life before the abuse; these may include a poor relationship with the primary caregiver because of neglect, which would have rendered the child more vulnerable to abuse. Until the child is permanently and safely settled after the abuse is discovered, the child may be more preoccupied with the question of where they are going to live and the nature of their future contact with their family and with the abuser. Indeed, while the child is "in transit", these latter concerns may stand in the way of the child feeling able to become fully involved in therapy concerned with the actual abuse. Some adolescents suffer from depression following a history of sexual abuse and others move on to drug abuse. Specific therapeutic work will be required here.

Some issues concern the abused child's feelings, thoughts, and experiences and are more appropriately addressed with the abused child, either individually or in a group.

Groups

The group setting is particularly liked by children and is an important component of the treatment process. Belonging to a therapy group reduces the child's sense of isolation, shame, and stigma. It offers the opportunity to share one's experiences in a safe setting. It is comforting to hear about the guilt and self-blame, which all group members experience to some degree, even if these feelings may not cease entirely. Learning about sexuality in a peer group reduces the embarrassment.

For victims over the age of six, groups are best offered to boys and girls separately and the work is most appropriately carried out with four to six children of a similar age and developmental stage within two- to three-year age bands. It is preferable for a group to include more than one child from any particular ethnic minority and children with common experiences such as living away from their biological family.

Groups require structure, a preplanned program and set of activities, and two co-therapists of whom at least one should be female. It is also extremely useful to run a parallel group for the children's carers. Group therapy, also serves as a forum for identifying those children who require further individual therapy.

Individual Therapy

Some children who are extremely troubled require individual therapy. Their difficulties may include a deep sense of guilt and low self esteem because of difficulties with attachment, the latter resulting from poor experiences of primary care. Other indications of the need for individual therapy include posttraumatic disorders and the experience of very prolonged, complex, and severe abuse. As with group therapy for children, it is important that the child's caregiver is seen separately.

The Mother

There are many troubling issues that arise for the mother or other nonabusive parent following discovery of sexual abuse. Chief among these are the following:

1. There are feelings of guilt at not having been aware of the abuse or for not having acted earlier on suspicions of abuse.
2. For various reasons, which include guilt, possible memories of previous own sexual abuse and the consequences of believing the child may lead the mother to have serious difficulties in believing the child's account.
3. If the abuser has been known to and is in any form of emotional or biological relationship with the nonabusing parent, and especially if the alleged abuser is denying the allegation of abuse, there is a serious dilemma for the mother or nonabusing carer(s) in deciding whose side to take.
4. If the mother believes and supports the child, this may be at considerable costs to her in losses of relationship and income and in fear of the abuser.

There is therefore a need to provide therapy for the nonabusing carer who is often, at the crisis time of disclosure, more needy than the child. Groups for mothers are often particularly helpful in enabling the participants to share their dilemmas and co-construct solutions. These groups are more than self-help groups and require the leadership of therapists.

There is empirical evidence indicating that the outcome for the child postdisclosure is significantly determined by the ability of the mother to actively support the child, not only by believing but also by supporting and protecting the child. The latter is measured by the mother's capacity to separate from the abuser, at least until the abuser takes responsibility and receives treatment.

The Abuser

For the protection of other children and particularly if the abuser is to return to an unsupervised relationship with the child, there is a need for therapy. Given the addictive nature of sexually abusive activity, the best that therapy can be expected to achieve is a full recognition by the abuser that he has the potential to abuse children and that he needs to guard indefinitely against this danger by a number of means.

Much therapy for abusers is carried out in groups, one of whose advantages is that group members are in a good position to discourage minimization and denial by other members while retaining a degree of mutual compassion and support. For therapy to commence, there is a prior requirement for the abuser to admit to the abuse. Therapy deals with various aspects of denial, which include denial of the list below:

- the extent of the abuse including the number of victims and frequency
- the abusive nature of the abuse including attempts at reframing it as education or love
- the discomfort, distress, and harm that the abuse has caused the victim
- responsibility for the abuse and the claim that the child invited it.

Therapy also explores antecedents to abuse in the abuser's own past childhood history. This is likely to include one or more harmful experiences including emotional, physical, or sexual abuse; exposure to parental/domestic violence; neglect; or disruption of primary attachments. Therapy for abusers needs to balance compassion for victimization with an insistence on the taking of full responsibility for the abuse. Therapy also needs to explore in detail the abusers' distorted thinking about abuse and the cycle of precipitating stimuli that lead to abuse as well as the grooming behavior that the abusers develop in relation to their child victims. Therapy then leads the abuser to develop alternate ways of responding to the temptation to sexually abuse children.

Therapy for Relationships

As well as dealing with individual issues, either in individual work or in groups of children, mothers, and abusers, there is also a need to facilitate therapeutic communication about many painful and unresolved issues between the various members of the family network surrounding the abuse. This includes the mother and child (where the child might question why the mother had not protected the child), the nonabusing parent and abuser, and, if the abuser is able to acknowledge responsibility for the abuse and the child is willing to meet him, therapeutic conversations between the child and the abuser. Whole family meetings are also important but can only usefully include the abuser if he has taken responsibility for the abuse. Following abuse, there is often an ongoing need for some therapeutic input for some time. Different needs arise at different stages in the child's development such as the onset of puberty. It is often not possible to rely on one comprehensive and definitive treatment program that will encompass and successfully anticipate all future therapeutic needs.

THE RECOVERED MEMORY CONTROVERSY

Recovered memories of previously forgotten, usually traumatic, experiences have latterly generated a great deal of heat due to the effects of family members being accused of abuse, which is totally denied by the alleged abuser. In clarifying this contentious issue, several points can be made.

- It is very rare to be able to recall reliably memories of experiences that occurred before the age of three and a half.
- There is debate about the existence of the phenomenon of repression.
- Some adults who, for the first time, describe previous childhood sexual abuse may never have forgotten the abuse but may merely not have talked about it to any one.
- In suggestible adults (and children), it is possible to induce false beliefs to which the person subsequently adheres.
- There are psychological techniques, which in the course of exploration of past experiences, induce false beliefs—thereafter called memories.
- The theoretical stance of therapists who use these techniques varies. For some, it is based on clinical encounters with patients/clients presenting with nonspecific symptoms who are subsequently found to have been sexually abused. For others, the practice is based on a strongly held belief that a history of child sexual abuse underlies much adult psychological symptomatology and that the active uncovering of the memories of this abuse is the route to recovery.
- Considerable caution is required in the use of, and interpretation of, the results of techniques used to recover memories.

CHILDREN AND ADOLESCENTS AS PERPETRATORS

Many sexual abusers describe the onset of their activity in adolescence. There is no indication to suggest that this is a temporary, developmental phase, which "burns out" with greater maturity. A significant proportion of adolescent abusers have learning difficulties. Moreover, adolescent abusers may be older siblings and family members or they may act as babysitters or neighbors and thus gain access to younger children. The nature of the abusive activity and its effects on victims is not significantly different from abuse by adults.

Abuse by adolescents poses particular dilemmas for parents when abuser and victim are siblings: parents face the difficult choices of whether to censor or condemn one child in order to protect and support another. Many parents resort to disbelief as a solution (albeit a maladaptive one).

Research has shown that sexual victimization is neither necessary nor sufficient to predict sexually abusive behavior. What distinguishes abusers is a history of exposure to violence, discontinuity of care, rejection by the family, and past sexual victimization of the mother.

Treatment for adolescent abusers is possible and, if supported by the young person's parents, can be effective in preventing subsequent further abuse by the adolescent. A balance is required between the recognition of the young person's incomplete maturation and their undoubted responsibility for the abuse.

There is accumulating information indicating that some genitalia-orientated activity is common in young children. They include looking at and touching their own and other children's genitalia and women's breasts and are assumed to be part of normal curiosity and exploration. These behaviors only give rise to concern when they are coercive, oft-repeated, or preoccupying to the child. Conversely, inserting fingers or objects into another child's vagina or anus and oral-genital contact is regarded as indicating inappropriate prior exposure of the child to sexual activity. This exposure could be by direct observation of sexual activity, either live or on the media/videos, or through having been sexually abused.

A minority of preadolescent children who have been severely and perversely abused, often in organized activity, may go on to repeatedly and coercively involve other children in sexual activity, which is perceived by the recipient child as disturbing and highly abusive. The behavior of these "abusing" children is often disturbed in other ways also. These children require very diligent surveillance and intensive therapy, sometimes in special residential settings.

PREVENTION

Prevention of sexual abuse is a complex issue, which can be approached from several directions. These include direct protection of children, reduction of precursors of abusive behavior, and facilitating the early reporting of, and subsequent belief in, reported abuse. There have been many initiatives teaching young children to value their body, to recognize "good" and "bad" touches, to be wary of strangers, and above all, to report unwanted or uncomfortable events and encounters, even when told to keep these a secret. It is difficult to evaluate these programs. Although children show subsequent retention of the learned knowledge, this cannot be equated with any certainty that such learned knowledge would, when needed, be translated into action. Abusers report that they tend to target vulnerable and isolated children who are less likely to be in a position to utilize protective measures. The main reliance on protection of children (young especially) must be placed on their primary carers.

As stated earlier, there are indicators of precursors of sexually abusive behavior. Psychosocial approaches to the reduction of adversity in childhood would very likely reduce sexual abuse of children, but these are nonspecific preventive measures.

Despite the now greater acknowledgment of child sexual abuse, denial and disbelief are still a frequent response to a report or disclosure of abuse by a child. Such responses are an active discouragement to disclosure and are, indeed, relied upon by abusers to ensure the continuation of sexual abuse with impunity. Continuing factual and unsensational public education about the known facts concerning child sexual abuse is therefore an important preventive measure.

Confidential telephone lines for children facilitate early reporting and are widely used by children. It is difficult to determine how effective they are in actually preventing or curtailing abuse.

RECOMMENDED READINGS

Finkelhor, D. (1984). Child Sexual Abuse. New Theory and Research. Free Press, New York.

Glaser, D., & Frosh, S. (1993). Child Sexual Abuse. 2nd ed. Macmillan, London.

Holmes, W.C., & Slap, J.B. (1998). Sexual abuse of boys: Definition, prevalence, correlates, sequelae & management. JAMA 280(21), 1855-1862.

Jones, D.P.H., & Ramchandari, P. (1999). Child Sexual Abuse: Informing Practice from Research. Radcliffe Medical Press Ltd., Jacksonville, Oregon.

Kendall-Tackett, K.A., Williams, L.M., & Finkelhor, D. (1993). Impact of sexual abuse on children: A review and synthesis of recent empirical studies. Psychol. Bull. 113, 164-180.

Parton, N. (1999). Child Sexual Abuse. John Wiley & Sons, New York.

Chapter 17

Hereditary and Acquired Mental Retardation

JAN STERN

Biological Psychiatry, pages 379-436.

ISBN: 1-55938-819-6

INTRODUCTION

The great majority of pregnancies end in the birth of a normal healthy child. However, 3% of children are born with major abnormalities. The CNS ranks fifth among the organ systems affected, and frequently these children will be mentally retarded. Mental retardation is a permanent impairment of the intellect sufficiently severe to prejudice normal existence in the community. It is a social as well as a psychopathological concept; it is a function both of the malfunction of the brain of an affected individual and of the nature and the complexity of the society in which he lives.

The terminology of mental retardation is confused because of attempts to provide accurate definitions free from stigma. Synonyms suggested have included "mental handicap", "subnormality", "intellectual disability", and, in Great Britain, increasingly "learning disability". For the sake of simplicity, we shall, in the main, refer to mental retardation. Often, children are classed as moderately retarded if their intelligence quotient (IQ) falls in the range of 50 to 70, and as severely retarded if they have an IQ of 50 or below. Psychologists are well aware of the danger of placing too much emphasis on this measure rather than on actual performance in school and in the community. In this chapter emphasis will be on biological mechanisms resulting in mental retardation.

Prevalence

In several recent studies prevalence of severe mental retardation in school-age children has been found to be between three and four per thousand, and that of moderate retardation between 20 and 30 per thousand, depending to some extent on definitions and methods of ascertainment. At birth, the incidence of some disorders with severe retardation is considerably higher, but many of these infants still die during the first few years of life sometimes before their intellectual deficit has been ascertained. Adult prevalence of severe retardation is increasing due to the improving expectation of life of individuals with disabilities, such as patients with Down syndrome.

Etiology

Genetic and environmental factors will permanently impair intelligence if they produce structural or metabolic defects in the brain that preclude normal function. Excess or deficiency of endogenous or exogenous metabolites may overwhelm the adaptive capacity of the brain and result in cell death, in mitotic delay, or in interference with histogenesis, with differentiation of cells, or with cellular migration. Mutations and chromosomal abnormalities often result in disorganization of cerebral development at both the macroscopic and micro-

Table 1. Classification of Causes of Mental Retardation

Prenatal
Genetic: chromosomal disorders, inherited syndromes of known or unknown cause, inherited metabolic or degenerative disorders
Environmental: exposure to teratogens or poisons, nutritional or hormonal insufficiencies, intrauterine infections
Perinatal
Injuries acquired late in uterine life, placental insufficiency, asphyxia, hypoxia–ischemia, infections
Postnatal
Physical abuse, infections, environmental deprivation
Unclassified
No known or dominant known cause or causes

Table 2. Examples of Disorders Causing Mental Retardation

Aminoacidurias	Phenylketonuria, nonketotic hyperglycinemia, homocystinuria, citrullinuria, argininosuccinic aciduria
Organic acidurias	Methylmalonic aciduria, propionic aciduria, multiple carboxylase deficiency, acyl-CoA dehydrogenase deficiencies
Mitochondrial disease	Pyruvate dehydrogenase deficiency, cytochrome oxidase deficiency
Lysosomal disorders	Mucopolysaccharidoses, mucolipidoses, sphingolipidoses, some leucodystrophies
Peroxisomal disorders	Zellweger syndrome, X-linked adrenoleucodystrophy
Other inborn errors	Galactosemia, Lesch-Nyhan disease, Wilson's disease, Menkes' disease
Hormonal disorders	Hypothyroidism, pseudoparathyroidism
Chromosomal disorders	Down syndrome, Prader-Willi syndrome, Angelman syndrome
Environmental hazards	Hypoxia–ischemia, bilirubin encephalopathy, mineral or organic poisons, burns encephalopathy, infections, head injury, malnutrition

scopic level. Traditionally the causes of mental retardation have been placed into four categories (Table 1).

In a dozen studies of the causes of mental retardation dated between 1966 and 1989 quoted by Prensky (1992), perinatal problems accounted for 10% to 12% of cases and postnatal cases for a comparable proportion. In roughly half the cases, prenatal causes were implicated leaving 25% to 30% of cases unclassified. The etiological pattern of mental retardation is not static. The contribution to prevalence by phenylketonuria, congenital hypothyroidism, and rubella embryopathy has been, or is being, eliminated. On the other hand, industrial pollution, iatrogenic disasters, and HIV infection have added to the causes of mental handicap. With improved diagnostic techniques in cytogenetics, biochemical genetics, and molecular biology, new disorders are being discovered and less severely affected patients with milder variants of known diseases identified, diminishing the proportion of unclassified cases (Table 2).

Pathogenetic Factors

At autopsies it is found that metabolic or chromosomal disorders sufficiently severe to result in permanent intellectual deficit are almost always associated with identifiable morbid anatomical changes in the nervous system. The distribution of lesions is often topologically uneven and most neuropathological findings are nonspecific: the repertory of structural changes in the brain is rather limited in relation to the great variety of adverse factors to which it may be exposed. Genetic and environmental factors may act via *common pathological pathways.* For example, Leigh's disease (subacute necrotizing encephalopathy) presents a specific neuropathological pattern of spongiform degeneration, vascular proliferation, gliosis, and demyelination, the basal ganglia and brain stem being affected worst. These changes may be the consequence of mutations in nuclear DNA resulting in deficiencies of the pyruvate dehydrogenase complex or of the neoglucogenic enzyme pyruvate carboxylase. This may be caused by mutations in nuclear or mitochondrial DNA resulting in deficiencies of complex I, IV, or V of the respiratory chain; or may be caused environmentally by the combined effect on pyruvate dehydrogenase of thiamine deficiency and acetaldehyde (from ingested alcohol) in Wernicke's encephalopathy.

Severe mental retardation may result from the effects of a single gene or of a well defined environmental hazard such as head injury. More often, the etiology of the mental defect and indeed that of psychiatric illness in general is multifactorial, the result of the interaction of several genetic and environmental factors as propounded by Eisenberg (1977). For example, the very low birth weight infant requiring intensive care for the first few months of life may well suffer periods of hypoxia and other metabolic stresses and is also at risk from intraventricular and periventricular hemorrhage. However, the premature birth may itself be due to multiple antenatal factors—maternal or environmental. It may then be difficult to apportion weight to individual adverse factors contributing to etiology. Often, it is difficult to define 'risk' levels for specific metabolites. The rate at which a metabolic disturbance develops, the developmental stage of the brain, the selective vulnerability of its formations at the time of the insult, the interaction of protective and exacerbating factors—both genetic and environmental—and, not least, the quality of the environment in which the child is subsequently reared will determine the outcome. Workers in basic research aim to isolate pathogenetic factors for study in controlled conditions. The task of the clinician is to utilize the knowledge thus gained in assessing the significance and contribution of interacting factors in individiual patients.

There is a well defined time course for the development of specific structures within the brain. If a genetic, metabolic, or toxic insult occurs at the time when a particular population of cells is supposed to form, migrate, differentiate, or in some way interact with surrounding structures, these processes may be interfered with. Often it will be the developmental stage of the brain, the severity and duration of the insult, and the effectiveness of protective mechanisms that will determine the outcome rather than the precise nature of the insult. During the first half of gestation, interference with neural tube formation, differentiation of cerebral vesicles, and generation and migration of neurons is associated with a wide spectrum of malformations: neural tube defects, holoprosencephalies, microcephaly, and defects of migration such as polymicrogyria, pachygyria, and lissencephaly ('smooth brain'). Most can be caused by mutations, chromosomal aberrations, teratogens, and other environmental factors acting singly or in combination. The period from the last trimester of pregnancy to 18 months or two years after birth has been called the 'brain growth spurt'. This is the period of glial multiplication, dendritic arborization, and synaptogenesis, and of a high rate of myelination when the brain is particularly vulnerable to malnutrition and hormonal imbalance. A detailed treatment of these topics will be found, for example, in Swaiman (1994) and Duckett (1995).

The genetic make-up of an individual may affect for better or for worse the effects of drugs or poisons; the clinical manifestations of an inborn error of metabolism are influenced by the environment and by other genes. Genetic heterogeneity profoundly affects the disease pattern of hereditary disorders. More than one mutation may occur at one locus, or a mutation may affect the pathway from gene to gene product. Often, this results in variable activity of the enzyme affected. Nearly always, the lower the residual activity of the enzyme the more severe the disorder and the greater the risk to the nervous system. Sometimes enzyme activity is reduced to a small fraction of normal but is still sufficient for normal development. Those affected may exhibit enhanced vulnerability to environmental hazards such as infection or inappropriate diet, which can unmask the underlying defect and produce a life-threatening crisis. Routine tests between attacks often fail to detect these disorders of which medium-chain acyl-CoA dehydrogenase deficiency is a good example.

The techniques of biochemical genetics and molecular biology can combine to provide insight into genotype–phenotype correlations. For a number of disorders, genes have been sequenced and disease-causing mutations identified. In some instances, mutation analysis in newborn infants can be used to predict the severity of the disease and to rationalize treatment. Unfortunately, while there is mostly correlation between mutational genotype and the biochemical phenotype, in some cases, neither correlates well with the clinical phenotype, particularly with respect to cognitive development. This is not surprising as the biochemical phenotype is more proximal to the mutation than the clinical genotype, which is more distant and, therefore, affected to a greater extent by other genetic and environmental factors.

PRESENTATION AND DETECTION

The diagnosis of mental retardation in a young infant may be very difficult. Damage sustained by the brain during early development may only become apparent months or years later as demands on the nervous system increase and ability for abstract reasoning is tested. The problem may be realized from failure to achieve relevant milestones or with the development of signs and symptoms such as skin lesions, microcephaly, or epilepsy. Early, sometimes presymptomatic diagnosis

Table 3. Some Examples of Neurological Signs in Mentally Retarded Patients

Ataxia
- Environmental (viral infections, hypoxia, antenatal, or postnatal cerebellar lesions, drugs, and poisons)
- Aminoacidopathies (late-onset urea cycle disorders, Hartnup disease)
- Organic acidurias (L-2-hydroxyglutaric aciduria, 4-hydroxybutyric aciduria)
- Respiratory chain disorders (mitochondrial encephalomyopathies)
- Lysosomal disorders (juvenile variants of some lipidoses and leucodystrophies)
- Peroxisomal disorders (Refsum's disease, adrenoleucodystrophy)
- Other metabolic disorders (vitamin E deficiencies, ataxia telangiectasia, Wilson's disease, cerebrotendinous xanthomatosis)

Extrapyramidal signs: dyskinesis, dystonia, choreoathetosis
- Aminoacidurias (biopterin synthesis or recycling deficiencies)
- Organic acidurias (glutaric aciduria type I)
- Lysosomal disorders (Krabbe's disease, metachromatic leucodynstrohy)
- Other metabolic disorders (Crigler-Najjar syndrome with kernicterus, Lesch-Nyhan disease, Wilson's disease, Huntington's chorea)

Hypotonia
- Unclassified mental retardation (very common, often variable, tends to get less severe with age)
- Down syndrome (some trisomic patients are hypertonic)
- Aminoacidurias (nonketotic hyperglycinemia, Lowe's syndrome)
- Organic acidurias (a common finding, e.g., in the lactic acidemias)
- Lysosomal disorders (observed in some lipidoses, mucopolysaccharidoses, leucodystrophies)
- Peroxisomal disorders (Zellweger syndrome, adrenoleucodystrophy)
- Other metabolic disorders (hypothyroidism, carbohydrate-deficient glycoprotein syndrome)

Hypotonia with muscle weakness
- Congenital myopathies (mitochondrial encephalomyopathies, Pompe's disease)
- Congenital myotonic dystrophy
- Congenital muscular dystrophy
- Duchenne muscular dystrophy

Peripheral neuropathies
- Abetalipoproteinemia
- Werdnig-Hoffmann disease
- Refsum's disease
- Familial dysautonomia
- A number of lysosomal and peroxisomal disorders
- Infectious and toxic polyneuropathies

of some disorders can can be achieved by comprehensive surveillance of all children and when indicated by use of appropriate screening tests.

Mentally retarded children often look 'different', but genuine physical signs must be distinguished from the impression conveyed by delayed growth and immature or inappropriate behavior. Somatic abnormalities may form a recognizable pattern as with Apert or the de Lange syndromes, two examples of disorders that may be obvious from birth. An excellent monograph by Jones (1988) for the diagnostic use of clinical signs also records useful laboratory findings. Winter and Baraitser (1993) provide details of over 2000 dysmorphic syndromes, and OMIM, the online version of McKusick (1994), provides details of genotypes and phenotypes in over 6000 entries. Mulvihill (1995) has pointed out that there are many more genes that can go wrong than there are ways for the human organism to be abnormal. There is a limited repertoire of recognizable syndromes.

Ataxia, Dyskinesia, and Dystonia

Ataxia must be distinguished from other forms of unsteadiness and the immature coordination often seen in retarded children. Several metabolic disorders are seen in patients with ataxia, in some cases with a clinical picture of spinocerebellar degeneration. Abnormal movements (dyskinesia) and abnormal tone (dysto-

Table 4. Some Ocular Findings in Mentally Retarded Patients

Cataracts
Galactosemia
Lowe's syndrome
Pseudohypoparathyroidism
Peroxisomal biogenesis defects
Cockayne syndrome
Marinesco-Sjögren syndrome
Hallerman-Streiff syndrome
Corneal opacities
Lysosomal disorders (frequent finding)
Wilson's disease (Kayser-Fleischer rings)
Tyrosinosis type II
Macular cherry red spot
GM_1 and GM_2 gangliosidoses
Niemann-Pick disease type A
Sialidosis, galactosialidosis
Retinal degeneration
Peroxisomal disorders (frequent finding)
Respiratory chain disorders (frequent finding)
Vitamin E deficiencies
Fatty alcohol oxidoreductase deficiency (Sjögren-Larsson)
Neuronal ceroid lipofuscinosis
Carbohydrate deficient glycoprotein syndrome
Laurence-Moon-Biedl syndrome
Cockayne syndrome
Retinitis pigmentosa and congenital deafness (Usher syndrome)
Dislocation of the lens
Homocystinuria
Sulphite oxidase deficiency

Table 5. Some Causes of Hearing Loss in the Mentally Retarded*

Environmental
Meningitis
Encephalitis
Ototoxic drugs (e.g., aminoglycosides)
Congenital infection (e.g., rubella)
Bilirubin encephalopathy
Tumors
Genetic
Mucopolysaccharidoses I, II and III
Mucolipidosis II
Mannosidosis
Some mitochondrial encephalopathies (e.g., MERFF)
Some peroxisomal disorders (e.g., X-linked adrenoleucodystrophy)
Biotinidase deficiency
Aspartoacylase deficiency
Cockayne syndrome
Some chromosomal disorders (e.g., trisomy 13)

Note: * Hyperacusis occurs in Tay-Sachs disease.

nia), signs of pyramidal dysfunction, are also seen in some neurometabolic disorders (Table 3).

Of special interest is the infant with hypotonia (Dubowitz, 1995). Hypotonia and weakness may result from a disorder of the motor unit (i.e., the anterior horn cell, peripheral nerves, and neuromuscular junction of muscle fibers) but may also be secondary to disorders affecting levels above the motor neuron. In the newborn, hypotonia is often seen following trauma, hypoxia–ischemia, and intracranial hemorrhage or infection, and in numerous genetic disorders, eventually giving way at a later date to cerebral palsy.

Sensory Organs

Ocular findings are common in severely retarded children. Nystagmus, strabismus, optic atrophy, and microphthalmia are frequently seen, while retrolental fibroplasia is rarely seen nowadays. Some neurometabolic disorders associated with ocular signs are shown in Table 4.

Hearing loss is probably as prevalent as blindness in the severely retarded. It may be acquired in the following ways: prenatally, as in maternal rubella and other infections; perinatally, when it may be associated with cerebral palsy or other evidence of an insult to the nervous system; and postnatally, for example, as sequelae to meningitis or the action of ototoxic drugs. Deafness is also a feature of a number of hereditary disorders (Table 5).

Table 6. Some Abnormalities of Skin and Hair in the Mentally Retarded

Adenoma sebaceum	
	Tuberous sclerosis
Café-au-lait patches	
	Neurofibromatosis type 1
Facial hemangioma	
	Sturge-Weber syndrome
Angiokeratoma	
	Fucosidosis
	Galactosialidosis
	β-mannosidosis
	Fabry's disease (intelligence mostly normal)
Subcutaneous nodules	
	Farber's disease
	Carbohydrate deficient glycoprotein syndrome
Rashes	
	Biotinidase deficiency (also alopecia)
	Hartnup disease (usually not retarded)
Brittle hair (trichorrhexis nodosa)	
	Menkes' disease (also alopecia)
	Argininosuccinic aciduria
Fine hair	
	Homocystinuria

Abnormalities of skin and hair are found in a number of disorders (Table 6).

Some organic acidurias are associated with peculiar odors, but there is wide variation among individuals in how these odors are perceived and described. Extensive documentation of clinical signs in mental retardation syndromes can be found in several reviews (Adams and Lyon, 1982; Nyhan and Sakati, 1987; Jones, 1988; Swaiman, 1994; Scriver et al., 1995).

Growth Disorders

Many patterns of malformation associated with growth retardation are the result of congenital hypoplasia in the skeletal system and in other organs, which often include the brain (Jones, 1988). The disorders may be caused by teratogens, chromosomal abnormalities, or mutant genes (Table 7).

In many cases, there is good correlation between the linear growth retardation and brain growth deficiency and/or mental retardation. Postnatally, there is usually a lack of catch-up growth due to irreversible antenatal impairment of brain and skeletal development. In mentally retarded children, secondary growth retar-

Table 7. Growth Disorders in the Mentally Retarded

Primary growth deficiency	
	Environmental (alcohol, drugs, poisons)
	Chromosomal imbalance (aneuploidy, deletions)
	Mutant genes (e.g., lysosomal storage disorders)
	Unknown causes (many dysmorphic syndromes)
Secondary growth deficiency	
	Environmental (malnutrition, radiotherapy, psychosocial deprivation)
	Metabolic (rickets, renal disease, heart disease, respiratory disease)
	Endocrinological (hypothyroidism, pseudohypoparthyroidism)
	Cerebral palsy
	Chronic severe infection
Tall stature	
	Chromosomal (XYY karyotype, often mentally normal)
	Beckwith-Wiedemann syndrome
	Cerebral gigantism (Soto's syndrome)
Megalencephaly	
	Neurocutaneous disorders
	Some lysosomal disorders (e.g., MPS I)
	Aspartoacylase deficiency
	Glutaric aciduria type I
Microcephaly	
	Seen in many syndromes; reduced brain weight is the commonest single abnormality in cases of mental retardation, cerebral palsy with epilepsy

dation is also common but, in contrast to primary growth deficiency, this can sometimes be prevented or reversed by appropriate treatment. On the other hand, severely impaired somatic growth during early postnatal life can be associated with subsequent impairment of mental abilities (Skuse et al., 1994). Tall stature is mostly genetically determined but is a feature of a few mental retardation syndromes (Table 7).

Behavior Disturbance and Autism

Recently, genotype–phenotype correlations have been extended to include behavior. Syndrome-specific observations can usefully complement psychiatric and psychometric assessments (Dykens, 1995). Inevitably, behavioral phenotypes show some within-syndrome variability as a result of the interaction of the gene affected, the individual's entire genetic endowment, and the environment. The behavior pattern may be striking, even bizarre, but there is also overlap across syndromes: the repertory of behavioral responses is not unlimited. Thus, self-mutilation is a feature of the de Lange as well as of the Lesch-Nyhan syndrome, and impaired speech is a feature of the Rett and Angelman syndromes. Behavior problems such as temper tantrums and stereotyped behavior are common in the retarded; often they amount to no more than behavior appropriate for the mental age of the patient. However, in a kindred studied by Brunner and colleagues (1993), the presence of a point mutation in the gene for monoamine oxidase A was associated with mental retardation and a propensity to aggressive behavior. Behavior problems have been attributed to food additives. Retarded children as a group show no such adverse behavioral reaction, although a few individuals may do so.

Infantile autism, as a specific syndrome, was first described more than 50 years ago by Kanner. Affected children fail to develop social relationships, exhibit ritualistic and repetitive behavior, and suffer from deficits in communication and perception, with onset in the first three years of life. The organic basis of autism is no longer disputed. In the severely retarded, the behavioral phenotype is often associated with known hereditary disorders such as tuberous sclerosis or some lipidoses whereas, in the intellectually normal or mildly retarded with so-called lesser variants, the etiology commonly appears to involve several genes (Rutter et al., 1994; Bolton et al., 1994). The behavioral phenotype in autism is thus the product of a range of pathogenetic mechanisms.

The Rett syndrome, with an incidence of one in 12,000 girls and onset at around ages one to three, also exhibits a characteristic behavioral phenotype that includes stereotypic hand movement, sleep disturbance, night laughing, crying spells, hyperventilation and apnea, absence of speech, and profound retardation. It may be the result of an X-autosomal translocation.

Acute Metabolic Encephalopathy

Some hereditary diseases produce a life-threatening crisis, which usually manifests itself during the first few days of life but may present at any age. Examples are urea cycle disorders, some organic acidurias, and galactosemia. Metabolic encephalopathies can also be caused by environmental factors (Table 8).

Acute episodes may be triggered or aggravated in some disorders by viral or bacterial infections, by fasting, by inappropriate diet, or by drugs. The neonate has a limited repertory of responses to severe illness: hypotonia, vomiting, diarrhea, lethargy, seizures, and coma (which are more often the result of environmental hazards). Suspicion of an inborn error of metabolism will be aroused when an infant deteriorates after a symptom-free period and does not respond to symptomatic therapy. Prompt and vigorous treatment is essential to prevent death or mental handicap. Details can be found in standard texts (e.g., Levene et al., 1995; Scriver et al., 1995).

Mass Screening

A number of disorders posing a threat of mental retardation can be diagnosed by presymptomatic screening of the newborn or by antenatal detection.

Table 8. Some Causes of Acute Encephalopathy with Risk of Mental Retardation

Trauma	Head injury, nonaccidental injury
Tumors	Including remote effect of tumors
Hypoxia–ischemia	Including near-miss sudden infant death syndrome
Intracranial hemorrhage	
Toxic shock encephalopathy	
Infections	Meningitis, encephalitis
Hemolytic uremic syndrome	
Reye's syndrome	
Inborn errors of metabolism	
Drugs and poisons	Lead, drug overdose, nonaccidental poisoning (Munchausen *by proxy*)

Mass screening is usually restricted to diseases that can be prevented or that respond to treatment (Anderson et al., 1994; Farriaux and Dhondt, 1994). In many countries, mass screening of the newborn has been successfully implemented for phenylketonuria and congenital hypothyroidism. Antenatal detection of Down syndrome and neural tube defects is also widely offered. In at-risk populations, screening for sickle cell disease gives the option for affected infants of prophylaxis against pneumococal meningitis and other hazards to their intellectual development. Opinion is still divided over the desirability of screening for galactosemia, biotinidase deficiency, maple syrup urine disease, and Duchenne muscular dystrophy.

Mild Mental Retardation

A recurrent problem in mental deficiency practice is the patient with mild mental retardation and no dysmorphic features. In some patients, no lesions can be demonstrated posthumously by routine neuropathological techniques. Most, however, show changes similar to, but less severe than, those of severely retarded patients. Some minor chromosomal abnormalities may produce mental retardation without dysmorphic features (Flint et al., 1995). The handicap associated with some monogenic disorders may span a wide range of mental disability. The same is true of environmental factors, which furthermore may have been acting over a limited period only and be no longer detectable when the patient is seen. Mild retardation, like severe retardation, is caused by one or more mutations, chromosomal abnormalities, and environmental factors acting singly or in concert. The concept of 'subcultural' mental retardation is no longer favored. Factors pointing to the involvement of a major gene in the etiology of mild mental retardation are listed in Table 9.

Table 9. Factors Pointing to the Involvement of a Major Gene in the Etiology of Mild Retardation

- Intelligence, while not grossly impaired, is significantly below that of parents and most members of the family
- Consanguinity in the parents
- More than one slow child in the family
- Behavior disorder, autistic features, or other psychiatric disturbance
- Neurological or physical signs
- Deteriorating performance at school
- Backwardness not explained
- Exposure to drugs or poisons excluded

Investigations

As in other branches of medicine, specialized investigations should be preceded by a careful history and physical examination and first-line tests. These should give an indication of how best to proceed to specialized investigations and definitive tests. Disorders may be missed in routine tests when the concentration of abnormal metabolites is too low for detection by standard screening tests, when the abnormality is not expressed in the body fluids normally examined, or when it only develops with time. When samples are referred, only centers with first-hand experience of the disorder sought should be approached. An example of a scheme for the investigation of mentally retarded patients is shown in Figure 1.

The traditional biochemical techniques of thin-layer chromatography, high pressure liquid chromatography, gas chromatography, and electrophoresis have been widely used. A recent advance in analytical technique, tandem mass spectrometry will detect metabolites in nanomole amounts and makes possible the detection of more than 20 metabolic disorders from blood spots collected on filter paper from neonates (Matsumoto et al., 1994). A definitive diagnosis is often possible by enzyme studies on plasma, the formed elements of blood, or fibroblasts. Cell culture studies in the form of crosscorrection experiments have been instrumental in identifying lysosomal disorders (Neufeld, 1991). Cultured cells also provide useful material for DNA studies.

Mutation analysis includes *diagnostic methods* for the detection of known mutations by such methods as allele-specific oligonucleotide (ASO) probing, amplification created restriction sites (ACRS), and amplification refractory mutation system (ARMS). *Scanning methods* determine whether a particular DNA fragment contains a mutation. They include single-strand conformation polymorphism (SSCP), denaturing gradient gel electrophoresis (DGGE), and chemical cleavage of mismatch (CCM) analysis (Cotton, 1993; Grompe, 1993; Forrest et al., 1995).

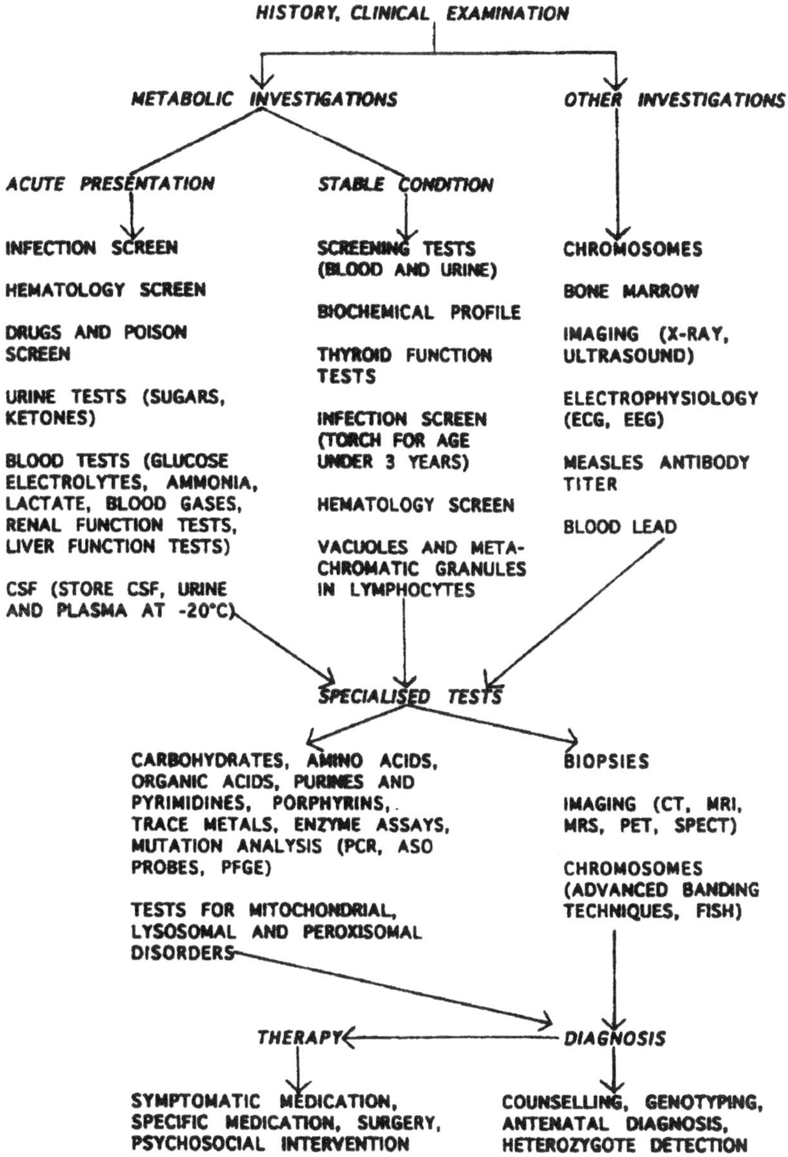

Figure 1. Scheme for the investigation and management of patients with mental retardation disorders. TORCH: toxoplasma, rubella, cytomegalovirus; CT: computerized tomography; MRI: magnetic resonance imaging; MRS: magnetic resonance spectroscopy; PET: positron emission tomography; SPECT: single-photon emission computerized tomography; PCR: polymerase chain reaction; ASO: allele-specific oligonucleotides; PFGE: pulsed field gel electrophoresis; FISH: fluorescent *in situ* hybridization.

Routine chromosome banding techniques have been refined by high resolution techniques and, more recently, subtle, submicroscopic lesions in chromosomes have become detectable by FISH, fluorescent *in situ* hybridization (Davies, 1993). *In situ* hybridization has also been used for chemical mapping of metabolic pathways in the brain by demonstrating where mRNA for a specific gene is expressed. Practical details of techniques in human biochemical genetics can be found in a review by Hommes (1991) and wider aspects of diagnosis in reviews by Adams and Lyon (1982), Nyhan and Sakati (1987), Levene and colleagues (1995), and Scriver and colleagues (1995).

Imaging techniques for the study of cerebral lesions include computed tomography (CT), magnetic resonance imaging and spectroscopy (MRI and MRS), neuronosonography, and Doppler ultrasound for the assessment of the cerebral circulation. Imaging techniques may be used to locate lesions such as intraventricular hemorrhage and periventricular leukomalacia in premature neonates and to monitor the evolution of such lesions. Imaging may be supplemented by electrophysiological investigations. The electroencephalogram (EEG) is diagnostic, for example, in subacute sclerosing panencephalitis and in nonketotic hyperglycinemia. Nerve conduction studies are helpful in the diagnosis of leucodystrophies. Evoked potentials (EP) obtained in response to repetitive sensory stimuli afford a means of assessing sensory as well as central nervous system function. Recently, the scope of imaging has been extended by techniques such as near infrared spectroscopy (NIRS), which can be used near the bedside, while positron emission tomograhy (PET) and functional magnetic resonance imaging (fMRI) can reveal changes in cerebral blood (CBF), which in turn is tuned to demands of neural activity. It is thus possible to identify activity and inactivity in the brain. (Swaiman et al., 1994; Levene et al., 1995).

ENCEPHALOPATHIES LARGELY OF ENVIRONMENTAL ORIGIN

Hypoxemia-Ischemia and Intraventricular Hemorrhage

Hypoxemic-ischemic brain injury is an important cause of mental handicap originating in the perinatal period (Swaiman et al. 1994; Levene et al. 1995). The brain can be deprived of oxygen by *hypoxemia,* a diminished oxygen level in the blood, or *ischemia,* a reduced perfusion of the tissue. Both may occur as a result of *asphyxia,* impairment of the respiratory exchange of oxygen and carbon dioxide. Hypoxia-ischemia is associated with increased utilization of glucose, increased glycolysis, and diminished production of high energy phosphates. Involvement of the white matter may occur in premature infants in the form of periventricular leukomalacia; neuronal necrosis may be seen both in premature and in fullterm infants. Accumulation of excess of cytotoxic excitatory amino

Table 10. Some Mechanisms of Anoxic Brain Damage

- Substrate depletion (hypoglycemia, failure of ketone body production)
- ATP depletion
- Excitatory amino acid toxicity with excessive entry of Ca^{2+} into neurons
- Damage by free radicals (oxygen, nitric oxide)
- Cerebral edema (vasogenic, cytotoxic)
- Cerebrovascular injury, disseminated vascular coagulation
- Disintegration of lysosomes

acids (Lipton and Rosenberg, 1994) and free radicals (Halliwell, 1994) play an important part in hypoxic-ischemic injury; cerebral edema both cytotoxic and vasogenic is a late complication in severely asphyxiated infants. Some mechanisms of anoxic brain damage are listed in Table 10.

In general, cognitive impairment reflects the severity of the insult. A classical disability of survivors of severe perinatal asphyxia is cerebral palsy, which may be associated with epilepsy and blindness in addition to mental retardation. However, in more than 80% of individuals with cerbral palsy there is no evidence of severe asphyxia. Prenatal hazards, brain malformations, chromosomal abnormalities and metabolic disorders account for the bulk of cases.

In the newborn, particularly the preterm infant, intraventricular hemorrhage, or more precisely germinal matrix hemorrhage-intraventricular hemorrhage, is now the most frequent and important form of intracranial hemorrhage. It may arise from birth trauma or a circulatory disturbance but is most likely to occur in babies with the respiratory distress syndrome. Provided the hemorrhage does not extent into the parenchyma and there is no ventricular dilatation, the prognosis appears to be no worse than for infants of similar birth weight but without hemorrhage. The association of low birth weight and mental handicap has long been known. Respiratory distress, ischemic lesions, and intraventricular hemorrhage are common in preterm babies, while infants with intrauterine growth retardation are more likely to develop hypoglycemia. The risk of cognitive deficits, weight for weight, is greater in this group, which includes infants with severe intrauterine malnutrition, chromosomal abnormalities, and congenital malformations. The low birth weight may then be the consequence of preexisting damage rather than its cause. A longterm study of a cohort of low birth weight infants has been published by Paneth and colleagues (1994). Improved management of neonates has greatly reduced the contribution of kernicterus and electrolyte imbalance to the prevalence of mental retardation. (For a detailed discussion of hypoxia and related topics, refer to Swaiman et al., 1994; Duckett, 1995; and Levene et al., 1995.)

Table 11. Some Causes of Hypoglycemia in Infants and Children

Environmental
Transient neonatal hypoglycemia
Hypoxia–ischemia, respiratory distress syndrom, hypothernia
Shock, hemorrhage, intracranial injury
Septicemia, meningitis
Reye's syndrome (environmental causes)
Hepatocellular failure
Intrauterine malnutrition
Ethanol, methanol, salicylate ingestion
Endocrine
Hyperinsulinism (maternal diabetes, nesidio-blastosis)
Deficiency of insulin antagonists (thyroid, adrenal pituitary)
Accidental or nonaccidental administration of insulin or hypoglycemic drugs
Hereditary
Defects in gluconeogenesis
Defects in glycogenolysis, glycogen synthetase deficiency
Galactosemia, fructose intolerance
Defects in oxidation of fatty acids
Defects in ketone body formation
Organic acidurias (propionic, methylmalonic, glutaric aciduria type II)
Reye's syndrome (hereditary causes)
Beckwith-Wiedemann syndrome

Hypoglycemia

Hypoglycemia has long been recognized as a cause of mental retardation. Some causes of hypoglycemia are listed in Table 11.

In the neonate, hypoglycemia is often associated with additional adverse factors such as anoxia, jaundice, hypothermia, hypocalcemia, acidosis, and septicemia. The individual contributions of these and other factors to symptoms and any permanent damage to the brain are difficult to disentangle. Neurological dysfunction has been demonstrated by brainstem auditory evoked potentials in infants at blood glucose levels between 2.6 and 2.0 mmol/l, in some cases in the absence of symptoms. Brain damage may ensue if dysfunction is prolonged or repeated. Before feeding is established, the infant depends on endogenous glycogen and fat. Low birth weight infants, particularly those small for gestational age, may have low reserves. The demand for glucose may not be met, particularly when gluconeogenesis and the production and utilization of ketone bodies, an alternative fuel for the brain, are functioning suboptimally (Cornblath and Schwartz, 1991; Hawdon et al., 1992). In inherited disorders of gluconeogenesis and

Table 12. Some Causes of Hyperammonemia

Inherited disorders of the urea cycle	
	N-acetylglutamate synthetase deficiency
	Carbamyl phosphate synthetase deficiency
	Ornithine transcarbamylase deficiency
	Argininosuccinate synthetase deficiency (citrullinuria)
	Argininosuccinate lyase deficiency (argininosuccinic aciduria)
	Arginase deficiency
Other hereditary disorders	
	Hyperornithinemia, hyperammonemia, homocitrullinuria (HHH syndrome)
	Lysinuric protein intolerance
	Fatty acid ß-oxidation defects
	Some organic acidurias
	Reye's syndrome (precipitated by a hereditary metabolic defect)
Environmental	
	Transient hyperammonemia of the newborn
	Reye's syndrome precipitated by viral infection, toxins, drugs
	Hypoxia–ischemia
	Shock, septicemia
	Drugs, notably valproate
	Proteus urinary infection with stasis

medium-chain acyl-CoA dehydrogenase deficiency, careful dietary control of blood glucose levels can greatly reduce the risk of mental retardation. In general, the risk of brain damage is higher in nonketotic hypoglycemia as, for example, in hyperinsulinism due to nesidioblastosis, than in ketotic hypoglycemia.

Hyperammonemia

Ammonia is a major product of nitrogen metabolism. It is generated in all organs including brain where it is formed in increased amounts during convulsions. Prolonged or recurrent hyperammonemia causes irreversible brain damage and may be fatal. Hyperammonemia may be hereditary, notably due to deficiency of a urea cycle enzyme, or it may be environmental (Table 12).

Urea cycle defects and transient hyperammonemia of the newborn carry the highest risk of death or mental retardation. However, three enzymes involved in the urea cycle, *N*-acetylglutamate synthetase, carbamyl phosphate synthetase, and ornithine transcarbamylase, are mitochondrial so that any disorder prejudicial to the metabolic integrity of this organelle can give rise to hyperammonemia.

In brain, detoxication occurs largely in astrocytes by amidation of glutamate to glutamine. Severe brain damage is associated with brain edema due to the osmotic effects of excess of glutamine in the astrocytes and the failure of the mitochondrial endothelial cells of the brain capillaries to maintain the integrity of the blood–brain barrier. Ammonia also affects neuronal excitability and acts on the chloride ionophore. These events precede significant depletion of energy stores of the brain. Postmortem, a characteristic finding are Alzheimer type II astrocytes with mitochondrial degeneration and lobulation of nuclei (Duckett, 1995).

Urea cycle deficiencies present most often as neonatal emergencies, but late onset cases also occur. In acute hyperammonemic episodes, hemodialysis is the recommended treatment. Subsequent management is by administration of phenylbutyrate or sometimes sodium benzoate to increase waste nitrogen excretion, restriction of dietary protein, and supplements of arginine. While lifesaving, this treatment cannot protect cases presenting as neonatal emergencies from mental retardation. On the other hand, more than half the cases treated prospectively can achieve normal intellectual development (Scriver et al., 1995). Inheritance of ornithine transcarbamylase deficiency is X-linked, whereas the other urea cycle enzyme deficiencies are inherited in an autosomal recessive manner. In heterozygous girl carriers of ornithine transcarbamylase deficiency, one of each pair of X chromosomes is inactivated randomly so that only one X chromosome is expressed. The abnormal gene is expressed in some liver cells, the normal gene in others. The relative proportion determines the severity of the presentation, which may be acute with neonatal or late onset and with a variety of neuropsychiatric symptoms, often with neurological sequelae. Even clinically normal heterozygotes have IQ scores six to 10 points lower than controls. A variety of techniques is offered for antenatal diagnosis and heterozygote detection. Experiments of gene transfer in animals and cultured cells look promising for the longterm. This topic has been reviewed by Nyhan and Sakati (1987), Swaiman and colleagues (1994), and Scriver and colleagues (1995). Very high levels of ammonia also occur in transient hyperammonemia of the newborn. The cause of this disorder has not been established, but with prompt and vigorous treatment, the prognosis is generally good.

Epilepsy

Epilepsy is relatively common in children with about six to eight per thousand being affected. The incidence is higher if febrile convulsions, breathholding attacks, and similar events are included. Epileptics as a group are somewhat less intelligent than nonepileptics, although some are not only intellectually normal but brilliant. Seizures are more frequent in the mentally retarded. About 10% of the mildly retarded and about a third of the severely retarded may be affected (Aicardi, 1994). In the mentally retarded, seizures may be the primary cause of the handicap. In some cases, seizures and retardation may both be the result of an underlying disease process, which is often a disorder of neuronal migration.

The view that convulsions can harm the brain is deep-rooted. Fits or their consequences such as cerebral anoxia, asphyxia, vascular disturbances, and associated biochemical changes can cause permanent injury, and this may occur in a previously well or retarded child. Sometimes a vicious circle is set up in which the lesions are produced by fits thus producing further fits. A disease process, perhaps unrecognized, may give rise to both fits and mental retardation. In the neurocutaneous disorders tuberous sclerosis and Sturge-Weber syndrome, it is the patients with epilepsy whose intelligence is most likely to be impaired. Myoclonic epilepsy may be in the forefront of the clinical signs of mitochondrial and lysosomal disorders; infantile spasms are observed in a number of hereditary disorders (Aicardi, 1994; Hopkins et al., 1995). Fits may also be iatrogenic, caused by sudden withdrawal of anticonvulsants or by administration of drugs that interfere with the metabolism of anticonvulsants. Overtreatment with anticonvulsants may have adverse effects on learning.

Excitatory amino acids play an important part in the genesis of neuronal injury in epilepsy. Activation of glutamate receptors is followed by excessive influx of calcium into the neuron, overstimulation of hydrolytic enzymes, and injury to the cell (Lipton and Rosenberg, 1994). Selective vulnerability of neurons may relate to their ability to buffer excess calcium.

Nutrition

The adult brain can withstand extreme and prolonged starvation without inevitable permanent effects on cognitive functioning. Malnutrition presents a greater risk to the developing brain, particularly during the brain growth spurt from the start of the third trimester of antenatal life to the second birthday. In humans, the greater part of this spurt takes place after birth, and intrauterine malnutrition should therefore be correctable to a significant extent in the postnatal period unless any damage is compounded by postnatal malnutrition or unless biological deficits are compounded by a poor environment and lack of stimulation. There have been numerous studies of malnutrition and dietary supplementation in pregnant women in poor populations, but so far we lack a clear understanding of the nature and magnitude of any effects on later infant and child behavior and cognition. A major difficulty is to separate the effets of malnutrition from those of poverty and poor education, two of the strongest predictors of poor mental performance (Susser, 1989).

Postnatally, malnutrition as a cause of mental retardation is probably rare in Western Europe except as a form of child abuse or because of neglect or ignorance. In the Third World, it still constitutes an important, preventable cause of educational underachievement, mental handicap, and death in childhood. Neurological abnormalities such as hyporeflexia, reduced nerve conduction velocities, and mental changes are sometimes seen in protein–calorie malnutrition. While usually insidious in onset, kwashiorkor may present with acute encephalopathy, and neurological symptoms may persist for some time after dietary correction. Many children so affected remain mentally retarded.

Studies of postnatal malnutrition also face the problem of isolating the effects of malnutrition as a specific cause of mental retardation from those of other adverse environmental factors. Chavez and colleagues (1995) conducted longitudinal studies on malnourished Mexican children. A group of children whose nutrition was supplemented before and after birth showed improved growth and mental and behavioral performance compared to a nonsupplemented control group. There was some catch-up on the part of the control group, but significant differences were still present at age 18 years. Early diet in preterm babies, notably its content of long-chain polyunsaturated acids, may affect their developmental status (Makrides et al., 1995).

In animals, malnutrition results in reduced brain size, which in itself is not of much significance. However, certain formations, for example, the cerebellum, are selectively affected. Irreversible changes are few, and they primarily involve cell number. Deficits in dendritic arborization, synaptogenesis, and development of enzyme systems can mostly be restored by nutritional rehabilitation, although distortion of the structural and biochemical development might have functional consequences.

Vitamins, Teratogens, Drugs, and Poisons

The role of vitamins as coenzymes suggests that deficiency may result in disturbance of neurological function. Beriberi caused by thiamine deficiency may present in infants as an acute encephalopathy; pellagra, due to nicotinic acid deficiency, may result in dementia; and pyridoxine deficiency may cause convulsions. Vitamin D deficiency may produce hypocalcemic fits in neonates. Vitamin E deficiency can be associated with neurological symptoms including ataxia. In B_{12} and folate deficiency, neuropsychiatric symptoms may include mental retardation. Postnatally, in advanced countries, inadequate vitamin intake is probably rare as a cause of overt vitamin deficiency except in alcoholic, psychiatric, and psychogeriatric patients. Patients on anticonvulsants may require vitamin D and folate supplements. The claim that vitamin supplements can improve the performance in psychometric tests and the maladaptive behavior of some children remains unproven. However, many countries are reducing their expenditure on welfare and, as a result, vitamin intake may now be inadequate in some groups of deprived children.

A few individuals have a constant specific requirement for a particular vitamin. They may require up to several hundred times the recommended intake: they exhibit *vitamin dependency* (Scriver et al., 1995). This is found in a small minority of patients with inborn errors. Untreated, it often carries a high risk of death or serious illness. Examples are variants of homocystinuria, some organic acidurias, and pyridoxine-dependent convulsions in neonates. These vitamin-dependent individuals are very rare; there is therefore no case for massive and indiscriminate vitamin supplements beyond the recommended norms. However, folate supplements when given before conception and during the first trimester of pregnancy can significantly reduce the incidence of neural tube defects. Treatment with large doses of vitamin E can prevent or ameliorate the ataxia and neurological symptoms in AVED (ataxia with vitamin E deficiency) due to deficiency of the liver α-tocopherol transfer protein, and in abetalipoproteinemia caused by deficiency of the microsomal triglyceride transfer protein (DiDonato, 1995).

Drugs and poisons can interfere with prenatal development by disturbing embryogenesis or by acting on individual fetal organs causing deformities. A single teratogen can be responsible for a wide range of defects; conversely, identical malformations may result from diverse environmental and genetic factors often acting synergistically. The timing of an insult may determine the form of the malformation as much as the agent responsible. The oocyte in the resting phase is probably relatively immune to toxic agents. On the other hand, consideration must be given to the "drugged sperm" as a potential mechanism for CNS abnormalities.The embryo, during the preimplantation phase of development, may be

Table 13. Some Antenatal Causes of Mental Retardation

Physical agents
- Radiation
- Trauma
- Nonaccidental injury

Maternal metabolic disorder
- Phenylketonuria
- Diabetes mellitus
- Malnutrition
- Iodine deficiency
- Folate deficiency

Maternal exposure to teratogens
- Carbon monoxide
- Anticonvulsants (valproate, phenytoin, carbamazephine)
- Antimetabolites (aminopterin)
- Isoretinoin
- Alcohol
- Tobacco
- Addictive compounds (marijuana, cocaine, heroin)
- Methyl mercury (Minamata disease)
- Lead

Infections
(See Table 14)

Genetic
- Chromosomal abnormalities
- Single-gene inheritance
- Multigenic inheritance

Multifactorial

exposed to drugs, and the fetus is at risk from eight weeks after conception. The role of the placenta must also be considered. Equilibration of drugs between mother and fetus may take from two minutes to two hours. The blood–brain barrier protects the CNS against many compounds, it matures around week 28 of gestation. Highly lipid-soluble drugs readily cross the blood–brain barrier and can affect the developing CNS. Neurons, once lost, cannot regenerate making the CNS particularly vulnerable. If a drug interferes with cell differentiation or migration, even minor resulting changes may have profound consequences for neurological and cognitive function (Duckett, 1995).

Table 13 lists some of the agents that may have adverse effects on the developing nervous system. Anticonvulsants increase the incidence of malformations in the offspring of epileptic women, but maternal seizures are almost certainly more dangerous to the fetus than anticonvulsants. Trimethadione, valproate, and multiple drug regimes are treatments of high risk. Children exposed to phenytoin *in utero* had an IQ 10 points lower than controls, while children exposed to carbamazepine did not differ from their controls (Scolnik et al., 1994).

Heavy consumption of alcohol during pregnancy is associated with the fetal alcohol syndrome in the offspring with microcephaly, stunted growth, a characteristic facies and mental retardation (Jones, 1988). In the United States, the incidence is probably higher than one in 6,000 live births. In the absence of the syndrome, any effects of alcohol on intelligence are of a lower order. In Sweden, nearly 10% of cases of mild mental retardation in school children were attributed to fetal alcohol effects. The pathogenetic process is not understood. Acetaldehyde and deficiency of retinoic acid have been suggested as agents that may be involved (van Baar et al., 1994; Levene et al., 1995). Alcohol abuse in pregnancy is often associated with other factors that have an adverse effects on the fetus. The mothers are often heavy smokers, and some are on psychotropic or addictive drugs. Most health authorities now recommend total abstinence during pregnancy.

Drugs taken in pregnancy often affect intelligence and behavior in the absence of recognizable malformations. Long-lasting functional disturbances in the developing brain may be produced by drugs that alter the balance of central neurotransmitter activity. This includes most psychoactive and addictive drugs. The longterm effects of alcohol and drugs taken during pregnancy on child development have recently been reviewed by van Baar and colleagues (1994) and Levene and colleagues (1995).

Infections

The principal stages in life when infection can lead to mental retardation are as follows: (1) prenatal (i.e., preconception to parturition), (2) perinatal, and (3) postnatal. Into this classification can be grouped a wide variety of infectious agents, which may infect or affect the nervous system and lead to mental retardation (Table 14).

Table 14. Some Examples of Infective Causes of Mental Retardation

Prenatal
- Cytomegalovirus (CMV)
- Rubella
- *Herpes simplex* (HSV)
- *Varicella zoster* (V-S)
- Human immunodeficiency virus (HIV)
- Listeriosis
- Syphilis
- Toxoplasmosis
- Malaria, trypanosomiasis, cysticercosis, cryptococcosis (in some parts of the world)

Perinatal
- Group B streptococci
- *Escherichia coli*
- Other Gram-negative infections
- Enteroviral infections
- *Herpes simplex*
- Cytomegalovirus
- Human immunodeficiency virus

Postnatal
- Bacterial meningitides (*H. Influenzae*, pneumococcal, meningococcal
- Viral encephalitides (*Herpes simplex*)
- Postinfectious encephalitides (measles, varicella, mumps)
- Viral and bacterial vaccines (measles, pertussis)

In acute infections, handicap may result from direct neuronal invasion or by toxic or hyperpyrexial cell damage. Respiratory arrest, for example, in hemophilus epiglottitis or pertussis may lead to hypoxia. Gastroenteritis may be complicated by hypernatremic dehydration and cerebrovascular accidents. Interference with the circulation or absorption of CSF may result in hydrocephalus. Cerebral edema may follow overenthusiastic fluid therapy. Sometimes an infection, in itself not necessarily serious, can be highly deleterious in a malnourished child or can aggravate the expression of an inborn error of metabolism.

Some infective agents that cross the placenta can cause fetal damage involving the CNS. Syphilis and listeriosis are nowadays rare, and they are preventable as causes of mental retardation as is rubella. Cytomegalovirus, herpes simplex virus, *Varicella zoster* virus, and human immunodeficiency virus can all injure the fetal brain and the infection can continue after birth. Although more than 90% of infants congenitally infected with cytomegalovirus grow up neurologically and developmentally normal, nearly three-quarters of those with neonatal neurological symptoms will have permanent sequelae including mental retardation. Maternal *Varicella zoster* infection carries a relatively low risk of embryopathy of about

2%. Herpes simplex infections, both pre- and perinatal, are fortunately rare but associated with high risk of mortality, psychomotor retardation, and neurological sequelae. Among perinatal infections, Lancefield Group B hemolytic streptococcal and *E. coli* septicemia and meningitis carry a high risk of death or permanent brain damage. Preterm babies requiring prolonged intensive care are most at risk. Postnatally, bacterial meningitis is still an important cause of mental retardation: late diagnosis, inadequate treatment, and the emergence of drug resistant strains of *N. meningitidis* or *H. influenzae* may adversely affect the outcome. Measles can cause mental retardation by acute encephalomyelitis at the time of eruption, and in immunosuppressed patients two to three months after an attack of measles, whereas subacute sclerosing panencephalitis (SSPE) can occur up to several years later. Vaccination has greatly reduced the incidence of encephalitides.

About 80% to 90% of cases of congenital toxoplasmosis are asymptomatic; the classical triad of hydrocephalus, chorioretinitis, and intracranial calcification is seldom in evidence. The contribution to mental handicap of this infection is therefore difficult to assess. Early maternal infection results in a lower rate of congenital infection than late maternal infection, but the risk of damage to the fetus is far greater.

The worldwide epidemic of the acquired immunodeficiency syndrome (AIDS) is expected to infect over 5 million children by the year 2000. Most affected children acquire human immunodeficiency virus (HIV) infection by vertical transmission from the mother, before, during, or after delivery. Neuropsychiatric presentation includes progressive encephalopathy, loss of motor milestones, and corticospinal tract abnormalities (Fowler, 1994). The dementia caused by HIV is termed AIDS dementia complex. It is the direct outcome of the virus entering the brain. Cerebrospinal fluid analysis shows the presence of HIV and protein p24. The incidence of dementia in AIDS patients is 6% (in adults) and 21% in children under 13 years of age. In a European study, three-quarters of infected children were still alive at age five, and most of these were free from neurological symptoms. A detailed presentation of this topic will be found in Swaiman (1994), Duckett (1995), and Levene and colleagues (1995).

HEREDITARY DISORDERS WITH MENTAL RETARDATION

Degenerative Diseases of the Nervous System

Loss of previously acquired milestones is seen in encephalopathies of diverse origin. Often it is difficult to decide whether or not a condition is progressive. Neurological examination of retarded children is notoriously difficult. Clinical signs and age of onset may narrow the diagnostic options (Adams and Lyon, 1982; Swaiman, 1994), but often longitudinal observations supplemented by psychological assessments and electrophysiological and imaging studies are

Table 15. Examples of Disorders Leading to Intellectual Deterioration

Tumors (medulloblastoma, glioma)
Infections (subacute sclerosing panencephalitis)
Autoimmune postinfectious disorders (disseminated encephalomyelitis)
Chronic poisoning (lead, organic mercury)
Hereditary disorders
Neurocutaneous disorders (tuberous sclerosis)
Spinal and spinocerebellar degeneration (Friedreich's ataxia)
Lysosomal storage disorders (sphingolipidoses, (mucopolysaccharidoses)
Leucodystrophies (metachromatic leucodystrophy, adrenoleucodystrophy
Spongiform encephalopathies (aspartoacylase deficiency)
Neuroaxonal dystrophy (Schindler disease)
Huntington's chorea
Wilson's disease
Childhood autism (some cases)
Rett syndrome

necessary to demonstrate slowing of acquisition of skills and eventually regression. Regression is not the same as a drop in IQ. Slow progress when related to age may appear as a drop in IQ. Pathological processes may be active before birth and result in perinatal complications, which can mask the underlying disease process. Loss of previously acquired milestones is seen in encephalopathies of diverse origin (Table 15).

Lysosomal Disorders

The lysosomes are intracellular organelles surrounded by a single membrane and are found in all cells other than mature erythrocytes. They possess a range of specific hydrolytic enzymes active at an acid pH for the degradation of macromolecules such as the glycosaminoglycans (mucopolysaccharides) of connective tissue and the sphingolipids (i.e., sphingomyelin, gangliosides, cerebroside, and sulfatide) of brain. Proper functioning of the lysosomal enzymes requires firstly that newly synthesized polypeptides acquire signals that direct them—via the endoplasmic reticulum and Golgi network—to the lysosomes and, secondly, the existence of receptors that recognize the signals (Tager et al., 1994). Loss of activity is followed by accumulation of the molecules that cannot be broken down. In the brain, storage leads to mechanical distortion of cells and interference with their metabolic activity ultimately resulting in their distruction. Storage material is seen in neurons, neuroglial cells, and macrophages and may appear on electron-microscopy as characteristic structures.

Lysosomal mutations have been found to involve transcription of genes, processing of mRNA , transfer of the gene product from its site of synthesis to the lysosome, and the synthesis of activator or stabilizer proteins. There is great variability in age of onset and progression of these disorders, determined in large

Table 16. Lysosomal Disorders

Mucopolysaccharidoses	
	MPS I (Hurler, Scheie)
	MPS II (Hunter), X-linked
	MPS III (Sanfilippo), 4 variants
	MPS IV (Morquio), 2 variants, not retarded
	Mps VI (Maroteaux-Lamy), usually not retarded
	MPS VII (Sly)
Lipidoses	
	GM_1-Gangliosidosis
	GM_2-Gangliosidosis, 3 variants
	Gaucher's disease, cerebral variants
	Niemann-Pick disease type A and B
	Niemann-Pick disease type C
	Wolman disease
	Farber disease
	Fabry disease, X-linked, not retarded
Leucodystrophies	
	Metachromatic leucodystrophy
	Krabbe's disease
Mucolipidoses	
	Mucolipidosis II (I-cell disease)
	Mucolipidosis III (pseudo-Hurler polydystrophy)
Glycoproteinoses	
	Sialidosis
	Galactosialidosis
	Mannosidosis, 2 variants
	Fucosidosis
	Aspartylglucosaminuria
	Schindler disease
Other disorders	
	Pompe's disease (GSD II)
	Batten's disease (neuronal ceroid lipofuscinosis)
	Salla disease (transport defect)

measure by genetic heterogeneity. Over 40 lysosomal disorders have been identified with incidences mostly in the range of one in 25,000 to one in 100000, suggesting an overall incidence greater than about one in 2000. The heterozygote frequency of some lysosomal diseases is high in the Ashkenazi Jewish population and is attributed to the founder effect and, in some cases, to selective advantage (Motulsky, 1995). Selected lysosomal disorders are listed in Table 16.

Disturbances of ganglioside metabolism are often associated with nerve cell destruction expressed by psychomotor retardation, a cherry red spot on the macula, and later spasticity and paralysis. Disorders of the metabolism of sulfatide and cerebroside, important constituents of myelin, affect the peripheral and central nervous system, giving rise to peripheral neuropathy, spasticity, and ataxia. Seizures may occur, but usually in the later stages. In the mucopolysaccharidoses, storage in the skeletal system gradually results in a characteristic appearance and bony changes (dysostosis multiplex). When a stored substance has a high turnover in liver or spleen, visceromegaly ensues. Some disorders combining the clinical and biochemical features of sphingolipidoses and mucopolysaccharidoses have been termed mucolipidoses. The leucodystrophies involve primarily the white matter. Their etiology is by no means exclusively lysosomal. Primary leucodystrophy is a feature of several peroxisomal disorders and of Canavan–Van Bogaert disease, while the metabolic defect is as yet unknown in others (Aicardi, 1993). Features of leucodystrophy (dysmyelination) are also seen secondary to untreated phenylketonuria and in some mitochondrial encephalopathies.

Diagnostic tools include neuroimaging, particularly MRI, EEG, auditory and visual evoked potentials, electroretinograms, and nerve conduction studies. Metachromatic inclusions or vacuoles in lymphocytes or abnormal cells in bone marrow occur in several lysosomal disorders. Urine tests need to be interpreted with care (Hommes, 1991). Histochemical and electronmicroscopic studies are helpful when the biochemical changes are ill-defined or unknown. For a definitive diagnosis, identification, characterization and assay of the deficient enzyme are essential. Leucocytes or fibroblasts are mostly used in these assays.

The techniques of molecular biology increasingly contribute to our understanding of these disorders (Cotton, 1993; Forrest et al., 1995). For example, the coding sequence of the gene for Gaucher's disease (acid β-glucosidase deficiency) has been established and it is now often possible to predict the severity of the disease by identification of mutations at the DNA level. Thus, patients homozygous for mutation L444P are likely to have neuronopathic disease and mental retardation while, in patients homozygous for mutation N370S, the nervous system is not affected and some may remain free from major symptoms(Scriver et al., 1995). Similar problems arise with metachromatic leucodystrophy caused by deficiency of arylsulfatase A encoded by a gene on chromosome 22 (Gieselmann et al., 1994). About a dozen mutations have been identified. Of the two most frequent alleles, Int2SD a splice donor site mutation gives rise to an early onset and severe course, while P426L ($Pro^{426} \rightarrow Leu$) encodes an unstable, but still active, enzyme, resulting in low but detectable enzyme activity and a juvenile or adult form of the disease. In 0.5% to 2.0% of the population, a substantially decreased level of enzyme activity is not associated with risk of developing disease, for the amounts of enzyme are obviously sufficient to sustain a normal phenotype. This so-called pseudodeficiency creates problems in both diagnosis and genetic counseling. These have been largely overcome by characterization of the mutations involved.

While the pathophysiological mechanisms in the lysosomal disorders are well understood, treatment has proven difficult. Cells can take up enzymes by endocytosis into pinocytotic vesicles, which fuse with lysosomes. An exogenous enzyme therefore can take part in the degradation of a stored substance. Lysosomal enzyme deficiencies have thus been prime candidates for enzyme replacement therapy. Enzymes have been administered in red cells or liposomes as carriers and are sometimes modified by linking them to recognition markers. Results have been disappointing. An exception is nonneuronopathic Gaucher's disease. In this disorder, glucocerebroside accumulates preferentially in macrophages. Glucocerebrosidase (from human placenta) can be targeted to macrophages by modifying its oligosaccharide chains to create a mannose terminal. This approach is promising for other nonneuronopathic conditions, but additional problems arise when storage occurs primarily in the brain, particularly when the disease process is already active *in utero* as, for example, in Tay-Sachs disease. Within the past decade bone marrow transplantation has been widely used for treating storage disorders (Hoogerbrugge et al., 1995; Krivit et al., 1995; Shapiro et al., 1995; Walkeley and Dobrenis, 1995). Results suggest that in patients with Hurler syndrome (MPS I), Krabbe's disease (globoid cell leucodystrophy), and the peroxisomal disorder adrenoleucodystrophy, this treatment can be effective in at least delaying dementia if started early enough in the disease. Patients in the preclinical phase of metachromatic dystrophy may also benefit, but the treatment is not recommended for the Sanfilippo and Hunter syndromes. It is likely that gene therapy will soon be tried for some lysosomal disorders as, in animal experiments, bone marrow cells have been reconstituted with transformed cells. Current emphasis is on prevention. Screening, carrier detection, and antenatal diagnosis are offered to high risk populations in many countries. For recent reviews of lysosomal disorders refer to Neufeld (1991), Swaiman (1994), Duckett (1995), and Scriver and colleagues (1995).

Peroxisomal Disorders

The peroxisomes are subcellular organelles surrounded by a single membrane that occur in all cells other than mature erythrocytes. They are involved in the β-oxidation of a variety of fatty acids that are poorly handled by mitochondria and also in the synthesis of bile acids and ether phospholipids, notably the plasmalogens, which are widely distributed in mammalian cell membranes and are particularly abundant in brain, especially in myelin.

At least 17 human disorders are linked to peroxisomal dysfunction, and 15 of these involve neuropsychiatric deficits (Fournier et al., 1994; Scriver et al., 1995). Peroxisomal function is impaired if the organelle fails to be formed or maintained; or if there is hereditary deficiency of one or more peroxisomal enzymes. In the former case, import of enzymes synthesized outside the organelle is defective; in single enzyme defects the structure of the peroxisome is intact and consequences depend, as in other inborn errors, on the role of the enzyme affected (Table 17).

In the normal human brain, young neurons migrate to the cortical plate at between 7 and 10 weeks of gestational age. Waves of late migration must pass through layers already formed. Neurons that do not migrate normally are at risk of stunting or early death. Defects in fatty acid oxidation in mitochondria and perox-

Table 17. Peroxisomal Disorders

Generalized loss of peroxisomal function (impaired biogenesis)
Zellweger syndrome (cerebrohepatorenal syndrome)
Neonatal adrenoleucodystrophy
Infantile Refsum disease
Hyperpipecolic acidemia
Multiple loss of peroxisomal function
Rhizomelic chondroplasia punctata
Zellweger-like syndrome
Single peroxisomal enzyme defects
X-linked adrenoleucodystrophy
Acyl-CoA oxidase deficiency
Peroxisomal thiolase deficiency (pseudo-Zellweger)
Bifunctional enzyme deficiency
Trihydroxycholestanoyl-CoA oxidase deficiency
Pipecolic acid oxidase deficiency
Glutaryl-CoA oxidase deficiency
Dihydroxyacetonephosphate acyltransferase deficiency
Refsum's disease (phytanic acid storage disease)
Without nervous system involvement
Acatalasemia
Hyperoxaluria type I

isomes appear to interfere with neuronal migration. Absence of peroxisomes results in striking and characteristic disorders of migration in Zellweger syndrome; migrational defects and heterotopias are more variable in other peroxisomal disorders with impaired fatty acid oxidation. Sudanophilic leucodystrophy is found in Zellweger syndrome and is often most striking in neonatal adrenoleucodystrophy (Moser, 1989).

The clinical presentation of patients affected by a particular peroxisomal disorder is highly variable and much dependent on the age of presentation. Even well defined entities show wide phenotypic variation. Patients may present with a variety of abnormalities:

- Neurological abnormalities (e.g., hypotonia, fits, hearing loss)
- Ocular abnormalities (e.g., cataracts, retinal degeneration)
- Craniofacial abnormalities
- Skeletal abnormalities (e.g., calcified stippling, shortening of proximal parts of the limbs)
- Hepatological abnormalities (e.g., hepatomegaly, fibrosis, cirrhosis).

All patients are mentally retarded, none of the other signs is obligatory. In the laboratory, assay of plasma very long chain fatty acids is generally used as the primary test, supplemented if necessary by assay of phytanic acid and bile acid intermediates in plasma and plasmalogen in red blood cells. Assay of plasma pipecolic acid may be useful in some cases.

The most common peroxisomal disorder is X-linked adrenoleucodystrophy. Its gene encodes a peroxisomal integral membrane protein probably involved in the activation of very long-chain fatty acids, their import into the peroxisomes, or both. In about half of the patients, the disease presents in childhood, often with behavior problems and learning difficulties at school rapidly progressing to visual and auditory disturbances, quadroplegia, seizures, total disability, and death. In the CNS, demyelination is associated with an inflammatory response. The second most common form is adrenomyeloneuropathy, with onset in early adulthood and a protracted course that involves mainly the spinal cord and shows little or no inflammatory response. In both phenotypes, symptoms of adrenal insufficiency may precede, coincide with, or follow the onset of neurological involvement. Yet other phenotypes occur. These include adolescent and adult onset cerebral forms, Addisonian adrenal involvement only, and longterm presymptomatic cases. Furthermore, it has been found that identical mutations can give rise to mild and severe phenotypes sometimes even in the same nuclear family. It is thought that an autosomal modifier locus may have a major role in determining phenotypic variability for a given mutation (Moser, 1995). Treatment with monounsaturated fatty acids (Lorenzo's Oil), which reduce the rate of synthesis of very long-chain fatty acids, while at the same time restricting dietary intake, has a mild slowing effect on the disorder (Moser, 1995). Bone marrow transplantation may be of some benefit if done early enough in the disease (Krivit et al., 1995).

Mitochondrial Diseases, Lactic Acidosis, and Organic Acidurias

The mitochondrion is the site of the major energy-generating reactions of the cell. Aerobic oxidation of carbohydrate and some glucogenic amino acids proceeds via conversion of pyruvate to acetyl-CoA in the mitochondrion. This compound also lies on major pathways of amino acid and fatty acid oxidation. Acetyl-CoA enters the tricarboxylic acid cycle (TCA) in which it is oxidized to CO_2. The respiratory chain transfers electrons to oxygen from $NADH_2$ and $FADH_2$ formed largely in the TCA cycle and β-oxidation pathway of fatty acids. The proton electrochemical gradient thereby created is used to drive ATP synthesis.

The majority of defects in mitochondrial energy production are due to mutations affecting the pyruvate dehydrogenase (PDH) complex, fatty acid transport and β-oxidation, or the electron transfer complexes (ETC). There are five of these complexes comprising between them at least 64 protein subunits. The majority are encoded by nuclear DNA. However, about 1% of total cellular DNA is contained within the mitochondrion in the form of a circular molecule of 16569 bp (mtDNA), which encodes 22 tRNAs and 2 rRNAs. The polypeptides are all subunits of the ETC and ATP synthetase. Most disorders of mitochondrial function, notably

Table 18. Mitochondrial Disorders*

Defects of substrate utilization
- Pyruvate carboxylase deficiency
- Defects in the pyruvate dehydrogenase complex
- Defects in the β-oxidation of fatty acids (short-chain, medium-chain, long-chain, short-chain 3-hydroxy, long-chain 3-hydroxy, and multiple acyl-CoA dehydrogenase deficiencies

Defects of the Krebs cycle
- Fumarase deficiency
- 2-ketaglutarate dehydrogenase deficiency

Defects of the respiratory chain
- Complex I (NADH-CoQ reductase)
- Complex II (succinate-CoQ reductase)
- Complex III (CoQ-cytochrome c oxidoreductase)
- Complex IV (cytochrome oxidase)
- Complex V (ATP synthase)

Defects in mitochondrial transport
- Carnitine palmityl transferase deficiencies I and II
- Carnitine acylcarnitine translocase deficiency
- HHH syndrome (transport of ornithine)
- Methylmalonyl-CoA mutase (one variant)

Acquired mitochondrial dysfunction
- Drugs and poisons (zidovudine, MPTP)
- Reye's encephalopathy

Note: *See also Tables 19 and 20.

defects of the PDH complex and of β-oxidation of fatty acids, result from nuclear gene mutations. Some deletions and duplications of mtDNA are also the consequence of a primary nuclear gene abnormality. However, in a group of disorders the primary mutation is in mtDNA and these conditions present some unique features (Wallace, 1992; Wallace, 1993; Scriver et al., 1995).

Mitochondrial DNA is transmitted via the ovum with no contribution from sperm—in other words, mtDNA is maternally inherited. In most individuals all mtDNA molecules are the same (*homoplasmy)* whereas patients with mtDNA mutations usually have a mixture of mutant and wild-type molecules (*heteroplasmy*). The phenotype will depend on the extent of the defect and the dependence of different organs on mitochondrial oxidative phosphorylation, which decreases in this order: brain, skeletal muscle, heart, kidney, liver. The proportion and proliferation of mutant mtDNA may vary in different tissues: rapidly dividing cells like cultured fibroblasts may tend toward homoplasmy. The mutation rate of mtDNA genes is much higher than that of the nuclear DNA genes; many cases are sporadic, and the effect on cognitive development is variable.

Oxidation of pyruvate begins with its transport into the mitochondrion and conversion to acetyl-CoA by the PDH complex. PDH deficiency, particularly of the

E1a subunit of the complex, is the most common enzyme defect in infants with primary lactic acidosis and is also a cause of severe neurodegenerative disease with developmental anomalies. These arise before birth and affect structures that are forming at the time when the enzyme normally appears during cerebral development (Brown, 1994; Brown and Squier, 1996).The disease is mostly due to new germline mutations arising in one of the parents. The underlying mutation has been identified in a number of cases and found to be different in almost every case.

Selected mitochondrial disorders with high risk of mental retardation are listed in Table 18.

Diseases of the mitochondrion, particularly those of the respiratory chain and PDH, are diagnostically taxing (Morris et al., 1995). An unusual combination of affected tissues is a frequent finding in mitochondrial disease. Myopathy combined with an unrelated symptom is characteristic, as is progressive external ophthalmoplegia and myoclonus epilepsy. Traditionally, lactic acidosis has been regarded as the hallmark of mitochondrial disease. It may be a primary consequence of an enzyme defect as in PDH deficiency or of impaired aerobic metabolism, when anaerobic lactate production is required partially to sustain ATP levels. In practice, lactic acidosis may, however, be absent in some patients with PDH deficiency and in general is neither a specific nor a sensitive diagnostic indicator (Table 19).

Biopsy specimens, usually muscle, are required for specialized biochemical and histochemical tests. Even then, results may be equivocal. For example, in disorders in which they are found, the frequency and distribution of *ragged red fibers* (subsarcolemmal aggregates of abnormal mitochondria seen on Gomori trichrome staining) may vary widely. Neuropathological changes in these disorders are also highly variable. In some syndromes, myoclonic epilepsy with ragged red fibers (MERFF); mitochondrial encephalomyopathy, lactic acidosis, and

Table 19. Some Causes of Lactic Acidosis

Environmental
Shock, cardiopulmonary disease
Uremia
Liver disease
Acute infection, septicemia
Diabetic ketoacidosis
Drugs and poisons
Hereditary
Disorders of gluconeogenesis
Disorders of the pyruvate dehydrogenase complex
Disorders of the Krebs cycle
Disorders of the respiratory chain
Disorders of fatty acid oxidation
Organic acidurias

strokelike episodes (MELAS); or neurogenic muscle weakness, ataxia, and retinitis pigmentosa (NARP) mutation analysis is useful but may be complicated by heteroplasmy for the mutant mtDNA. Overall, diagnosis of these disorders is currently still imprecise in many cases, the pathophysiology not well understood, the natural history often unpredictable, and the treatmemt largely ineffective.

The outlook is more favorable for patients with defects in the β-oxidation of fatty acids. These disorders often present in childhood with recurrent attacks of hypoketotic hypoglycemia and acute encephalopathy, which may be complicated by hypertrophic cardiomyopathy. Abnormal organic acid and acylcarnitine patterns are found in the body fluids, but these may be much less pronounced between attacks (Hoffmann and Bremer, 1994; Pollitt, 1995). Fast atom bombardment tandem mass spectrometry (FAB-MS/MS) is useful in identifying abnormal patterns in these circumstances. Patients are managed by strict avoidance of fasting and, where appropriate, a low fat diet. Medium-chain triglycerides benefit patients with long-chain acyl-CoA and long-chain hydroxyacyl-CoA deficiencies.

Medium-chain acyl-CoA dehydrogenase deficiency (MCAD) is the most important of these disorders, with a particular mutation (A856G) being found in 90% of mutant alleles so far examined. The carrier frequency is high in Europe, being between one in 50 and one in 100, but only about 10% of homozygotes are affected by acute episodes. These may be fatal and, of the survivors, more than a third have neuropsychiatric sequelae (Scriver et al., 1995). Pathogenesis probably involves

- decreased capacity to produce energy from fatty acids
- impaired ketone body formation
- deficit of acetyl-CoA due to sequestration of CoA as acyl-CoA esters
- toxic effects of acyl-CoA intermediates
- neurotoxicity of free fatty acids, notably octanoic acid (Pollitt, 1995; Scriver et al., 1995)

Propionic acidemia, deficiency of propionyl-CoA carboxylase, a biotin dependent enzyme, is in its classical form only very rarely compatible with normal intellectual development. Propionyl-CoA, which accumulates rapidly, is highly toxic. It interferes with the citric acid and urea cycles, inhibits pyruvate dehydrogenase, leads to accumulation of odd-chain fatty acids, and, in most patients, leads to hyperammonemia and hyperglycinemia ("ketotic hyperglycinemia"). Most patients present in the first few days of life with severe ketoacidosis, but some have been mentally retarded without acute episodes, and a few individuals with severe enzyme defects have been found in family studies who were essentially asymptomatic (Scriver et al., 1995). *Methylmalonic acidemia*, deficiency of the B_{12}-dependent enzyme methylmalonyl-CoA mutase, can arise from an apoenzyme deficiency (mut^- and mut^0 variants), or from a defect in the transport or processing of its B_{12}-derived cofactor adenosylcobalamin (cbl variants). Clinically, the disorder resembles propionic acidemia in its variability from overwhelming neonatal illness in the majority of patients to freedom from major symptoms. Methylmalonyl-CoA has been implicated in pathogenesis. Some patients with cbl mutations respond to treatment with hydroxycobalamin. *Maple syrup urine disease* is caused by deficiency of branched-chain α-ketoacid dehydrogenase, a multienzyme complex located on the mitochondrial inner membrane. Mutations occur in at least four subunits giving rise to varying amounts of enzyme and resulting in presentation that may be as follows: acute neonatal ("classic"); chronic, with neurological symptoms and mental retardation; or intermittent mostly with normal intelligence. Many patients are compound heterozygotes, that is, genotype–phenotype correlations are not straightforward. In all three disorders, hemodialysis is the treatment of choice in the acute phase. Subsequent management is by restriction of the intake of pathogenic precursor amino acids, nutritional supplements, and antibiotics as appropriate. If treatment is started in the first few days of life, a minority of patients achieve near-normal intellectual development. However, mortality is high, and longterm complications are common. The treatment is arduous and creates major psychosocial problems in the families. Liver transplantation appears to be a promising alternative (Leonard, 1995).

The outlook is much better for patients with *multiple carboxylase deficiency* affecting all four mammalian carboxylases caused either by deficiency of holocarboxylase synthetase or biotinidase. In both variants, the nervous system is severely affected but most symptoms improve on treatment with biotin. Presymptomatic screening for biotinidase deficiency is operated in some countries. *Aspartoacylase deficiency* is a neurodegenerative disorder characterized by spongy degeneration of the brain and extensive breakdown of myelin, often referred to as Canavan disease. *N*-acetylaspartate, the substrate of the enzyme, is only formed in brain, and its function is not understood. Levels in brain and body fluids are greatly increased in the disorder. The enzyme has been purified and human cDNA and genomic clones isolated. Several mutations have been identified but the pathogenesis of the phenotype is still unclear (Matalon et al., 1993; Scriver et al.,1995). *Succinic semialdehyde dehydrogenase deficiency* on the metabolic pathway of the neurotransmitter GABA results in elevated levels of 4-hydroxybutyric acid in the body fluids and varying degrees of mental retardation and neurological impairment. Inhibition of the formation of 4-hydroxybutyric acid by the GABA transaminase inhibitor vigabatrin has been beneficial in some patients (Jakobs et al., 1993). *Glutaryl-CoA dehydrogenase deficiency* (glutaric aciduria type I) may be of acute or insidious onset with progressive head enlargement, dyskinesia and dystonia, cerebral atrophy, and basal ganglia pathology. The gene has been cloned and a number of mutations identified. Phenotypes vary in patients with the same mutation and clinical manifestations do not correlate with enzyme activities or metabolite levels (Scriver et al., 1995). Other organic acid disorders are listed in Table 20.

Table 20. Selected Organic Acidurias with Mental Retardation

Involving gluconeogenesis and pyruvate metabolism
(*See also* Tables 18 and 19)
Glucose-6-phosphatase deficiency (three genotypes, mental retardation preventable)
Fructose-1, 6-diphosphatase deficiency (mental retardation preventable)
Pyruvate carboxylase deficiency (two variants)
Phosphoenolpyruvate carboxykinase deficiency
Disorders of the pyruvate dehydrogenase complex
Involving amino acid derivatives
Maple syrup urine disease (branched chain amino acids, several genotypes and phenotypes)
Isovaleric acidemia (leucine)
3-Methylcrotonyl-CoA carboxylase deficiency
3-Methylglutaconic aciduria (leucine; multiple etiology)
3-Hydroxy-3-methylglutaryl-CoA lyase deficiency (leucine)
Mevalonic aciduria (leucine)
2-Methylacetoacetyl-CoA thiolase (isoleucine)
Acetoacetyl-CoA thiolase deficiency (isoleucine; mitochondrial, cytosolic, peroxisomal)
3-Hydroxyisobutyryl-CoA deacylase deficiency (valine)
Propionic acidemia (valine, isoleucine; pccA, pccB, and pccC mutants)
Methylmalonic acidemia (mut^0, mut^-, cblA, cblB, cblC, cblD, cblF mutants)
2-Ketoadipic aciduria (lysine; no causal link to mental retardation established)
Glutaric aciduria type I (lysine, tryptophan)
4-Hydroxybutyric aciduria (glutamic acid)
2-Oxoprolinuria (pyroglutamic aciduria; glutathione)
Miscellaneous
Multiple carboxylase deficiency (defects of holocarboxylase synthetase or biotinidase)
L-2-Hydroxyglutaric aciduria
Fumarase deficiency
2-Ketoglutarate dehydrogenase deficiency
Aspartoacylase deficiency
D-Glyceric aciduria (no causal link to mental retardation established)
Succinyl-CoA:3-ketoacid-CoA transferase deficiency (mental retardation preventable by treatment)
Orotic aciduria
Involving fatty acid oxidation
Mitochondrial β-oxidation of fatty acids (Table 18)
Peroxisomal oxidation of fatty acids: pipecolic acid, very long chain fatty acids, phytanic acid, pristanic acid, trihydrocholestanoic acid (*see also* Table 17)
No significant involvement of CNS
Primary oxaluria type I and II
Tyrosinemia type I (hepatorenal tyrosinemia)

Hyperphenylalaninemia and Other Aminoacidurias

The metabolic block in phenylketonuria (PKU) is in the hydroxylation of the essential amino acid phenylalanine to tyrosine by an enzyme complex consisting of the liver enzyme phenylalanine hydroxylase, a cofactor, tetrahydrobiopterin, and the enzyme dihydropteridine reductase, which is required to recycle the oxidized cofactor. As a result of the block, phenylalanine accumulates in the body fluids and some is diverted into minor metabolic pathways to form phenylpyruvic acid (hence the synonym "phenylketonuria") and other aromatic compounds. More than 98% of hyperphenylalaninemias are due to deficiency of the liver enzyme phenylalanine hydroxylase and have an incidence of about one in 10,000. The structure of the gene has been determined and over 200 mutations identified (Güttler and Zetterström, 1994; Scriver et al., 1995), most patients are compound heterozygotes. There is correlation between some genotypes and the metabolic phenotype. For example, the mutations R408W and IVS-12nt1 are associated with a high blood phenylalanine level at diagnosis, a low dietary tolerance for phenylalanine, and a high risk of severe mental retardation, whereas patients with mutation Y414C on one of their chromosomes have a lower blood phenylalanine at diagnosis, higher tolerance of dietary phenylalanine, and a somewhat lower risk of severe mental retardation. However, neither the mutation genotype nor the metabolic phenotype always correlate with the clinical phenotype as determined by cognitive development, as this also involves interaction with other genes and with the environment.

The pathogenesis of phenylalanine hydroxylase deficiency is not fully understood. Phenylalanine itself probably plays the leading role in pathogenesis. There is evidence that brain function is adversely affected by decreased dopamine synthesis. Cognitive abilities dependent on the prefrontal cortex appear specially vulnerable. Myelin metabolism is impaired in oligodendroglial cells, particularly the synthesis of sulfatide, with an increased turnover of myelin and a propensity to demyelination (Güttler and Zetterström, 1994; Scriver et al., 1995).

The devastating effects of untreated maternal phenylketonuria are well documented. They include microcephaly, mental retardation, and multiple congenital malformations. The maternal genotype is an environmental factor for the fetus; phenylalanine is a teratogen. The higher the maternal blood phenylalanine level, the greater the risk to the fetus. The risk is much lower for the offspring of mothers with blood phenylalanine levels below 400 μmol/l (Levy et al., 1994).

Neonatal screening and prompt treatment with a diet low in phenylalanine have dramatically reduced the number of phenylketonuric children who are mentally retarded. In Great Britain, current recommendations are that all infants with blood levels significantly above 360 μmol/l (six times the mean normal) should be treated, that treatment must start as soon as possible after birth and not later than age 21 days, and that blood levels should be held, as far as possible, within the range 120 to 360 μmol/l., though somewhat higher levels may have to be tolerated as patients approach adolescense. Phenylketonuric mothers with blood phenylalanine levels above 300 μmol/l should be treated, and levels should be controlled ideally within the the range 60 to 250 μmol/l with treatment planned to start, if possible, before conception. In all cases, whether severe or mild, outcome is closely related to the age at which treatment started, to the mean blood phenylalanine level during the first few years of life, and to the extent to which levels fluctuate about the mean (Beasley et al., 1994; Costello et al., 1994; Fisch et al., 1994; Ris et al., 1994). While early treated children are generally of normal intelligence, their IQ scores fall on the average half a standard deviation below that of their unaffected sibs. In addition, many have learning disabilities and psychiatric and behavioral problems also later in life. How factors other than the diet affect the outcome and the extent to which minor neuropsychiatric symptoms can be reversed is not yet clear. Sophisticated neuropsychological tests and noninvasive imaging techniques are now available to monitor the progress of patients.

Most patients with hyperphenylalaninemia lack the liver phenylalanine hydroxylase. However, 1% to 2% of patients have a deficiency of the cofactor tetrahydrobiopterin. This co-factor is also required for the hydroxylation of tyrosine and tryptophan to the neurotransmitter precursors 3,4-dihydroxyphenylalanine (dopa) and 5-hydroxytryptophan (5-HTP). The deficiency may be produced by impaired synthesis of the cofactor, most frequently by deficiency of pyruvoyl tetrahydrobiopterin synthase, or by deficiency of dihydropteridine reductase. Untreated patients develop progressive neurological disease with severe mental retardation, not arrested by the phenylalanine low diet. Replacement therapy is with dopa and 5-HTP protected by the (extracerebral) decarboxylase inhibitor carbidopa, with a phenylalanine low diet to protect patients from the effects of excess of phenylalanine, and with folinic acid supplements for patients with reductase deficiency and tetrahydrobiopterin for patients with defects in cofactor synthesis. The treatment is of some benefit but remains to be fully evaluated. Some mildly affected patients with synthase activity are heterozygotes with considerably lower enzyme activity than the expected value of half of the mean value for normal homozygotes, an example of *negative allelic complementation.*

By germline mutagenesis with *N*-ethyl-*N*-nitrosourea mouse models, ENU-1, ENU-2, and ENU-3 of phenylalanine hydroxylase deficiency have been created. ENU-1 has the characteristics of the milder forms of hyperphenylalaninemia, while ENU-2 and ENU-3 have those of the severe forms (Scriver et al., 1994). These models will be valuable in experimental studies of pathogenesis and the development of effective somatic gene therapy.

Excessive excretion of homocystine (*homocystinuria*) is found in several inborn errors of metabolism of which *cystathionine β-synthase deficiency* is the most important (Kraus, 1994; Scriver et al., 1995). Patients have fine, fair hair, a malar flush, skeletal abnormalities reminiscent of Marfan's syndrome, and characteristic dislocation of the lenses. The defect results in a block in the transsulfuration pathway from methionine to cystine, accumulation of methionine, homocysteine, and their metabolites in the body fluids, and excessive excretion of homocystine and other sulfur amino acids in the urine. Homocystinuria carries a high risk of thromboembolic crises, which often occur after minor surgery. Fewer than half the patients with homocystinuria are severely retarded. Epilepsy and a variety of psychiatric disorders are found in many patients including some who are only mildly retarded or intellectually normal.

The disorder is genetically heterogeneous, in nearly half the patients, the biochemical abnormality can be corrected to a large extent by pyridoxine (100-500 mg/day). The cDNA of the gene has been cloned, the gene sequenced, and 14 mutations identified (Kraus, 1994). Most patients are compound heterozygotes. Unexpectedly, some patients with undetectable enzyme activity are pyridoxine responsive. Again, there is lack of correlation between genotype and phenotype, and an identical genotype does not result in the same phenotype even within a family.

Cerebral thromboses probably contribute significantly to the neuropsychiatric symptomatology. Homocysteic acid is present in excess in the body fluids. It is an excitatory amino acid that may act in the nervous system by overstimulation of the *N*-methyl-D-aspartic acid (NMDA) receptor (Lipton and Rosenberg, 1994). Homocysteic acid is also implicated in the thromboembolic crises. The risk of these crises is greatly increased in patients who also carry the factor V Leiden mutation.

Treatment of the disorder with pyridoxine in responders and with a diet low in methionine supplemented with cystine, folate and betaine in nonresponders is beneficial. The effects of the treatment on the longterm sequelae of the disorder, particularly dislocation of the lenses, thromboembolisms, and osteoporosis, are not yet known.

In man, hyperglycinemia occurs with ketosis in some organic acidurias. This hyperglycinemia is called ketotic. Hyperglycinemia without ketosis, *nonketotic hyperglycinemia*, is associated with deficiency of the glycine cleavage system consisting of four protein components and being situated at the inner mitochondrial membrane of liver, brain, kidney, and placenta. It is the almost complete absence of enzyme activity in brain that results in extreme hypotonia, myoclonic fits, apneic episodes, and a characteristic electroencephalogram with paroxysmal bursts on an almost flat record. Symptoms start within hours or days of birth; more than half the affected infants die in the first month. Survivors are nearly always severely retarded. A significant finding is an elevation of the ratio of the glycine level of CSF to that of plasma. A few atypical milder and late onset cases have been described.

The role of glycine as an inhibitory neurotransmitter in the brain stem and spinal cord is well established. However, glycine is also an excitatory allosteric modulator for the *N*-methyl-D-aspartic acid glutamate receptor channel so that high levels of glycine may enhance excitotoxic events (Tada and Kure, 1993; Lipton and Rosenberg, 1994; Scriver et al., 1995). Antagonists of the NMDA receptor such as dextromethorphan have been used with some success in reducing seizure frequency and normalizing the EEG. Benzoate, which conjugates with glycine to form hippurate, has also been used. No treatment has as yet made any impact on the developmental progress of patients. The pathogenetic process is probably

Table 21. Some Disorders of Amino Acid Metabolism in Relation to Mental Retardation

Untreated disorder generally associated with mental retardation
Phenylketonuria
Homocystinuria (cystathionine β-synthase)
Methylenetetrahydrofolate reductase deficiency with homocystinuria
Methionine synthetase deficiency with homocystinuria (cobalamin E or G disease)
Urea cycle enzymes (see Table 12)
HHH syndrome (see Table 12)
Nonketotic hyperglycinemia
Hypervalinemia (transamination defect)
Hyperleucine-isoleucinemia (transamination defect)
Sulphite oxidase deficiency
Glutathione synthetase deficiency (5-oxoprolinuria)
γ–glutamyl transpeptidase deficiency
ß-alaninemia
Lowe's syndrome (transport defect)
No causal relationship to mental retardation established
Histidinemia
Sarcosinemia
Cystathioninuria
Hyperprolinemia type I and type II
Hydroxyprolinemia
Carnosinemia, homocarnosinosis
Hyperlysinemia
Hartnup disease (transport defect)

active before birth when the NMDA channel is particularly active, enhancing the vulnerability of the perinatal brain.

Other disorders of amino acid metabolism in relation to mental retardation are shown in Table 21.

It should be pointed out that, while in disorders such as *histidinemia* or *Hartnup disease,* a causal relationship to mental retardation has not been established, those affected appear more *vulnerable* to metabolic or nutritional hazards.

Purines and Pyrimidines

The *Lesch-Nyhan syndrome*, an X-linked recessive disorder, is characterized by compulsive self-mutilation, choreoathetosis, spasticity, gout, and severe mental retardation. The enzyme affected, hypoxanthine-guanine phosphoribosyl transferase (HPRT), plays a vital part in purine salvage, the recycling of purine nucleotides from purine bases. Its deficiency results in overproduction and urinary excretion of purines (e.g., uric acid), urolithiasis, and ultimately renal failure. Treatment with allopurinol alleviates most of the symptoms of the disorder except those affecting the nervous system. The gene has been sequenced and many mutations identified. There is broad correlation between the severity of the symptoms and the extent of the enzyme defect. At one end of the spectrum are patients with very low enzyme levels and the full syndrome; at the other end are those with relatively high enzyme levels (>8% of normal) solely with hyperuricemia and its consequences; those with intermediate enzyme levels have good mental function and behavior but neurological deficits (Nyhan and Sakati, 1987; Gresser, 1994; Scriver et al., 1995). Assay of the enzyme, antenatal diagnosis, and identification of carriers are all well established.

The pathogenesis of the involvement of the nervous system is not understood. There is some evidence that the nigrostriatal dopaminergic pathways are implicated. It may be significant that synthesis of serotonin and dopamine precursors requires a purine derived cofactor. The disorder is a prime candidate for gene replacement therapy. Progress has been made in animal and *in vitro* experiments, including the transfer of the entire genomic HPRT gene contained within a yeast artificial chromosome into mouse cells. Patients will in due course benefit from these advances.

Adenylosuccinase deficiency is an autosomal recessive disorder of *de novo* purine synthesis and the purine nucleotide cycle in which two compounds, succinyladenosine and succinyl-aminoimidazole carboxamide ribotide, accumulate in greatly increased amounts. Most patients have shown severe psychomotor retardation, epilepsy, and autistic features, whereas some have been less severely retarded. Two out of four documented inborn errors of pyrimidine metabolism can affect the nervous system. Mild mental retardation is an inconstant feature of *orotic aciduria*, or uridine-5-monophosphate synthetase deficiency. Patients treated with uridine have remained apparently well, free from the obstructive uropathy and macrocytic anemia characteristic of this disorder. However, the possibility of late neuropsychiatric changes has not yet been excluded. *Dihydropyrimidine dehydrogenase deficiency* is accompanied by urinary excretion of large amounts of thymine, uracil, and 5-hydroxymethyl uracil. The phenotype is as yet ill-defined, but many patients present with epilepsy, psychomotor retardation, and autistic features, singly or in combination, while others are asymptomatic. Dihydropyrimidine deficiency may trigger or aggravate the neurotoxic effects of fluorouracil, a pyrimidine antagonist used in the treatment of cancer. (For reviews, see Gresser, 1994; Scriver et al., 1995.)

Dynamic Mutations—Unstable DNA

Until recently, anticipation, or the worsening of disease phenotype in succeeding generations, was dismissed as due to ascertainment bias (Teisberg, 1995). Classical genetics was predicated on stable and inheritable mutations. It has been known for some years that, in the human genome, runs of simple sequence repeats of DNA occur that are widely used as genetic markers and for finger printing. Normally, these sequences produce no harmful effect, but genetic instability results from runs of triplet repeats, which increase in size between generations, a phenomenon termed "dynamic mutations". In general, the larger the number of repeats the more severe the disease or the earlier the onset. The *fragile X syndrome* has a frequency of about one in 1250 males and one in 2500 females. It is recognized as the commonest genetic cause of mental retardation after Down syndrome. A fragile site on the X chromosome, FRAXA, was first observed in chromosome preparations more than 25 years ago. It is associated with a mutation of a gene, *FMR1*, at the FRAXA site. This mutation involves an increase in length of a stretch of CGG tandem repeats in the 5' untranslated region of the gene. The trinucleotide repeat number is highly polymorphic. In normal individuals, it varies from six to 60 repeats. If the number of repeats exceeds 60, it is unstable when transmitted to the next generation. Repeat numbers between 60 and 200 constitute a premutation, which occurs in unaffected transmitting males and unaffected carrier females. Amplifications to more than 200 repeats result in a full mutation with appearance of the cytogenetic abnormality and the clinical phenotype, which presents varying degrees of mental retardation, minor dysmorphic features affecting particularly the face, connective tissue dysplasia, and macroorchidism. More than half the females who carry the full mutation are mentally retarded. The length of a full mutation is also unstable during cell divisions so that mosaicism may occur in the cells of affected individuals.

Affected males and to a lesser extent affected females exhibit behavior problems, which may include gaze aversion, hyperactivity, attention deficit disorders, and social dysfunction. The behavioral phenotype thus resembles that of autism, albeit with subtle differences (Dykens, 1995). The degree of mental impairment is variable, and there is a significant inverse correlation between IQ and both CGG repeat number and percent methylation of the CpG island promoter of the FMR1 gene. Methylation of the promoter is associated with transcription suppression and absence of the encoded protein FRMP. This protein occurs in at least four to five isoforms. High expression is found in brain, in neurons rather than in glial cells, and in several epithelia and spermatogonia. The function of FRMP is so far unknown, but progress is being made with molecular diagnostic techniques for family studies and reliable antenatal diagnosis. (For recent reviews of this syndrome, refer to Symposium, 1994; Scriver et al., 1995.)

The X chromosome occupies a unique position in mental retardation studies. Half the genes assigned to the X chromosome are disease-related compared to less than a third for a comparable autosome, and most diseases caused by an X-linked mutant gene include mental retardation as one of their manifestations. At a recent count (Symposium, 1994), there were no fewer than 127 X-linked mental retardation syndromes and, of these, 22 had mental retardation as the only consistent finding. Identifying these genes and their products should assist in discovering how genetic factors affect intelligence.

Myotonic dystrophy is an autosomal dominant disorder characterized by myotonia, a wide spectrum of severity of manifestation, and anticipation, with earlier onset and more severe symptoms in successive generations. The gene is expressed in brain and muscle. It codes for an unidentified protein kinase. The mutation is the expansion of a CTG repeat near the 3' end of the gene. Normal individuals have fewer than 30 copies of this repeat and minimally affected patients have 50 to 80 copies. Typically affected adult onset patients who may be mildly mentally retarded have 100 to 500 copies, while patients with the severe congenital form of the disorder with hypotonia, delayed motor development, and mental retardation have 500 to 2000 copies (Scriver et al., 1995).

It is not widely known that *Huntington's disease* may present in children with progressive mental retardation as a prominent clinical sign. The mutation involves amplification of a CAG trinucleotide repeat at the 3' end of the gene on chromosome 4. There is a significant inverse correlation between the number of repeats and the age of onset.

OTHER METABOLIC DISORDERS

Galactosemia

Classical galactosemia, caused by galactose-1-phosphate uridyl transferase deficiency, presents with failure to thrive, acute liver disease, and cataracts, often aggravated by *E. coli* septicemia. Survivors are severely retarded. Timely treatment with a galactose-free diet can correct these complications and is often lifesaving. However, even when treated early, patients with the most common mutation, Q188R, develop significant cognitive deficits, neurological signs, and neuropathological abnormalities, which are detectable by MRI (Gitzelmann and Steinmann, 1995). These deficits do not correlate with the age at diagnosis nor with blood levels of galactose-1-phosphate, widely used to monitor dietary control. It has recently been demonstrated that exclusion of galactose from the diet is frustrated by endogenous galactose synthesis, which occurs in patients at the same rate as in normals (Holton and Leonard, 1994; Berry et al., 1995). The response to treatment of some patients with mutations other than Q388R and with some residual enzyme activity is more favorable. Genetic counseling and antenatal diagnosis can offer at-risk couples the benefit of informed choice in planning their families.

Smith-Lemli-Opitz Syndrome

Smith-Lemli-Opitz syndrome is an autosomal recessive disorder with multiple congenital anomalies and mental retardation showing considerable clinical variability. Common manifestations include microcephaly, ptosis, characteristic facies, limb abnormalities, cataracts, and hypogenitalism. These patients have a deficiency of 7-dehydrocholesterol reductase on the cholesterol synthetic pathway leading to elevated plasma 7-dehydrocholesterol, decreased plasma cholesterol levels, and abnormalities in urinary bile acids (Opitz and de la Cruz, 1994). Cholesterol has a key role in the formation of the plasma membrane and outer mitochondrial membrane of all cells, and in myelination. The normal fetus can synthesize cholesterol. In man, myelination starts in the third trimester of pregnancy and subcortical myelination after birth. However, the fetus must require cholesterol much earlier during morphogenesis and histogenesis for the formation of membranes in rapidly dividing cells. It is still debated how and to what extent maternal cholesterol can cross the placenta and blood–brain barrier and compensate for the failure of the fetus to synthesize this key metabolite. Trials of dietary treatment with supplements of cholesterol and bile acids are in progress (Salen et al., 1996).

Pelizaeus-Merzbacher Disease

Proteolipid protein (PLP) is the major myelin protein of the CNS. Mutations in the gene encoding PLP occur in several species and have severe pleiotropic effects on myelination. In man, they give rise to Pelizaeus-Merzbacher disease, an X-linked heredodegenerative disease exhibiting dysmyelination and abnormal myelin periodicity. Age of onset, severity, rate of intellectual deterioration, and neuropathological expression are variable (Griffith et al., 1995; Harding et al., 1995). A spectrum of mutations within the PLP structural gene has been described. It includes missense mutations, frame shifts, and deletion of the entire gene. The majority of cases of Pelizaeus-Merzbacher disease are associated with lack of the functional PLP gene, but duplication of the region containing the PLP gene can also cause the disease implying that increased gene dosage too can be associated with dysmyelination (Harding et al., 1995). A form of X-linked spastic paraplegia (SPG2) is also associated with point mutations of the PLP gene, while mutation of the peripheral myelin protein gene PMP22 on chromosome 17 gives rise to the peripheral neuropathy Charcot-Marie-Tooth disease, which also appears to be caused by too much *or* too little of the myelin protein involved.

Muscular Dystrophy

Dystrophin, a large cytoskeletal protein, is encoded by a gene on the short arm of the X chromosome. Deletions, duplications, and point mutations predominantly result in lack of dystrophin. If they involve a frameshift, they generally result in severe *Duchenne muscular dystrophy* (DMD). The allelic, milder *Becker muscular dystrophy* usually involves a reduction in size or quantity of dystrophin produced in most cases by in-frame mutations of the gene. Dystrophin occurs in five isoforms. At least two of these are expressed in brain where they may have a role in neurotransmission. Psychometric studies have consistently shown mean IQ levels of 80 for patients with Duchenne muscular dystrophy. The retardation is not progressive. The incidence of DMD is one in 3500, and nearly one-third of cases are new mutations. The muscle protein utrophin can replace dystrophin in some functions in the muscle cell and is not affected in DMD. Upregulating of the endogenous utrophin gene has been suggested as a means of influencing the progress of the disease; somatic gene therapies are being developed.

The congenital muscular dystrophies are a heterogeneous group of disorders. About half the patients have a deficiency of merosin, a protein on the outside of the muscle membrane encoded by a gene on chromosome 6. The gene is also expressed in oligodendroglia and Schwann cells. The phenotype is severe. It affects skeletal and cardiac muscle and the central and peripheral nervous system. Mental retardation is often present and at times is quite severe. (The muscular dystrophies are reviewed by Dubowitz, 1995; Scriver et al., 1995.)

Trace Metals

Menkes' disease is a very rare X-linked disorder of copper transport. Clinical signs include abnormal ("kinky") hair and facies, changes in bone and blood vessels, hypopigmentation, rapid neurological deterioration, and severe mental retardation. The disorder is produced by deficiency of a P-type ATPase, which transports copper across cell membranes (Vulpe et al., 1993). Absorption of copper from the intestinal tract is impaired, as is intracellular transport of copper leading to low levels in mitochondria. In brain, copper is trapped in astrocytes and fails to reach the neurons (Kodama, 1993). Most clinical features are attributable to impaired formation of the copper-containing enzymes cytochrome oxidase, lysyl oxidase, and tyrosinase. Treatment has been largely ineffective, but antenatal diagnosis is well established.

Wilson's disease is a recessive disorder with an incidence of about one in 100,000 in which copper accumulates in liver, brain, kidney, heart, and cornea where it produces the classic Kayser-Fleischer rings. Patients may present with liver disease, hemolytic anemia, psychiatric problems, or commonly neurological symptoms largely attributable to dysfunction of the basal ganglia. Biliary excretion of copper and incorporation into ceruloplasmin are severely impaired. The accumulation of copper in the liver leads to progressive liver damage and overflow of copper to the brain, cornea, and other organs. The order in which organs are affected and the response to the toxic excess of copper varies greatly from patient to patient. The Wilson's disease gene also encodes a copper transporting P-type ATPase. More than 25 mutations have been identified They include small insertions and deletions, and missense, nonsense, and splice mutations (Thomas et al., 1995). In most families, normal, heterozygous, and homozygous individuals can now be reliably identified. Phenotypes vary widely, for even identical twins may not be concordant for neurological or hepatic presentation and age of onset (Walshe, 1995). Most patients' symptoms can be reversed by achieving and monitoring a negative copper balance. For this, penicillamine is widely used as chelating agent. Triethylene tetramine dihydrochloride ('trientine') is a superior, less toxic chelating agent but not widely available (Walshe,1995).

The enzymes sulfite oxidase, xanthine oxidase and aldehyde dehydrogenase share a cofactor in which molybdenum is linked to a pterin, molydopterin. *Molybdenum cofactor* deficiency is associated with severe mental retardation, fits, dislocation of the lenses, and xanthine stones. The neurological and neuropathological findings closely resemble those seen in isolated sulfite oxidase deficiency.

Mental retardation, as the result of an acute *lead encephalopathy,* is rare in the developed world. There is, however, continuing concern about impaired intellectual development and disturbed behavior as a result of low level lead exposure associated with blood lead levels below 20 μg/dl. Taylor (1991) concluded from a metaanalysis of 19 studies that there is a real effect on intelligence, albeit small in comparison to that of other environmental factors determining intelligence. The timing, prenatal or postnatal, of the exposure, and the synergistic action of other heavy metals may be important in determining the effect on childhood cognition (Lewis et al., 1992). The upper limit of blood lead levels considered acceptable has been lowered to 10 μg/dl (0.48 μmol/l). Meso-2,3-dimercaptosuccinic acid ("succimer") is likely to become the drug of choice for the treatment of low level plumbism (Norman and Boardley, 1995). A multicentered, randomized, placebo controlled, double-blind intervention trial is now on the way in the United States to establish the effects on cognition and behavior of treating with succimer children with blood levels below 45μg/dl (2.17 μmol/l).

Congenital Hypothyroidism

In the mentally retarded, abnormal endocrine function is more often part of a mental retardation syndrome rather than the cause of the mental defect. An exception is congenital hypothyroidism. In experimental animals, thyroid deficiency *in utero* or early postnatal life adversely affects brain growth, myelination, dendritic arborization, and the development of neurotransmitter systems. In man, congenital hypothyroidism may be caused by maldevelopment or maldescent of the thyroid, inborn errors of metabolism affecting the thyroid hormones or their receptors, ingestion of goitrogens, or iodine deficiency (Burrow et al., 1994; Scriver et al., 1995). Endemic cretinism due to iodine deficiency is by far the most common cause of mental retardation in the Third World; the result of the combined effects of maternal and fetal hypothyroidism. Both, are in principle, preventable by enriching the diet with iodized salt or oil (Hetzel, 1994). In economically advanced countries, the incidence of congenital hypothyroidism is about one in 4000 births. Agenesis or dysgenesis of the thyroid account for more than 85% of cases, inborn errors of metabolism, and defects at the level of the pituitary or hypothalamus make up the remainder (Scriver et al., 1995). Clinical signs of hypothyroidism are not obvious in the majority of affected infants. In the absence of screening, the diagnosis is, therefore, easily missed.

Neonatal screening and prompt and vigorous treatment have reduced intellectual impairment. However, before birth there is only limited transfer of thyroid hormones from mother to fetus, affording only incomplete protection from the effects of fetal hypothyroidism. Even with adequate treatment, patients with plasma levels for thyroxine at diagnosis below a threshold value of between 40 and 50 nmol/l show a mean deficit of about 10 IQ points at school age. Minor neurological signs are also common in these children (Kooistra et al., 1994; Tillotson et al., 1994). The optimum conditions for effective replacement therapy are not clearly defined. A higher dose of L-thyroxine may be superior for intellectual development but may result in behavioral problems, and periods of sensitivity to thyroid hormone for verbal and nonverbal skills may not be identical (Grant, 1995; Rovet and Ehrlich, 1995).

CHROMOSOMAL DISORDERS

Chromosomal anomalies are a major cause of severe mental retardation (Table 22). These anomalies may be numerical or structural. Structural changes in chromosomes may be brought about by a variety of agents including radiation, toxic substances, and viral infections. They occur as *deletions* involving the loss of a chromosome fragment following chromosome breakage. Deletions may be terminal or interstitial, or may result in a ring chromosome. *Inversions* result from two breaks on the same chromosome and inversion of the intervening segment; they can be detected by banding techniques. *Translocations* follow the break-up of two chromosomes with transfer and fusion of the fragments to each other. A translocation is balanced if the amount of genetic material is presumed to be identical to that in a normal cell. *Robertsonian translocations* occur between two acrocentric chromosomes resulting in the formation of a new metacentric chromosome. Among abnormalities of chromosome number, only trisomies play an important part in the etiology of mental retardation. Even so, they contribute more to fetal loss than to live births with mental retardation. About three-quarters of cases of trisomy 21 (Down syndrome) are miscarried or stillborn, whereas the proportion of other aneuploidies lost by miscarriage is 90% with the notable exception of the trisomies of the sex chromosomes 47,XXX; 47,XXY; and 47,XYY.

Virtually all disorders involving loss or addition of a whole or of a significant part of a chromosome are associated with moderate or severe mental retardation. In addition, chromosomal disorders generally present a pattern of dysmorphism affecting several organs, growth disorders, and endocrine abnormalities. While each syndrome is associated with a recognizable pattern of phenotypic features, not all features are present in every patient, and some are found in more than one chromosome-linked disorder (Jones, 1988). Chromo-

Table 22. Selected Chromosomal Abnormalities

Major autosomal abnormalities
Trisomy 8 (Warkany syndrome)
Trisomy 13 (Patau syndrome)
Trisomy 18 (Edwards syndrome)
Wolf-Hirschon syndrome (4p-)
Cat's cry syndrome (cri-du-chat syndrome, 5p-)
Sex chromosomal abnormalities
Monosomy X (Turner syndrome), rarely retarded
XXX, mildly retarded or normal
XXY (Klinefelter syndrome), mildly retarded or normal
XXXY, more severely retarded
XYY, normal or mildly retarded
Contiguous gene syndromes
Langer-Giedion syndrome (8q-), intelligence very variable
Prader-Willi syndrome (15q-), paternal origin or uniparental maternal disomy
Angelman syndrome (15q-), maternal origin or uniparental paternal disomy
Rubinstein-Taybi syndrome (16p-)
Miller-Dieken syndrome (17p-)
Smith-Magenis syndrome (17p-)
Williams-Beuren syndrome (7q-)
DiGeorge syndrome (22q-), most patients mildly or moderately retarded

some banding techniques allow identification of individual chromosomes and the detection of most structural abnormalities, including translocations, duplications, and deletions (Rooney and Czepulkowski, 1992). Recently, banding techniques have been refined to allow the detection of rearrangements down to 1 to 2 megabases.

In studies of mildly retarded patients, the frequency of chromosomal abnormalities has ranged from 3% to 19%. In some cases, the causal relationship between the abnormalities and the mental retardation was not certain (Göstason et al., 1991). Mental retardation is mild or absent in the most common disorders of the sex chromosomes—the 45,X; 47,XXX; 47,XXY; and 47,XYY syndromes— and is more severe in the polysomies of the X and Y chromosomes. There has long been a suspicion that subtle submicroscopic chromosomal abnormalities may account for a proportion of "unclassified" mental retardation. Flint and colleagues (1995) postulated that the terminal region may be enriched for cryptic chromosomal abnormalities. Using highly informative variable tandem repeat probes, they surveyed 28 subtelomeric regions and found three deletions in a group of 99 patients with varying degrees of mental retardation but with no obvious chromosomal abnormalities as judged by routine karyotype analysis. This suggests that the contribution of chromosomal disorders to the prevalence of mental retardation is currently underestimated.

Down Syndrome

Down syndrome is the most common genetic cause of mental retardation. It results from a complete or partial triplication of chromosome 21. Maternal nondisjunction appears to be responsible for 95% of cases, nearly 80% occurring in meiosis I. Robertsonian translocations, 21q isochromosomes, and mosaics are much rarer. The causes of nondisjunction have not yet been conclusively identified. A variety of developmental abnormalities include a characteristic facies, microbrachycephaly, epicanthic folds, Brushfield's spots on the irides, transverse palmar creases, and congenital defects of gut and heart. Hypotonia is a constant finding. Brain weight is reduced, particularly that of the cerebellum and brain stem. Histopathological and morphological changes have been described but are not consistently found. Most reported metabolic abnormalities have been neither pronounced nor persistently present. Recurrent infections have been attributed to impaired immune responses. An increased incidence of leukemia and of self-remitting "leukemoid" states is associated with the disorder as is an increased incidence of congenital hypothyroidism and of thyroid autoantibodies. Brains of patients older than 30 to 40 years show histological changes also seen in Alzheimer's disease, namely, neurofibrillary tangles and senile plaques, suggesting that premature or accelerated aging is part of the pathogenetic process.

Theoretically, in the absence of gene dosage compensation, all genes on chromosome 21 could contribute to the Down phenotype. From investigations of patients in whom only part of chromosome 21 is triplicated it appears that most features of the Down phenotype can be replicated by a region in band 21q22, the so-called "DS critical region", which may contain up to 100 genes. The gene for amyloid precursor protein and a gene for a rare familial form of Alzheimer disease have been assigned to chromosome 21 as has the gene for the calcium-binding protein S100β, which is involved in neuronal growth and differentiation. The role of these proteins in the pathogenetic process has not yet been elucidated.

The mental retardation of patients is primarily determined by the aneuploidy but is also affected by parental genes, the environment in which the patient is reared, and the early onset of neuronal degeneration of Alzheimer type. This makes the assessment of the longterm effects of any treatment difficult. The pattern of cognitive deficits in Down syndrome differs from that seen in other disorders with mental retardation, especially in the development of language. Insight into the nature of the cognitive impairment might be gained from a study of exceptional cases who do not show common domain-specific deficits (Rondal and O'Connor, 1995) and from animal models such as the Ts65Dn mouse with segmental trisomy at dosage imbalance for genes corresponding to those on human chromosome 21q21-22.3, which includes the "DS critical region". These mice exhibit impaired performance in complex learning tasks (Reeves et al., 1995).

The association between Down syndrome risk and increasing maternal age has long been known, but most Down syndrome births are to mothers aged under 35 years. Antenatal screening by assay of α-fetoprotein, β-human chorionic gonadotrophin, unconjugated estradiol, pregnancy-associated plasma protein, or a selection of these serum markers supplemented by expert ultrasonography will identify the great majority of affected fetuses. Tests on fetal cells harvested from the maternal circulation will in due course provide a highly sensitive and specific screen. (For recent reviews of Down syndrome and related problems, refer to Epstein, 1986; Symposium, 1990; Scriver et al., 1995.)

Contiguous Gene Syndromes

The term "contiguous gene syndrome" (Schmickel, 1986), refers to clinically recognizable disorders involving several unrelated genes sufficiently physically close to each other on a given chromosomal segment for their function to be affected by deletions or duplications. These may be submicroscopic, detectable only by molecular cytogenetic techniques such as fluorescent *in situ* hybridization (FISH). Disease genes may be interspersed with other genes whose dosage has no effect on the phenotype. Most of these disorders occur sporadically and, in some cases, mutations of individual genes located in the segment involved occur, giving rise to disorders with a Mendelian mode of inheritance.

X-linked contiguous gene syndromes have been described on both the short and long arms of the chromosome. The Xp22.3 and Xq21 syndromes include mental retardation genes. Mental retardation is also a frequent finding in the Xp21 syndrome, which includes among others the genes for adrenal hypoplasia congenita, glycerol kinase, Duchenne muscular dystrophy, and ornithine transcarbamylase. The *Smith-Magenis syndrome* includes brachycephaly, dysmorphic features of face and limbs, speech delay, behavior problems, and mental retardation. Patients have deletions at 17p11.2. In the *Rubinstein-Taybi syndrome*, which is constituted by mental retardation, maxillary hypoplasia and other facial defects, broad thumbs and toes, and cryptorchidism, deletions at 16p13.3 are responsible for at least a significant proportion of cases. The *Miller-Dieken syndrome* is a multiple malformation syndrome with lissencephaly. Patients have deletions at 17p13.3 and one gene within this region may be involved in a signal transduction pathway crucial for cerebral development. It is likely that the *Williams-Beuren syndrome*, which involves supravalvular aortic stenosis, facial anomalies, transient infantile hypercalcemia, and mental retardation, but with preservation of some social and verbal skills, is also a contiguous gene syndrome. A deletion at 7q11.2 involving the elastin gene has been implicated.

If a deletion occurs in a region where one or more genes are differentially expressed ("imprinted"), different syndromes may result from similar deletions on the maternally or paternally derived chromosomes of a pair. *Prader-Willi syndrome* features obesity, short stature, hypogonadism, and mental retardation, and the *Angelman syndrome* presents epilepsy, ataxia, paroxysms of laughter, and mental retardation. The microdeletions that appear in the 15q11-13 region in two-thirds of patients with Prader-Willi syndrome are detectable by FISH. The deletions always occur on the paternally derived chromosome, normally as a sporadic event. In most of the remaining cases, both number 15 chromosomes are inherited from the mother—a phenomenon called unimaternal disomy. Angelman syndrome is caused by similar microdeletions on the maternally derived chromosome 15 or by unipaternal disomy 15. Presumably, a gene or group of genes is expressed only from the maternal chromosome resulting in Angelman syndrome when silenced, whereas another gene or group of genes is expressed only from the paternal chromosome resulting in Prader-Willi syndrome when repressed. Methylation is involved in these processes and parent-of-origin–specific methylation has in fact been used diagnostically. When the Prader-Willi syndrome is caused by maternal disomy, both 15 chromosomes are inherited from the mother (heterodisomy). Why the two maternal chromosomes, while functioning normally in the mother, produce Prader-Willi syndrome in the offspring is puzzling. Presumably, in this instance, complementary contributions from both parents are necessary for normal growth and development. (Clinical and cytogenetic aspects of the contiguous gene syndromes are discussed by Borgaonkar, 1994; Swaiman, 1994; and Scriver et al., 1995.)

CONCLUSIONS

Mental impairment is caused by damage to the brain. In this chapter, emphasis has been placed on the pathophysiological processes resulting in retardation and on how nature and nurture interact to shape intelligence and behavior (Dykens, 1995; Editorial, 1995). Diverse mutations or chromosomal abnormalities may act via common neuropathological, psychiatric, or behavioral pathways. Some mutations are expressed only at certain developmental stages of the brain, some selectively affect certain formations of the brain, and some affect specific cognitive functions, language or memory (Goldman-Rakic, 1995; Tully, 1995). Phenotypes in complex organisms have multiple determinants. As a rule, they reflect the expression of a major gene, of modifier genes, and of their interaction with each other and with the environment. Not only the magnitude of a metabolic disturbance but also the speed at which it develops may be influenced by random (stochastic) events with implications for the vulnerability of the brain. Cognitive phenotypes are subject to a wider range of modifying influences than metabolic phenotypes, which are closer to the gene product of the major gene involved. This may result in significant differences in cognitive function in patients with an identical mutation of a major gene. Psychologists are studying the relationships and disparities between domain-general and domain-specific abilities as well as the persistence of any differences with development in normal individuals and in those with mental retardation syndromes such as the Beuren-Williams or Down syndromes (Karmiloff-Smith, 1995; Rodan and O'Connor, 1995; Slater, 1995). The resilience, plasticity, and recuperative power of the brain is well recognized. Nevertheless, even with sophisticated investigative techniques, it is frequently not possible to differentiate learning disabilities that are reversible and essentially equate with delayed development from those that are irreversible with limited possibilities of catch-up.

Advances in molecular biology have greatly expanded the range of techniques for the diagnosis of hereditary disorders with the prospect of correction of the underlying defect by somatic gene therapy. Sadly, while significant progress has been recorded in animal and *in vitro* experiments, low transduction efficiencies in some delivery systems and inflammatory and immunological reactions to viral vectors have so far largely frustrated clinical applications of somatic gene therapy. This is not the first instance of hopes unfulfilled. The President's Committee (1972) made this statement: "Using present knowledge and techniques from the biomedical and behavioral sciences, it is possible to reduce the occurrence of mental retardation by 50% by the end of the century", a goal unlikely to be attained. However, it cannot be gainsaid that there have been significant advances in biomedical amelioration and prevention, not only by treatment of disorders such as phenylketonuria and congenital hypothyroidism and by a reduction of the prevalence of the likes of rubella embryopathy and bilirubin encephalopathy, but also by important early intervention programs for at-risk infants (Blair et al., 1995).

Ethical problems arise in all medical research and new forms of treatment directly involving patients, more so in the case of mentally retarded patients who are unable to give informed consent. To what extent should they be expected to take part in research projects, which may well benefit future generations but will do nothing to improve their own condition? Diagnosis of a disorder, antenatal or postnatal, may set a limit to the expectation of life or to the intellectual potential of a fetus or child. Except on religious grounds, few would oppose a request for aborting a fetus with an incurable, rapidly fatal storage disease. Circumstances are different in the case of a fetus with Down syndrome. In a caring environment, Down syndrome patients can look forward to a tolerable, if limited, existence. Should the right to live be determined by the likely score in an IQ test? For whose benefit would the termination of pregnancy be carried out? In this quandary, the mother has rights to be considered in arriving at a decision, but so has the fetus, so have any unaffected siblings whose prospects might be adversely affected by a handicapped child in the family, and so has society, which is responsible for the legal framework within which its citizens exist and serves as guardian of their collective conscience. These issues have been inconclusively debated by physicians, lawyers, philosophers, and theologians. In the real world, the most effective safeguard for patients and public alike is access to competent doctors, nurses, carers, and administrators with compassion, common sense, and flexibility of mind. They are more likely to arrive at decisions acceptable to their clients and the community if they are not impeded by unreasonable financial curbs, rigid rules, and legislative constraints all of which are only too often inspired by prejudice, dogma, or fear.

REFERENCES

Adams, R.D. & Lyon, G. (1982). Neurology of Hereditary Metabolic Disease of Children. McGraw-Hill, London.

Aicardi, J. (1993). The inherited leucodystrophies: A clinical overview. J. Inher. Metab. Dis. 16, 733-743.

Aicardi, J. (1994). Epilepsy in Children, 2nd ed. Raven Press, New York.

Anderson, L.B., Fullerton, J.E., Holtzman, N.A., & Motulsky, R.G. (1994). Assessing Genetic Risks: Implementation for Health and Social Policy. National Academy Press, Washigton, D.C.

van Baar, A., Koppe, J., Boer, K., & Zetterström, R. (Eds.)(1994). Drug and Alcohol Dependency during Pregnancy, Long-term Effects on Child Development. Acta Paediatr. 83 (Suppl. 404), 1-77.

Beasley, M.G., Costello, P.M., & Smith, I. (1994). Outcome of treatment in young adults with phenylketonuria detected by routine neonatal screening between 1964 and 1971. Quart. J. Med. 87, 155-160.

Berry, G.T., Nissim, I., Lin, Z., Mazur, A.T., Gibson, J.B., & Segal, S. (1995). Endogenous synthesis of galactose in normal men and patients with hereditary galactosaemia. Lancet 346, 1073-1074.

Blair, C., Ramey, C.T., & Hardin, J.M. (1995). Early intervention for low birth weight premature infants: participation and intellectual development. Am. J. Ment. Retard. 99, 542-554.

Bolton, P., Macdonald, H., Pickles, A., Rios, P., Goode, S., Crowson, M., Bailey, A.L., & Rutter, M. (1994). A case-control family history study of autism. J. Child Psychol. Psychiatr. 35, 877-900.

Borgaonkar, D.S. (1994). Chromosomal Variation in Man, 7th ed. Wiley-Liss, New York.

Brown, G.K. (1994). Metabolic disorders of embryogenesis. J. Inher. Metab. Dis. 17, 448-458.

Brown, G.K. & Squier, M.V. (1996). Neuropathology and pathogenesis of mitochondrial diseases. J. Inher. Metab. Dis. 19, 553-572.

Brunner, H.G., Nelen, M., Breakfield, X.O., Ropers, H.H., & van Oost, B.A. (1993). Abnormal behavior associated with a point mutation in the structural gene for monoamine oxidase. Science 262, 578-580.

Burrow, G.N., Fisher, D.A., & Larsen, P.R. (1994). Maternal and fetal thyroid function. N. Engl. J. Med. 331, 1072-1078.

Chavez, A., Martinez, C., & Soberanes, B. (1995). Effects of early malnutrition on late mental and behavioral performance. Dev. Brain Dysfunct. 8, 90-102.

Cornblath, M. & Schwartz, R. (1991). Disorders of Carbohydrate Metabolism in Infancy, 3rd ed. Blackwell Scientific Publications, Boston.

Costello, P.M., Beasley, M.G., Tillotson, S.L., & Smith, I. (1994). Intelligence in mild atypical phenylketonuria. Eur. J. Pediatr. 153, 260-263.

Cotton, R.G.H. (1993). Current methods of mutation detection. Mutat. Res. 285, 125-144.

Davies, K.E. (Ed.)(1993). Human Genetic Disease: A Practical Approach, 2nd ed. IRL Press, Oxford, England.

DiDonato, S. (1995). Can we avoid AVED? Nature Genet. 9, 106-107.

Dubowitz, V. (1995). Muscle Disorders in Childhood. W.B. Saunders, London.

Duckett, S. (Ed.)(1995). Pediatric Neuropathology. Williams & Wilkins, Baltimore, Maryland.

Dykens, E.M. (1995). Measuring behavioral phenotypes: provocations from the "New Genetics". Am. J. Ment. Retard. 99, 522-532.

Editorial (1995). The Question of Intelligence. Nature Genet. 9, 97-98.

Eisenberg, L. (1977). Development as a unifying concept in psychiatry. Br. J. Psychiatr. 131, 225-237.

Epstein, C.J. (1986). The Consequence of Chromosome Imbalance. Principles, Mechanisms, and Models. Cambridge University Press, New York.

Farriaux, J.P. & Dhondt, J.L. (Eds.)(1994). New Horizons in Neonatal Screening. Proceedings of the 2nd meeting of the International Society of Neonatal Screening, Lille, September, 1993. Elsevier, Amsterdam The Netherlands.

Fisch, R.O., Chang, P.N., Weisberg, S., Guldberg, P., Güttler, F., & Tsai, M.Y. (1995). Phenylketonuric patients decades after diet. J. Inher. Metab. Dis. 18, 347-353.

Flint, J., Wilkie, A.O.M., Buckle, V.J., Winter, R.M., Holland, A.J., & McDermid, H.E. (1995). The detection of subtelomeric chromosomal rearrangements in idiopathic mental retardation. Nature Genet. 9, 132-140.

Forrest, S., Cotton, R., Landegren, U., & Southern,S. (1995). How to find all these mutations. Nature Genet. 10, 375-376.

Fournier, B., Smeitink, J.A.M., Dorland, L., Berger, R., Saudubray, J.M., & Poll-The, B.T. (1994). Peroxisomal disorders: A review. J. Inher. Metab. Dis. 17, 470-486.

Fowler, M.G. (1994). Pediatric HIV infection: Neurologic and neuropsychologic findings. Acta Paediatr. 83 (Suppl. 400), 59-62.

Gieselmann, V., Polten, A., Kreysing, J., & von Figura, K. (1994). Molecular genetics of metachromatic leucodystrophy. J. Inher. Metab. Dis. 17, 500-509.

Gitzelmann, R. & Steinmann, B. (Eds.)(1995). Galactosemia. Eur. J. Pediatr. 154 (Suppl. 2), S1-S106.

Goldman-Rakic, P.S. (1995). Cellular basis of working memory. Neuron 14, 477-485.

Göstason, R., Wahlström, J., Johannison, T., & Holmquist, D. (1991). Chromosomal aberrations in the mildly mentally retarded. J. Ment. Defic. Res. 35, 240-246.

Grant, D.B. (1995). Congenital hypothyroidism: Optimal management in the light of 15 years experience of screening. Arch. Dis. Child. 72, 85-89.

Gresser, U. (Ed.)(1993). Molecular Genetics, Biochemistry and Clinical Aspects of Inherited Disorders of Purine and Pyrimidine Metabolism. Springer-Verlag, Berlin.

Griffiths, I.R., Montague, P., & Dickinson, P. (1995). The proteolipid protein gene. Neuropath. Appl. Neurobiol. 21, 85-96.

Grompe, M. (1993). The rapid detection of unknown mutations in nucleic acids. Nature Genet. 5, 111-117.

Güttler, F. & Zetterström, R. (Eds.)(1994). Phenylketonuria—Past, Present, Future. Acta Paediatr. 83 (Suppl. 407), 1-129.

Halliwell, B. (1994). Free radicals, antioxidants and human disease: Curiosity, cause or consquence. Lancet 344, 721-724.

Harding, B., Ellis D., & Malcolm, S. (1995). A case of Pelizaeus-Merzbacher disease showing increased dosage of the proteolipid protein gene. Neuropath. Appl. Neurobiol. 21, 111-115.

Hawdon, J.M., Ward Platt, M.P., & Aynsley-Green, A. (1992). Pattern of metabolic adaptation for preterm and term infants in the first neonatal week. Arch. Dis. Child. 65, 341-343.

Hetzel, B. (1994). Iodine deficiency and fetal brain damage. N. Engl. J. Med. 331, 1770-1771.

Hoffmann, G.F. & Bremer, H.J. (Eds.)(1994). Selective screening for inborn errors of metabolism. Eur. J. Pediatr. 153 (Suppl. 1), S1-S100.

Holton, J.B. & Leonard, J.V. (1994). Clouds still gathering over galactosaemia. Lancet 344, 1242-1243.

Hommes, F.A. (Ed.)(1991). Techniques in Diagnostic Human Biochemical Genetics—A Laboratory Manual. Wiley-Liss, New York.

Hoogerbrugge, P.M., Brouwer, O.F., Bordigoni, P. Ringden, O., Kapaun, P., Ortega, J.J., O'Meara, A., Cornu, G., Souillet, G., Frappaz, D., Blanche, S., & Fischer, A. (1995). Allogenic bone marrow transplantation for lysosomal storage disease. Lancet 345, 1398-1402.

Hopkins, A., Shorvon, S., & Cascino, G. (Eds.)(1995). Epilepsy, 2nd ed. Chapman & Hall, London.

Jakobs, C., Jaeken, J., & Gibson, K.M. (1993). Inherited disorders of GABA metabolism. J. Inher. Metab. Dis. 16, 704-715.

Jones, K.L. (1988). Smith's Recognizable Patterns of Human Malformations, 4th ed. W.B. Saunders, Philadelphia.

Karmiloff-Smith, A. (1995). The extraordinary cognitive journey from foetus through infancy. J. Child Psychol Psychiatr. 36, 1293-1313.

Kooistra, L., Schellekens, J.M.H., van der Meere, J.J., & Kalverboer, A.F. (1994). Motor and cognitive development in children with congenital hypothyroidism: a long-term evaluation of the effects of neonatal treatment. J. Pediatr. 124, 903-909.

Kraus, J.P. (1994). Molecular basis of phenotype expression in homocystinuria. J. Inher. Metab. Dis. 17, 383-390.

Krivit, W., Lockman, L.A., Watkins, P.A., Hirsch, J., & Shapiro, E.G. (1995). The future for treatment by bone marrow transplantation for adrenoleucodystrophy, metachromatic leucodystrophy, globoid cell leucodystrophy and Hurler syndrome. J. Inher. Metab. Dis. 18, 398-412.

Leonard, J.V. (1995). The management and outcome of propionic and methylmalonic acidaemia. J. Inher. Metab. Dis. 18, 430-434.

Levene, M.I., Lilford, R.J., Bennett, M.J., & Punt, J. (Eds.)(1995). Fetal and Neonatal Neurology and Neurosurgery, 2nd ed. Churchill Livingstone, Edinburgh, Scotland.

Levy, H.L., Waisbren, S.E., Lobregt, D., Allred, E., Schuler, A., Trefz, F.K., Schweitzer, S.M., Sardharwalla, I.B., Walter, J.H., Barwell, B.E., Berlin, C.M., & Leviton, A. (1994). Maternal mild hyperphenylalaninaemia: an international survey of offspring outcome. Lancet 344, 1589-1594.

Lewis, M., Worobey, J., Ramsay, D.S., & McCormack, M.K. (1992). Prenatal exposure to heavy metal: Effect on childhood cognition, skills and health state. Pediatrics 89, 1010-1015.

Lipton, S.A. & Rosenberg, P.A. (1994). Excitatory amino acids as a final common pathway for neurological disorders. N. Engl. J. Med. 330, 613-622.

McKusick, V.A. (1994). Mendelian Inheritance in Man, 11th ed. Johns Hopkins University Press, Baltimore.

Makrides, M., Neumann, M., Simmer, K., Pater, J., & Gibson, R. (1995). Are long-chain polyunsaturated fatty acids essential nutrients in infancy? Lancet 345, 1463-1468.

Matalon, R., Kaul, R., & Michals, K. (1993). Canavan disease: Biochemical and molecular studies. J. Inher. Metab. Dis. 16, 744-752.

Matsumoto, T., Caprioli, R., Gross, M.L., & Seyama, Y. (Eds.)(1994). Biological Mass Spectrometry: Present and Future. John Wiley, New York.

Morris, A.A.M., Jackson, M.J., Bindoff, L.A., & Turnbull, D.M. (1995). The investigation of mitochondrial respiratory chain disease. J. Roy. Soc. Med. 88, 217P-222P.

Moser, H.W. (1989). Peroxisomal diseases. Adv. Pediatr. 36, 1-38.

Moser, H.W. (1995). Adrenoleucodystrophy: Natural history, treatment and outcome. J. Inher. Met. Dis. 18, 435-447.

Motulsky, A.G. (1995). Jewish diseases and origins. Nature Genet. 9, 99-101.

Mulvihill, J.J. (1995). Craniofacial syndromes: No such thing as a single gene disease. Nature Genet. 9, 101-103.

Neufeld, E.F. (1991). Lysosomal storage diseases. Ann. Rev. Biochem. 60, 257-280.

Norman, E.H. & Bordley, W.C. (1995). Lead toxicity intervention in children. J. Roy. Soc. Med. 88, 121-124.

Nyhan, W.L. & Sakati, N.O. (1987). Diagnostic Recognition of Genetic Disease. Lea & Febiger, Philadelphia.

Opitz, J.M. & de la Cruz, F. (1994). Cholesterol metabolism in the RSH/Smith-Lemli-Opitz syndrome: Summary of an NICHD Conference. Am. J. Med. Genet. 50, 326-338.

Paneth, N., Rudelli, R., Kazam, E., & Monte, W. (Eds.)(1994). Brain Damage in the Preterm Infant. Cambridge University Press, Cambridge, England.

Pollitt, R.J. (1995). Disorders of mitochondrial long-chain fatty acid oxidation. J. Inher. Metab. Dis. 18, 473-490.

Prensky, A.L. (1992). Mental retardation. In: Diseases of the Nervous System. (Asbury, A.K., McKhann, G.M., & McDonald, W.I., Eds.), pp. 612-626. W.B. Saunders, Philadelphia.

President's Committee on Mental Retardation (1972). Entering the Era of Human Ecology. DHEW publ. no. (05) 72-7. Washington, D.C.

Reeves, R.H., Irving, N.G., Morgan, T.H., Wohn, A., Kitt, C., Sisodia, S.S., Schmidt, C., Bronson, R.T., & Davisson, M.T. (1995). A mouse model for Down syndrome exhibits learning and behaviour deficits. Nature Genet. 11, 177-183.

Ris, M.D., Williams, S.E., Hunt, M.M., Berry, H.K., & Leslie, N. (1994). Early treated phenylketonuria: adult neuropsychologic outcome. J. Pediatr. 124, 388-392.

Rondal, J.A. & O'Connor, N. (1995). Exceptional Language Development in Down Syndrome. Cambridge University Press, Cambridge, England.

Rooney, D.E. & Czepulkowski, B.H. (1992). Human Cytogenetics: A Practical Approach, 2nd ed. Oxford University Press, New York.

Rovet, J.F. & Ehrlich, R.M. (1995). Longterm effects of L-thyroxine therapy for congenital hypothyroidism. J. Pediatr. 126, 380-386.

Rutter, M., Bailey, A., Bolton, P., & Le Couteur, A. (1994). Autism and known medical conditions: Myth and substance. J. Child Psychol. Psychiatr. 35, 311-322.

Salen, G., Shefer, S., Batta, A.K., Tint, G.S., Xu, A., & Honda, A. (1996). Abnormal cholesterol biosynthesis in sitosterolaemia and the Smith-Lemli-Opitz syndrome. J. Inher. Metab. Dis. 19, 391-400.

Schmickel, R.D. (1986). Contiguous gene syndrome: A component of recognizable syndromes. J. Pediatr. 109, 231-233.

Scolnik, D., Nulman, I., Rovet, J., Gladstone, D., Gardner, H.A., Gladstone, R., Ashby, P. Weksberg, R., Einarson, T., & Koren, G. (1994). Neurodevelopment of children exposed *in utero* to phenytoin and carbamazepine monotherapy. J. Am. Med. Assn. 271, 767-770.

Scriver, C.R., Beaudet, A.L., Sly, W.S., & Valle, D. (Eds.)(1995). The Metabolic and Molecular Basis of Inherited Disease, 7th edn. McGraw Hill, New York.

Scriver, C.R., Eisensmith, R.C., Woo, S.L.C., & Kaufman, S. (1994). The hyperphenylalaninemias of man and mouse. Annu. Rev. Genet. 28, 141-165.

Shapiro, E.G., Lockman, L.A., Balthazor, M., & Krivit, W. (1995). Neuropsychological outcome of several storage diseases with and without bone marrow transplantation. J. Inher. Metab. Dis. 18, 413-429.

Skuse, D., Pickles, A., Wolke, D., & Reilly, S. (1994). Postnatal growth and development: Evidence for a "sensitive" period. J. Child Psychol. Psychiatr. 35, 521-545.

Slater, A. (1995). Individual differences in infancy and later IQ. J. Child Psychol. Psychiatr. 36, 69-112.

Susser, M.W. (1989). The challenge of causality: Human nutrition, brain development and mental performance. Bull. NY. Acad. Med. 65, 1032-1049.

Swaiman, K.F. (Ed.)(1994). Pediatric Neurology, 2nd ed. Mosby, St. Louis. Symposium (1990). Trisomy 21 (Down syndrome). Am. J. Med. Genet. (Suppl. 7), 1-330.

Symposium (1994). X-linked Mental Retardation-6. Am. J. Med. Genet. 51, 281-622.

Tada, K. & Kure, S. (1993). Nonketotic hyperglycinaemia: Molecular lesion, diagnosis, pathophysiology. J. Inher. Metab. Dis. 16, 691-703.

Tager, J.M., Aerts, J.M.F.G., van den Bogert, C., & Wanders, R.J.A. (1994). Signals on proteins, intracellular targeting and in-born errors of organellar metabolism. J. Inher. Metab. Dis. 17, 459-469.

Taylor, E. (1991). Developmental neuropsychiatry. J. Child Psychol. Psychiatr. 32, 3-47.

Teisberg, P. (1995). The genetic background of anticipation. J. Roy. Soc. Med. 88, 185-187.

Thomas, G.R., Forbes, J.R., Roberts, E.A., Walshe, J.M., & Cox, D.W. (1995). The Wilson's disease gene: Spectrum of mutations and their consequences. Nature Genet. 9, 210-217.

Tillotson, S.L., Fuggle, P.W., Smith, I., Ades, A.E., & Grant, D.B. (1994). Relation between biochemical severity and intelligence in early treated congenital hypothyroidism: a threshold effect. Br. Med. J. 309, 440-445.

Tully, T, (1995). Dissection of memory formation: From pharmacology to molecular genetics. Trends Neurosci. 18, 212-217.

Vulpe, C., Levinson, B., Whitney, S., Packman, S., & Gitschier, J. (1993). Isolation of a candidate gene for Menkes disease and evidence that it encodes a copper-transporting ATPase. Nature Genet. 3, 7-13.

Walkeley, S.V. & Dobrenis, K. (1995). Bone marrow transplantation for lysosomal disease. Lancet 345, 1382-1383.

Wallace, D.C. (1992). Disease of the mitochondrial DNA. Annu. Rev. Biochem. 61, 1195-1215.

Wallace, D.C. (1993). Mitochondrial diseases: Genotype versus phenotype. Trends Genet. 9, 128-133.

Walshe, J.M. (1995). Copper: Not too little, not too much, but just right. J. Roy. Coll. Physicians (London) 29, 280-288.

Winter, R.M. & Baraitser, M. (Eds.)(1993). Multiple Congenital Anomalies. Oxford University Press, Oxford.

Chapter 18

Autism

CHRISTOPHER GILLBERG

Biological Psychiatry, pages 437-450.

ISBN: 1-55938-819-6

INTRODUCTION

The term Autism originally derived from the Greek "autos" (self) was meant to depict the egocentric thinking characteristic of the thought disorder in schizophrenia but is nowadays often used to refer to childhood autism or autistic disorder. Egocentric thinking also appears to be a fundamental feature of childhood autism, and so the autism "label" is a reasonable concept in this disorder as well as in some cases of schizophrenia

Clear-cut examples of childhood autism are on record as early as the late eighteenth and early nineteenth centuries. Young children with tuberous sclerosis and autism were described in the early 1930s, but the behavioral problems were regarded as "psychosis". It was not until the 1940s that childhood autism (infantile autism) was firmly established as a medical diagnosis. In 1943, Leo Kanner described 11 children (8 boys and 3 girls) with "autistic disturbances of affective contact", most of whom were (at least functionally) mentally retarded. A year later, Hans Asperger described a group of patients with "autistic psychopathy" most of whom were of normal or superior intelligence.

Childhood autism is commonly regarded as being multifactorial in etiology and a distinct syndrome with no precisely defined boundaries. Most authorities agree that it exists as a part of a spectrum with autisticlike conditions. So does Asperger's "autistic psychopathy", now recognized as Asperger's syndrome (Asperger, 1944; Wing, 1981).

DIAGNOSTIC CRITERIA

The diagnostic criteria currently in use for autistic disorder and those for Asperger's syndrome, which closely resemble the criteria for childhood autism, can be distinguished by taking into account the features of Asperger's as set forth in Table 1.

Most children who are diagnosed as having autistic disorder fail to meet the diagnostic criteria of Asperger's syndrome and vice versa. There is a built-in check to make certain that an individual is not diagnosed as having both disorders: language development in autism is always abnormal but it is normal in people diagnosed with Asperger's syndrome. However, as it turns out, most children who otherwise conform to Asperger's classical description of the syndrome have some abnormality of language development (e.g., some delay compared to siblings, semantic-pragmatic problems, odd prosody, problems in comprehension of spoken language as a means of communication). Clinically, the boundaries between autism and Asperger's syndrome are not distinct and there are individuals who conform to the classic autism picture in early childhood but later meet the Asperger criteria perfectly. More rarely, the reverse holds true. There are also individuals who do not meet the criteria for any of the two syndromes who show

Table 1. Diagnostic Criteria for Asperger's Syndrome

Severe impairment in social interaction (extreme egocentricity) as manifested by at least two of the following four:
- decreased ability to interact with elderly peers
- lack of desire to interact with peers
- lack of appreciation of social cues
- socially and emotionally inappropriate behavior

All-absorbing narrow interest, as manifested by at least one of the following three:
- exclusion of other activities
- repetitive adherence
- more rote than meaning

Imposition of routines and interests, as manifested by at least one of the following two:
- imposition on self
- imposition on others

Speech and language problems, as manifested by at least three of the following five:
- delayed development of language
- superficially perfect expressive language
- formal, pedantic language
- odd prosody, peculiar voice characteristics
- impairment of comprehension including misinterpretations of literal/implied meanings

Nonverbal communication problems, as manifested by at least one of the following five:
- limited use of gestures
- clumsy/gauche body language
- limited facial expression
- inappropriate expression
- peculiar, stiff gaze

Developmental coordination problems, as manifested by
- motor clumsiness, as documented by poor performance on neurodevelopment examination

After Gillberg & Gillberg (1989)

marked "autistic features". Some of these have the triad of social, communicational, and behavioral restriction in combination with severe mental retardation, while others are normally intelligent and show a number of autistic disorder/Asperger syndrome symptoms but do not meet full diagnostic criteria for these disorders. In the latter group are several of those with deficits in attention, motor control, and perception (DAMP), a majority of whom also meet diagnostic criteria for attention deficit hyperactivity disorder (ADHD). Thus it is convenient but reasonable to lump these groups under the heading of autism spectrum disorders (ASD).

According to recent models, Asperger syndrome is seen as a mild variant of autism, autism in more intellectually able individuals, or a disorder involving overlap, though not necessarily having identical dysfunctional neural circuits. It should be added that there is as yet no universal agreement that autistic disorder and the Asperger syndrome belong to the same spectrum of disorders. However, it is a matter of diagnostic importance that a diagnosis of autistic disorder can be made with confidence in most cases around age three, whereas the diagnosis of Asperger's syndrome can be made usually only after age five. Unfortunately, many with this syndrome do not attend clinics until well into school age, and some (perhaps even the majority) do not apply for help at all in childhood or adolescence.

DIFFERENTIAL DIAGNOSIS

The disorders that most often arouse diagnostic discussion (and even confusion) in the field of autism include childhood schizophrenia, childhood disintegrative disorder, developmental language delay, severe and profound mental retardation with attention deficits and stereotypes, depression, emotional deprivation, and various sensory deficits. There are also a number of medical conditions that can cause autism (vidé infra), but these should not be regarded in the differential diagnoses. Instead, in a multiracial classification system in which the behavioral syndrome is coded on one axis, such disorders should be coded on a separate axis for associated medical disorders.

The status of childhood schizophrenia is still unclear. (Asarnow, 1994). Most authors consider it a rare disorder usually with onset in the preadolescent and early adolescent periods and altogether separated from autism. The diagnosis should be made only when full diagnostic criteria for schizophrenia are met.

Childhood disintegrative disorder (Volkmar, 1992) is a rare condition commonly onsetting in the three- to five-year-old age range and affecting children with normal or almost normal development until that age. After a period of restlessness, language deterioration, and confusion, the clinical picture in the school age is often inseparable from that of autistic disorder. There are sometimes considerable problems distinguishing disintegrative disorder from Landau-Kleffner syndrome (Russo et al., 1994). In this syndrome of "acquired aphasia/auditory agnosia", there are often severe behavior problems, which may sometimes be inseparable from the syndrome of autism or disintegrative disorder. Most authorities regard disintegrative disorders as being a part of the autism spectrum.

Developmental receptive language disorders (Cantwell and Baker, 1987) may be difficult to separate from autism. The similarities between these two classes of disorders are sometimes striking, and there appears to be an area of overlap in which the boundary is completely blurred.

Children with profound mental retardation can be sociable and unsociable (Wing, 1989). In the latter case, a diagnosis of autistic disorder is often considered. Full diagnostic criteria are not always met in such cases (because of the limited repertoire shown by individuals with such severe intellectual handicap). A diagnosis of autisticlike condition/atypical autism is often helpful in such children, because it suggests the need for certain types of intervention.

Depressed and emotionally deprived children, particularly when very young, can show several symptoms suggestive of autism. In such patients, there is usually some major negative psychosocial factor the removal of which leads to recovery or marked overall improvement rather rapidly. (This may also happen even if the psychosocial environment is ameliorated in other ways.) To be sure, some children with autism are subjected to psychosocial deprivation and improvement in environmental change in such instances also follows. A few months later, however, plateauing occurs and any further improvement is much slower.

Hearing and visual impairments are important parts of the differential diagnoses for several reasons: (1) children with autism spectrum disorders often have both impairments with the visual being more rare (Steffenburg, 1991), (2) hearing impairment may be diagnosed in autism when, in fact, there is none, (3) sensory impairments may be missed altogether in autism because of the typical lack of cooperation during examination, at least in the youngest age group, and (4) children who are deaf, blind, or both have a higher-than-average rate of autism, but the diagnosis may be missed and falsely attributed to the sensory deficit per se or regarded as an integral part of it (which it is not, in most cases).

PREVALENCE

Autistic disorder is more common than previously believed. It affects 8-12 per 10,000 children born (Gillberg, 1994). Its prevalence is relatively stable across cultures. Boys are affected much more often than girls at ratios of 2-3:1 in most of the population-based studies (Wing, 1993). Mental retardation (full-scale IQ under 70) occurs in 65-90% of all cases (Gillberg, 1990). Epilepsy is associated with autism in 25-35% of all cases. About half of the epileptics are seen in the first few years of life. Most of the remaining ones have their onset around adolescence (Gillberg, 1991).

Asperger's syndrome, once believed to be very rare, has been found to be several times more frequent than autistic disorder. Thus, at least 36 and possibly as many as 71 in 10,000 school-age children may be affected (Ehlers and Gillberg, 1993). Mental retardation is seldom seen in this syndrome and epilepsy affects fewer than 5% in childhood and adolescence.

SEX RATIOS

Boys are much more often affected by autism and its spectrum disorders than are girls. The most commonly cited figure for male to female ratios is 3:1, but this may be an overestimate given that, in the population studies, sex ratios have tended toward 2:1 to 2.5:1 instead. Male to female ratios appear to be lowest in the most severely handicapped, most profoundly mentally retarded groups and higher in those who are of normal intelligence. In clinical practice, Asperger's syndrome is usually diagnosed 10-fold more often in boys than in girls, but the only available population study related a male to female ratio of only 4:1 (Ehlers and Gillberg, 1993).

SOCIAL FACTORS

Autism occurs in all social classes at a rate that corresponds to the population norms (Wing, 1993). Asperger's syndrome might be slightly more common in upper social classes, but this finding might be a reflection of clinical referral bias—only those cases from the upper social classes consulting clinics—or the effect of IQ being higher in so-called Asperger's syndrome (and the parents also having a higher IQ and coming from higher social class) than in classic autism.

Autism is more common among children born to first-generation immigrants (Wing, 1993). This could be taken to mean that such children are more often exposed to intrauterine infections such as rubella, which can cause autistic symptoms or that fathers (or mothers, albeit rarely) with Asperger's syndrome or autistic features seek spouses in cultures where their own social impairments are not as readily obvious.

ASSOCIATED IMPAIRMENTS AND MEDICAL DISORDERS

Classic autism is accompanied by other impairments in a vast majority of the patients. Thus, for example, mental retardation is present in about 75%, epilepsy in 35%, and moderate-to-severe hearing and visual impairments in 10% to 20% each. The rate of motor impairment in an unselected group of young people with autism has not yet been properly studied. Moreover, several retardation disorders are connected with autism and are considered to be the underlying cause of the brain dysfunction leading to the autistic symptoms. About one in four of a population of children with autism have such an associated medical disorder (Gillberg and Coleman, 1996). Some of the disorders so far reported are listed in Table 2.

Among these, the most important are tuberous sclerosis and other neurocutaneous disorders as well as the fragile X syndrome and other chromosomal disorders,

Table 2. Related Medical Disorders and Subgroups in Autism

Fragile X syndrome	Hypomelanosis of Ito
Other sex chromosome anomalies	Goldenhar syndrome
Partial trisomy/tetrasomy 15q11-13	Rett syndrome
Other chromosome anomalies	Moebius syndrome
Tuberous sclerosis	PKU
Neurofibromatosis	Lactic acidosis
Rubella embryopathy	Cytomegalic virus
Herpes simplex encephalitis	Williams syndrome
	Duchenne muscular dystrophy

which together constitute more than half the group with a "dual diagnosis" (autism *plus* another disorder).

In a child psychiatry clinic, it is likely that the rate of associated medical disorders is lower than in the general population. Though it may seem surprising, a likely explanation for this is that if a medical disorder such as tuberous sclerosis has been diagnosed, the additional problem of autism will go unrecognized or be seen (appropriately) as a consequence of the medical disorder, and hence, be regarded (inappropriately) as a problem of secondary importance. The result could then be that referral to a specialized child psychiatry service might not be sought.

BRAIN DAMAGE

There is convincing evidence that brain damage can lead to autism (Folstein and Rutter, 1977). The mere fact that previously normal children who contract herpes encephalitis may develop the full syndrome of autism (Gillberg, 1986) supports the idea that damage may cause autism. However, it is less obvious to what extent pre- or perinatally acquired damage to the brain plays a part in the spectrum of causes leading to the autistic syndromes. Even though reduced optimality in the pre-, peri-, and neonatal periods is more common in autism than in normal children and in children with mental retardation without autism, it remains unclear that such reduced optimality is a clear-cut and reliable measure of brain damage. For it could easily be a reflection of some genetic vulnerability in the fetus, which in itself leads to various "nonoptimal" events during pregnancy or the intrapartum or newborn periods.

GENETIC FACTORS

From Table 2, it is clear that genetic factors are one important cause of autism. Such factors also appear to play a very important role in Asperger's syndrome (Frith, 1991). In the remaining disorders that fall within the autism spectrum, the influence of genetic factors is much less clear.

Twin studies have shown autism to be a genetic disorder. Identical twins are usually concordant for the syndrome, whereas nonidentical twin pairs are concondant in only a few percent of all individuals. This and the observation of an increased rate of autism in siblings (about 5%) favor a genetic model for the development of autism in such patients. However, it is still unknown as to what proportion of these children with autism has the disorder because of purely genetic factors. In other words, the nature of exactly what is inherited has yet to be established: Is it a general/specific cognitive/language deficit (as reflected in speech–language disorders and dyslexia in relatives) or a more specific social/social-cognitive deficit (as reflected in Asperger's syndrome or Asperger traits in relatives)? The twin studies, and genome scan studies of sibling pairs in which both sibs are affected by autism, suggest that several (5-10?) susceptibility genes may act in concert to cause or predispose to the full syndrome of autism.

NEUROPSYCHOLOGY/COGNITION

People with autism and those with Asperger's syndrome usually have a very characteristic neuropsychological test profile (Frith, 1991). For instance, on the so-called Wechsler Intelligence Scale for Children (WISC), children with autism tend to score low on full-scale and verbal IQ, but relatively better on the performance parts of the test. Those diagnosed as suffering from Asperger's syndrome usually have normal (or superior) overall IQ (particularly verbal IQ). A relatively low result on picture arrangement is characteristic of both autism and Asperger's syndrome. This subtest measures empathy in general. People with autism and young children with Asperger's syndrome score poorly on the comprehension subtest, which measures "common sense". With age, the Asperger individual may well score better on the comprehension scale because he or she will have learned, perhaps through formal education (rather than from social experience), what would be an appropriate response to a question such as "What would you do if there was a fire?". Most individuals with autism score above their own average—or indeed in the superior range—on the block-design subtest. This is a test of visual constructive ability, an area in which many individuals with autism excel. Most people with Asperger's syndrome too will do well on this subtest, but some may fail because of their slow motor performance. Object assembly (a jigsaw puzzle–type task guided by the overall outline of the depicted object), on the other hand, may be an area of relative failure in the Asperger group, possibly because of the failure on the part of such individuals to attend to wholes rather than detail.

The neurocognitive profile in addition to experimental evidence coming from the laboratory suggests that people with autism have a much reduced capacity for empathy—the so-called "theory of mind" (Frith, 1991). This is seen in the problem they confront in executive functions (e.g., planning functions, which generally reflect frontal lobe dysfunction). They appear to be lacking in their drive for central coherence (Happe, 1994).

NEUROBIOLOGY

Few consistent findings have been proved by the study of autism neurobiology (which is compatible with the notion of autism as a behaviorally defined syndrome with multiple etiologies). However, the fact remains that brainstem, cerebellar, temporal lobe, and frontal lobe dysfunctions have been reported by many groups (Gillberg and Coleman, 1992). Autopsy findings are consistent with abnormalities in the cerebellum and amygdala (Bauman, 1991). Neurochemical findings include evidence of imbalance of dopamine/norepinephrine and hyperserotonemia and excess levels of markers for glial and synapse activity and endorphin dysfunction (Gillberg and Coleman, 1992). One theory has suggested imbalance in the dopamine system (which is closely linked to the endorphin system) projecting from the brainstem to the temporal and mesial frontal lobes and that this could be the primary neurochemical explanation of the autistic syndrome (Damasio and Maurer, 1978; Gillberg and Coleman, 1992).

WORK-UP

All children with autism and Asperger's syndrome require a proper and thorough work-up that includes the following:

a. Tests for autistic behavior such as the DISCO test (Wing and Gould, 1994) to quantify the degree to which the individual is affected
b. A neuropsychological test, such as the WISC or the Leiter test to evaluate the level of general cognitive functioning
c. An examination by a highly skilled neuropsychiatrist who has had considerable training in child neurology/neuropsychiatry (including a search for skin changes that could signal the presence of an underlying neurocutaneous disorder)
d. A laboratory assessment tailored to meet the requirements for an adequate work-up in the individual being examined. This should include, in all instances, evaluation of vision and hearing, a chromosomal culture and DNA-analysis for the FMR-1 gene, and blood and urine screening for various metabolic disorders. In addition, neuroimaging, EEG, Auditory Brainstem Response (ABR), and cerebrospinal fluid examinations may all be indicated, depending on the age of the individual, associated IQ medical history, and the result of the clinical examination.

INTERVENTIONS AND TREATMENT

There is no specific treatment that can cure or alleviate the symptoms of autism. Those who have epilepsy will need a proper evaluation by a skilled team of neuropsychiatrists/neurologists/neuropsychologists in order to arrive at the best possible recommendation for pharmacotherapy, neurosurgery, and psychosocial interventions. Some children with autism and epilepsy do better overall if allowed the occasional seizure rather than being completely seizure-free and suffering from major side-effects of drug treatment. Others benefit greatly from antiepileptic drug treatment both in terms of seizure activity and behavior problems (Gillberg, 1991).

Hyperactivity and general, unspecific behavioral problems are sometimes amenable to treatment with neuroleptics, fenfluramine, or vitamin B6 plus magnesium, but the pros and cons of such treatment have to be weighed carefully in each individual and there is no rationale for generalized guidelines regarding these drugs. The neuroleptics obviously have some positive effects on core autistic symptoms, but the hazards of such treatment (including a high rate of tardive dyskinesia in this patient group) preclude their widespread use in the whole population with autism.

Psychoeducative methods and behavior modification in autism have been used with considerable success (Schopler, 1989; Lovaas et al., 1989), though claims in this regard often represent considerable overstatements. All individuals with autism, including those diagnosed with Asperger's syndrome, need an individualized education plan encompassing a high degree of structure and external goal criteria. Few skills come easy to people with autism. Because of their particular inability to seek out information and new knowledge from other people (in a learning situation), they fail to show the initiative and spontaneity of children without autism. They need the highly structured guidance of both teachers and trainers familiar with the basic handicaps intrinsic to autism. A well integrated, multidisciplinary approach to treatment/intervention is necessary for each patient.

OUTCOME

Outcome in autism is highly variable (Gillberg, 1991a). Depending on whether or not autism and Asperger's syndrome constitute phenotypically different expressions of the same underlying pathology, outcome from the psychosocial point of view in autism can be said to be anything from very poor to excellent.

In the present state of our knowledge (and given the types of therapy currently provided), IQ under 50 around age five- to seven-years-old, predicts extremely low levels of independence in adult age. On the other hand, about 50% of those with IQ above 70 have a fair-to-good outcome and may well live on their own and hold down a job even though "completely normal" social functioning is uncommon. Outcomes in the typical Asperger group have been subjected to little systematic study, but it is clear that they are generally much better than in the classic autism group. However, the prevalence of psychiatric disorder leading to *adult* psychiatric consultation is probably much elevated as compared with that in the general population. Adult psychiatrists are still often unfamiliar with Asperger's syndrome, and patients may be misdiagnosed as having borderline conditions, atypical depression, paranoid disorder, type II schizophrenia, or psychosis NOS.

Lastly, mortality appears to be much increased in classical autism, most probably because of the high incidence of associated severe medical disorders, epilepsy, and accidents. In Asperger's syndrome, the incidence of suicide attempts, and possibly of completed suicides, is high (Wing, 1981).

SUMMARY

Childhood autism is currently conceived as a behaviorally defined syndrome of multiple neurobiological (including genetic) etiologies. It comprises a triad of social, communicational, and behavioral restriction and exists in a spectrum together with nonclassical variants of the disorder such as Asperger's syndrome. In its classical form, however, autism is usually associated with mental retardation and occurs in about one in 1000 children with a male-to-female ratio between 2:1 and 3:1. Asperger's syndrome, by comparison, is associated with higher IQ and good verbal ability, occurs in about four in 1000 children, and has a male-to-female ratio that is probably higher than in classical autism. Brainstem (and cerebellar), temporal and frontal lobe dysfunction have been proposed to account for autistic symptomatology. Particular neuropsychological profiles exist that probably reflect specific mental deficiencies in autism spectrum problems. The outcome is psychosocially disappointing in classical variants of the disorder but much better in highly functioning individuals, including those diagnosed with Asperger's syndrome. Rational cures are mostly unavailable, but psychoeducative measures and behavior modification, while not dramatically altering the basic problems, may go a long way in alleviating some of the most severe consequences of the disorder.

REFERENCE

American Psychiatric Association. (1987). Diagnostic and Statistical Manual of Mental Disorders (3rd ed.). Washington, D.C.

American Psychiatric Association. (1994). Diagnostic and Statistical Manual of Mental Disorders (4th ed.). Washington, D.C.

Asarnow, J.R., Tompson, M., Hamilton, E.B., Goldstein, M.J., and Guthrie, D. (1994). Family-expressed emotion, childhood-onset depression, and childhood-onset schizophrenia spectrum disorders: Is expressed emotion a nonspecific correlate of child psychopathology or a specific risk factor for depression. J. Abnormal Child Psychol. 22, 129-146.

Asperger, H. (1944). Die Autistischen Psychopathen im Kindesalter. Archiv fur Psychiatric und Nervenkrankheiten 117, 76-136.

Akefeldt, A. & Gillberg, C. (1991) Hypomelanosis of Ito in three cases with autism and autistic-like conditions. Develop. Med. Child Neurology. 33, 737-743.

Bauman, M.L. (1991) Microscopic neuroanatomic abnormalities in autism. Pediatrics 87, 798-743.

Bleuler, E. (1919). Dementia Praecox or the Group of Schizophrenias. Vienna: (Zinkin, J., Trans.), International University Press, New York.

Bolton, P. & Rutter, M. (1990). Genetic influences in autism. Int. Rev. Psychiatry 2, 67-80.

Campbell, M. (1989). Pharmacotherapy in autism: An overview. In: Diagnosis and Treatment of Autism. (Gillberg, C. Ed.), Plenum, New York.

Campbell, M. & Malone, P. (1991). Mental retardation and psychiatric disorders. Hosp. Comm. Psychiatry, 42, 374-379.

Cantwell, D.P. & Baker, L. (1987). Psychiatric symptomatology in language-impaired children: A comparison. Dev. Med. Child Neurology 2, 128-133.

Chess, S., Korn, S.J. & Fernandez, P.B. (1971). Psychiatric Disorders of Children with Congenital Rubella. Brunner/Mazel, New York.

Coleman, M. & Blass, J.P. (1985). Psychiatric symptomatology in language-impaired children: A comparison. Neurology 2, 128-133.

Critchley, M. & Earl, C.J.C. (1932). Tuberose sclerosis and allied conditions. Brain 55, 311-346.

Damasio, A.R. & Maurer, R.G. (1978). A neurological model for childhood autism. Arch. Neurology, 35, 777-786.

Delong, G.R., Beau, S.C., & Brown, F.R. (1981). Acquired reversible autistic syndrome in acute encephalopathic illness in children. Arch. Neurology 38, 191-194.

Ehlers, S. & Gillberg, C. (1993). The epidemiology of Asperger syndrome. A total population study. J. Child Psychol. Psychiatry 34, 1327-1350.

Folstein, S. & Rutter, M. (1977). Infantile autism: A genetic study of 21 twin pairs. J. Child Psych. Psychol. 18, 297-321.

Friedman, E. (1969). The autistic syndrome and phenylketonuria. Schizophrenia 1, 249-261.

Frith, U. (Ed.)(1991). Autism and Asperger Syndrome. Cambridge University Press, Cambridge, England.

Frith, U. (1989). Autism and "theory of mind". In: Diagnosis and Treatment of Autism. (Gillberg, C., Ed.), pp. 33-52. Plenum Press, New York.

Gillberg, C. (1986). Brief report: Onset at age 14 of a typical autistic syndrome. A case report of a girl with herpes simplex encephalitis. J. Aut. Dev. Disorders 16, 369-375.

Gillberg, C. (1989). Asperger syndrome in 23 Swedish children. Dev. Med. Child Neurology 31, 520-531.

Gillberg, C. (1990). Autism and pervasive developmental disorders. J. Child Psychol. Psychiatry 31 (1), 100-119. [Published erratum appears in J. Child Psychol. Psychiatry 1991, 32(1), 213.]

Gillberg, C. (1991a). Outcome in autism and autistic-like conditions. J. Am. Acad. Child Adol. Psychiatry 30, 375-382.

Gillberg, C. (1991b). The treatment of epilepsy in autism. J. Aut. Dev. Disorders 21, 61-77.

Gillberg, C. (1992). The Emanuel Miller Memorial Lecture 1991: Autism and autistic-like conditions: Subclasses among disorders of empathy. J. Child Psychol. Psychiatry 33, 813-842.

Gillberg, C., (1994). The prevalence of autism and autism spectrum disorders. In: The Epidemiology of Child and Adolescent Psychopathology. (Verhulst, F.C. & Koot, H.M., Eds.), Oxford University Press, Oxford, England.

Gillberg, C. & Coleman, M. (1992). The Biology of the Autistic Syndromes, 2nd ed. Cambridge University Press, New York.

Gillberg, C. & Coleman, M. (1996). Autism and possibly related medical disorders. A review of the pertinent literature. Dev. Med. Child Neurology 38, 191-202.

Gillberg, C. & Forsell, C. (1984). Childhood psychosis and neurofibromatosis—more than a coincidence? J. Aut. Dev. Disorders 14, 1-8.

Gillberg, C., Steffenburg, S. & Eeg-Olofsson, O. (1991). Autism associated with marker chromosome. J. Am. Acad. Child Adol. Psychiatry 30, 489-494.

Gillberg, I.C. & Gillberg, C. (1989). Asperger syndrome—some epidemiological considerations: A research note. J. Child Psychol. Psychiatry 30(4), 631-638.

Gillberg, I.C., Gillberg, C., & Kopp, S. (1992). Hypothyroidism and autism spectrum disorders. J. Child Psychol. Psychiatry. 33, 531-542.

Hagerman, R.J. (1989). Chromosomes, genes and autism. In: Diagnosis and Treatment of Autism. (Gillberg, C., Ed.). Plenum Press, New York.

Happe, F.G.E. (1994). An advanced test of theory of mind: Understanding of story characters' thoughts and feelings by able autistic, mentally handicapped, and normal children and adults. J. Aut. Dev. Disorders 24, 129-154.

Hellgren, L., Gillberg, C., & Gillberg, I.C. (1994). Children with deficits in attention, motor control and perception (DAMP) almost grown up: The contribution of various background factors to outcome at age 16 years. Eur. Child Adol. Psychiatry 1-15.

Howlin, P. & Rutter, M. (1987) Treatment of Autistic Children. John Wiley & Sons, Chichester, England.

Hunt, A. & Dennis, J. (1987). Psychiatric disorder among children with tuberous sclerosis. Dev. Med. Child Neurology. 29, 190-198.

Kanner, L. (1943). Autistic disturbances of affective contact. Nervous Child, 2, 217- 250.

Komoto, J., Udsui, S., Otsuki, S., & Terao, A. (1984). Infantile autism and Duchenne Muscular dystrophy. J. Aut. Dev. Disorders 14, 191-195.

Krug, D.A., Arick, J., & Almond, P. (1980). Behavior checklist for identifying severely handicapped individuals with high levels of autistic behavior. J. Child Psychol. Psychiatry 21, 221-229.

Landgren, M., Gillberg, C., & Stromland, K. (1992). Goldenhar syndrome and autistic behavior. Dev. Med. Child Neurology 34, 1000-1005.

Lovaas, I., Calouri, K., & Jada, J. (1989). The nature of behavioral treatment and research with young autistic persons. In: Diagnosis and Treatment of Autism. (Gillberg, C., Ed.) Plenum, New York.

McKenna, K., Gordon, C.T., Lenane, M., Kaysen, D., Fahey, K., & Rapoport, J.L. (1994). Looking for childhood-onset schizophrenia: The first 71 cases screened. J. Am. Acad. Child Adol. Psychiatry. 33, 636-644.

Ornitz, E.M., Guthrie, D., & Farley, A.J. (1977). The early development of autistic children. J. Aut. Child. Schiz. 7, 207-229.

Reiss, A.L., Feinstein, C., Rosenbaum, K.N., Borengasser-Caruso, M.A., & Goldsmith, B.M. (1985). Autism associated with Williams syndrome. Pediatrics 106, 247-249.

Russo, M., Perry, R., Gillberg, C., & Kolodny, E. (1994). Heller's syndrome in a preschool boy: Proposed medical evaluation, and hypothesized pathogenesis. Eur. Child. Adol. Psychiatry 5, 172-177.

Rutter, M., Bailey, A., Bolton, P., & Le Couteur, A. (1994). Autism and known medical conditions: Myth and substance. J. Child Psychol. Psychiatry 35, 311-322.

Schopler, E. (1989). Principles for directing both educational treatment and research. In: Autism. Diagnosis and Treatment of Autism. (Gillberg, C., Ed.), Plenum, New York.

Steffenburg, S. (1991). Neuropsychiatric assessment of children with autism: A population-based study. Dev. Medicine Child Neurology 33 495-511.

Steffenburg, S., Gillberg, C., Hellgren, L., Anderson, L., Gillberg, I.C., Jakobsson, G., & Bohman, M. (1989). A twin study of autism in Denmark, Finland, Iceland, Norway and Sweden. J. Child Psychol. Psychiatry 30, 405-416.

Stubbs, E.G. (1976). Autistic children exhibit undetectable hemagglutination-inhibition antibody titers despite previous rubella vaccination. J. Aut. Child. Schiz. 6, 269-274.

Szatmari, P. (1992). The validity of autistic spectrum disorders: A literature review. J. Aut. Dev. Disorders 22 (4), 583-600.

Volkmar, F.R. (1992). Childhood disintegrative disorder: Issues for DMS-IV. J. Aut. Dev. Disorders 22, 625-642.

Wing, L. (1981). Asperger's syndrome: A clinical account. Psychological Medicine 11, 115-129.

Wing, L. (1989). The diagnosis of autism. In: Diagnosis and Treatment of Autism. (Gillberg, C., Ed.), Plenum Press, New York.

Wing, L. (1993). The definition and prevalence of autism: A review. Eur. Child Adol. Psychiatry 2, 61-74.
Wing, L. & Gould, J. (1979). Severe impairments of social interaction and associated abnormalities in children: Epidemiology and classification. J. Aut. Dev. Disorders 9(1), 11-29.
Wing, L. & Gould, J. (1994). Diagnostic Interview for Social and Communication Disorders (DISCO). Third complete revision of the Handicaps, Behavours, and Skills Schedule.
World Health Organization (WHO)(1992). International Classification of Diseases and Disorders (ICD-10). Geneva, Switzerland.
World Health Organization (WHO)(1993). The ICD-10 Classification of Mental and Behavioral Disorders. Diagnostic Criteria for Research. Geneva, Switzerland.

RECOMMENDED READINGS

Frith, U. (Ed.)(1991). Autism and Asperger Syndrome. Cambridge University Press, Cambridge, England.
Gillberg, C. & Nordin, V. (1998). Long-Term course of autistic disorders: Update on follow up studies. Acta. Psychiatr. Scand. 97(2), 99-108.
Schopler, E., Mesibor, G.B., & Kunce, L. (Eds.) (1998). Asperger's Syndrome and High Functioning Autism. Plenum, New York.

Chapter 19

Personality Disorders

PETER TYRER and KATHRYN EVANS

Biological Psychiatry, pages 451-461.

ISBN: 1-55938-819-6

INTRODUCTION

Personality disorders constitute a set of conditions that are frequently encountered in psychiatric practice and, indeed, in everyday life. They are common conditions but are not necessarily associated with any other psychiatric disorder. It is only recently, however, that the diagnosis of the group has become sufficiently reliable to allow them to take their place alongside other mental health disorders. Despite this, there is much argument over the status of these conditions and there are many aspects of their classification that are unsatisfactory.

One of the reasons for this nonconsensus is that the term has become an extremely unattractive psychiatric label. No one likes being tagged with it as it includes many negative attributes. These include aggression, major problems in relationships, lack of consistency, excessive "help-seeking behavior", antisocial acts, and an apparent total inability to learn from past mistakes. There is therefore a danger that people who are not liked by the doctor (or by another therapist) will receive this diagnosis. This attribution is almost invariably followed by another—the notion that personality disorders are untreatable and consequently not worthy of therapist time.

Table 1. Main Characteristics of Personality Disorders (World Health Organization, 1992) and the Diagnostic and Statistical Manual for Mental Disorders (DSM-IV) (American Psychological Association 1994)

Persistent pattern of abnormal behavior	Impaired ways of perceiving and interpreting events from people, abnormal affectivity (emotional arousal and response), poor control over impulses, impairment of interpersonal relationships.
Dysfunctioning	Personality disorders show behavior that is inflexible and maladaptive and occur in a wide range of personal social situations
Suffering	Both the person and the environment suffer as a consequence of the persistent abnormal behavior shown in these disorders
Onset	Onset in late childhood or adolescence with relative stability of the condition subsequently
Relationship to mental state	Condition is not a consequence of a mental state disorder or a consequence of organic brain disease or injury

Note: (adapted from ICD-10)

WHAT IS PERSONALITY DISORDER ?

There is considerable argument over the definition of personality and therefore it is not surprising that the same will extend to personality disorders. The main difference between normal personality and disorder is that the former is adaptable and can adjust to the prevailing environment, whereas personality disorder is deviant and maladaptive so that there is usually a misfit with the environment. The other distinction that needs to be made is between a temporary disorder of mental state such as depression, which may apparently affect the personality but not in a long-standing way, and personality disorder itself, which is enduring. In the United States, personality disorders are given a separate axis of classification (Axis II) compared to the mental state disorders (Axis I)(American Psychiatric Association, 1994). However, in the world classification of psychiatric disorders, ICD-10 (Chapter V)(World Health Organizsation, 1992), personality disorder is grouped with the other mental and behavioral disorders on the same axis. The main characteristics of personality disorder are shown in Table 1. These confirm the picture of a persistent disorder, manifest mainly by problems of behavior and relationships leading to social dysfunction, which is independent of any mental state disorder.

CATEGORIES AND SEVERITY OF PERSONALITY DISORDER

There are hundreds of ways of describing normal personality and, although these are reduced for personality disorder, the validity of each separate condition remains an open question. Table 2 lists personality disorders by subdivision into clusters according to their comparative classifications in DSM IV and ICD-10.

Although each of the diagnoses has face validity, in practice, the phenomenon of co-morbidity (i.e., the simultaneous presence of two or more disorders) is the rule rather than the exception. For example, the most commonly diagnosed personality disorder in the United States and in most other countries of the world is borderline personality disorder (Loranger et al., 1994), but this condition alone is found in less than one in 10 of all those with personality disorder—the others including a mixture of co-morbidity with other personality disorders as well as mental state disorders (Fyer et al., 1988). It is generally found that the greater the severity of disorder, the more likely are the criteria for many different personality disorders to be satisfied (Oldham et al., 1992).

There is also a strong argument in favor of using a dimensional system of classification for personality disorders rather than the simple dichotomous classification that currently exists between "no personality disorder" and "personality disorder" (Widiger and Frances, 1985; Widiger, 1991). One advantage of a dimensional classification is that it would reduce the current pejorative use of the

Table 2. DSM IV and ICD-10 Personality Disorder Classification

	DSM IV	*ICD-10*
Cluster A	Paranoid Schizoid Schizotypal	Paranoid Schizoid
Cluster B	Antisocial Borderline Histrionic Narcissistic	Dissocial Emotionally unstable: • Impulsive type • Borderline type Histrionic
Cluster C	Obsessive–Compulsive Avoidant Dependent Not otherwise specified	Anakastic (obsessive–compulsive) Anxious (avoidant) Dependent Unspecified

diagnosis. Terms such as personality accentuation and difficulty (Tyrer et al, 1990) represent a lower level of severity and allow greater flexibility (as well as accuracy) in achieving a stable and reliable diagnosis (Clark, 1992). Because of the considerable overlap between different personality disorders, it is now common convention to divide the group into three major groups or clusters—the odd group or eccentricity, which is extremely withdrawn and is characterized by isolation and eccentricity (cluster A), the flamboyant cluster in which personality abnormality shows itself through prominent and dramatic behavior (cluster B), and the anxious or fearful group in which individuals are inhibited, frightened, and timid (cluster C) (Reich and Thompson, 1987).

The distinguishing features of the main three clusters of personality disorder are shown in Table 3. Presently, the most fashionable approach to classifying these conditions is using operational criteria, a technique that has been extremely effective in identifying other mental disorders but is less satisfactory for personality disorders. This is because personality disorders are, by definition, longterm, and are more reliably described in terms of tendencies and traits rather than clearly defined behavioral acts. In each of these conditions the main criteria listed in Table 1 also have to apply. This is an important issue because personality disorder, unlike all other mental diagnoses, is dependent to some extent on the culture from which it emanates.

It is recognized that the form of presentation of a specific disorder of mental state may be influenced greatly by cultural factors, but its incidence and prevalence are not. In the case of personality disorders, if the society from which the individual comes is itself unusual, certain abnormalities of personality may actually thrive and be successful. Thus, the antisocial personality disorder in Germany at the time of the Third Reich and possibly obsessive–compulsive disorder in

Table 3. Clinical Features of Personality Disorder

	DSM IV	*Main Features*	*ICD-10*
Cluster A (Withdrawn)	Paranoid	Suspiciousness, refusal to trust people, fear of conspiracy, persistent blaming of others, persistently bears grudges, morbid jealousy	Paranoid
	Schizoid	Detachment from social relationships and voluntary isolation, restricted range of expression, ignorance of social conventions	Schizoid
	Schizotypal	Social withdrawal, eccentricities, magical thinking, paranoid ideation, excessive social anxiety which does diminish with familiarity	
Cluster B (Flamboyant)	Antisocial	Short-tempered with a tendency to react with physical violence, insensitivity to feelings of others, failure to conform to social norms, blaming others, excessive irritability	Dissocial
		Tendency to act on impulse followed by regrets because of negative consequences, failure to plan ahead or to anticipate consequences, inability to maintain tasks	Emotionally unstable/ Impulsive
	Borderline	Instability of self image emotional liability, patterns of unstable and intense interpersonal relationships, recurrent suicidal behavior	Borderline
	Histrionic	Excessive emotionality and attention seeking tendency toward dramatization and manipulative behavior, suggestible, egocentric and labile	Histrionic
	Narcissistic	Grandiosity, need for admiration, lacks empathy, envious of others	
Cluster C (Fearful)	Obsessive–Compulsive	Excessive, conscientiousness in everyday activity, tendency to plan everything in great detail, unable to alter plans to suit changing needs	Anakastic
	Avoidant	Persistent tension and anxiety, social inhibition, feelings of inadequacy, hypersensitivity to negative evaluation, tendency to avoid social contacts	Avoidant
	Dependent	Excessive need to be taken care of, fear of separation, difficulty adjusting to negative events, excessive dependence and submissiveness to others	Dependent

Japan at the present day are adaptive (and therefore not classified as disorder). This is because personality disorder has to be recorded in the context in which it occurs. Because societies vary, personality disorder also may vary although perhaps not quite to the same extent that may be implied at first sight, as the maladaptive behavior of the personality disorder shows itself across all areas of social functioning and not just in the person's general relationship with society.

CAUSES OF PERSONALITY DISORDER

Both genetic and environmental factors are instrumental in causing personality disorder. A discrimination is made between temperament, a largely inherited set of personality dispositions (Buss and Plomin, 1986), and adult personality disorders, which have much less, but still significant, genetic contribution (Torgersen, 1984). There is now a great deal of evidence that major adverse experiences in childhood, particularly those involving inconsistent parenting and repeated sexual or physical abuse, lead to personality disorder in adult life (Herman et al., 1989) and, in many ways, such conditions can be thought of as delayed expressions of posttraumatic stress.

By contrast, those who, through temperamental attributes, may be predisposed to personality disorder can be protected from its expression by consistent caring and attachment figures in childhood. Environmental influences appear to be particularly influential in the etiology of the flamboyant (cluster B) personality disorders. Genetic influences are more important in the schizoid, schizotypal, and paranoid (cluster A) disorders (Kendler et al, 1993), and roughly equal genetic and environmental influences give rise to the anxious-fearful group (cluster C).

EPIDEMIOLOGY

In the past, personality disorder was considered to be a fairly uncommon condition affecting no more than around 3% of the population. With the introduction of more formal methods of assessment and the recognition that personality disorder often co-exists with mental state disorders (another form of co-morbidity), the evidence now suggests that approximately one in 10 members of a typical population suffers from a disorder of personality (de Girolamo and Reich, 1993). There is an unfortunate tendency to describe individuals as *either* suffering from a personality disorder *or* a mental illness whereas, in clinical practice, the two are often seen together.

The most common associations between the Axis I and Axis II disorders of the DSM classification are illustrated in Table 4.

In strict terms, this association should not really be described as co-morbidity because co-morbidity implies a presence of two separate "diseases" occurring at the same time in one person. For many of these conditions, the association is a genuine association rather than a chance one. In other words, the personality abnormality and the mental state disorder seem likely to be part of the same condition, although this still has to be confirmed. The most common association is between the cluster C personality disorders and the neurotic ones (sometimes called the general neurotic syndrome; Tyrer, 1985), the cluster B (flamboyant) personality disorders and substance misuse, and, to a much lesser degree, the eccentric cluster (A) and a schizophrenic disorder. This association has led some

Table 4. Summary of Relationships between Axis I and Axis II Disorders (DSM Classification)

Axis I disorder	*Axis II disorder (cluster)*	*Degree of association*	*Source*
Substance use	B	High	Brooner et al., 1993; Dejong et al., 1993
Schizophrenia	A and B	Low	Cutting et al., 1985
Mood disorders	B and C	Moderate	Alnaes and Torgersen, 1988 Gunderson and Phillips, 1991
Stress disorders	B and C	Moderate	Tyrer et al., 1988
Neurotic disorders	C	High	Tyrer, 1985 Flick et al., 1993
Eating disorders	B and C	Moderate	Dowson, 1992; Fahy et al., 1993
Somatoform disorders	C	High	Stern et al., 1993

Note: Association definition: High = >50%
Moderate = 25–50%
Low = 15–25%

authorities to judge that personality and mental-state disorders are best viewed on a spectral continuum with personality tendencies at one extreme and mental state disorders at the other (Akiskal, 1987; Siever and Davies, 1991).

TREATMENT AND COURSE OF PERSONALITY DISORDERS

Effective treatments for personality disorder are not yet unequivocally available. Much of the literature and investigations of personality disorder are studies of its natural history in the absence of proven interventions. The natural history is different for the three clusters and so they are worth discussing separately.

Cluster A (Odd Group)

These disorders tend to begin in late adolescence and to persist relatively unchanged to late life. There has been much debate about the relationship between this group of personality disorders and what used to be called childhood schizophrenia but now is called Asperger's syndrome. This condition was characterized by social isolation, lack of feeling for others, and introspective behavior unaffected by social conventions and was felt to be an enduring personality disorder (Asperger, 1944). What is clear is that there is no relationship between the condition of early infantile autism and this group of personality disorders but that some similar conditions that develop later in adolescence could be the early manifestations of this group of personality disorders (Wolff, 1993).

Cluster B (Flamboyant)

These conditions characteristically manifest themselves at an earlier age in other personality disorders, often in early adolescence. There is now abundant evidence dating from the pioneering studies of Lee Robins (1966) that antisocial and delinquent behavior in childhood is followed by similar behavior in adult life. On the positive side, however, these conditions tend to resolve at an earlier age than another personality disorders, and this may be as soon as the late twenties in some individuals. It is possible to view this group of disorders in developmental terms to regard them as immature personality disorders as opposed to most of the others, which are mature personality disorders that persist for most of adult life (Tyrer et al., 1991). This conclusion is, to some extent, supported by electroencephalographic studies that indicate a greater proportion of immature brain wave frequencies (theta rhythm: 5–8 Hz) in people with this group of disorders.

Cluster C (Anxious/Fearful)

These conditions manifest themselves in late adolescence or sometimes earlier and show some improvement later in life but at a later stage than the flamboyant group of personality disorders. There is some evidence that one group of personality traits, obsessional ones, become more prominent later in life (Tyrer and Seivewright, 1988), but this remains to be confirmed.

TREATMENT OF PERSONALITY DISORDER

As mentioned in the beginning of this chapter, there is a general notion throughout psychiatric practice that personality disorders are untreatable and to be avoided by those with a therapeutic bent. This view is slowly being eroded but remains predominant. One consistent finding is that disorders of mental state (Axis I) tend to have a better prognosis when they are not associated with a personality disorder than when they are (Reich and Green, 1991; Murphy and Tyrer, 1991; Tyrer et al., 1992), and so it is easy to see how the idea of untreatability has developed.

This is to some extent unfair because in most of these examples of "co-axial" diagnoses, the personality disorder has not been treated; it has merely been recognized as a sort of black shadow hanging over the disorder of mental state. When the personality disorder has been addressed as a treatable condition in its own right, the results are not all gloomy. A long time scale has to be taken before one can be certain that a treatment is effective, but there is now a range of evidence suggesting that certain drug treatments, particularly, low doses of antipsychotic drugs but also compounds such as carbamazepine and antidepressants such as phenelzine and fluoxetine, are beneficial in the cluster B group (Soloff et al., 1986; Cowdry and Gardner, 1988), and that tricyclic antidepressants help with the cluster C group (Tyrer et al., 1993). There is also a similar consensus developing over the value of psychotherapeutic intervention in borderline personality disorder (Higgitt and Fonagy, 1992) and of an interesting form of behavior therapy combined with cognitive elements called dialectical behavior therapy (Linehan et al., 1987). This is the first psychosocial intervention to have reduced the repetition rate of deliberate self harm (parasuicide) in borderline personality disorder and is being widely taught as part of managed care policies in the United States.

SUMMARY

Personality disorder constitutes a range of important conditions, which is as yet imperfectly defined, and represents a large part of psychiatric disorders and cannot be ignored by anyone involved in the management of mental illness. It is now realized that personality disorder is more common than even mental state disorders and affects approximately 10% of the adult population. It leads to considerable suffering, both to the individuals concerned and to others, and tends to adversely affect response to treatment and longterm outcome of mental disorders. However, in the last few years there have been optimistic reports supported by controlled clinical trials that these disorders can be helped by several forms of treatment, and this is beginning to alter their status as the diagnostic lepers in the community of psychiatric disorders.

REFERENCES

Akiskal, H.S. (1987). The boundaries of mood disorders: Implications for defining temperamental variants, atypical subtypes, and schizoaffective disorder. In: Diagnosis and Classification in Psychiatry: A Critical Appraisal of DSM-III. (Tischler, G., Ed.) Cambridge University Press, Cambridge.

Alnaes, R. & Torgersen, S. (1988). The relationship between DSM-III symptom disorders (Axis I) and personality disorders (Axis II) in an outpatient population. Acta Psychiat. Scand. 78, 348-355.

American Psychiatric Association (1987). Diagnostic and Statistical Manual of Mental Disorders. American Psychiatric Association, Washington, D.C.

American Psychiatric Association (1994). Diagnostic and Statistical Manual of Mental Disorders, 4th revision. American Psychiatric Association, Washington, D.C.

Asperger, H. (1944). Die autistischen Psychopathen im Kindesalter. Arch. Psychiat. Nervenkrank. 177, 76-137.

Brooner, R.K., Greenfield, L., Schmidt, C.W., & Bigelow, G.E. (1993). Antisocial personality disorder and HIV infection among intravenous drug abusers. Am. J. Psychiatry 150, 53-58.

Buss, A.H. & Plomin, R. (1986). The EAS approach to temperament. In: The Study of Temperament: Changes, Continuities and Challenges. (Plomin, R. & Dunn, J., Eds.) Lawrence Erlbaum, Hillside, New Jersey.

Clark, L.A. (1992). Resolving taxonomic issues in personality disorders. J. Personal. Dis. 6, 360-376.

Cowdry, R.W. & Gardner, D.L. (1988). Pharmacotherapy of borderline personality disorder: Alprazolam, carbamazepine, trifluoperazine and tranylcypromine. Arch Gen. Psychiatry 45, 111-119.

Cutting, J., Cowen, P.J., Mann, A.H., & Jenkins, R. (1986). Personality and psychosis: Use of the standardized assessment of personality. Acta Psychiat. Scand. 73, 87-92.

de Girolamo, G. & Reich, J.H. (1993). Personality disorders. World Health Organization, Geneva, Switzerland.

DeJong, C.A.J., Van den Brink, W., Harteveld, F.M., & van der Wielen, E.G.M. (1993). Personality disorders in alcohol and drug addicts. Compr. Psychiatry 34, 87-94.

Dowson, J.H. (1992). Associations between self-induced vomiting and personality disorder in patients with a history of anorexia nervosa. Acta Psychiat. Scand. 86, 399-404.

Fahy, T.A., Eisler, I., & Russell, G.F.M. (1993). Personality disorder and treatment response in bulimia nervosa. Br. J. Psychiatry 162, 765-770.

Flick, S.N., Roy-Byrne, P.P., Cowley, D.S., Shores, M.M., & Dunner, D.L. (1993). DSM-III-R personality disorders in a mood and anxiety disorders clinic: Prevalence, comorbidity and clinical correlates. J. Affect. Dis. 27, 71-79.

Fyer, M.R., Frances, A.J., Sullivan, T., Hurt, S.W., & Clarkin, J. (1988). Co-morbidity of borderline personality disorder. Arch Gen. Psychiatry 45, 348-352.

Goldberg, S.C., Shulz, S.C., Shulz, P.M., Resnick, R.J, Haymer, R.M., & Friedel, R.O. (1986). Borderline and schizotypal personality disorders treated with low dose thiothixene versus placebo. Arch Gen. Psychiatry 43, 680-686.

Gunderson, J.G. & Phillips, K.A. (1991). A current view on the interface between borderline personality disorder and depression. Am. J. Psychiatry 148, 967-975.

Herman, J.L., Perry, J.C., & Van der Kolk, B.A. (1989). Childhood trauma in personality disorder. Am. J. Psychiatry 146, 490-495.

Higgitt, A. & Fonagy, P. (1992). Psychotherapy in borderline and narcissistic personality disorder. Br. J. Psychiatry 161, 23-43.

Kendler, K.S. & Eaves, L.J. (1986). Models for the joint effects of genotype and environment on liability to psychiatric illness. Am. J. Psychiatry 143, 279-289.

Linehan, M. (1987). Dialectical behavior therapy for borderline personality disorder: Theory and method. Bulletin of the Menninger Clinic 51, 261-276.

Livesley, W.J. & Jackson, D.N. (1993). The internal consistency and factorial structures of behaviors judged to be associated with DSM-III personality disorders. Am. J. Psychiatry 143, 1473-1474.

Loranger, A.W., Sartorius, N., Andreoli, A., Berner, W., Buchheim, P., Channabasavanna, S.M., Coid, B., Dahl, A., Diekstra, R.F.W., Jacobsberg, L.B., Mombour, W., Ono, Y., Regier, D.A., Tyrer, P., & von Cranach, M. (1994). The International Personality Disorder Examination: The WHO/ADAMHA International Pilot Study of Personality Disorders. Arch Gen. Psychiatry 51, 215-224.

Murphy, S.M. & Tyrer, P. (1991). A double-blind comparison of the effects of gradual withdrawal of lorazepam, diazepam, and bromazepam in benzodiazepine dependence. Br. J. Psychiatry 158, 511-516.

Oldham, J.M., Skodol, A.E., Kellman, H.D., Hyler, S.E., Rosnick, L., & Davies, M. (1992). Diagnosis of DSM-III-R personality disorders by two semistructured interviews: Patterns of comorbidity. Am. J. Psychiatry 149, 213-220.

Reich, J. & Thompson, W.D. (1987). DSM-III personality disorder clusters in three generations. Br. J. Psychiatry 150, 471-475.

Reich, J.H. & Green, A.I. (1995). Effect of personality disorders on outcome of treatment. J. Nerv. Ment. Dis. 179, 74-82.

Robins, L.N. (1966). Deviant Children Grown Up. Williams & Wilkins, Baltimore.

Shea, M.T., Pilkonis, P.A., Beckham, E., and et al., (1990). Personality disorders and treatment outcome in the NIMH treatment of depression collaborative research program. Am. J. Psychiatry 147, 711-718.

Shea, M.T., Widiger, T.A., & Klein, M.H. (1992). Comorbidity of personality disorders and depression: Implications for treatment. J. Consulting Clin. Psychol. 60, 857-868.

Siever, L.J. & Davis, K.L. (1991). A psychobiologic perspective on the personality disorders. Am. J. Psychiatry 148, 1647-1658.

Soloff, P.H., George, A., Nathan, R.S., Schulz, P.M., Ulrich, R.F., and Perel, J.M. (1986). Progress in pharmacotherapy of personality disorders: A double blind study of amitriptyline, haloperidol, & placebo. Arch Gen. Psychiatry 43, 691-697.

Stern, J., Murphy, M., & Bass, C. (1993). Personality disorders in patients with somatisation disorder: A controlled study. Br. J. Psychiatry 163, 785-789.

Stone, M.H. (1993). Long-term outcome in personality disorders. Br. J. Psychiatry 162, 299-313.

Tyrer, P. (1985). Neurosis divisible? Lancet 1, 685-688.

Tyrer, P., Casey, P., & Ferguson, B. (1991). Personality disorder in perspective. Br. J. Psychiatry 159, 463-471.

Tyrer, P., Casey, P., & Gall, J. (1983). The relationship between neurosis and personality disorder. Br. J. Psychiatry 142, 404-408.

Tyrer, P. & Seivewright, H. (1988). Studies of outcome. In: Personality disorders: Diagnosis, Management, and Course. (Tyrer, P., Ed.), pp. 119-136. Butterworth, London (Wright).

Tyrer, P., Seivewright, N., Ferguson, B., Murphy, S., Darling, C., Brothwell, J., Kingdon, D., & Johnson, A.L. (1990). The Nottingham Study of Neurotic Disorder: Relationship between personality disorder and symptoms. Psychol. Med. 20, 423-431.

Tyrer, P., Seivewright, N., Ferguson, B., Murphy, S., & Johnson, A.L. (1993). The Nottingham Study of Neurotic Disorder: Effect of personality status on response to drug treatment, cognitive therapy, and self-help over two years. Br. J. Psychiatry 162, 219-226.

Tyrer, P., Seivewright, N., Ferguson, B., & Tyrer, J. (1992). The general neurotic syndrome: A coaxial diagnosis of anxiety, depression and personality disorder. Acta Psychiat. Scand. 85, 201-206.

Widiger, T.A. (1991). Personality disorder dimensional models proposed for DSM-IV. J. Personal. Dis. 5, 386-398.

Widiger, T.A. & Frances, A.J. (1985). The DSM-III personality disorders: Perspectives from psychology. Arch Gen. Psychiatry 42, 615-623.

World Health Organisation (1992). ICD-10: Classification of Mental and Behavioral Disorders. World Health Organization, Geneva, Switzerland.

Chapter 20

Alcoholism, Alcohol Abuse, and Alcohol Dependence

DAVID H. OVERSTREET and
ALEXEY B. KAMPOV-POLEVOY

Biological Psychiatry, pages 463-485.

ISBN: 1-55938-819-6

INTRODUCTION

Many individuals in the U.S. and other countries are still consuming alcohol at rates that put their own health and frequently the safety of others in jeopardy. Although these individuals may still be classified as alcoholics in lay terms, workers in the field now use two more clearly recognizable terms to describe problems with alcohol drinking. Alcohol abuse is commonly defined as "repetitive patterns of drinking in harmful situations with adverse consequences, including impaired ability to fulfill responsibilities. "(U.S. Dept. Of Health and Human Services, 1997, p. 2). Alcohol dependence refers to a more severe condition characterized by "impaired control over intake, withdrawal from alcohol, tolerance, and drinking despite problems " (U.S. Dept. of Health and Human Services, 1997, p. 1). In order to treat these conditions, which can occur in over 10% of the populations in some countries, it is important to know more about the patterns of alcohol use and how this commonly abused drug exerts its effects in the brain.

This chapter will briefly survey recent trends in alcohol consumption, including changes in the number of individuals who abuse or are dependent upon alcohol, as a background to outlining some of the key effects alcohol has on neurotransmitters and behaviors. Finally, a brief survey of treatment/management approaches to alcohol abuse and dependence will be presented.

EPIDEMIOLOGY OF ALCOHOL USE

Trends in Alcohol Consumption

The use of alcohol has been declining nationwide over the past 16 years, with the value in 1993 being as low as the value in 1964 (Williams et al., 1995). This trend in declining consumption is attributable mostly to a fairly large decrease in the consumption of spirits (Williams et al., 1995); the consumption of wine and beer has decreased to a lesser degree. The decline in alcohol consumption in the United States has been mirrored by the majority of other countries in the world for which reliable statistics have been gathered (Edwards et al., 1994). Thus, there has been a worldwide decrease in per capita consumption of alcohol in recent years.

Recent evidence suggests that the decrease in consumption does not apply to all subgroups. Midanik and Clark (1994) presented a much more detailed analysis of changes in drinking patterns between 1984 and 1990. They found that there were no changes in % of current drinkers, but there were decreases in the % of weekly drinking (for both males and females) and the percent of drinkers of more than five drinks (males only). The decreases in heavy drinking were seen only in individuals with incomes above the median, fully employed individuals, married individuals, or individuals with a high school education (Midanik and Clark, 1994). Thus, the decline in alcohol consumption seems to be confined to the higher socioeconomic groups.

Adverse Consequences of Drinking

Excessive drinking can have a number of adverse consequences, both for the individual drinker (e.g., liver cirrhosis) and for friends or even unrelated individuals (e.g., alcohol-related car crashes). This section will present a brief survey of these two topics to illustrate the extent of the problems with excessive alcohol drinking.

Cirrhosis is characterized by the gradual replacement of normal liver tissue with abnormal tissue, leading to scarring, liver failure, and death. This irreversible condition was the 11th leading case of death in 1992, accounting for 25,407 deaths (DeBakey et al., 1995). Even so, the number of deaths from cirrhosis has steadily declined from a peak in 1973 (DeBakey et al., 1995). Although this decrease parallels the decline in alcohol consumption over the same period, it is not clear whether these two variables are directly related. It is known that the rate of cirrhosis-related deaths is increased in heavy drinkers (DeBakey et al., 1995) and that the rate of heavy drinking has declined (see above), but factors such as nutrition may have also contributed. Nevertheless, cirrhosis is still a major health consequence to the individual who drinks excessive amounts of alcohol.

Alcohol has been implicated in 40% to 45% of fatal traffic accidents over the past 20 years (Campbell et al., 1995). A determination of the involvement of alcohol is made based on one of three factors: the judgement of the investigating officer, a reading on blood alcohol concentration, or an arrest for driving under the influence (DUI). If any one of these is checked, then the accident is considered alcohol-related. Although the number of alcohol-related fatal accidents has been declining over the past several years, there is still a concern that the younger age group (16 to 24 years old) who represent 15% of the driving population are involved in a disproportionate number of the alcohol-related (28%) fatalities (Campbell et al., 1995). Another uncertainty in this area is how many of the deaths are caused by individuals who are dependent on alcohol and how many are caused by social drinkers. Since alcohol treatment programs are typically aimed at alcohol-dependent individuals, it is also unclear whether these treatment programs will have a significant impact on alcohol-related traffic accidents. Such uncertainty has led to the creation of education/awareness programs at several college campuses, including the University of North Carolina at Chapel Hill, where the negative consequences of binge drinking are discussed. It will require several years to assess the impact of such programs.

Alcohol Abuse and Alcohol Dependence

The prevalence of alcohol abuse and alcohol dependence has been examined by a number of groups, and there are differences among these surveys due to differences in diagnostic criteria and other factors. A comprehensive study of the 1992 population has concluded that the overall prevalence rate of alcohol dependence and abuse is 7.41% (Grant et al., 1994). However, the prevalence rates vary tremendously according to age (the 18 to 29 age group has more than double the overall prevalence rate), gender (the prevalence rate for males is more than twice that for females), and race (the prevalence rate for nonblacks is higher than that for blacks).

CORRELATES/MARKERS FOR ALCOHOL ABUSE/DEPENDENDCE

Studies conducted in humans have convincingly demonstrated that alcoholism runs in certain families. Evidence suggests that as many as 80% of alcoholics may have close biological relatives with a history of alcohol-related problems. The average risk for developing alcoholism is seven times greater among first degree relatives of alcoholics than among controls. Male relatives of male alcoholics appear to be at particular risk for the disease. Adoption and twin studies also indicate that the risk for alcoholism may be determined at least partially by the genetic factors.

Heterogeneity of Alcoholism

The large Swedish adoption study of Cloninger, Bohman, and colleagues deserves special mention because of its evidence for a genetic heterogeneity of alcoholism. Based on that study, it was suggested that genetic transmission was involved in the development of at least two forms of alcoholism. One form, referred to as Type 2, was found to be limited to males and associated with early-onset drinking and antisocial characteristics. The other, called Type 1, was not limited to males, was associated with late-onset drinking without antisocial features, and revealed evidence for a gene/environment interaction. Though methodological problems have been noted with the design employed in this impressive study (e.g., the use of medical registrations to identify alcoholism as opposed to direct interview), the data strongly support a genetic role in the transmission of more than one form of alcoholism suggesting that different phenotypic expressions should be detectable across subtypes as well.

The evidence from genetic studies that alcoholism is heterogeneous bolstered earlier phenomenological observations of more than one form of alcoholism. The concept of the heterogeneity of alcoholism has a rich history. Jellinek proposed one of the first typologies of alcoholism that captured the notion that the origin and course of alcoholism may be influenced by cultural, psychological, and physiological factors. Recently, interest in the phenomenology of alcoholism has been rekindled in an attempt to understand the heterogeneity of the illness and its relevance for prognosis, treatment, and pathophysiology.

The efforts to identify meaningful subtypes of alcoholism based on clinical phenomenology have utilized a variety of populations, assessment methods, and research designs. Therefore, it is difficult to directly compare findings across studies. Furthermore, there is controversy in the field in at least two areas: (1) Should individuals with antisocial personality disorder and alcohol problems represent a subgroup of alcoholics or rather a primary personality disorder with secondary alcoholism? (2) Do subgroups of alcoholics represent distinct diagnostic groups or rather clusters of extremes on one or more phenomenological/temperamental dimensions?. Despite these unanswered questions, a degree of convergence is evident in recent studies of the clinical phenomenology of alcoholism.

Genetic Markers for Alcohol Dependence

The discovery of the role of heredity in the development of alcoholism stimulated a search for genetic markers of this disease. Review of the literature shows that, over the last two decades, a number of markers of the inherited vulnerability to alcoholism have been identified. These markers can be grouped into three categories:

1. Electrophysiological markers: (a) an excess of fast electroencephalographic (EEG) activity (beta activity); (b) a greater power in the fast EEG alpha range; (c) a reduced amplitude of P300 waveform of event related potentials.
2. Biochemical markers: (a) low platelet MAO activity, at least in Type 2 alcoholics; (b) low platelet adenylate cyclase activity after stimulation with cesium fluoride and prostaglandin E_1; (c) higher platelet serotonin uptake rate in high risk individuals.
3. Differences in reactions to alcohol. (a) a higher alcohol-induced increase in baseline heart rate in individuals at high risk for alcoholism compared to individuals at low risk; (b) reduced intoxication and motor impairment to alcohol in individuals at high risk for alcoholism compared to individuals at low risk.

Evidence for a Relationship between Intake of Sweets and Alcohol

In Animals

One of the most intriguing recent findings in animals is an association between consumption of sweets and alcohol. Alcohol-preferring (P) rats were shown to consume greater amounts of sucrose solutions compared to alcohol-nonpreferring (NP) rats. It was also shown that the association between consumption of sweets and alcohol is based not on the high caloric value of the sweet substance but on its hedonic (pleasurable) value. Therefore, consumption of the artificial sweetener such as saccharin may also be associated with subsequent alcohol intake. Since the first reportings of these findings in 1978, the results have been replicated in a variety of mouse and rat strains/lines. The correlation between consumption of saccharin solution and alcohol in these studies was reported to be as high as 0.8. This makes saccharin consumption a reliable predictor of alcohol intake in animals.

In animals, the high consumption of sweet solutions in alcohol-preferring animals has been attributed to two major characteristics. One is a tendency to consume sweet solutions far beyond the limits of normal daily fluid intake; the other is the preference for more concentrated sweet solutions.

In Humans

The review of the clinical literature does not provide the direct indication of association between alcoholism and sweet-liking. However, a number of researchers reported a high level of co-morbidity between eating disorders and substance abuse. In a survey of the literature, Jonas found that alcohol abuse occurs in 14% to 60% of subjects with eating disorders, and other forms of drug abuse occur in 10% to 31%. Overall, the data suggest that between 24% and 65% of individuals with eating disorders will have a present or past diagnosis of some form of substance abuse. Conversely, 20% to 40% of substance abusers were found to have a past or present history of eating disorders. More recent studies support this hypothesis and suggest that manifestations of bulimia and alcohol/drug abuse are not only related, but they also can substitute for each other. Some authors even suggest that alcoholic bulimia patients form a distinct clinical subgroup. There are indications that bulimics with the selective craving for carbohydrates ("sweet bulimia") are especially prone to alcoholism and substance abuse. These findings support the hypothesis regarding potential usefulness of sweet-liking as a marker for alcoholism.

The relationship between sweet-liking and alcoholism was more directly examined recently at the University of North Carolina. The subjects were given five concentrations of sucrose solution in random order and were asked to rate how sweet it was and how much they liked the solution. Sixty-five percent of alcoholics preferred the higher sucrose concentration (0.83M) compared to just 16% of the control group (Kampov-Polevoy et al., 1997). Therefore, sweet-liking and/or consumption is highly associated with high alcohol consumption in both animals and humans.

INTERACTIONS OF ALCOHOL WITH NEUROMODULATORS

In recent years, how alcohol interacts with neuromodulators in the brain has been substantially clarified. It seems clear now that alcohol does not produce its intoxicating or sedative effects by some nonspecific actions such as a change in the fluidization of the cellular membrane. Rather, alcohol selectively interacts with specific neurotransmitter receptors to produce many of its effects. A detailed description of these interactions is beyond the scope of this chapter. A brief synopsis of alcohol's interactions with several key neurotransmitters is given below. A more comprehensive list of alcohol's targets in the brain is given in Table 1 (See Department of Health and Human Services, 1997).

GABA

The inhibitory neurotransmitter γ-aminobutyric acid (GABA) is widely distributed throughout the nervous system. Alcohol may produce its sedative effects, in part, by interacting with subtypes of the GABA-A receptor. Its interaction with these receptors facilitates the action of GABA itself, leading to a greater conductance of chloride into the cell and an inhibitory effect. However, by applying alcohol directly to specific brain regions, it has been found that the variation in sensitivity to alcohol in these regions is correlated with the subtypes of GABA-A receptors that exist in those regions (Criswell et al., 1993).

Table 1. Brain Targets for Alcohol

Target	*Alcohol's Effects*	*Potential Behavioral Effects*
GABA-$_A$ Receptor	Stimulates Function	Sedation; intoxication
NMDA Receptor	Blocks Function	Sedation; intoxication
5-HT_3 Receptor	Stimulates Function	Rewarding Properties
Nicotinic Acetylcholine Receptor	Mixed	Uncertain
ATP Receptor	Blocks Function	Uncertain
Calcium Channels	Blocks Function	Sedation?
G-Protein-Coupled Receptors	Mixed	Uncertain

Glutamate

The excitatory neurotransmitter, glutamate, is also widely distributed throughout the nervous system. Recent investigations of the interaction of alcohol with the receptors for glutamate have indicated that alcohol may selectively block one subtype of receptor—the one for *N*-methyl-D-aspartate (NMDA). This blockade is believed to contribute to the intoxicating effects of alcohol because other drugs known to block these receptors also produce intoxicating effects. As for GABA-A receptors, there is a regional sensitivity of NMDA receptors to alcohol, and this might be related to different subtypes in the different brain regions (Simson et al., 1993).

Serotonin

Some of the effects of serotonin (5-hydroxytryptamine, or 5-HT) are mediated via the 5-HT3 receptor, which like the GABA-A and NMDA receptors, is a ligand-gated ion channel. Therefore, it is not surprising that alcohol also interacts with the 5-HT3 receptor. Alcohol interacts directly with the 5-HT3 receptor to increase the conductance of sodium and potassium, but it also potentiates the effects of agents that also interact with the 5-HT3 receptor (Peoples et al., 1996). Alcohol will also increase the release of serotonin as well as dopamine in certain brain regions, but it is not clear whether this effect has any relationship to alcohol's interaction with 5-HT3 receptors. Several investigators have demonstrated that the increase in dopamine release induced by alcohol appears to be dependent upon 5-HT3 receptors, and it has been suggested that these receptors may be involved in the rewarding effects of alcohol.

Dopamine

Unlike the neurotransmitters described above, alcohol does not interact with dopamine receptor subtypes. Instead, alcohol induces an increase in the release of dopamine in key brain regions such as the nucleus accumbens. It is commonly believed that this increase in the release of dopamine is the mediator of the rewarding properties of alcohol because it seems to be involved in the rewarding effects of cocaine and other drugs of abuse (Koob, 1992).

Other Targets

As indicated in Table 1, alcohol may also interact with a number of other targets in the brain. The reader is referred to the 1997 report and Alcohol and Health for further information and references (Department of Health and Human Services, 1997).

NEUROBEHAVIORAL EFFECTS OF ALCOHOL

Through its interaction with specific membrane proteins in the brain, alcohol produces a variety of behavioral effects. The behaviors selected for discussion in this section represent some of the most carefully characterized or the most problematic effects of alcohol.

Intoxication

The induction of intoxication is probably the best known action of alcohol, in part because it is commonly depicted in the movies. However, the effects of alcohol on motor behavior is more complex than the typical popularized account of alcohol intoxication. As the dose of alcohol is increased, the following changes in behavior progressively occur. Initially, at low doses, alcohol may produce disinhibition such that the individual is more talkative and the experimental animal is more active. At intermediate doses, there is a mixture of stimulatory and inhibitory effects such that the individual's pressured speech becomes slurred and the experimental animal's activity becomes discoordinated. Finally, at higher doses, clear-cut sedative effects are seen in both animals and humans. A variety of biological and environmental factors are known to affect intoxicating responses to alcohol. For example, it has been possible to develop strains of rats and mice that are differentially sensitive to the sedative effects of alcohol by selective breeding techniques (e.g., Dietrich, 1993). The fact that sons of alcoholics tend to be less sensitive to the intoxicating effects of alcohol (Schuckit, 1994) is further evidence to support the view that genetic factors play an important role in influencing the intoxicating effects of ethanol.

Reinforcement

Although it is well known that many humans and experimental animals will readily self-administer alcohol, there are tremendous individual differences. As indicated above, data from 1992 indicate that 7.4% of people in the United States were suffering from alcohol dependence and/or abuse. As many as 25% of the population are not current drinkers in most surveys. In a typical laboratory rat population, the majority of rats do not readily drink much alcohol, but a certain number readily drink substantial quantities. This observation has led several investigators to develop lines of rats that drink very high amounts of alcohol by selective breeding approaches (Li et al., 1993). It has also been recognized that inbred mouse strains vary in the quantities of alcohol they will self-administer, and investigators are actively engaged in genetic studies to detect genes (quantitative trait loci; QTL) that influence alcohol drinking behavior in these mice (Phillips et al., 1994). Finally, it has been recently recognized that a mouse knockout for the 5-HT1B gene self-administers alcohol at a higher rate, suggesting that the 5-HT1B receptor might influence alcohol intake (Crabbe et al., 1996).

There are also many developments in the study of genetic contributions to alcohol dependence and abuse in humans. Because alcoholism is a heterogeneous disorder that is most likely influenced by a variety of genes, progress in detecting genes that influence alcoholism has been slow and distracted by controversy. Perhaps the most controversial area is the failure of many investigators to replicate an early report, which suggested that a particular polymorphism of the dopamine 2 receptor was associated with alcoholism (Gelertner et al., 1993).

Aggression

The aggression-promoting properties of alcohol consumption in humans has been frequently popularized in the entertainment industry, so almost everyone is aware that there is a connection. However, it should be stressed that the large majority of individuals who use and even abuse alcohol do not become aggressive or violent. The mechanisms underlying the aggressive behavior induced by alcohol in a subset of individuals have not been clearly established, but there is some evidence that an underlying, possibly genetic deficiency of serotonin may contribute (Virkkunen et al., 1994).

A number of preclinical investigations have suggested that the effects of alcohol on aggressive behavior in animals may be mediated by some of the same neurotransmitters with which alcohol is known to interact. Firstly, it must be emphasized that alcohol tends to stimulate aggressive behavior in animals at low to intermediate doses (Miczek et al., 1994), the same doses that tend to stimulate locomotor activity. These proaggressive effects of alcohol can be blocked by GABA-A receptor antagonists and potentiated by GABA-A agonists. It has also been demonstrated that increases in serotonin may reduce the stimulatory effects of alcohol on aggressive behavior in animals. Thus, individuals who have deficiencies in their central serotonergic systems (Virkkunen et al., 1994) may be more susceptible to the proaggressive effects of alcohol. However, more study is needed on these interrelationships because it is not clear how the serotonin elevation that is produced by alcohol itself, as described above, might be involved in the proaggressive effects of alcohol.

Learning and Memory

Both acute and chronic alcohol consumption and/or treatment can impair learning and memory, although possibly through different mechanisms. Since the early work of Donald Overton, it has long been known that alcohol can produce a phenomenon referred to as state-dependent learning. For example, a rat that learns a specific task while under the influence of alcohol, exhibits no evidence of having learned the task when given a vehicle. Conversely, a rat that learns a task when given the vehicle may not perform well when it is administered alcohol. These findings could be interpreted to suggest that acutely administered alcohol might affect both learning and memory processes. However, a number of more recent studies using more sophisticated behavioral tasks (e.g., see Melchior et al., 1993) have suggested that alcohol interferes with memory processes selectively.

The mechanisms underlying the acute effects of alcohol on memory might involve some of the neurotransmitters described above. The most likely candidate underlying alcohol's memory-disruptive effects is glutamate, the excitatory amino acid. This neurotransmitter, via the NMDA receptors, is believed to mediate the form of neuronal plasticity known as longterm potentiation (Lister et al., 1987). Because alcohol is known to act as an antagonist at the NMDA receptor and to inhibit longterm potentiation, it is probable that a blockade of NMDA receptors may contribute to alcohol's effects on memory.

So far, the available evidence suggests that a different neurotransmitter system may be involved in the learning and memory deficits, which are associated with chronic alcohol abuse. Firstly, it should be emphasized that chronic alcoholics may have nutritional deficiencies in addition to their alcohol abuse and that brain damage is commonly seen in these individuals, in contrast to those individuals/animals that take alcohol acutely. Regions of the cerebral cortex and the hippocampus have exhibited degenerative processes, and it has been suggested that cholinergic neurons are among those nerve cells, which have degenerated, because it has been possible to improve the memory of animals chronically exposed to alcohol by giving either pharmacological treatments that increase acetylcholine, brain tissue transplants that contain cholinergic neurons, or neurotrophic facts that stimulate acetylcholine (Beracochea et al., 1992; Bruckner and Arendt, 1992). The applicability of these techniques to improving the memory of chronic alcoholics has not been established as yet. It is probably better to focus on treatments that will lead to a reduction in drinking among alcohol abusers so that the memory and cognitive deficits can be prevented. Some of the treatment approaches used will be discussed later.

Tolerance and Physical Dependence

Individuals who drink or animals given large amounts of alcohol chronically become tolerant to the effects of alcohol. Consequently, they can consume or be given larger amounts of alcohol. For example, it has been determined that alcoholics who have developed tolerance may appear sober at blood alcohol concentrations of 230 to 460 mg/dl, 23 to 46 times the level used in many states for driving under the influence determinations (Charness et al., 1989). Because tolerance does develop to many of the effects of alcohol, it is likely that chronic abusers increase their drinking so that the levels of alcohol reach toxic levels for some regions or organs.

Tolerance to the effects of alcohol most likely involves multiple mechanisms. It is clear that the liver enzymes responsible for the metabolism of alcohol increase in activity; therefore, more alcohol is needed to reach the same blood levels. However, neuronal adaptations must also be involved because, as reported above, tolerant individuals can appear to function very well at blood alcohol levels, which would severely impair the behavior of an alcohol-naïve subject. There has been a tremendous amount of research on the neuronal mechanisms underlying tolerance to alcohol; in fact, the whole of this chapter could have easily been devoted to this topic itself. This section will only consider GABA and glutamate mechanisms because they have been the most thoroughly examined.

Before discussing these systems, another aspect of alcohol's chronic effects must be mentioned. It is well known that chronic alcohol consumption also leads to physical dependence, a state of quasi-normality that requires alcohol to be present. When alcohol is withdrawn from an alcohol-dependent person or animal, several characteristic withdrawal signs can be noted. These symptoms include agitation, tremulousness, and autonomic hyperactivity characterized by sweating, increased blood pressure, and elevated heart rate (Charness et al., 1989) and are normally observed in eight to 12 hours after withdrawal from alcohol. The more severe withdrawal sign, delirium tremens, which includes the above symptoms and hallucinations, may appear within two days after alcohol withdrawal and last up to a week.

It will be noted that the constellation of symptoms observed in the alcohol-withdrawn individual or animal is opposite in nature to the sedative effects observed after acute alcohol administration. Because both GABA and glutamate systems appear to mediate alcohol's acute sedative effects, it is possible that adaptations in these neurotransmitter systems are involved in the development of tolerance to and physical dependence upon alcohol. Indeed, investigators have found that treatment with benzodiazepines, which interact with the same GABA-A receptors as alcohol, will reduce many of the alcohol withdrawal symptoms, and so will administration of the NMDA receptor agonist MK-801. Thus, the excess excitability of the alcohol-withdrawn brain is the result of both an adaptation of the glutamate system toward more excitation to overcome the blockade induced by alcohol and an adaptation of the GABA system to less inhibition to overcome the chronic stimulation induced by alcohol. Changes in GABA-A or NMDA receptor subtypes may play a role in these adaptive responses.

TREATMENT/MANAGEMENT OF ALCOHOLISM

As indicated above, there are not any successful treatment strategies to control or reverse some of the more tragic consequences of chronic alcohol abuse and dependence. Neither cirrhosis of the liver nor the impairments of cognitive behavior associated with brain damage are amenable to correction given our current knowledge about alcohol's actions. Although the extensive research in progress might lead to certain successful palliative therapies, it might be better to focus our attention on developing preventative and therapeutic strategies to create a reduction in alcohol drinking in alcohol abusers. The present sections considers some of those therapies, some currently in use and others that might have promise in the future.

Behavioral Therapies

A wide variety of behavioral techniques have been used in an attempt to influence the alcohol-dependent individual to reduce or control his/her drinking, including self-help groups such as Alcoholics Anonymous (AA) and extensive individual counseling with a therapist specializing in drug abuse counseling.

Brief Interventions

The use of brief interventions in an attempt to reduce drinking in individuals who abuse alcohol has been more successful than control groups, which have been simply screened for alcohol use, and is generally nearly as successful as more intense interventions (Bien et al., 1993). Brief interventions are different from most other treatments for alcohol abuse because of their short duration, primary care setting, performance by general counselors, focus on individuals at risk for alcohol dependence, and goal of moderate drinking rather than abstinence.

The fact that brief interventions are modestly successful (Bien et al., 1993) supports the use of this strategy as a first option in the management of alcohol abuse and dependence. Other characteristics of brief intervention, which make it a particularly useful strategy, include the following: (1) many people who are not dependent upon alcohol still drink at dangerous levels and need assistance; (2) it is usually relatively inexpensive and, as indicated above, does not require a specialized drug abuse counselor; and (3) brief intervention need not be regarded as the exclusive treatment received by the patient, as other options can follow later.

The effectiveness of brief intervention is only modest, as suggested above. It appears that men may benefit more from this procedure than women (Babor and Grant, 1992), but it is not clear why this is so. Other investigators identified six variables (Miller and Sanchez, 1994) seemingly important to influencing the effectiveness of brief intervention:

1. feedback
2. responsibility
3. advice
4. menu
5. empathetic style
6. self-efficacy

Thus, a great deal is known about the effectiveness of brief intervention in reducing alcohol intake in individuals who abuse alcohol; it has modest shortterm benefits. Nevertheless, there are often barriers to its use because of reluctance on the part of general practitioners, inattention to screening for alcohol problems, and lack of incentives by health insurance companies.

Alcoholics Anonymous

This self-help approach to maintain sobriety in individuals who formerly abused alcohol has been very popular as indicated by a recent survey. Approximately 9% of those people surveyed had attended an AA meeting at sometime during their life, and 3.6% had attended an AA meeting within the last year (Room and Greenfield, 1993). Attendance is more likely at AA meetings than other therapy group sessions, including other 12-step programs.

Although an association with AA seems to work well for some people, it does not work well for others. The reasons for these differences and the basis for the success of AA as a treatment intervention are not known at this time. In recent years, investigators have begun to explore the reasons for AA's success and it has been reported that individuals who attended AA meetings regularly reported a greater use of behavioral change mechanisms (Snow et al., 1994).

Despite its relative success (67% of Finnish AA members who maintained sobriety for one year continued to go to AA meetings), the emphasis on sobriety might make AA an unattractive choice for some individuals. There is some evidence that controlled drinking can be a treatment option for certain patients, particularly those with less severe dependence symptoms who reject abstinence as a treatment goal (Miller et al., 1992).

Relapse Prevention

The goal of any alcohol-treatment program, including the brief intervention strategy mentioned above, is to prevent relapse or substantially lengthen the time of sobriety before relapse occurs. Since alcohol abuse and dependence is a complex, multifactorial disorder, it will not be surprising that the treatment strategies employed to prevent relapse are also multifaceted and complex. While space does not permit a full discussion of these strategies, we will call attention to just a few of the key issues in developing a strategy to prevent relapse in alcohol-dependent individuals.

Over the years, it has been recognized that the success of any treatment program may depend as much on the characteristics of the patient and the counselor as it does on the specific counseling technique. For example, counselors who use a confrontational approach tend to be less successful because they engender resistance in their patients. Because of this recognition, a comprehensive multisite study has been conducted in an attempt to match the patient characteristics with the type of therapy. This approach, known as Project MATCH, has been ongoing for many years, and the first set of comparative results already has been published (Project MATCH Research Group, 1997).

The Project MATCH study randomly assigned 952 outpatient clients and 774 clients with aftercare (i.e., following inpatient or intensive care hospital treatments) to one of three treatment approaches: twelve step facilitation therapy, cognitive behavioral therapy, and motivational enhancement therapy. The results indicated that all three treatment approaches were similarly successful, with one exception. Clients low in psychiatric severity benefitted more from the twelve step facililtation therapy than the cognitive behavioral therapy. While other client attributes such as motivational readiness, network support for drinking, alcohol involvment, gender, and sociopathy were predictive of drinking outcomes over time, they did not interact with treatment strategy (Project MATCH, 1997). Thus, except for psychiatric severity, it does not seem to matter which treatment is applied to a specific client.

Pharmacotherapies

Medications, which have been used to influence the individuals who are dependent upon alcohol, can be divided into three categories:

1. Relief of withdrawal symptoms
2. Increase in the toxicity of alcohol
3. Reduction in alcohol drinking (anti-craving medications)

Litten and colleagues (1996) have recently completed a comprehensive survey of these medications and the reader is referred to that article for details. This chapter will focus on the anti-craving medications partly because a new anti-craving medicine was approved for use in the treatment of alcohol abuse in 1994.

Treating Alcohol Withdrawal

The benzodiazepines have been widely used to treat alcohol withdrawal for many years, but concerns about abuse liability have led researchers to search for either better benzodiazepines or drugs working through other mechanisms. For example, antagonists of the NMDA receptors have been shown to reduce alcohol withdrawal symptoms in experimental animals; however, their use in humans may be limited by potential toxic effects (Litten et al., 1996). Treatment of alcohol withdrawal symptoms with the antiepileptic carbamazepine and with calcium channel blockers has also shown some promise (Litten et al., 1996), but more work is necessary to identify even more successful compounds.

Increasing the Toxicity of Alcohol

Disulfiram and carbimide are two drugs that inhibit the liver enzyme aldehyde dehydrogenase, which breaks down the acetaldehyde that is formed when alcohol dehydrogenase catabolizes alcohol. When an individual subsequently drinks alcohol, the acetaldehyde levels build up and the person may experience unpleasant side effects such as palpitations, difficulty in breathing, headache, and nausea. These drugs can be useful in maintaining abstinence in certain individuals who abuse alcohol, but there is a problem with compliance (Litten et al., 1996).

Reducing Alcohol Drinking

There is a wide variety of drugs that reliably may reduce alcohol drinking in experimental animals. Some of these have also proven beneficial in reducing the drinking in humans. This section will provide a brief overview of the drugs already tested (Table 2). (See the review by Litten and colleagues (1996) for specific details and references.)

Naltrexone

Animal literature had demonstrated that alcohol intake or administration could lead to the release of endogenous opiates and that administration of opiate antagonists, including naltrexone, could reduce the amount of alcohol drinking in experimental animals (Litten et al., 1996). These findings suggested that the opiate mechanism may be involved in voluntary alcohol drinking and led to a series of human studies, which have confirmed the efficacy of naltrexone in the treatment of alcohol abuse and dependence and led to the approval of naltrexone (ReVia) for the treatment of alcohol abuse and dependence. Some studies suggest

Table 2. Drugs used to Reduce Alcohol Intake

		Effects in	
Drug or Class	*Mechanism of Action*	*Animals*	*Humans*
Naltrexone	Opiate Antagonist	Reduction	Therapeutic
Naloxone	Opiate Antagonist	Reduction	Untested
Nalmafene	Opiate Antagonist	Reduction	Therapeutic
Acamprosate	Uncertain, NMDA?	Reduction	Therapeutic
SSRIs	Block 5-HT reuptake	Reduction	Mixed results
5-HT3 antagonist	Block 5-HT3 Receptor	Reduction	Mixed results.
5-HT2 antagonist	Block 5-HT2 Receptor	Reduction	Negative
5-HT1A agonist	Stimulate 5-HT1A Rec	Reduction	Mixed results.
Bromocriptine	Stimulates DA Receptor.	Reduction	Mixed results
Tiapride	Blocks DA Receptor	Reduction	Shows Promise
TA-0910	Stimulates TSH Release	Reduction	Untested

that naltrexone reduces the stimulatory effects of alcohol while increasing the sedative effects (Litten et al., 1996). Although it seems to be an effective treatment, a variety of side effects, including nausea (10%), have been reported, and investigators are currently examining the effects of other narcotic antagonists as well (Litten et al., 1996).

Acamprosate

Several European studies have suggested that alcohol abuse and dependence might be treated by acamprosate (calcium acetylhomotaurine). The doses used in these studies are much higher than those typically used for naltrexone, but there appear to be very few side effects and it does not appear to have abuse potential (Litten et al., 1996). The European studies gave placebo or acamprosate for up to one year and reported that the lengths of abstinence periods were longer and the percentage of retention was higher for the acamprosate-treated groups (Litten et al., 1996). Trials are underway in the United States to confirm the efficacy of acamprosate. Although the biochemical mechanisms of action of acamprosate have not been established, there is evidence that interactions with the GABA-ergic or NMDA systems could be involved. The involvement of opiate systems has not been ruled out either, but if they are, then we could have two relatively effective compounds that work via different mechanisms. Since alcohol abuse and dependence is a heterogeneous disorder, it may be that both drugs may be needed to manage effectively all cases of the disorder.

Selective Serotonin Reuptake Inhibitors

The serotonergic system has long been implicated in alcohol drinking, as described above. Since a deficiency of 5-HT is commonly associated with elevated voluntary drinking, it is logical to propose that selective serotonin reuptake inhibitors (SSRIs) may reduce excessive alcohol drinking. Indeed, there is an impressive animal literature demonstrating that fluoxetine, citalopram, and other SSRIs significantly reduce voluntary alcohol consumption (Litten et al., 1996). However, the effect of SSRIs on alcohol intake in humans is much less impressive. Some studies show a decrease of up to 20% in the first week of treatment, and no differences from placebo after longer treatments. Other studies suggest that the SSRIs may be more effective in the heaviest drinkers (Litten et al., 1996). In conclusion, SSRIs have not been impressive in reducing alcohol intake in humans, but they might still be beneficial in some subtypes.

5-HT3 Antagonists

As indicated above, alcohol appears to stimulate the release of 5-HT, which interacts with 5-HT3 receptors to stimulate the release of dopamine. If dopamine underlies the rewarding properties of alcohol, then drugs that break the cycle of alcohol-induced rises in dopamine might be efficacious in reducing voluntary alcohol drinking. A number of experimental animal studies have demonstrated that 5-HT3 antagonists do indeed induce a reduction in voluntary alcohol drinking, but the only human trial to date has been negative (Litten et al., 1996). A problem with this class of drugs is the apparent inverted U-shaped dose–response function whereby intermediate doses are the most effective. Carefully controlled human studies may be extremely difficult to carry out.

5-HT2 Antagonists

As with the 5-HT3 antagonists, a wide variety of animal studies have demonstrated the effectiveness of 5-HT2 antagonists in reducing voluntary alcohol drinking (Litten et al., 1996). Ritanserin, a 5-HT2C antagonist, has been tested for efficacy in human alcohol abuse and found to be relatively ineffective. Amperozide, a 5-HT2A antagonist, has been recently confirmed to be very effective in reducing voluntary alcohol drinking (Overstreet et al., 1997), but neither this compound nor other 5-HT2A antagonists have been tested in humans as yet.

5-HT1A Agonists

A variety of other serotonergic agents have been tested for their efficacy in reducing alcohol intake in experimental animals. These include 5-hydroxytryptophan, the precursor of 5-HT, and fenfluramine, a 5-HT releaser. However, it is not known whether these agents work by acting as antagonists at the 5-HT2 or 5-HT3 receptors or by stimulating the 5-HT1A receptors. It is known that drugs that are selective for the 5-HT1A receptors, such as 8-OH-DPAT, buspirone, and ipsapirone, also induce a decrease in voluntary alcohol drinking in experimental animals (Litten et al., 1996). Buspirone has had some success in treating individuals with alcohol abuse and co-morbid anxiety disorders, as discussed below.

GABAergic Agents

Studies on the interaction of GABAergic agents and voluntary alcohol intake have been less common and the results have been mixed, with some agents reducing alcohol intake and others increasing it (Litten et al., 1996). γ-hydroxybutyrate, a metabolite of GABA, has been shown to reduce alcohol intake both in experimental animal models and in clinical trials (Litten et al., 1996). Further work is necessary because of the abuse potential of many GABAergic agents such as the benzodiazepines, and the abuse liability of γ-hydroxybutyrate, and other potentially therapeutic GABAergic agents must be examined.

Dopaminergic Agents

As indicated above, dopamine is released by the injection of alcohol, and it is believed that dopamine may underlie the rewarding properties of alcohol. This model would predict that drugs that block dopamine's effects, such as dopamine receptor antagonists, would reduce voluntary alcohol drinking. Surprisingly, dopamine antagonists have had very little effect on voluntary alcohol consumption in experimental animals (Litten et al., 1996). Although bromocriptine, a dopamine agonist, has been reported to reduce alcohol intake in animals, it has been ineffective in the human studies completed to date (Litten et al., 1996). Finally, tiapride, a dopamine D2 receptor antagonist, has shown promise in reducing alcohol intake in early clinical trials (Litten et al., 1996). A problem that these dopamine antagonist studies have is that the compounds are very potent in reducing locomotor activity, so it may be difficult to see a selective effect of alcohol intake.

Treatment of Comorbid Disorders

Alcohol abuse and dependence may be present in individuals who suffer from a variety of psychiatric disorders, including both anxiety and depressive disorders. Alcohol abuse and dependence may also be present in individuals who are simultaneously abusing other drugs, including nicotine. If there is a direct relationship between the two co-morbid disorders, it might be possible to modify one condition by treating the other. Thus, treating anxiety or depression may have beneficial effects on the individual's incidence of drinking.

Depressive disorders, including mania, were detected in 53% of the patients being treated for alcohol abuse and dependence. It is not clear in many cases whether the individual has both disorders independently or whether the alcohol abuse or depressive disorder led to the other condition. In any case, treatment of the depressive disorder in such individuals might well lead to the improvement of both conditions. Indeed, early studies with tricyclic antidepressants and SRRIs have shown that the excessive drinking is reduced in individuals with depressive symptoms, but not necessarily in those individuals who were not depressed (Litten et al., 1996). It is recommended that further trials on the efficacy of SSRIs in reduced alcohol drinking should be confined to depressed individuals.

Anxiety disorders were detected in 18% of individuals being treated for alcohol abuse and dependence. Since benzodiazepines have abuse liability themselves, they are not recommended to reduce anxiety in these individuals. Instead, serotonergic agents have been used. Buspirone, a partial 5-HT1A receptor agonist, has been demonstrated to be effective in reducing alcohol drinking in some studies but not others (Malec et al., 1996). One difference between these studies is that the positive study included a behavioral counseling component, while the negative study did not. This outcome emphasizes a key point about the pharmacotherapy of alcohol abuse and dependence. Some form of behavioral therapy should always be combined with drug treatment.

Nicotine is perhaps the most common other drug abused by those individuals with alcohol abuse and dependence, with the prevalence rate for smoking ranging from 75% to 90%. There is also a direct relationship between smoking and drinking, with individuals who drink excessively also smoking excessively. There has been only a limited effort to date to treat the nicotine addiction as well as the alcohol abuse in these individuals; this is partly because it was believed that being abstinent from both drugs would increase the risk for relapse. Those programs, which have recently been implemented, have been inconclusive (Toneatto et al., 1995).

Alcohol abuse is also occasionally associated with the abuse of other drugs, but there is relatively little information about the treatment prospects for these polysubstance abusers. From a theoretical standpoint, the evidence suggesting that dopamine mesolimbic pathways may be involved in the rewarding properties of many drugs with a potential for abuse (Koob, 1992) implies that there might be pharmacological treatments, which will be effective in both alcohol and cocaine abusers as well as those individuals who are abusers of both substances. The fact that naltrexone appears to help in the management of both opiate and alcohol abuse (Litten et al., 1996) supports this contention. Only further research with additional medications will permit a confirmation of this view.

SUMMARY AND CONCLUSIONS

Despite recent trends demonstrating a reduction in the per capita rate of alcohol consumption and of alcohol-related traffic deaths, the excesssive drinking of alcohol remains a significant problem in the United States, with more than 7% of the population still being diagnosed with alcohol abuse and/or dependence. Much progress has occurred in identifying the targets in the brain with which alcohol interacts; the GABAergic, serotonergic, dopaminergic, and NMDA systems appear to be key targets for alcohol's neurobehavioral actions. There has also been substantial improvement in treatment strategies, with the recent approval of naltrexone (ReVia), the first new drug for the treatment of alcoholism in 40 years. Several other medications have shown promise and it is likely that the testing of pharmacotherapies will remain a very active research area over the next decade. Given the heterogeneity of alcoholism, it is not impossible to envision a Project MATCH2 in which the patient characteristics are taken into account when prescribing a medication for treatment. The interaction of pharmacotherapies with behavioral therapies may also be an active area of investigation over the next decade. We can look to the next decade with optimism about the prospects of developing even more effective therapies for alcohol abuse/dependence.

REFERENCES

Babor, T.F. & Grant, M. (1992). Programme on Substance Abuse: Project on Identification and Management of Alcohol-Related Problems. Report on Phase II: A Randomized Clinical Trial of Brief Interventions in Primary Health Care. World Health Organization, New York.

Beracochea, D., Micheau, J., & Jaffard, R. (1992). Memory deficits following chronic alcohol consumption in mice: Relationships with hippocampal and cortical cholinergic activities. Pharmacol. Biochem. Behav. 42, 749-753.

Bien, T.H., Miller, W.R., & Tonigan, J.S. (1993). Brief interventions for alcohol problems. A review. Addiction 88, 315-336.

Bruckner, M.K. & Arendt, T. (1992). Intracortical grafts of purified astrocytes ameliorate memory deficits in rat induced by chronic treatment with alcohol. Neurosci. Lett. 141, 252-254.

Campbell, K.E., Zobeck, T.S., & Bertolucci, D. (1995). Trends in Alcohol-Related Fatal Traffic Crashes, United States 1977-1993. Surveillance Report No. 34. National Institute of Alcohol Abuse and Alcoholism, Rockville, MD.

Charness, M.E., Simon, R.P., & Greenberg, D.A.. (1989). Ethanol and the nervous system. N. Engl. J. Med. 321, 442-454.

Crabbe, J.C., Phillips T.J., Feller, D.J., Hen, R., Wenger, C.D., Lessov, C.N., & Schafer, G.L. (1996). Elevated alcohol consumption in null mutant mice licking 5-HT1B serotonin receptors. Nature Genetics 14, 98-101.

Criswell, H.E., Simson, P.E., Duncan, G.E., McCown, T.J., Herbert, J.S., Morrow, A.L., & Breese, G.R. (1993). Molecular basis for regionally specific action of ethanol on GABA A receptors. Generalization to other ligand-gated ion channels. J. Pharmacol. Exp. Ther. 267, 522-537.

Debakey, S.F., Stinson, F.S., Grant, B.F., & Dufour, M. (1995). Liver cirrhosis mortality in the United States, 1970-1992. Surveillance Report No. 37. National Institute on Alcohol Abuse and Alcoholism, Rockville, MD.

Dietrich, R.A. (1993). Selective breeding for initial sensitivity to ethanol. Behav. Genet. 23, 153-162.

Edwards, G., Anderson, P., Babor, T.F., Casswell, S., Ference , R., Giesbrecht, N., Godfrey, C., Holder, H.D., Lemmens, P., Makela, K., Midanik, L.T., Norstrom, T., Osterberg, E., Romelsjo, A., Room, R., Simpura J., & Skog, O.J. (1994). Alcohol Policy and the Public Good. Oxford University Press, New York.

Gelertner, J., Goldman, D.., & Risch N. (1993). The A1 allele at D2 receptor gene and alcoholism. JAMA 269, 1673-1677.

Grant, B.F., Harford, T.C., Dawson, D.A., Chou, P., Dufour M., & Pickering, R. (1994). Prevalence of DSM-IV alcohol abuse and dependence: United States, 1992. Alcohol Health Res. World 18, 243-248.

Kampov-Polevoy, A.B., Garbutt, J.C., & Janowsky, D. (1997). Evidence of preference for a high-concentration sucrose solution in alcoholic men. Am. J. Psychiatry 154, 269-270.

Koob, G.F. (1992). Drugs of abuse: Anatomy, pharmacology, and function of reward pathways. Trends Pharmacol. Sci. 13, 177-184.

Li, T.-K., Lumeng, L., & Doolittle, D.P. (1993). Selective breeding for alcohol preference and associated responses. Behav. Genet. 23, 163-170.

Lister, R.G., Eckhardt, J.J., & Weingartner, H. (1987). Ethanol intoxication and memory. Recent developments and new directions. In: Recent Developments in Alcoholism.(Galanter, M., Ed.), Vol. 5, pp. 111-126. Plenum Press, New York.

Litten, R.Z., Allen, J., & Fertig, J. (1996). Pharmacotherapies for Alcohol Problems: A review of research with focus on developments since 1991. Alcohol. Clin. Exp. Res. 20, 859-876.

Malec, T.J., Malec, E.A., & Dongier, M. (1996). Efficacy of buspirone in alcohol dependence: A review. Alcohol. Clin. Exp. Res. 20, 853-858.

Melchior, C.L., Glasky, A.J., & Ritzmann, R.F. (1993). A low dose of ethanol impairs working memory in mice in a win-shift foraging paradigm. Alcohol 10, 491-493.

Miczek, K.A., DeBold, J.F., & van Erp, S.M.M. (1994). Neuropharmacological characteristics of individual differences in alcohol effects on aggression in rodents and primates. Behav. Pharmacol. 5, 407-421.

Midanik, L.T. & Clark, W.B. (1994). Demographic distribution of U.S. drinking patterns in 1990: Description and trends from 1984. Am. J. Pub. Health 84, 1218-1222.

Miller, W.R. & Sanchez, V.C. (1994). Motivating young adults for treatment and lifestyle change. In: Alcohol Use and Misuse by Young Adults. (Howard, G.S. & Nathan, P.E., Eds.), pp. 55-81. University of Notre Dame Press, South Bend, IN.

Miller, W.R., Leckman, A.L., Delaney, H.D., & Tinkcom, M. (1992). Long-term follow-up of behavioral self-control training. J. Stud. Alcohol 53, 246-261.

Overstreet, D,H, McArthur, R.A., Rezvani, A.H., & Post, C. (1997). Selective inhibition of alcohol intake in diverse alcohol-preferring rat strains by the 5-HT2A antagonists amperozide and FG 5974. Alcohol. Clin. Exp. Res. 21, 1448-1455.

Peoples, R.W., Li, C., & Weight, F.F. (1996). Lipid versus protein theories of alcohol action in the nervous sysem. Ann. Rev. Pharmacol. Toxicol. 36, 185-201.

Phillips, T.J., Crabbe, J.C., Metten, P., & Belknap, J.K. (1994). Localization of genes affecting alcohol drinking in mice. Alcohol. Clin. Exp. Res. 18, 931-941.

Project MATCH Research Group. (1997). Matching alcoholism treatments to client heterogeneity: Project MATCH posttreatment drinking outcomes. J. Stud. Alcohol 58, 7-29.

Room R. & Greenfield, T. (1993). Alcoholics Anonymous, other 12-step programs, and psychotherapy in the U.S. population, 1990. Addiction 88, 555-562.

Schuckit, M.A. (1994). Low level of response to alcohol as a predictor of future alcoholism. Am. J. Psychiatry 151, 184-189.

Simson, P.E., Criswell, H.E., & Breese, G.R. (1993). Inhibition of NMDA-evoked electrophysiological activity by ethanol in selected brain regions: Evidence for ethanol-sensitive and ethanol-insensitive NMDA-evoked responses. Brain Res. 607, 9-16.

Snow, M.G., Prochaska, J.O., & Rossi, J. (1994). Processes of change in Alcoholics Anonymous: Maintenance factors in long-term sobriety. J. Stud. Alcohol. 55, 362-371.

Toneatto, T., Sobell, L.C., Sobell, M.B., & Kozlowksi, L.T. (1995). The effect of cigarette smoking on alcohol treatment outcome. J. Subst. Abuse 7, 245-252.

U.S. Department of Health and Human Services. (1997). Ninth Special Report to the U.S. Congress on Alcohol and Health. June, Washington, D.C.

Virkkunen, M., Kallio, E. Rawlings, R., Tokola, R., Poland, R.E., Guidotti, A., Nemeroff, C., Bissette, G., Kalogeras, K., Karonen, S.-I., & Linnoila, M. (1994). Personality profiles and state aggressiveness in Finnish alcoholic, violent offenders, fire setters, and healthy volunteers. Arch. Gen. Psychiatry 51, 28-33.

Williams, G.D., Stinson, F.S., Stewart, S.L., & Dufour, M.C. (1995). Apparent Per Capita Alcohol Consumption: National, State, and Regional Trends, 1977-1992. Surveillance Report No. 35. National Institute of Alcohol Abuse and Alcoholism, Rockville, MD.

RECOMMENDED READINGS

Hannigan, J.H., Spear, L.P., Spear, N.E., & Goodlett, C.R. (Eds.) (1999). Alchohol and Alcoholism: Effects on Brain and Development. pp. 282, Lawrence Erlbaum Associates, Mahwah, N.J.

Litten, R.Z., Allen, J., & Fertig, J. (1996). Pharmacotherapies for Alcohol Problems: A review of research with focus on developments since 1991. Alcohol. Clin. Exp. Res. 20, 859-876.

Project MATCH Research Group. (1997). Matching alcoholism treatments to client heterogeneity: Project MATCH posttreatment drinking outcomes. J. Stud. Alcohol 58, 7-29.

U.S. Department of Health and Human Services. (1997). Ninth Special Report to the U.S. Congress on Alcohol and Health. Washington, D.C.

Chapter 21

Homelessness and Mental Health

JAN SCOTT

INTRODUCTION

The increase in the number of homeless people in the United States of America and Europe since the 1970s has been accompanied by marked changes in the

Biological Psychiatry, pages 487-498.

ISBN: 1-55938-819-6

demographics of this population. The median age of the homeless population now lies between 29 and 39 years (Fischer et al., 1986; Morse and Calsyn, 1986; Sargent, 1989). Just less than half (47%) of all homeless people are single adults, 33% are youths aged 16 to 21 years, and families comprise 30%. When "hidden homeless" groups are included, 10% to 25% of homeless people are female (Burt and Cohen, 1989; George et al., 1991). Compared with the domiciled population, individuals from ethnic minorities are overrepresented among the homeless (Kroll et al., 1986; Morse and Calsyn, 1986; Burt and Cohen, 1989).

The past two decades have also seen a disproportionate, possibly twofold (Tantum, 1991) increase in the number of mentally ill people among the homeless (Scott, 1993). Research on this group has expanded significantly, but controversy over what actually constitutes homelessness and obstacles to reliable and valid assessments of psychopathology have made the information obtained difficult to interpret. This chapter draws on the most robust studies undertaken to identify the size of the problem and to describe why mentally ill people may be at risk of becoming homeless. Other sections review data on the nature and extent of mental health problems among different subpopulations of homelessness people, highlight the physical health problems faced by these individuals, and identify the key elements of effective services for homeless mentally ill people.

PREVALENCE OF HOMELESS AND MENTAL ILLNESS

It is virtually impossible to give accurate figures on the number of homeless mentally ill people. Pressure groups may overestimate figures whereas official statistics tend to underestimate the number of people affected. A critical problem is the definition of homelessness employed. Official definitions are usually restricted to "those without adequate shelter". In the USA on any given day, this "literal" homeless population numbers about 500,000 to 600,000 people (Manderscheid and Rosenstein, 1992) of whom about 200,000 may have a major mental disorder. Although these "street people" may be the most visible group, they represent only one component of the "home to homelessness" continuum (Austerberry and Watson, 1986). Austerberry and Watson (1986) identified four categories or levels of homelessness according to geographic location:

1. Street people or those sleeping rough
2. Residents of shelters and hostels for the homeless
3. Residents of hotels or bed and breakfast accommodation—often families placed in "temporary" accommodations for prolonged periods.
4. Other unique situations such as staying with others—people without their own accommodations (e.g., those currently staying with family or friends).

If the "hidden homeless" populations (groups 3 and 4) are included in the calculations, the local homeless population in the USA may be 2.5 million people (Caton, 1990). Less is known about the prevalence and nature of the problems faced by these additional groups, but estimates suggest that 30% to 80% of individuals may have some form of mental health problem ranging from psychosis to less severe reactive disorders.

PATHWAYS TO HOMELESSNESS

Arguments rage as to whether homelessness among mentally ill people is fundamentally an economic problem or a health problem. Pathways to homelessness are often complex. Both general and specific factors that may put people with mental illness at risk are discussed below.

General Factors

Economic Factors

Unemployment, poverty, and shortage of affordable housing contribute significantly to the homelessness problem (Schutt and Garrett, 1992). Mentally ill people are particularly vulnerable to such influences (Scott, 1993). Although about 50% of homeless people are high school graduates, there are significant differences in skill levels between the homeless and the domiciled population (Sargent, 1989). The reduced demand for low skilled laborers puts these individuals at greater risk of unemployment (Wright and Weber, 1987). Few homeless people have a paid job (Scott, 1993) and 66% do not receive any financial benefits (Schutt and Garrett, 1992). In the United States, the reduced availability of low cost housing has been compounded by the closure of 50% of single-room-occupancy (SRO) accommodation between 1970 and 1980 (Cohen, 1994).

Social Problems

Homeless men and women are disaffiliated from their families and society. Such isolation may have its origins early in life. Childhood histories reveal that experiences of abuse and placements away from the family home in institutional settings are frequent features of homeless populations (Bassuk et al., 1986; Susser et al., 1987). At least 50% of homeless adults and 60% to 90% of homeless mentally ill adults have no social contacts (Scott, 1993). This marginalization may be both a cause and a consequence of homelessness (Cohen et al, 1984; Stark et al., 1989).

Intentional Homelessness

Although President Reagan once remarked that "the homeless are homeless...by choice" research suggests that less than 4% of homeless people report that they chose this lifestyle (Baxter and Hopper, 1984; Start et al., 1989).

Specific Factors

The Nature of Mental Illness

The symptoms of severe mental illness (which may include bizarre behavior), the sufferers' reaction to the illness, and the response of family members or other individuals may directly or indirectly lead to mentally ill people becoming homeless (Connelly and Crown, 1994; Schutt and Garrett, 1992). Symptoms of illness such as social withdrawal may reduce the availability of potential support networks, while the level of disability experienced may prevent the individual coping with independent living. The sufferers' lack of insight into their problems may lead them to reject treatment or accommodation that is offered. In addition, individuals with mental health problems are frequently rejected by their immediate social network and are stigmatized by the community at large.

Adverse Reactions to Homelessness

While mental illness may contribute to the process of becoming homeless, it is also clear that the adversity associated with homelessness increases the risk of suffering from mental health problems. A third of homeless women in one survey reported having been raped and a significant number suffered posttraumatic disorders (Breakey et al, 1989). Also, the stress of living in temporary accommodation has been found to be associated with an increased prevalence of depressive disorders and acute distress reactions (Linn et al., 1990; Tacchi and Scott, 1996).

Changes in Mental Health Service Provision

Deinstitutionalization and preclusive admission policies have been associated with the increase in the size of the homeless mentally ill population. However, it is not the policy of deinstitutionalization per se, but its inadequate implementation, that is likely to make this an important contributory factor (Thornicroft and Bebbington, 1989). It has been calculated that there is an interval of 30 months from the time of hospital discharge into adequate accommodation with outpatient support to the time of becoming homeless.

Health and welfare organizations continue to struggle to provide effective community aftercare for people who were previously long-stay hospital residents. However, many homeless mentally ill people have never been admitted to a mental hospital. Engaging these individuals and newly diagnosed mentally ill people (particularly young males) with services that may rescue them from homelessness is an unresolved problem.

THE NATURE AND EXTENT OF MENTAL ILLNESS

Estimates of the nature, extent, and severity of mental illness and substance abuse among homeless people vary considerably (Sargent, 1989). The lack of consistency in the findings of earlier studies reflects problems in sample selection, methods of investigation, and case definition (Susser et al., 1990). The most recent studies show greater uniformity both nationally and internationally and suggest that 50% of the total homeless population may have some form of mental disorder (Breakey et al., 1989; Herrman et al., 1989; Stark et al., 1989), with 70% to 80% receiving lifetime diagnoses (Fischer et al., 1986; Herrman et al., 1989). Fischer and colleagues (1986) estimated that twice as many homeless as domiciled people have psychiatric disorders and that these problems are of a more severe nature and tend to worsen if the state of homelessness continues. The suicide rate in homeless people is about four times higher than the expected population rate (Connelly and Crown, 1994).

Studies on hostel and shelter populations that used both trained researchers and structured clinical interviews have been reviewed by several authors (Sargent, 1989; Scott, 1993; Schutt and Garrett, 1992). These papers, mainly referring to homeless males, suggest that the prevalence of functional psychoses is about 33%. The majority of individuals in this group suffer from schizophrenia, but about one in five suffer from bipolar disorder. Among mentally ill individuals, 30% to 80% have a history of previous psychiatric hospitalization, and many demonstrate high levels of associated psychosocial morbidity, which significantly impedes their day-to-day functioning (Stark et al., 1989).

There is less information on women who are homeless, but evidence suggests that they also suffer high rates (40–60%) of mental disorder and substance abuse (Marshall and Reed, 1992; Smith et al., 1993; Tacchi and Scott, 1996). However, studies show that, compared with homeless men, homeless women have more variable rates of severe mental disorders (from 6–50%), have higher levels of depressive and anxiety disorders (10–40%), and have less experience of institutional living in either hospitals or prisons (Burt and Cohen, 1989; Tantum, 1991).

Substance abuse is a significant problem among street and hostel residents, occurring in 9% to 63%. The rates of alcohol and drug abuse reported in homeless populations are at least five times higher than those in the general population (Wright and Weber, 1987; Breakey et al., 1989). In one study undertaken in Los Angeles, 30% of the 400 homeless people interviewed showed evidence of long-term substance abuse without the presence of any other mental disorder (Koegel and Burman, 1988).

The median prevalence of alcohol abuse in homeless males and females is 38%, and the rate in males (47%) is about three times that of females (16%). Older homeless adults (> 50 years old) show a prevalence of alcohol abuse of 43%, about twice that of younger; (< 30 years old) homeless people (Wright and Webber, 1987; Breakey et al., 1989). Not surprisingly, homeless alcohol-dependent people (Koegel and Burman, 1988). Co-morbidity of alcohol problems and major mental illness is also a significant finding in at least 20% hostel residents (Scott, 1993).

Drug abuse is more common in America than in Europe, and it is becoming a particularly frequent feature in younger homeless people on both sides of the Atlantic. The median prevalence of drug abuse among homeless people is about 15%. There is a suggestion that the rates of drug abuse in homeless females are comparable with those in homeless males. (Breakey et al., 1989). Intravenous drugs are used twice as often by homeless as by domiciled individuals (Gelberg et al., 1990). HIV infection rates are four to five times greater in homeless adults using drugs than in homeless adults who are not using drugs (Connelly and Crown, 1994). The prevalence of co-morbidity of drug abuse with major mental disorders is 10% to 20%.

The prevalence of personality disorders has not been investigated in all studies, but rates in both homeless males and females are about 50% (Breakey et al., 1989). Studies of homeless families suggest that the prevalence of Axis II disorders may be particularly high in young female single parents who head the family (Bassuk, 1990). A third of these women reported abuse and neglect during childhood and two thirds reported major family disruptions (Bassuk et al., 1986).

In these studies, looking at acute as well as longterm mental health problems (Fischer et al., 1986; Stark et al., 1989; Tacchi and Scott, 1996), 30% to 50% of the individuals interviewed reported current psychological distress on the General Health Questionnaire (GHQ; Goldberg, 1972). Linn and colleagues (1990) found that 80% of the homeless sample of 214 people reported significant levels of acute distress compared with 49% of a control sample ($n = 250$) drawn from the general population. Distress in the homeless was particularly correlated with unemployment, greater alcohol intake, poor physical health, lack of social support, and anxieties related to difficulties in gaining access to medical services. Burt and Cohen (1989) found that the rate of deliberate self harm in the homeless ranged from 14% to 26%, seven times that of the general population. The rate was lowest in homeless women with children.

Relatively few studies have been undertaken on homeless teenagers. The evidence suggests that rates of psychosis are low but that behavioral difficulties are frequent (Caton, 1990). The most common problems identified in homeless adolescents are school difficulties, antisocial behavior, depression, and deliberate self-harm. The latter occurred in about 15% of boys and 30% of girls. Family disputes and disorganization were frequent accompaniments of these problems, and 60% of homeless adolescents stated that they did not wish to live with their parents.

The problems of young children in homeless families requires investigation. In one of the few studies available, Bassuk and colleagues (1986) assessed 151 children living with their mothers in hostels for the homeless in Boston. Among those younger than five years old, 47% showed at least one developmental delay and 33% showed two or more delays. At school, a third were attending "special classes". More than half were positive on the Beck childhood depression scale and 48% showed high levels of anxiety.

PHYSICAL HEALTH PROBLEMS

The presence of any physical disorder may exacerbate or precipitate mental health problems or significantly complicate the treatment of the latter. Co-morbidity for physical illness and alcohol problems or physical and mental illness occurs in about a third of homeless adults (Scott, 1993). Alcohol-dependent street dewellers exhibit extreme levels of morbidity, with one study suggesting a 98% prevalence of physical illness (Blumberg et al., 1973). HIV infection in homeless people is also becoming a major public health concern, with some reports putting HIV positive rates at 25% to 30% (Schutt and Garrett, 1992; Connelly and Crown, 1994).

Homeless people attending casualty departments report a disproportionate prevalence of traumatic injuries (14–33%), infections (with 5–15% having active tuberculosis), scabies and lice (20%), and cellulitis (Brickner, 1984; Kelly, 1985; Wright and Weber, 1987). Those living on the street show high rates (10–23%) of peripheral vascular disease, leg ulcers, and frostbite (Brickner, 1984). Sexually transmitted diseases affect about one in 10 homeless people (Breakey et al. 1989). Mortality rates are increased at least threefold in homeless as opposed to domiciled populations (Alstrom et al., 1975).

MENTAL HEALTH SERVICES FOR HOMELESS PEOPLE

This section briefly summarizes previous reviews of service provision to homeless mentally ill people (Goldfinger and Chafetz, 1984; Caton, 1990; Schutt and Garrett, 1992; Scott, 1993). In terms of good practice, a number of common themes emerge: the need to engage the individual with the services, the need for assertive multidisciplinary input through outreach and day care programs provided in places where homeless people congregate, and the need to offer a variety of housing options.

The homeless mentally ill are a multiproblem group (Breakey, 1987) who present with a complex combination of psychological, physical, and social problems. These difficulties can be organized into a hierarchy of needs, but for most homeless people, basic requirements (especially housing, food, and clothing) take precedence over the acquisition of treatment for mental health problems. It should hardly come as any surprise that someone trying to survive life on the streets may not be a regular attender at outpatient clinic appointments or may not see taking medication as prescribed as their first priority. Gross mental disturbances, suspiciousness, and paranoia may make homeless mentally ill people reluctant to accept input, but previous negative experiences of the mental health services is also an important predicator of nonengagement (Eagle and Caton, 1990). However, several researchers report that the majority of homeless people will accept mental health service input and that more management plans fail because of lack of availability of appropriate services or failure of providers to negotiate health and welfare priorities collaboratively with the sufferer. The provision of accommodation is often the crucial first step and usually dictates engagement of the individual with the services.

The above principles apply to homeless people suffering from the whole range of mental health problems but, to date, most of the specialist services developed have focused on the needs of those with severe mental illnesses living in shelters. The most effective services appear to make selective use of hospitalization with multidisciplinary team input and case managers working to maintain the individual in the community. While 60% of the therapy undertaken focuses on case management or support activities, 40% involves crisis work and counseling (Breakey, 1987). Specialist teams can also link with other agencies to provide support and advice to other professionals and voluntary staff involved in the care of homeless mentally ill people. This may help reduce rejection of difficult-to-manage clients. Eventually, the goal of most specialist teams is to return their clients to the mainstream services to try to avoid permanent marginalization.

In the United States, Baxter and Hopper (1984) describe a three-tiered system of housing to meet the short and longer term needs of the homeless mentally ill. They emphasize that tier one (emergency shelters) must be backed up by transitional accommodations (tier 2) and longterm supportive residence in the community (tier 3). The need for the coordination between all components is also stressed. A major issue to be addressed is how to prevent temporary shelters from becoming the longterm residencies of many homeless individuals because of lack of available alternatives or stigmatization of the homeless mentally ill applying for accommodation (Alisky and Iczkowski, 1990). Shortterm solutions such as cold weather initiatives need to be backed up by longterm housing strategies if a reduction in the number of homeless mentally ill people is to be achieved (Schutt and Garrett, 1992).

SUMMARY AND CONCLUSIONS

This chapter suggests that 30% to 50% of homeless people may be suffering from major mental illness or substance abuse disorders. Functional psychoses are particularly common among hostel residents, but in families in temporary accommodations, personality disorders and acute nonpsychotic disorders may predominate. Drug abuse is a significant problem in younger adults who are homeless. Co-morbidity of substance abuse and mental illness, or substance abuse and personality disorder affects at least a fifth of those interviewed. Physical illnesses are also prevalent. Personal histories reveal that family disorganization and early disadvantage are commonly reported by adults who later become homeless, and many homeless teenagers have run away from home on several previous occasions.

Although most research has focused on single adults residing in shelters or living on the streets, the homeless population is heterogeneous with a multiplicity of health and social needs. There are at least three important subgroups within the homeless mentally ill population with differing service requirements:

1. those with a previous history of inpatient treatment who have severe mental illness and become homeless some time after discharge from hospital
2. younger adults who have avoided or failed to receive any psychiatric treatment and whose mental illness (varying from personality dysfunction, substance abuse, or psychosis) may have contributed to social drift
3. those whose (nonpsychotic) mental health problems are a consequence of becoming homeless or have acted as an important intervening variable in increasing the risk of the development of social problems that may cause homelessness

To a large extent, the first group epitomizes the problems of providing community care to those with complex chronic difficulties (Bachrach, 1984). The failure to provide adequate alternative accommodation and the failure of aftercare has led to many expatients becoming homeless (Talbott and Lamb, 1984). As described, the limited resources available for specialist provision has so far focused on this subgroup. Other groups will require the development of new initiatives, as it is unlikely that there is a "single service" solution (Breakey, 1987). For younger homeless people with mental illness, engaging them with the services seems to be the primary issue that must be addressed. Less is known about the needs of the third subgroup, but this population includes homeless families, and offering them some input might help prevent the transmission of a disadvantaged lifestyle and deprivation to the children involved.

Second generation research studies must focus more specifically on discovering why some mentally ill people become homeless and others do not. How does mental illness contribute to the causation of homelessness? Do other specific factors operate in mentally ill populations that are not found in other (nonpsychiatric) homeless groups? Prospective follow-up of those discharged from hospitals may also make it possible to predict future housing instability among the severely mentally ill. The dynamics of entry into and exit from the homeless population also need to be explored (Schutt and Garrett, 1992). What types of input will increase the chance of a previously homeless person achieving stable independent living?

Finally, it should be noted that the "literal" homeless compose only 3% of the population living below the "poverty line" (Rossi et al., 1987). Many of these extremely poor people require more systematic service input as they suffer the same types of deprivation and disadvantage as homeless people and experience similar mental health problems as a cause or consequence of this adversity. This population also represents an important "prehomeless" group, which may benefit from interventions aimed at preventing them from joining the ranks of the homeless.

REFERENCES

Alisky, J.M. & Iczkowski, B.A. (1990). Barriers to housing for deinstitutionalised psychiatric patients. Hosp. Comm. Psych. 41, 93–95.

Alstrom, C.H., Lindelius, R., & Salum, I. (1975). Mortality among homeless men. Br. J. Addiction 70, 245–252.

Austerberry, H. & Watson, S. (1986). Housing and Homelessness. Routledge & Kegan Paul, London.

Bachrach, L. (1984). The homeless mentally ill and the mental health services. In: The Homeless and Mentally Ill. (Lamb, H.R., Ed.), pp. 11–54. American Psychiatric Association, Washington, D.C.

Bassuk, E.L.. (1990). Who are the homeless families? Characteristics of sheltered mothers and children. Comm. Mental Hlth. J. 26, 421–430.

Bassuk, E.L., Rubin, L., & Lauriat, A.S. (1986). Characteristics of Sheltered Homeless Families. Am. J. Public Health, 76, 1097–1102.

Baxter, E. & Hopper, K. (1984). Shelter and housing for the homeless mentally ill (Lamb, H.R. Ed.). In: The Homeless Mentally Ill. pp. 103–139. (Lamb, H.R., Ed.). American Psychiatric Association, Washington, D.C.

Blumberg, L (1973). What to do about the men on skid row: Report of the Greater Philadelphia Movement to the Redevelopment Authority of the City of Philadelphia. Greater Philadelphia Movement, Philadelphia.

Breakey, W.R. (1987). Treating the homeless. Alcohol Hlth. Res. World 3, 11–42.

Breakey, W.R., Fischer, P., Kramer, M., Nestadt, G., Romanoski, A., Ross, A., Royal, R., & Stine, O. (1989). Health and mental health problems of homeless men and women in Baltimore. J. Am. Med. Ass. 262, 1352–1357.

Bricker, P.W. (1984). Medical aspects of homelessness. In: The Homeless Mentally Ill. pp. 243–259. (Lamb, H.R. Ed.). American Psychiatric Association, Washington, D.C.

Burt, M.R. & Cohen, B.E. (1989). Differences among homeless single women, women with children, and single men. Social Problems 36, 508–524.

Caton, C. (1990). Solutions to the homeless problems. In: Homeless in America. (Caton, C., Ed), pp. 174–190. Oxford University Press, Oxford, England.

Cohen, C. (1994). Down and out in New York and London: A cross-national comparison of homelessness. Hosp. Comm. Psych. 45, 769–776.

Cohen, N.L., Putnam, J.F., & Sullivan, A.M. (1984). The mentally ill homeless: Isolation and adaptation. Hosp. Comm. Psych. 35, 922–924.

Connelly, J., Crown, J. (1994). Homelessness and Ill Health. pp. 23-67. Royal College of Physicians, London.

Eagle, P.F. & Caton, C.L.M. (1990). Homeless and mental illness. In: Homeless in America. (Caton, C. Ed), pp. 59–75. Oxford University Press, Oxford, England.

Fischer, P., Shapiro, S., Breakey, W.R., Anthony, J., & Kramer, M. (1986). Mental health and social characteristics of the homeless: A survey of mission users. Am. J. Public Hlth. 76, 519–523.

Gelberg, L., Linn, L., Usatine, R., & Smith, M. (1990). Health, homelessness and poverty: A study of clinic users. Arch. Inter. Med. 150, 2325–2330.

George, S., Shanks, N., & Westlake, L. (1991). Census of homeless people in Sheffield. Br. Med. J. 302, 1387–1389.

Goldberg, D.P. (1972). GHQ: The Detection of Psychiatric Illness by Questionnaire. Oxford University Press, Oxford, England.

Goldfinger, S.M. & Chafetz, L. (1984). Developing a better service delivery system for the homeless mentally ill. In: The Homeless Mentally Ill. (Lamb, H.R. Ed.), pp.91–108. American Psychiatric Association, Washington, D.C.

Herrman, H., McGorry, P. Bennett, P. van Reil, R., & Singh, B. (1989). Prevalence of severe mental disorders in disaffiliated and homeless people in inner Melbourne. Am. J. Psych. 146, 1179–1184.

Kelly, J.T. (1985). Trauma: With the example of San Francisco's shelter program. In: Health Care of Homeless People.(Brickner, P., Scharer, L., Conanan, B., Elvy, A., & Savarese, M., Eds), pp. 72–85. Springer Verlag, New York.

Koegel, P. & Burman, M. (1988). The prevalence of specific psychiatric disorders among homeless individuals in the inner city of Los Angeles. Arch. Gen. Psych. 45, 1085–1092.

Kroll, J., Carey, K., Hagedorn, D., Fire Dog, P., & Banvides, E. (1986). A survey of homeless adults in urban emergency shelters. Hosp. Comm. Psych. 37, 283–286.

Linn, L., Gelberg, L., & Leake, B. (1990). Substance abuse and mental health status of homeless and domiciled low-income users of a medical clinic. Hosp. Comm. Psych. 41, 306–310.

Marshall, J. & Reed, J. (1990). Psychiatric morbidity in homeless women. Br. J. Psych. 160, 761–765.

Manderscheid, R. & Rosenstein, M. (1992). Homeless persons with mental illness and alcohol or other drug abuse: Current research policy and prospects. Curr. Opin. Pysch. 5, 273–278.

Morse, G. & Calsyn, R. (1986). Mentally disturbed homeless people in St Louis: Needy, willing, but underserved. J. Mental Hlth. 14, 74–94.

Rossi, P., Wright, J., Fisher, G., & Willis, G. (1987). The urban homeless: Estimating composition and size. Science 235, 1336–1341.

Sargent, M. (1989). Update on programmes for the homeless mentally ill. Hosp. Comm. Psych. 40, 1015–1016.

Schutt, R. & Garrett, G. (1992). Responding to the Homeless. pp. 21–105. Plenum Press, New York.

Scott, J. (1993). Homelessness and mental illness. Br. J. Psych. 162, 314–324.

Smith, E., North, C., & Spitznagel, E. (1993). Alcohol, drugs and psychiatric comorbidity among homeless women: An epidemiological survey. J. Clin. Psych. 54, 82–87.

Stark, C., Scott, J. Hill, M., & Morgan, W. (1989). A Survey of the "Long-Stay" Users of DSS Resettlement Units. A Research Report. Department of Social Security, London.

Susser, E., Conover, S., & Struening, E. (1990). Mental illness in the homeless: Problems of epidemiological method in surveys of the 1980s. Comm. Mental Hlth. J. 26, 387–410.

Susser, E. Struering, E., & Conover, M. (1987). Childhood experiences of homeless men. Am. J. Psych. 144, 1599–1601.

Tacchi, M. & Scott, J., (1996). Homeless women in government direct access hostels in London. Psych. Services. 47, 196-198.

Talbott, J. & Lamb, R. (1984). Summary and recommendations. In: The Homeless Mentally Ill. (Lamb. H.R., Ed), pp. 1-10. American Psychiatric Association, Washington, D.C.

Tantum, D. (1991). High risk groups: The homeless and ethnic minorities. Curr. Opin. Psych. 4, 295–303.

Thornicroft, G. & Bebbington, P. (1989). Deinstitutionalisation–from hospital closure to service development. Br. J. Psych. 155, 739–753.

Wright, J. & Weber, P. (1987). Homelessness and Health., pp. 20–108. McGraw-Hill, New York, NY.

RECOMMENDED READINGS

Schutt, R. & Garrett, G. (1992). Responding to the Homeless. pp. 21–105. Plenum Press, New York.

Scott, J (1993). Homelessness and mental illness. Br. J. Psych. 162, 314–324.

Chapter 22

Eating Disorders:
Anorexia Nervosa and Bulimia Nervosa

David S. Goldbloom

Biological Psychiatry, pages 499-509.

ISBN: 1-55938-819-6

INTRODUCTION

The eating disorders anorexia nervosa (AN) and bulimia nervosa (BN) have been the focus of increasing clinical attention over the past three decades both within psychiatry and throughout medicine. More so than any other psychiatric disorders, they encompass perturbations in virtually every physiological system studied and provide a paradigmatic example of the complex interplay between the body and the mind. Additionally, they reflect the powerful influence of cultural values and social pressures on the expression of distress and disorder. Unfortunately, AN and BN are, like many psychiatric disorders, sometimes trivialized by the media and even within the medical profession, obscuring the significant mortality and morbidity of these disorders. Commonly misperceived as a late twentieth century disorder, AN was well characterized and named more than a hundred years ago—long before current cultural preoccupation with thinness. While AN has been the subject of extensive scientific inquiry (Szmukler, Dare, and Treasure, 1995), BN has been characterized more recently (Russell, 1979).

DEFINITIONS

Diagnostic criteria for these disorders have evolved considerably on the basis of empirical research and expert consensus, and current standard definitions are provided in Table 1 (American Psychiatric Association, 1994). For AN, it is important to note that weight loss is necessary per se but not on its own pathognomonic for the disorder; in young women, other psychiatric disorders such as depression, schizophrenia, and conversion disorder as well as a variety of medical disorders may produce weight loss. Rather, this often critical degree of weight loss is accompanied by an intense fear of weight gain or fatness and a distortion in the way one perceives one's body. Typically, body weight becomes the yardstick by which all self-worth is measured, and the severity of the emaciation is denied. Finally, the cessation of menstrual function, which accompanies the weight loss, may at times precede it—reflecting as yet unknown connections between psychological states, nutrition, and neurohormonal regulation of menstrual function (Goldbloom, 1993). In BN, the psychological preoccupations with body weight and shape parallel those of AN but most commonly occur in people at a relatively normal weight. The central behavioral event in BN is the binge eating episode, the core characteristics of which are a profound sense of loss of control accompanied by the rapid and nonpleasureful consumption of a large amount of food—often thousands of calories—in a brief period of time. Then, driven by psychological preoccupation with shape and weight, efforts are made to counteract the effects of the ingested calories via purging through self-induced vomiting and abuse of purgative medications, exercise, or severe caloric restriction. Finally, this is typically not an occasional disorder, but rather one that recurs

Table 1. Diagnostic Criteria

Anorexia Nervosa

a. Refusal to maintain body weight at or above a minimally normal weight for age and height (e.g., weight loss leading to maintenance of body weight less than 85% of that expected; or failure to make expected weight gain during a period of growth, leading to body weight less than 85% of that expected)
b. Intense fear of gaining weight or becoming fat, even though underweight
c. Disturbance in the way in which one's body weight or shape is experienced, undue influence of body weight or shape on self-evaluation, or denial of the seriousness of the current low body weight
d. In postmenarcheal females, amenorrhea (i.e., the absence of at least three consecutive menstrual cycles; a woman is considered to have amenorrhea, if her periods only occur following administration of hormone—e.g., estrogen)

Restricting Type: during the current episode of Anorexia Nervosa, the person has not regularly engaged in binge eating or purging behavior (i.e., self-induced vomiting or the misuse of laxatives, diuretics, or enemas)
Binge-Eating/Purging Type: during the current episode of Anorexia Nervosa, the person has regularly engaged in binge-eating or purging behavior

Bulimia Nervosa

a. Recurrent episodes of binge eating, as characterized by both of the following:
1. eating, in a discrete period of time (e.g., within any two-hour period), an amount of food that is definitely larger than most people would eat during a similar period of time and under similar circumstances

1. a sense of lack of control over eating during the episode (e.g., a feeling that one cannot stop eating or control what or how much one is eating)
b. Recurrent inapproriate compensatory behavior in order to prevent weight gain (e.g., self-induced vomiting, misuse of laxatives, diuretics, enemas or other medications, fasting, or excessive exercise)
c. The binge eating and inappropriate compensatory behaviors both occur, on average, at least twice a week for three months
d. Self-evaluation is unduly influenced by body shape and weight
e. The disturbance does not occur exclusively during episodes of Anorexia Nervosa

Purging type: during the current episode of Bulimia Nervosa, the person has regularly engaged in self-induced vomiting or the misuse of laxatives, diuretics, or enemas

Nonpurging type: during the current episode of Bulimia Nervosa, the person has used other inappropriate compensatory behaviors (e.g., fasting or excessive exercise) but has not regularly engaged in self-induced vomiting or the misuse of laxatives, diuretics, or enemas

Note: Adapted from information in: American Psychiatric Association (1994). *Diagnostic and Statistical Manual of Mental Disorders*. 4th ed. American Psychiatric Association, Washington, D.C.

typically several times per week; people with BN have typically suffered the disorder for several years prior to seeking therapeutic attention.

ETIOLOGY

Along formal linear lines of causality as might be applied to physics or infectious diseases, the etiology of AN and BN is unknown. Most clinicians and researchers embrace a multidimensional model of pathogenesis (Garfinkel and Garner, 1982) that acknowledges risk factors at the level of the individual, the family, and society.

At the individual level, the most common proximate cause—or mere antecedent—of AN or BN is dieting behavior. However, dieting behavior is common and eating disorders are fortunately far less common. Thus, there are individual factors beyond the dieting behavior that must be considered. These typically include perfectionism, a need for a sense of control, a lack of awareness of internal physiological and psychological states, and a sense of ineffectiveness globally that may turn affected individuals toward their body weight as the one element of their lives they can regulate. Being an adolescent or young adult female is clearly a risk factor, as is having a personal history of obesity or chronic medical illness such as diabetes mellitus; a history of sexual abuse has also been identified as a risk factor although nonspecific for a number of psychiatric disorders. AN has also been construed as a phobic avoidance of the psychosocial and sexual demands of adolescence. Additionally, there is evidence that the biological perturbations of dieting behavior affect men and women differentially with more pronounced central neurotransmitter changes in women; these changes affect neurotransmitters implicated in the hypothalamic regulation of hunger and satiety (Goldbloom and Kennedy, 1993).

At the level of the family, there is intriguing evidence for genetic contributions as reflected by monozygotic versus dizygotic twin studies (Treasure and Holland, 1995). Precise heritability patterns or genetic markers for eating disorders await elucidation, but increased vulnerability in the context of an affected family member is documented. Beyond genetics is an important contribution of families toward the pathogenesis of these disorders. While there is no single type of family that promotes eating disorders, families may contribute through shared preoccupation with food, shape, and weight. Family dynamics that exacerbate difficulties in the development of normative autonomy during adolescence may create a stage for the development of eating disorders. Furthermore, a family history of mood disorder and/or substance abuse may amplify risk.

Finally, at the level of society, eating disorders are largely culture-bound phenomena of food-abundant, weight-preoccupied groups and nations. The idealization of the thin female form over the past three decades may account for greater awareness of and/or occurrence of eating disorders without explaining phenomena in their entirety. At the same time, increasing pressures on women to perform, achieve, and please others may exacerbate risk—as evidenced in the overrepresentation of eating disorders in female dancers and competitive athletes (Brownell, 1995).

EPIDEMIOLOGY

The lifetime prevalence for AN among women is estimated across studies at 0.5% to 1%. The incidence of AN over the past 40 years has increased. Less than 10% of patients with AN are males but, from the earliest descriptions of AN, men have been recognized to suffer this disorder. The lifetime prevalence of BN has been demonstrated in the largest community epidemiological study to date to be 1.1% of females and 0.1% of males, with partial syndromes being evident in an additional 1.3% (Garfinkel et al., 1995). The typical age of onset for AN is bimodal at 14 years and 18 years and that of BN is usually 18 years.

CLINICAL FEATURES

The ubiquity of dieting behavior, especially among young females in our society, does not typically lead to AN. The course of dieting that leads to AN includes continually lowering weight goals, weight loss that leads to increased criticism of one's body, and social isolation. Denial of the weight loss—even at the point of frank emaciation—escalates as the weight decreases. Preoccupation with food through cooking (for others), thinking about, reading about, and even dreaming about food as well as highly ritualized behavior toward food and weight characterize the withdrawal from food. Specific foods become morally valenced, and a pattern of dichotomous thinking—good or bad, fat or thin—governs self-appraisal and behavior.

In addition to a pattern of dietary restriction that may include fasting for days at a time, there may be a schedule of intensive exercise devoted, not to fitness or pleasure, but a purging of body fat. About 50% of AN sufferers will develop binge eating and/or purging behaviors as seen in BN in the context of their starvation and weight loss. The bodily symptoms of AN may be conceptualized as medical complications of the disorder, and are discussed below.

For BN, in contrast to the highly visible sequelae of AN, both the behaviors and the consequences may be highly secretive. Because binge eating episodes represent, for many BN sufferers, a response to overwhelming affects and serve as a source of subsequent shame, guilt, and self-loathing, they typically occur in a private context. Purging follows commonly through self-induced vomiting (either manually or through the use of chemical emetics such as ipecac), abuse of diet pills, laxatives, or diuretics (sometimes in quantities that can cause significant fluid and electrolyte imbalance). Unfortunately, when severe caloric restriction follows a binge it may ultimately set up intractable hunger urges, which precipitate the next binge, locking the sufferer into a vicious cycle.

CLINICAL COURSE

AN is largely a chronic disease. A longitudinal outcome study conducted in Toronto following patients some five to 14 years after their original consultation indicated that, even though approximately 38% of subjects had recovered and another 27% had made substantial improvements, 27% remained actively ill and 8% were deceased (Toner et al., 1986). While numbers vary, mortality rates as high as 18% have been documented in longterm outcome studies of AN (Ratnasuriya et al., 1991). This disturbing figure is as high as the rates of death, predominantly from suicide, in disorders such as schizophrenia and depression. Another important dimension to the clinical course is the longitudinal vulnerability to the development of other psychiatric disorders regardless of whether the AN is active or in remission. There is particular risk for the development of mood disorders and anxiety disorders. Because of both the occurrence of AN during peak reproductive years and its deleterious effects on neuroendocrine function, there is evidence of low pregnancy rates among AN sufferers and for increased rates of eating disorder among women undergoing evaluation for infertility. For women with AN who do conceive, the pregnancy and development of the fetus may be compromised by severe caloric restriction or bulimic behaviors (Goldbloom, 1993).

In contrast to AN, the clinical course of BN is not as well documented—perhaps since the disorder was only described formally in 1979. No reliable findings across outcome studies have emerged, although shortterm outcomes related to specific therapeutic interventions have actually been better documented for BN than for AN (Garfinkel and Goldbloom, 1993). In this regard, five-year follow-up data from psychotherapy trials for BN represents the best available data (Fairburn et al., 1995). While significant and sustained improvement resulted from a variety of psychotherapies including an abstinence rate of almost half the sample, 46% met criteria for AN, BN, or a partial syndrome, and 28% met criteria for other psychiatric disorders, predominantly mood and anxiety disorders. There was no evidence of BN-related mortality.

COMPLICATIONS

Characterization of the medical complications of AN has been a consistent feature of clinical descriptions for more than a century and at times has led to biological hypotheses of etiology and efforts at treatment. With the evolution of new assessment measures in such diverse areas as immunology and neuroimaging, further evidence of the protean complications emerges (Goldbloom and Kennedy, 1995).

AN is associated with a range of both central and peripheral neurological disturbances. Magnetic resonance imaging studies have documented decreases in grey and white matter and increases in cerebrospinal fluid volumes compared to controls (Katzman et al., 1996). At a functional brain level, positron emission tomography (PET) has documented glucose hypermetabolism in the caudate region. Generalized muscle weakness and loss of muscle tissue commonly occurs.

Like the central neurological manifestations, disturbances in normal bone growth and metabolism may appear early in the course of AN and lead to pathological fractures, growth retardation, and osteoporosis. Cardiovascular complications reflect both decreased caloric availability for metabolic tasks and changing cardiac architecture with emaciation. Sinus bradycardia, low blood pressure, and acrocyanosis are common; cardiac imaging reveals changes in cardiac chamber size and muscle mass, with an increased propensity for mitral valve prolapse. If the emetic ipecac is used repeatedly to induce vomiting, this may result in a cardiomyopathy as a direct toxic effect of one of its ingredients. Unexplained deaths in AN may result from sudden cardiac events. The vulnerability of this cardiac status is also reflected in occurrence of congestive heart failure in the context of aggressive refeeding.

The gastrointestinal response to caloric restriction is a marked delay in gastric emptying time, which yields problems with abdominal bloating and pain. Constipation is also a common sequela both from decreased intake and laxative abuse. Dehydration resulting from vomiting or laxative or diuretic abuse may lead to irreversible renal tubular damage and renal calculi. Metabolic sequelae of vomiting include the development of a hypochloremic hypokalemic metabolic alkalosis. Parotid gland enlargement and elevation of salivary amylase can result from recurrent vomiting. Reversible decreases in red blood cells, white blood cells, and platelets are also seen. Dermatological complications include the development of lanugo hair—a fine, downy growth—and, in individuals who induce vomiting manually, a callus on the dorsum of the hand caused by friction against the teeth. Finally, the acidic effects of vomitus lead to erosion of dental enamel on the lingual aspect of the upper teeth. Beyond the somatic sequelae of AN and BN are the nonspecific cognitive and affective aspects of starvation; they include impairment in concentration and mood lability.

PREVENTION AND TREATMENT

Primary prevention involves efforts to prevent the onset of the disorder. Reducing cultural preoccupation with a thin, unachievable female form as presented through the media would be an example of a significant effort at primary prevention. Education about the nature and risk of eating disorders to high-risk populations, such as preadolescent female dancers, would be a focused primary preventative effort. However, no data exists to support the efficacy of such interventions and, despite any intuitive appeal, they should not be implemented without empirical evaluation and support. A secondary prevention program is concentrated on individuals at high risk and/or those who have already shown early signs of the disorder. Here, the goal is the prevention of chronicity and the reduction of morbidity associated with it. Educating students, parents, teachers, and physicians about the early warning signs of eating disorders and implementing screening programs may lead to earlier detection and intervention. While there is no direct evidence for the efficacy of these programs, some follow-up studies indicate that a longer duration of symptoms prior to intervention worsens outcome. Finally, tertiary prevention is aimed at minimizing the longterm complications of these disorders including death; this is often the role of clinicians in the longterm management of eating patients with the eating disorder.

Treatment of AN and BN is typically multimodal, involving a trusting therapeutic relationship, nutritional rehabilitation, structure, attention to intrapsychic and interpersonal issues, and, in some cases, hospitalization and medication. A detailed consideration of the research supporting each dimension of treatment is beyond the scope of this chapter but is elaborated in numerous recent textbooks and clinical practice guidelines (American Psychiatric Association, 1993).

A trusting relationship with a clinician is a cornerstone to treatment and is a difficult task, especially in the case of the AN patient: (1) the patient is typically ambivalent about treatment, particularly when it includes weight gain; (2) implicit in this relationship is the relinquishment of some control at a time when the patient may feel her weight is the only area of her life she controls; and (3) it is easy for the patient to see the clinician acting purely in *loco parentis*—typically in the wake of unsuccessful efforts by the family to reverse the weight loss. In BN, the difficulties in the therapeutic relationship may relate to the shame and guilt experienced by the patient over highly secretive behaviors, coupled with a wish to please the therapist; this may lead to concealment of symptomatology.

For both AN and BN, normalization of eating behavior and weight is an essential component of treatment. This is best accomplished through meal planning and food diaries for subsequent monitoring. In AN, a detailed meal plan is often constructed to insure adequacy of caloric and macronutrient intake as well as to introduce foods to which the patient has displayed phobic avoidance. For AN patients, this meal plan of three meals plus snacks per day may typically start at 1500 kilocalories per day, increasing by 300 kilocalorie increments weekly up to 3000 to 3500 kilocalories per day. Prior to treatment, patients may be literally starving or subsisting in a catabolic state on a few hundred kilocalories per day. Instituting, supervising, and enforcing this meal plan may require hospitalization and will often feel overwhelming for the patient who will require much support to go through it. The goal is to establish a rate of weight gain of 1-2 kg per week.

In BN, a food diary that is kept for a few weeks prior to treatment can be most revealing. It often reflects patterns of daytime dietary restriction followed by evening binge patterns in women at a statistically normal body weight (although they may have previously been overweight and can maintain their current weight only through severe restriction) as well as the occurrence of binge episodes as a sequela to recurring and characteristic interpersonal events and intolerable affects. The therapeutic task in nutritional rehabilitation then becomes one of normalizing eating patterns to prevent intractable hunger at the end of the day and making eating less of a secretive event and introducing avoided foods outside the context of binges (patients typically binge on precisely those foods they avoid at other times because of their perceived "badness" with regard to caloric and/or fat content). Additionally, both the therapist and the patient learn of patterns in binge eating behavior with regard to both antecedents and consequences, laying the groundwork for developing alternative coping mechanisms.

The issue of body weight must be addressed in AN and BN. In AN, a detailed history must include determination of premorbid weight, weight at which menstrual function became disrupted, current weight, and perceived ideal weight. Then a target weight is established that is usually about 90% of average values for height and frame and is compatible with both menstrual function and maintenance without dieting. It is important to establish a range of target weight within about 2 kg, both to reflect the natural tendency of body weight fluctuation and to decrease preoccupation with a single weight value as being all-important.

In BN, acceptance of modest weight gain to reach a weight that is compatible with relinquishing dieting may be a necessary tradeoff to arrest the cycle of binge eating and purging.

It is important to understand, at the outset, the context—both historical and proximate—in which the disorder arose and the adaptive functions it may serve; it is the recognition of the latter that forms part of the basis for shared goals between the patient and the therapist beyond simple weight gain or binge cessation. At the same time, the maladaptive elements must be seen as outweighing the adaptive ones and, over the course of treatment, newer coping strategies for dealing with the past and the present must be developed for the future. This is the task of psychotherapy.

Additionally, psychoeducation plays an important role in the treatment of these disorders. Despite seemingly encyclopedic knowledge about food, these patients often lack an understanding of the relationship between their eating behavior and the social and psychological sequelae, and they frequently feel isolated in their suffering. For less severe forms of BN, psychoeducation alone—delivered in a group setting—represents an innovative and effective form of treatment.

Pharmacotherapy of eating disorders has been a fruitful area of inquiry for BN but not for AN (Kennedy and Goldbloom, 1995). There is no evidence from randomized, controlled clinical trials that medications are of any benefit in AN; the drug of choice is food. Nevertheless, on compassionate grounds, some medications are occasionally used for specific and transient purposes: gastric prokinetic agents to reduce the delayed gastric emptying and thus relieve bloating in the early stages of refeeding; anxiolytic medications to reduce the psychological distress prior to meals in the early stages of refeeding; antidepressants to reduce dysphoria once the contributing effect of nutritional deprivation has been reversed.

In BN, numerous clinical trials have indicated the shortterm efficacy of various antidepressant drugs in the diminution and cessation of binge eating episodes and associated purging. The most frequently studied of these is fluoxetine, the selective serotonin reuptake inhibitor that has been shown to have an antibulimic effect at doses of 60 mg per day. How these drugs exert an antibulimic effect is unknown—but what is obvious is that they do not work through the relief of underlying depression. Rather, it may be that their central effects on the various neurotransmitters that regulate hunger and satiety centrally contribute to their therapeutic effects. It is also unclear what is the longterm role and benefit of these medications, and clinically, their use typically occurs against a backdrop of other simultaneous interventions.

Clinical trials of various forms of psychotherapy for BN are numerous and provide validation for the utility of a variety of approaches, including cognitive–behavioral therapy, interpersonal therapy, and psychodynamic psychotherapy—either in an individual or a group format (Garfinkel and Goldbloom, 1993). The scientific study of psychotherapy for AN is surprisingly limited but does provide support for the use of family therapy in weight-recovered AN patients who are under 18 years of age and have not been chronically ill.

CONCLUSION

The evolution of understanding and treatment of AN and BN is paradigmatic of changes in psychiatry over the last century, integrating neuroscience with psychology, pharmacotherapy with psychotherapy, and individual with culture. Their prevalence, morbidity, and mortality all point to the need for more research in this area of human behavior.

REFERENCES

American Psychiatric Association (1993). Guidelines for Eating Disorders. Am. J. Psychiat. 150, 212-228.

American Psychiatric Association (1994). Diagnostic and Statistical Manual of Mental Disorders, 4th ed. American Psychiatric Association, Washington, D.C.

Fairburn, C.G., Norman, P.A., Welch, S.L., O'Connor, M., Doll, H.A., & Peveler, R.C. (1995). A prospective study of outcome in bulimia nervosa and the long-term effects of three psychological treatments. Arch. Gen. Psychiat. 52, 304-312.

Garfinkel, P.E. & Garner, D.M. (1982). Anorexia Nervosa: A Multidimensional Perspective. Brunner/Mazel, NY.

Garfinkel, P. & Goldbloom, D. (1993). Bulimia nervosa: A review of therapy research. J. Psychother. Practice Res. 2, 38-50.

Garfinkel, P., Lin, E., Goering, P., Spegg, C., Goldbloom, D., Kennedy, S., Kaplan, A., & Woodside, D. (1995). Bulimia nervosa in a Canadian sample: Prevalence and a comparison of subgroups. Am. J. Psychiat. 152, 1052-1058.

Goldbloom, D. (1993). Menstrual and reproductive function in the eating disorders. In: Medical Issues and the Eating Disorders: The Interface, pp. 165-175. Brunner/Mazel, NY.

Goldbloom, D. & Kennedy, S. (1993). Neurotransmitter and neuropeptide disturbances in the eating disorders. In: Medical Issues and the Eating Disorders: The Interface. (Kaplan, A.S. & Garfinkel, P.E., Eds.), pp. 123-143. Brunner/Mazel, New York.

Goldbloom, D. & Kennedy, S. (1995). The medical complications of anorexia nervosa. In: Comprehensive Textbook of Eating Disorders and Obesity. (Brownell, K.D. & Fairburn, C.G., Eds.), pp. 266-270. Guilford Press, New York.

Katzman, D., Lambe, E., Ridgley, J., Goldbloom, D., Mikulis, D., & Zipursky, R. (1996). Cerebral gray matter and white matter volume deficits in adolescent females with anorexia nervosa. J. Pediatrics 129, 794-803.

Kennedy, S. & Goldbloom, D. (1995). Psychopharmacologic treatments for anorexia nervosa and bulimia nervosa. In: Treatments of Psychiatric Disorders: The DSM-IV Edition. (Gabbard, G.O., Ed.), pp. 2153-2162. American Psychiatric Press, Washington, D.C.

Ratnasuriya, R.H., Eisler, I., Szmukler, G.I., & Russell, G.F.M. (1991). Anorexia nervosa: Prognosis and outcome after twenty years. Br. J. Psychiat. 158, 495-502.

Russell, G.F.M. (1979). Bulimia nervosa: An ominous variant of anorexia nervosa. Psychol. Med. 9, 429-448.

Szmukler, G., Dare, C., & Treasure, J. (Eds.)(1995). Handbook of Eating Disorders: Theory, Treatment, and Research. John Wiley & Sons, New York.

Toner, B.B., Garfinkel, P.E., & Garner, D.M. (1986). Long-term follow-up of anorexia nervosa. Psychosom. Med. 48, 520-529.

Treasure, J. & Holland, A. (1995). Genetic factors in eating disorders. In: Handbook of Eating Disorders: Theory, Treatment, and Research. (Szmukler, G., Dare, C., & Treasure, J., Eds.), pp. 65-82. John Wiley & Sons, New York.

Chapter 23

Antidepressant and Anxiolytic Drugs

SHARON C. CHEETHAM and DAVID J. HEAL

Biological Psychiatry, pages 511-567.

ISBN: 1-55938-819-6

INTRODUCTION

Electroconvulsive shock, the first effective treatment for severe affective disorders was introduced in the 1930s (Cerletti and Bini, 1938). However, it was not until the late 1950s that effective drugs for the treatment of depressive illness and anxiety emerged. Since that time some significant improvements have been made. However, there have been no major breakthroughs in the treatment of these disorders. As a consequence, antidepressants and anxiolytics still remain the subject of intense research.

ANTIDEPRESSANT DRUGS

Clinical Features of Depression

Literary and clinical descriptions of depression date back to antiquity. The old testament story of King Saul describes a depressive syndrome as does the story of Ajax's suicide in Homer's "Iliad". The Ancient Greeks used the term "melancholia" to describe depression in the belief that it was caused by an excess of bile [Greek: melas, black; kholé, bile].

The word depression is used to describe both a clinical condition or illness and the brief, mild downward mood swings that we all experience as part of daily living. In the clinical context, the term depressive illness refers to a prominent and persistent disturbance of mood, either depressed, elevated (manic), or mixed that is clearly distinguishable from prior functioning, which may be associated with cognitive, psychomotor, psychophysiological, interpersonal, and behavioral difficulties. Patients with depressed mood exhibit loss of energy and interest, difficulty in concentrating, sleep and appetite disturbance, change in weight, psychomotor agitation or retardation, feelings of guilt and worthlessness, and thoughts of death or suicide. Patients with elevated mood (mania) exhibit expansiveness or irritability, flight of ideas, distractibility, pressured speech, hyperactivity, excessive involvement in activities without the use of judgement, decreased sleep, heightened self-esteem, and grandiose ideas.

In an attempt to achieve a reliable and objective system of psychiatric diagnosis and classification, the American Psychiatric Association proposed the now well accepted Diagnostic and Statistical Manual of Mental Disorders DSM-III (1980), which was later modified to DSM-III-R (1987) and more recently updated to DSM-IV (1994). Within this classification, the mood disorders are subdivided into the depressive disorders ("unipolar depression") and the bipolar disorders. The depressive disorders are characterised by depressed mood, whereas the bipolar disorders involve the presence (or history) of mania, mixed (rapidly alternating manic and depressed episodes), or hypomanic episodes. However, the boundaries between depression and sadness and depressive illness and anxiety states still remain to some extent arbitrary and ill defined (Tyrer, 1992). Furthermore, there is a high degree of comorbidity between depressive illness and anxiety states.

Historical Perspective

The discovery of the first drugs with antidepressant properties was entirely accidental. During clinical trials of the antituberculosis drug, iproniazid, some patients were noted to exhibit elevated mood and euphoria (Selikoff et al., 1952). Almost simultaneously iproniazid was shown to be a potent inhibitor of the enzyme monoamine oxidase (MAO) (Zeller et al., 1952). However, the link between the elevated mood and any possible benefit in depressed patients was not made until later (Loomer et al., 1957; Kline, 1958).

In 1958, Kuhn described the antidepressant effect of imipramine. This compound (which is structurally related to the phenothiazines, e.g., chlorpromazine) was initially found to be ineffective for the treatment of schizophrenia, but a number of schizophrenics with depressive features showed some improvement in the latter symptoms. As a consequence, the drug was evaluated in depressed patients with clear-cut and dramatic benefit. However, it was not until the 1960s that imipramine and other tricyclic antidepressants were shown to potently inhibit the uptake of the monoamine neurotransmitters, noradrenaline (Glowinski and Axelrod, 1964) and 5-hydroxytryptamine (serotonin, 5-HT) in the brain (Carlsson et al., 1969).

Around the same time a proportion of the patients treated with the antihypertensive drug, reserpine, were reported to develop symptoms resembling those of depression (Freis, 1954; Muller et al., 1955). Animal studies demonstrated that reserpine caused marked depletion of nerve terminal stores of noradrenaline (Holzbauer and Vogt, 1956), 5-HT (Pletscher et al., 1956) and dopamine (Bertler, 1961). These observations combined with the pharmacological actions of the monoamine oxidase inhibitors (MAOIs) and the tricyclic antidepressants to increase monoamine concentrations at central synapses by blockade of their degradation or reuptake led to the formulation of the monoamine hypothesis of depression. This states that depressive illness results from a functional deficit of noradrenaline (Bunney and Davis, 1965; Schildkraut, 1965) or 5-HT (Coppen, 1967; Lapin and Oxenkrug, 1969) in the brain and that the MAOIs and monoamine reuptake inhibitors produce their therapeutic effects by reversing this deficit.

Developments in Antidepressant Drugs

Serendipity has played a role in the discovery of many therapeutic agents, not only antidepressant drugs. This route has profound implications for novel drug development. Only when the pharmacological actions of the "first generation" MAOIs and monoamine reuptake inhibitors were determined could the "second generation" of drugs with improved side-effects be developed. Antidepressant drugs based on new pharmacological approaches were not possible until a

Table 1. The Characteristics of an "Ideal" Antidepressant Drug

1.	Rapid onset of clinical efficacy
2.	Efficacy in all depressed patients, irrespective of the various subcategories of this mental disorder
3.	No adverse side-effects
4.	Nontoxic when taken in overdose
5.	No drug dependence or rebound withdrawal effects

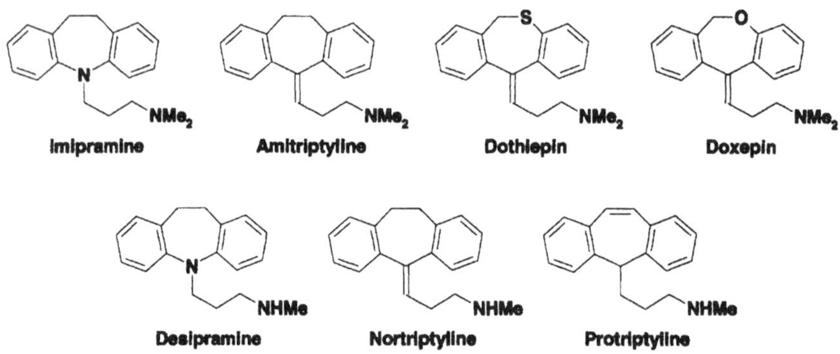

Figure 1. The structures of several first generation tricyclic antidepressants

hypothesis explaining the biochemical basis of the etiology of depression was proposed. However, in developing any new drug, it is important to initially define the properties an "ideal" drug would possess; these are summarized in Table 1.

Several new antidepressant drugs have recently been introduced into clinical practice with the aim of improving therapeutic efficacy, side-effects, and toxicity profile. To put these new drugs into perspective, the strengths and shortcomings of all classes of antidepressant drug will be reviewed in relation to efficacy, onset of action, side-effects, and toxicity.

Monoamine Reuptake Inhibitors

First Generation—Tricyclic Antidepressants

Until recently, tricyclic antidepressants were the most widely used treatment for depressive illness. They all possess the same basic chemical structure and are closely related to the phenothiazines, the main structural difference being the presence of an ethylene group in the central ring rather than a sulphur (Figure 1). The tricyclics inhibit the reuptake of noradrenaline and 5-HT (e.g., imipramine [Geigy], amitriptyline [Merck Sharpe and Dohme], and dothiepin [Knoll]) or of predominantly noradrenaline (e.g., desipramine [Geigy] and protriptyline [Merck Sharpe and Dohme]), with little effect on dopamine (Hyttel, 1982; Richelson and Pfenning, 1984; Heal et al., 1992; Table 2).

It is well established that the tricyclic antidepressants produce clinically significant improvements in depressed patients and are superior to placebo (reviewed by Brotman et al., 1987). However, only 65% to 70% of patients respond to treatment with these drugs (Klerman and Cole, 1965; Morris and Beck, 1974; Davis, 1975; Kupfer and Detre, 1978). The onset of clinical action is not immediate; a significant global improvement usually is demonstrated only after two to three weeks of continuous treatment, although some improvement in individual symptoms occurs quite rapidly in drug-responsive patients (Frazer, 1994).

The tricyclics produce a range of unwanted side-effects due to their interaction with muscarinic, histamine H_1, and α_1-adrenergic receptors (Shein and Smith, 1978; Tang and Seeman, 1980; Richelson and Nelson, 1984; Heal et al., 1992). The potency of the tricyclics in blocking muscarinic cholinergic receptors produces the commonly quoted side-effects of dry-mouth, blurred vision, constipation and urinary retention. Cholinergic blockade can also result in sinus tachycardia and short-term memory impairment (Tollefson, 1991). Sedation and drowsiness may result from their affinity for histamine H_1 receptors. As a consequence of blockade of α_1-adrenoceptors, tricyclics produce postural (orthostatic) hypotension, which occurs in as many as 20% of patients (Tollefson, 1991). The

Table 2. Potency of Various Tricyclic Antidepressants as Inhibitors of Monoamine Uptake in Vitro

	K_i values (nM)[a]		*Selectivity ratio*
	NA	*5-HT*	*NA : 5-HT*[b]
Desipramine	0.9	340	378
Protriptyline	1	280	280
Nortriptyline	4	260	65
Imipramine	13	42	3
Doxepin	19	280	15
Amitriptyline	24	66	3
Dothiepin	34	110	3

Notes: Values are taken from Richelson and Pfenning (1984).
[a] The smaller the number, the more potent the drug is as an inhibitor of noradrenaline (NA) or 5-HT uptake *in vitro*.
[b] The larger the number, the more selective the drug is in blocking the reuptake of NA.

combination of anticholinergic activity, inhibition of monoamine reuptake, and the direct depressant effects of the tricyclics can result in mild tachycardia. However, abnormalities in cardiac conduction can occur. These include prolongation of PR, QRS, or QT intervals or flattening or inversion of T waves due to slowing of both atrial and ventricular depolarization. The slowing of depolarization can lead to atrioventricular or bundle branch block or premature ventricular contractions. These cardiac effects are cause for concern but do occur more commonly in patients with preexisting cardiac problems (Roose et al., 1987). Less common side-effects include weight-gain (Cole and Brodkin, 1990), convulsions (Skowron and Stimmel, 1992), and disturbances of sexual function. These side-effects become particularly dangerous when tricyclics are taken in overdose as the convulsions, cardiac sequelae, respiratory depression, and coma can prove fatal. Death from tricyclic overdose is primarily due to cardiac arrest (Crome and Newman, 1979). The toxicity in overdose of the tricyclic antidepressants is of great concern as approximately 15% of patients with major depression die by suicide (Guze and Robins, 1970). Poisoning accounts for about 20% of all suicides with the tricyclics being the most commonly used drugs (Kapur et al., 1992). However, while this property of the tricyclics has been the subject of considerable criticism (Leonard, 1986; Cassidy and Henry, 1987; Henry, 1989) and should not be treated with complacency, it should be viewed within the context of the overall prescribing of these useful therapeutic agents.

Drug dependence is not a problem with the tricyclics. However, abrupt cessation of treatment can result in a mild withdrawal syndrome including nausea, vomiting, cramps, and general malaise in the case of imipramine (Davis, 1975; Kupfer and Detre, 1978). These symptoms can be prevented by gradual withdrawal, which also helps prevent relapse into depression.

Second Generation—Monoamine Reuptake Inhibitors

There are many new drugs that are marketed as "second generation" monoamine reuptake inhibitors. These drugs have diverse chemical structures; therefore, they will be herein categorized on the basis of their main pharmacological actions.

Selective serotonin reuptake inhibitors. The first selective serotonin reuptake inhibitor (SSRI) to be successfully introduced into the depression market was fluoxetine (Prozac; Lilly). Four other SSRIs have been approved for the treatment of major depressive disorder, namely paroxetine (Smith Kline Beecham), sertraline (Pfizer), citalopram (Lundbeck) and fluvoxamine (Solvay) (Figure 2). The main pharmacological actions of the SSRIs, as their name implies, is the potent and selective inhibition of the reuptake of 5-HT (Hyttel, 1982; Wong et al., 1983; Richelson and Pfenning, 1984; Bolden-Watson and Richelson, 1993; Table 3).

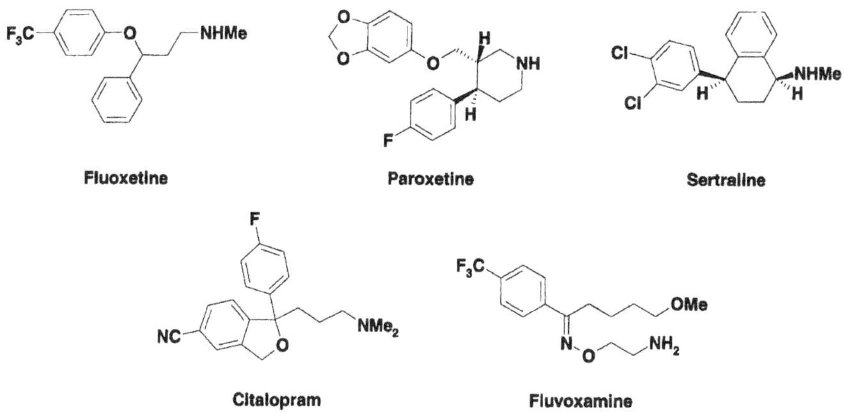

Figure 2. The structures of various selective serotonin reuptake inhibitors

Table 3. Potency of Various Second Generation Antidepressants as Inhibitors of Monoamine Uptake in Vitro

	K_i values (nM)[a]		*Selectivity ratio*[b]	
	5-HT	*NA*	*5-HT : NA*	*NA : 5-HT*
Fluoxetine	12	280	23	–
Paroxetine	0.7	33	47	–
Sertraline	3.4	220	65	–
Citalopram	1.3	400	307	–
Fluvoxamine	7	500	71	–
Maprotiline	3300	7.4	–	446
Amoxapine	470	4.4	–	107
Lofepramine	2400	1.9	–	1263
Venlafaxine	39	210	5	–
Trazodone	490	9500	19	–
Nefazodone	137	570	4	–

Notes: Values are taken from Richelson and Pfenning (1984) and Bolden-Watson and Richelson (1993).
[a] The smaller the number, the more potent the drug is as an inhibitor of 5-HT or noradrenaline (NA) uptake *in vitro*.
[b] The larger the number the more selective the drug is in blocking the reuptake of 5-HT or NA.

The SSRIs have clearly shown superior efficacy to placebo and comparable efficacy to the tricyclics in major depressive disorder, but they are neither more efficacious, nor faster acting than tricyclics (Rickels and Schweizer, 1990; Dechant and Clissold, 1991; Grimsley and Jann, 1992; Murdoch and McTavish, 1992; Kasper et al., 1992; Finley, 1994; Lane et al., 1995). There is, however, evidence indicating that SSRIs may be more effective in treating the anxiety associated with depression than the tricyclics (Montgomery and Fineberg, 1989). In addition, the SSRIs have been shown to be effective in obsessive-compulsive disorder (in which other antidepressants are ineffective with the exception of clomipramine, which is also a potent 5-HT reuptake inhibitor), panic disorder, dysthymia, social phobia, and bulimia (Finley, 1994; Lane et al., 1995).

The SSRIs do not induce the anticholinergic, cardiovascular, or sedative side-effects associated with the tricyclic antidepressants. This improvement is almost certainly because the SSRIs are highly selective 5-HT reuptake inhibitors with little affinity for receptors associated with other neurotransmitters including muscarinic, histamine H_1, and α_1-adrenergic receptors, with the exception of paroxetine, which exhibits moderate affinity for muscarinic receptors (Wong et al., 1983; Richelson and Nelson, 1984; Cusack et al., 1994). However, it would be incorrect to conclude that the SSRIs cause fewer adverse effects than the tricyclics. The main drawback of the SSRIs is a higher incidence of nausea (15–35%) and gastrointestinal disturbances (e.g., vomiting, diarrhea) than produced by the tricyclics (Rickels and Schweizer, 1990; Grimsley and Jann, 1992; Wagner et al., 1992); although these side-effects tend to lessen over time. Other side-effects include central nervous system symptoms (e.g., insomnia, agitation, headache), sexual dysfunction (anorgasmia or delayed ejaculation), and tremor (Finley, 1994). The critical issue is whether the side-effect profile of the SSRIs improves patient compliance. Between 0% and 14% of depressed patients taking placebo drop out of clinical trials because of side-effects (median: 5%), as compared to 7% to 23% (median: 15%) of those taking the SSRIs and 7% to 44% (median: 21%) of those taking tricyclics (Cookson, 1993). Thus, there is an improvement in the rate of compliance between the SSRIs and the tricyclics, but this increase is fairly modest (around 6%). The SSRIs have a greatly reduced risk of toxicity in overdose relative to the tricyclics (Office of Population Census Survey, 1990) due to their lack of cardiotoxic effects.

Many clinicians consider the SSRIs to be a substantial improvement over the tricyclics. However, their clinical efficacy and onset of action is equivalent to that of the tricyclics. Although the SSRIs are perceived to be considerably better tolerated than the tricyclics, there is only a modest increase in patient compliance compared to the tricyclics. Thus, the major advantage of the SSRIs is the greatly reduced toxicity in overdose. As a consequence, the SSRIs now dominate the antidepressants market both in the United Kingdom and United States.

However, there are concerns relating to the use of the SSRIs, in particular fluoxetine (Prozac), which is becoming a "lifestyle" drug. This does not relate to its use in depressed patients but to its use to brighten the mild downward mood swings that everyone experiences as part of daily living and to alter personality thus producing more optimistic, competitive, bold, and confident people. Whether the emotional manipulation of nondepressed subjects is beneficial to society is, at

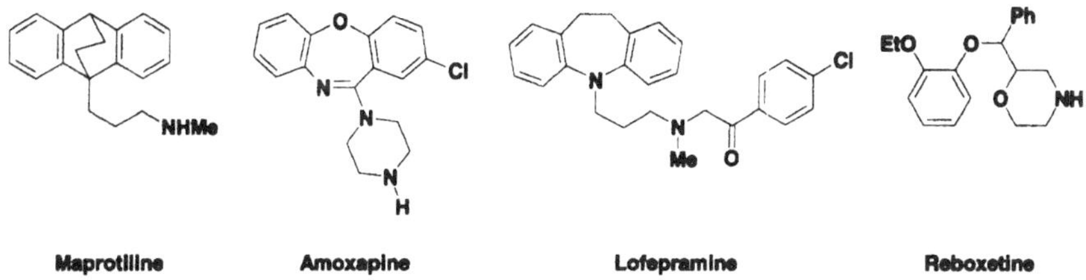

Figure 3. The structures of the second generation selective noradrenaline reuptake inhibitors.

this stage, undetermined; however, it does pose some very serious ethical and moral questions (Kramer, 1993).

Selective noradrenaline reuptake inhibitors. Until very recently, all of the selective noradrenaline reuptake inhibitors available—maprotiline (Ciba-Geigy), amoxapine (Lederle), and lofepramine (Merck)—were structurally related to the tricyclics (Richelson and Pfenning, 1984; Bolden-Watson and Richelson, 1993; Table 3; Figure 3). However in July 1997, Pharmacia and Upjohn launched reboxetine, a structurally novel selective noradrenaline reuptake inhibitor (NARI) in the UK (Riva et al., 1989; Figure 3).

Maprotiline, amoxapine, and lofepramine exhibit efficacy and a time-course of clinical improvement equivalent to the earlier tricyclics (Mathur, 1975; Middleton, 1975; Feighner et al., 1982a; Gruter and Poldinger, 1982; Jue et al., 1982; Kinney and Evans, 1982; Rickels et al., 1982a). Reboxetine is efficacious in both the short- (4–8 weeks) and longterm (up to 12 months) treatment of depression and is at least as efficacious as the tricyclics and SSRIs (Montgomery, 1997). However, reboxetine has been reported to be superior to fluoxetine in severely depressed patients (Montgomery, 1997). Furthermore, reboxetine showed significant advantages over fluoxetine in terms of social functioning, positively affecting patients' self-perception and motivation toward action (Dubini et al., 1997; Montgomery, 1997). There is no evidence that reboxetine is more rapidly acting than the tricyclics or SSRIs.

All these drugs exhibit significantly lower affinity for muscarinic receptors than the tricyclics (Richelson and Nelson, 1984; Batra and Björklund, 1986; Riva et al., 1989), a phenomenon associated with a lowered propensity to produce anticholinergic side-effects (e.g., dry mouth, blurred vision, constipation; Lancaster and Gonzalez, 1989; Ad Sitsen and Montgomery, 1994; Mucci, 1997). However, the affinity of amoxapine and maprotiline for histamine H_1 and α_1-adrenergic receptors is comparable to that of the tricyclics (Richelson and Nelson, 1984), accounting for their cardiovascular (orthostatic hypotension) and sedative side-effects (Gruter and Poldinger, 1982; Kinney and Evans, 1982). Both drugs also produce other serious side-effects including seizures with maprotiline (Skowron and Stimmel, 1992) and movement disorders with amoxapine due to its affinity for D_2 receptors (Jue et al., 1982; Thornton and Stahl, 1984). Lofepramine and reboxetine exhibit a significantly lower incidence of cardiovascular and sedative side-effects than the tricyclics (Peet, 1988; Mucci, 1997). Compared with the SSRI fluoxetine, reboxetine is equally well tolerated, but the profile of side-effects is qualitatively different (Mucci, 1997).

There is insufficient information to draw any firm conclusions regarding the toxicity of amoxapine and maprotiline in overdose. However, there is little evidence to suggest that they would be less toxic than the tricyclics. Lofepramine appears to be much safer in overdose than the tricyclics (Cassidy and Henry, 1987; Henry, 1989). Data relating to toxicity in overdose for reboxetine is awaited. However, its pharmacological profile suggests that it will be safer than the tricyclics.

Therefore, in terms of efficacy and onset of action, maprotiline, amoxapine, and lofepramine are not significantly advantaged over the tricyclics. For maprotiline and amoxapine, this comment also applies to their side-effect profile, whereas lofepramine is clearly better tolerated and less toxic in overdose than the first generation tricyclics. Reboxetine is clearly as efficacious as the tricyclics and the SSRIs but has been reported to be more efficacious in severely depressed patients and to have a significantly stronger effect on social functioning than the SSRI fluoxetine. These "beneficial" effects of reboxetine are being used in an attempt to create a "niche" for reboxetine within the SSRI-dominated antidepressant market. Whether these initial findings are robust enough to allow reboxetine to make a significant impact within the market place remains to be seen. In terms of side-effect profile, reboxetine is advantaged over the tricyclics and is likely to be less toxic in overdose. Reboxetine has no such advantages over the SSRIs.

Serotonin noradrenaline reuptake inhibitors. Venlafaxine (Wyeth; Figure 4) is described as the first of a new class of antidepressants observed to inhibit neuronal 5-HT and noradrenaline reuptake (serotonin noradrenaline reuptake inhibitor, or SNRI) (Muth et al., 1986; Bolden-Watson and Richelson, 1993). However, its uptake inhibition profile *in vitro* clearly shows that venlafaxine is only a weak noradrenaline reuptake inhibitor, being predominantly a 5-HT reuptake inhibitor (Table 3). Pierre-Fabre recently launched the SNRI, milnacipran, in France (July 1997). Milnacipran is a more "balanced" inhibitor of 5-HT and noradrenaline reuptake than venlafaxine (Briley and Moret, 1997; Figure 4).

Clinical studies have shown venlafaxine and milnacipran to be significantly more effective than placebo in patients with major depressive disorder (Khan et al., 1991; Schweizer et al., 1991; 1994; Mendels et al., 1993; Briley and Moret, 1997; Puech et al., 1997) and at least as efficacious as the tricyclics and the SSRIs (Samuelian et al., 1992; Clerc et al., 1994; Schweizer et al., 1994; Briley and Moret, 1997; Puech et al., 1997). There is evidence to suggest that venlafaxine

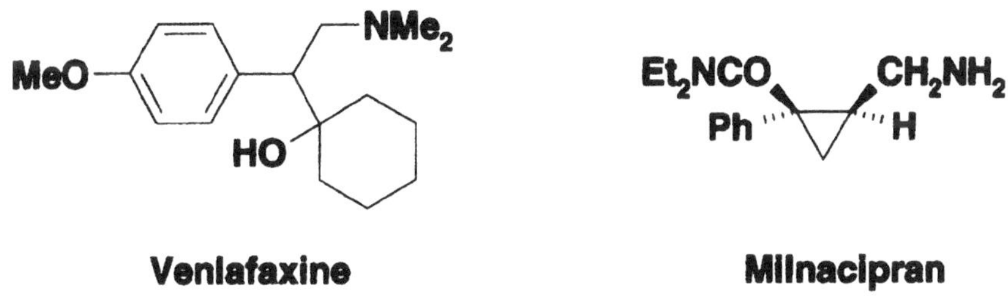

Figure 4. The structures of the serotonin noradrenaline reuptake inhibitors venlafaxine and milnacipran

and milnacipran have efficacy superior to the SSRIs, especially in more severe depression or when melancholia is present (Clerc et al., 1994; Dierick et al., 1996; Lopez-Ibor et al., 1996). The superior efficacy of venlafaxine and milnacipran is similar to that reported for clomipramine (Danish University Antidepressant Group, 1986; 1990). Thus, it has been proposed that combined inhibition of serotonin and noradrenaline reuptake may be associated with superior efficacy compared with selective serotonin reuptake inhibition alone (Briley and Moret, 1997). However, there is no evidence that venlafaxine and milnacipran are more efficacious than the tricyclics.

The results of a limited number of clinical studies in which the dose of venlafaxine was rapidly titrated (≥200 mg) suggest that venlafaxine may have a rapid onset of clinical action, albeit at the expense of a high incidence of side-effects (Rickels, 1991; Rudolph et al., 1991; Derivan et al., 1995; Guelfi et al., 1995). Although venlafaxine has been shown to be superior to placebo treatment according to the Hamilton Depression Rating Scale (HAM-D) total at week 1, this is a result of significant improvement in some, but not all, symptoms associated with major depressive disorder. A similar rapid improvement in certain symptomatologies has also been noted in depressed patients receiving high doses of amitriptyline and imipramine (Katz et al., 1987). These data raise the question of whether rapid attainment of optimum doses of antidepressants in general could initiate a more rapid clinical improvement in some symptoms of depression. To date, a more rapid onset of clinical action for venlafaxine compared to the tricyclics and SSRIs has not been established. A meta-analysis of the major comparative studies with milnacipran has shown no difference in onset of action of milnacipran compared to the tricyclics and SSRIs (Briley and Moret, 1997).

The adverse effects of venlafaxine resemble those of the SSRIs, particularly fluoxetine, which is not unexpected based on the uptake inhibition profile of venlafaxine (Table 3). The most common side-effects are nausea, headache, drowsiness, insomnia, dry mouth, sexual dysfunction, sweating, dizziness, nervousness, and constipation (Rickels, 1991; Montgomery, 1993; Holliday and Benfield,

1995). Venlafaxine has no affinity for muscarinic, histamine H_1 and α_1-adrenergic receptors *in vitro* (Muth et al., 1986; Montgomery, 1993). Thus, it has an improved side-effect profile compared with the tricyclics. There has been no evidence to suggest that venlafaxine affects cardiac conduction; however, there have been several reports of increases in blood pressure, particularly at high doses (Rickels, 1991; Montgomery, 1993; Holliday and Benfield, 1995). The implications of this side-effect in depressed patients with preexisting cardiovascular disease is, as yet, unclear. Milnacipran has considerably improved tolerability over the tricyclics, which probably arises from its pharmacological selectivity, in particular, its lack of affinity for a range of receptor subtypes (Moret et al., 1985; Briley et al., 1996). Milnacipran and the SSRIs possess comparable tolerability, but the profile of side-effects are qualitatively different. Nausea and diarrhea were more common with the SSRIs, and headache, dry mouth, and dysuria more common with milnacipran (Montgomery et al., 1996). The dysuria seen with milnacipran requires men with prostatic hyperplasia to be treated with caution and for the drug to be contraindicated where dysuria already exists.

Although the amount of data available are limited, venlafaxine and milnacipran appear to present few toxicity problems in overdose compared with the tricyclics (Holliday and Benfield, 1995; Puech et al., 1997).

Venlafaxine and milnacipran are no more efficacious or faster-acting than the tricyclics but, in terms of side-effect profile and toxicity in overdose, they show significant advantage. A limited number of studies do, however, suggest that venlafaxine and milnacipran exhibit superior efficacy than the SSRIs, especially in more severe depression, and show a tolerability profile globally similar. The potential of venlafaxine to produce increased blood pressure, especially at high doses, is a distinct disadvantage compared with the SSRIs. Thus, venlafaxine and milnacipran may provide a useful alternative to the tricyclics. At present, they are unlikely to make a substantial impact on the now SSRI-dominated antidepressant market. However, should their efficacy in more severely depressed patients prove to be a consistent finding, they may find a "niche" in the market place through the treatment of patients where the SSRIs are thought to be less effective.

Dopamine reuptake inhibitors. Bupropion (Burroughs Wellcome; Figure 5), is an antidepressant in the aminoketone class that acts as a relatively weak inhibitor of the neuronal uptake of noradrenaline serotonin, and dopamine. However, it does not inhibit monoamine oxidase. Presumably its antidepressant action is mediated by dopaminergic and noradrenergic mechanisms.

Bupropion has also proven to be no more effective or faster-acting than the tricyclics in the treatment of major affective disorder (Workman and Short, 1993). It lacks affinity for muscarinic, histamine H_1 and α_1-adrenergic receptors (Richelson and Nelson, 1984). Therefore, bupropion exhibits none of the anticholinergic, cardiovascular, or sedative side-effects common to the tricyclics. It also seems to be relatively safe in overdose (Van Wyck Fleet et al., 1983). The only

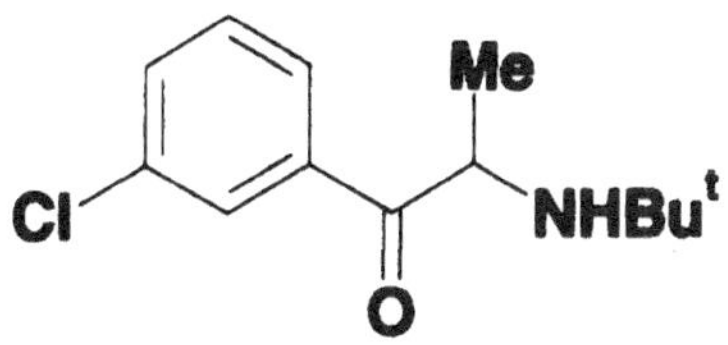

Bupropion

Figure 5. The structure of the dopamine reuptake inhibitor bupropion

serious adverse effect of bupropion is convulsions, especially in patients receiving high doses (>450mg/day; Peck et al., 1983; Skowron and Stimmel, 1992). Early clinical trials of bupropion indicated that the incidence of seizures at doses ≤450mg/day fell within the accepted range for antidepressant drugs, but the incidence significantly increased at doses ≤600mg/day (Peck et al., 1983; Johnston et al., 1991). However, in a trial of bupropion for the treatment of bulimia a high incidence of seizures was reported (Horne et al., 1988). As a result of the apparent high risk in this population and the discrepancy with the previously observed incidence, marketing of bupropion was delayed while further clinical data was obtained to clarify the situation. The earlier seizure estimates were confirmed with a significant proportion of those experiencing seizures having some predisposing factor such as history of head trauma or seizure, present seizure disorder, or concomitant use of drugs, which lower seizure threshold (Johnston et al., 1991). As a consequence, the maximum recommended dose for bupropion is 450mg/day. Furthermore, bupropion is not recommended for patients predisposed to seizures. Recently bupropion has been used with some success to treat attention deficit–hyperactivity disorder (ADHD) and boost success in quitting smoking while receiving patches.

Monoamine Oxidase Inhibitors

Irreversible Monoamine Oxidase Inhibitors

Within the central nervous system, the enzyme MAO exists in two forms, MAO_A and MAO_B (Johnston 1968; Knoll and Magyar, 1972). These isoenzymes act preferentially on different monoamines; 5-HT and noradrenaline are largely metabolized by MAO_A, while MAO_B more actively deaminates phenylethylamine, and both enzymes act on dopamine (Murphy et al., 1987). It is the inhibition of MAO_A that is associated with antidepressant activity (Murphy et al., 1987; Mann et al., 1989).

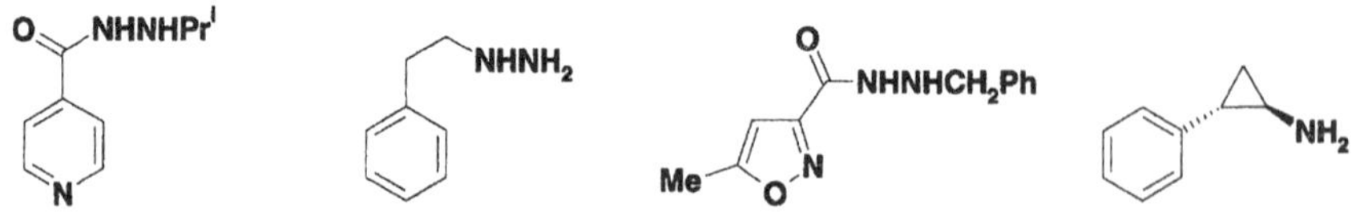

Figure 6. The structures of various first generation irreversible monoamine oxidase inhibitors

The first generation MAOIs, for example, iproniazid (Roche; discontinued), phenelzine (Parke-Davis), isocarboxazid (Roche), and tranylcypromine (SmithKline Beecham), are nonselective irreversible inhibitors of both enzymes (Youdim and Finberg, 1985). These drugs possess a phenylethylaminelike structure with a chemically reactive group, which enables the inhibitor to bind covalently to the enzyme resulting in a long lasting inhibition. In iproniazid and phenelzine, this reactive moiety is a hydrazine group; in tranylcypromine, it is a cyclopropylamine group (Figure 6). Recovery of MAO activity following cessation of treatment takes several weeks—the time required for synthesis replacement of MAO.

The efficacy of the MAOIs in the treatment of major depression has been controversial. While some clinical trials have demonstrated efficacy for the MAOIs in depressed patients that is equivalent to that of the tricyclics, others have reported that these drugs have little or no benefit (Morris and Beck, 1974; Kupfer and Detre, 1978; Tyrer, 1979; Murphy et al., 1987). It has been suggested that the poor performance of MAOIs is due to the administration of inadequate doses (Kupfer and Detre, 1978; Paykel and Hale, 1985). Attempts have been made to identify a subgroup of depressed patients who respond particularly well to the MAOIs. These patients have been described as having "atypical depression". Although the assessment of this issue has been confounded by the variety of criteria used to categorize this atypical group, there is evidence indicating that there is a subgroup of patients that are preferentially responsive to phenelzine (Liebowitz et al., 1988; Quitkin et al., 1991). These patients show mood reactivity, rejection sensitivity, hyperphagia, hypersomnolence, and lethargy. Furthermore, there are reports claiming that MAOIs are effective in tricyclic nonresponders (Nolen and Haffmans, 1989; Thase et al., 1992; McGrath et al., 1993).

The time-course for clinical improvement for the MAOIs is generally accepted to be several weeks. However, there is evidence indicating that the clinical response to these drugs, particularly tranylcypromine, can be extremely rapid (Lingl, 1964; Kupfer and Detre, 1978; Tyrer, 1979).

Many of the unwanted side-effects of the MAOIs result from their primary mode of action. Postural hypotension is a common side-effect, which is perhaps surprising in view of the increase in catecholamine levels and the lack of affinity of the MAOIs for α_1-adrenoceptors. A possible explanation is that inhibition of MAO allows amines such as dopamine and octopamine to accumulate within peripheral sympathetic nerve terminals and displace noradrenaline from the storage vesicles, thereby reducing noradrenaline release associated with sympathetic activity. Tremors, insomnia, and agitation can occur as a result of overstimulation of the central nervous system (CNS). Dry mouth, blurred vision, and urinary retention are observed but are less common than reported for the tricyclics and are not a consequence of affinity for muscarinic receptors. Other side-effects include weight gain, sexual dysfunction, and, very rarely, severe hepatotoxicity with MAOIs of the hydrazine type. However, the most publicized side-effect of the MAOIs is hypertensive crisis more commonly known as the "cheese effect" (Blackwell et al., 1967). This effect results from an interaction between MAOIs and tyramine, which is prevalent in certain foods (e.g., cheese), concentrated yeast extracts, and pickled herrings (Cooper, 1989). Tyramine is an indirectly acting sympathomimetic. Under normal circumstances, tyramine is metabolized in the gastrointestinal tract and thus little enters the circulation. However, if MAO_A is inhibited, tyramine enters the circulation causing the release of greater-than-normal amounts of noradrenaline and adrenaline from sympathetic nerve terminals, resulting in increased blood pressure (Da Prada et al., 1988; Cooper, 1989). This is associated with flushing, occipital headache, nausea, vomiting, photophobia, and raised intracranial pressure, which can occasionally lead to the more serious or even fatal subarachnoid hemorrhage (Kupfer and Detre, 1978; Blackwell, 1981). In order to avoid such episodes, patients receiving MAOI therapy must abide by certain dietary restrictions, a requirement that increases the likelihood of noncompliance. In addition, numerous over-the-counter cold and sinus medications contain indirectly-acting sympathomimetic amines; therefore, their use should also be avoided by patients treated with MAOIs. Thus, the irreversible MAOIs have a very poor side-effect profile. It was the publicity generated by the fatalities resulting from hypertensive crisis, along with the concerns relating to the efficacy of the MAOIs in major depressive disorder that resulted in the unpopularity of the MAOIs with general practitioners.

Other serious adverse reactions can occur when MAOIs are used with other drugs including narcotics, antihypertensives, insulin, anesthetics, tricyclics and SSRIs (Kurtz, 1990; Montgomery, 1991).

Toxicity in overdose is also a major concern for the MAOIs (Cassidy and Henry, 1987; Henry, 1989). Early symptoms include faintness, anxiety, flushing, sweating, headache, tachypnea, tachycardia, and tremor. More serious symptoms include muscular hyperactivity, coma, seizures, profound hypotension, and cardiac arrest (Meredith and Vale, 1985).

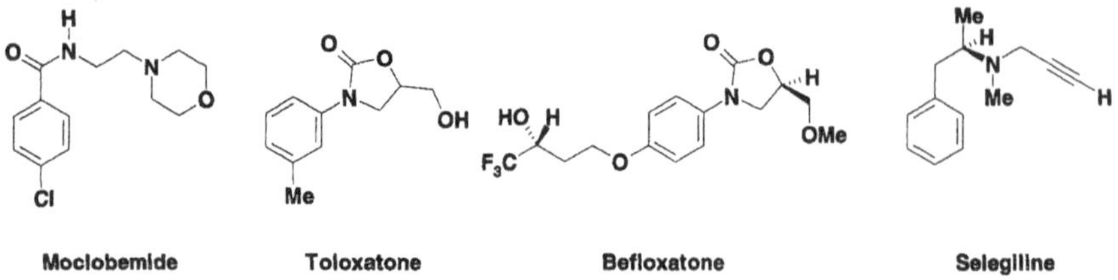

Figure 7. The structures of several second generation monoamine oxidase inhibitors

Although there is no evidence of drug dependence or rebound withdrawal effects with the MAOIs as these drugs are irreversible inhibitors of MAO, a wash-out period of four to six weeks for enzyme resynthesis is required before the prescribing of other psychoactive drugs. Therefore, the irreversible MAOIs are not used as first line therapy because nonresponders have to wait four to six weeks before an antidepressant drug with a different mode of action can be tried. This situation can have very serious implications for depressed patients, particularly those at risk of suicide.

"Second Generation" Monoamine Oxidase Inhibitors

In an attempt to circumvent the problems associated with the irreversible MAOIs, reversible inhibitors of MAO_A such as moclobemide (Roche), toloxatone (Synthélabo), and befloxatone (in phase II–III clinical trials; Synthélabo) and an irreversible inhibitor of MAO_B, namely, selegiline (Asta Medica), have been developed (Figure 7).

Reversible MAO_A inhibitors. The reversible inhibitors of MAO_A (RIMAs) are significantly more effective than placebo in patients with major depressive disorder and as efficacious as the tricyclics (Laux et al., 1989; Versiani et al., 1989; Lecrubier and Guelfi, 1990; Bakish et al., 1992). As the irreversible MAOIs were often administered at subtherapeutic doses because of the side-effect risk, RIMAs are considered more efficacious in real terms than the first generation MAOIs. However, there is no evidence to indicate that RIMAs have a more rapid onset of clinical action than the tricyclics (Laux et al., 1989).

The major improvement with the RIMAs is undoubtedly in side-effect and adverse reaction profile. The most common side-effects of moclobemide are dizziness, headache, dry mouth, insomnia, nausea, somnolence, and tachycardia (Baunhackl et al., 1989; Laux et al., 1989; Versiani et al., 1989; Bakish et al., 1992). Thus, its side-effect profile shows a closer similarity to that of the SSRIs rather than the tricyclics. This is due, at least in part, to the lack of affinity of moclobemide for muscarinic, histamine H_1 and α_1-adrenergic receptors (Da Prada et al., 1989). However, postural hypotension has been reported in some studies (Baunhackl et al., 1989). Moclobemide also produces significantly less potentiation of the effect of tyramine to raise systolic blood pressure than the irreversible MAOIs (Simpson and de Leon, 1989). Thus, moclobemide-treated patients have little risk of hypertensive crisis (Laux et al., 1989; Versiani et al., 1990).

Toxicity in overdose does not appear to be the serious problem it is for the tricyclics and the irreversible MAOIs, with deep sedation to coma being the most common effect (Hetzel, 1992). However, when taken in combination with tricyclics, the effects are much more severe (Myrenfors et al., 1993).

Despite clear evidence of efficacy in major depressive disorder and the improvement in the side-effect, adverse reaction, and toxicity profiles over the first-generation MAOIs, the reversible MAO_A inhibitors (RIMAs) have not been received favorably by general practitioners. As a consequence, they have not made a significant impact on the antidepressant markets.

Irreversible selective MAO_B inhibitors. The MAO_B inhibitor, selegiline, has been evaluated in depressed patients, but the results obtained have been equivocal (Mann and Gershon, 1980; Mendis et al., 1981). This is not unexpected as it is inhibition of MAO_A that is associated with antidepressant activity (Murphy et al., 1987; Mann et al., 1989). Selegiline was found to be efficacious in an open clinical trial when given at high dose (Quitkin et al., 1984). However, it was suggested that this effect was due to the loss of MAO_B selectivity (Liebowitz et al., 1985).

5-HT_{1A} Agonists

Buspirone (Bristol-Myers Squibb) is the only 5-HT_{1A} partial agonist to be extensively marketed for the indication of anxiety; tandospirone (Sumitomo) is registered in Japan, gepirone (Fabre-Kramer) and flesinoxan (Solvay/Duphar) are in the late stages of clinical development, and ipsapirone (Bayer-Troponwerke) has recently been discontinued (Figure 8; see "Anxiolytic Drugs" section for details). However, only flesinoxan is currently in development for depression. The 5-HT_{1A} agonist/5-HT_{2A} antagonist flibanserin (BIMT 17; Boehringer Ingelheim) is also in clinical development for depression (Figure 9).

Analysis of depressive symptoms in patients with generalized anxiety disorder (GAD) revealed an improvement with buspirone (Goldberg and Finnerty, 1979; Feighner et al., 1982b). As a consequence, several clinical trials were initiated to evaluate 5-HT_{1A} agonists in patients with major depressive disorder. The results of these studies looked promising with statistically significant, although often moderate, antidepressant effects being reported (Heller et al., 1990; Jenkins et al., 1990; Rausch et al., 1990; Robinson et al., 1990; Amsterdam, 1992; Ansseau et al., 1993). However, the impetus for the development of the 5-HT_{1A} agonists as

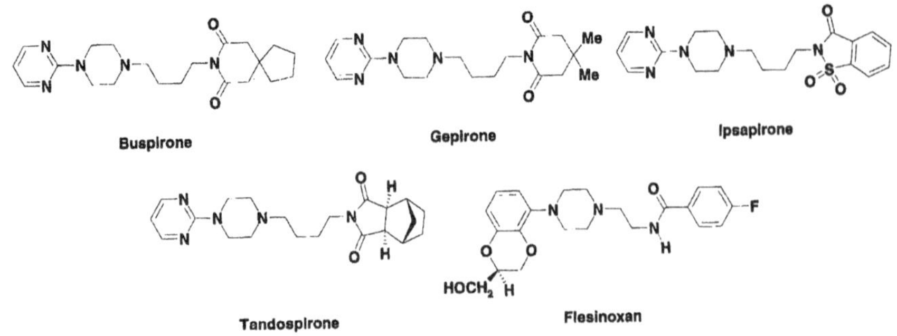

Figure 8. The structures of a selection of 5-HT_{1A} agonists

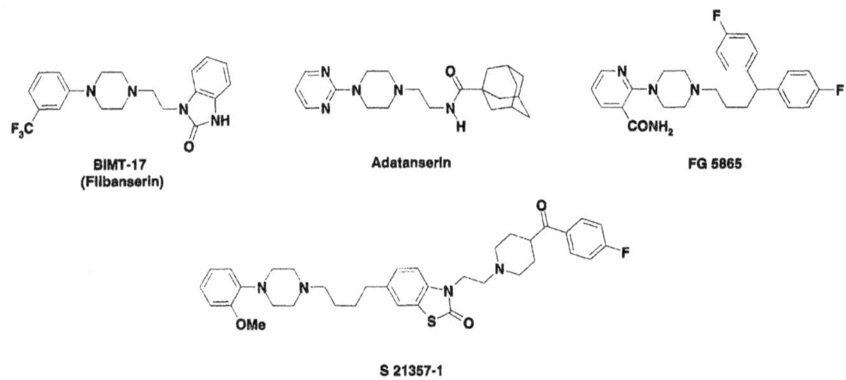

Figure 9. The structures of various 5-HT_{1A} agonists and 5-HT_{2A} antagonists

antidepressants appears to have lost momentum. This may be due to the emergence of the SSRIs. The initial clinical studies suggest that the 5-HT_{1A} agonists are not as efficacious as the SSRIs or tricyclics, although large scale studies versus reference compounds are required to address this issue. The 5-HT_{1A} agonists require one to three weeks to alleviate anxiety (Feighner and Boyer, 1988) and there is no evidence to suggest that these drugs will act more rapidly against depression. Side-effects are similar to those induced by the SSRIs: nausea, diarrhea, dizziness, light-headedness, and occasional headache (Feighner and Boyer, 1988). However, it has been suggested that 5-HT_{1A} full agonism or high efficacy partial agonism is required for antidepressant efficacy (De Vry, 1996). Flesinoxan, a high efficacy partial agonist (Ahlenius et al., 1991; Schipper et al., 1991), is in phase III trials. Flibanserin, a 5-HT_{1A} full agonist and 5-HT_{2A} antagonist (Borsini et al., 1995; 1997) is currently in phase II trials for depression.

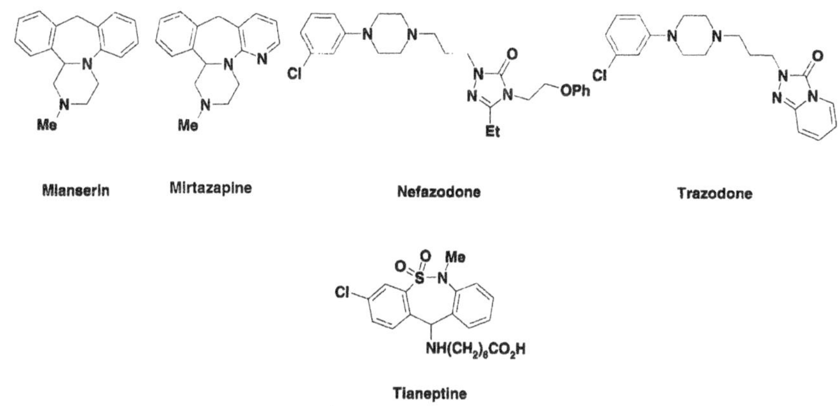

Figure 10. The structures of several miscellaneous antidepressants

Miscellaneous

Mianserin (Organon; Figure 10) is a tetracyclic antidepressant and is therefore structurally distinct from the classical tricyclics. When mianserin was introduced in 1975, its mode of action was unknown. It has subsequently been shown to be a potent α_2-adrenoceptor antagonist, which enhances noradrenergic neurotransmission by inhibiting presynaptically located α_2-adrenoceptors (Baumann and Maitre, 1977). The clinical efficacy of mianserin has been established in several placebo-controlled trials (see Carman et al., 1991; Vinar et al., 1991). There is no evidence of rapid onset of action. Unlike the tricyclics, mianserin is devoid of anticholinergic activity; however, it exhibits affinity for a wide range of other receptors including histamine H_1, α_1-adrenergic, 5-HT_1, 5-HT_2, and 5-HT_3 receptors (Tang and Seeman, 1980; Richelson and Nelson, 1984; Wander et al., 1986; Hoyer et al., 1988a; 1988b; Schmidt and Peroutka, 1989). The contribution, which its affinity for histamine H_1, 5-HT_2, and 5-HT_3 receptors makes to its therapeutic mode of action or side-effect profile, is unclear.

Much of the popularity of mianserin has been due to its safety in overdose (Cassidy and Henry, 1987; Henry, 1989), but reports that this drug can induce potentially fatal blood dyscrasias and may induce bone marrow depression (Chaplin, 1986; Rudorfer and Potter, 1989) have markedly reduced its use. The α_2-antagonist approach has been adopted by several other pharmaceutical companies resulting in idazoxan (Reckitt and Colman), imiloxan (Syntex), efaroxan (Reckitt and Colman), fluparoxan (Glaxo), and mirtazapine (Organon; Figure 10). Of these, only mirtazapine has been registered for the treatment of depression; the others have been abandoned during development. Mirtazapine is structurally related to mianserin and was developed to overcome the problem of blood dyscrasias. Mirtazapine has been shown to exhibit efficacy and a time-course of clinical improvement that is similar to that of the tricyclics (Smith et al., 1990; De Jongh, 1993; Bremner, 1995; Zivkov and De Jongh, 1995). The side-effect profile includes dry mouth, blurred vision, dizziness, decreased libido, tiredness, sedation, somnolence, and increased appetite (Claghorn and Lesem, 1995; Van Moffaert et al., 1995; Zivkov and De Jongh, 1995). Thus, mirtazapine has an improved side-effect profile over the tricyclics and mianserin, which is likely to be due to its low affinity for muscarinic cholinergic, dopamine, and α_1-adrenergic receptors (De Boer and Ruigt, 1995). Mirtazapine has affinity for histamine H_1 receptors (De Boer et al., 1988), which explains the observed sedation and drowsiness. Its affinity for 5-HT_{2A}, 5-HT_{2C}, and 5-HT_3 receptors may account for both its improved side-effect profile and therapeutic mode of action (De Boer and Ruigt, 1995). However, mirtazapine has been associated with agranulocytosis, which may be idiopathic, and with neutropenia. Therefore, mirtazapine is unlikely to make a significant impact in the antidepressants market.

Trazodone (Roussel; Figure 10) is structurally unrelated to the tricyclics and was introduced onto the UK market as an antidepressant in 1980. Trazodone exhibits high affinity for 5-HT_1, 5-HT_2, α_1-, and α_2-adrenergic receptors and is a weak inhibitor of 5-HT reuptake (Richelson and Nelson, 1984; Richelson and Pfenning, 1984; Wander et al., 1986; Table 3). However, it has also been suggested that m-chlorophenylpiperazine, the major metabolite of trazodone, which is a potent and nonselective 5-HT agonist, contributes to its therapeutic effect. Trazodone has been shown to be as effective as the tricyclics in the treatment of major depression with a similar time-course of clinical improvement (Feighner and Boyer, 1988; Fabre, 1989, 1990). Trazodone has an improved side-effect profile over the tricyclics in terms of anticholinergic effects (Rudorfer and Potter, 1989) and has improved safety in overdose (Cassidy and Henry, 1987; Henry, 1989). However, trazodone is sedative and can cause orthostatic hypotension, which almost certainly results from affinity for histamine H_1 and α_1-adrenergic receptors, respectively (Richelson and Nelson, 1984). Trazodone can also produce intractable priapism, which, although rare, is very serious, requiring corrective surgery or causing permanent loss of erectile function in some male patients (Scher et al., 1983).

Nefazodone (Bristol-Myers Squibb) is an analogue of trazodone (Figure 10). Furthermore, it shares a common metabolite with trazodone, m-chlorophenylpiperazine. Therefore, not unexpectedly, the pharmacology of nefazodone exhibits some similarities to trazodone, that is, that of a potent 5-HT_2 antagonist and a modest inhibitor of 5-HT reuptake (Table 3). However, nefazodone exhibits lower affinity for α_1-adrenoceptors than trazodone, and it has little affinity for muscarinic, 5-HT_{1A}, α_2-, and β_1-adrenergic receptors (Eison et al., 1990). Nefazodone has clearly shown superiority to placebo in depressed patients (D'Amico et al., 1990; Mendels et al., 1992) and equivalent efficacy to imipramine (Feighner et al., 1989; Van Moffaert et al., 1992; Rickels et al., 1994). Nefazodone treatment is associated with side-effects typical of the SSRIs (i.e., nausea, headache, and dizziness, and some sedation). Thus, there is no evidence to indicate that nefazodone is either more efficacious or faster acting than the tricyclics. However, it does show an improved side-effect profile and is unlikely to be toxic in overdose. Whether nefazodone exhibits any clear advantages over the SSRIs remains to be seen but, on the basis of current information, this is unlikely.

Probably one of the most paradoxical compounds to enter the antidepressant market is tianeptine (Servier; Figure 10), which is currently only available in France despite its launch in 1988. Tianeptine has been shown to increase presynaptic 5-HT reuptake (Wilde and Benfield, 1995), which is clearly at odds with the hypothesis that an increase in central 5-HT and/or noradrenaline function is required for antidepressant efficacy. Tianeptine is more efficacious than placebo in depressed patients (Costa e Silva and Ruschel, 1994; Staner et al., 1994; Wilde and Benfield, 1995; Ginestet, 1997) and exhibits equivalent efficacy to the tricyclics and SSRIs (Guelfi, 1992; Alby et al., 1993; Staner et al., 1994; Cassano et al., 1996). There is no evidence that it is more efficacious or rapidly acting than the tricyclics or SSRIs. Tianeptine appears to be well-tolerated with no effects on heart-rate, blood pressure, cardiac conductance or ventricular function. The main side-effects are dry mouth, insomnia and nightmares, headache, constipation, dizziness, and drowsiness (Guelfi, 1992; Wilde and Benfield, 1995). However, several cases of hepatotoxicity have been associated with tianeptine (Le Bricquir et al., 1994; Rifflet et al., 1996), which is almost certainly due to its structural similarity to amineptine, which is well known for inducing hepatotoxicity. This is likely to limit the success of this compound.

Iprindole (Wyeth) was introduced onto U.K. market in 1967. The therapeutic mode of action of iprindole remains obscure; however, it is not a monoamine reuptake inhibitor or a MAOI (Roth and Gillis, 1975; Zis and Goodwin, 1979). The efficacy of iprindole in patients with major depression is equivocal (Zis and Goodwin, 1979; Rudorfer and Potter, 1989) and this probably accounts for its decline in use. It does, however, produce only mild anticholinergic effects (Rudorfer and Potter, 1989) and is not toxic when taken in overdose (Cassidy and Henry, 1987; Henry, 1989).

Comparative studies with classical benzodiazepines in major depression indicate that these drugs do not alleviate the core symptoms of depression, but they do have effects on sleep and anxiety. In contrast, the triazolobenzodiazepines alprazolam (Upjohn; Fig 11) and adinazolam (Upjohn) have been reported to be effective in mild to moderate depression at approximately twice the dose used to treat anxiety; although the positive effects again appeared to be mainly on anxiety and sleep with minimal effects on the core symptoms of depression. These drugs are also inferior to the tricyclics in patients with melancholic or severe major depression (reviewed by Birkenhäger et al., 1995). The triazolobenzodiazepines do not induce anticholinergic effects and have a good margin of safety when taken in overdose (Rudorfer and Potter, 1989). However, these compounds are highly

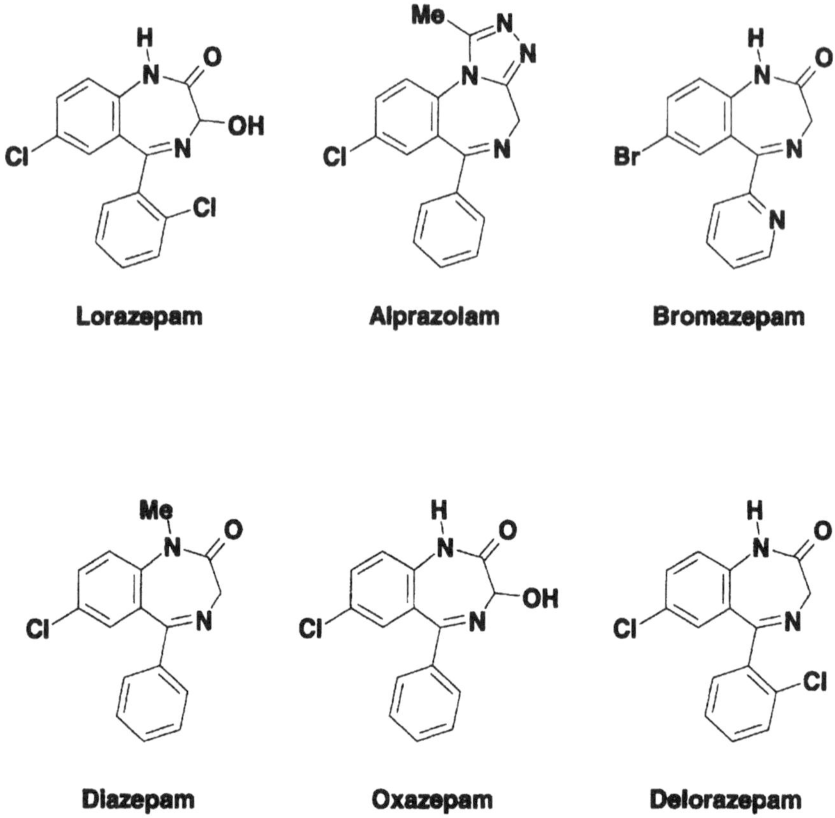

Figure 11. The structures of a selection of benzodiazepine anxiolytics

sedative and several patients receiving adinazolam have suffered seizures (Abuzzahab, 1989). The development of rapid physical dependence and withdrawal effects are also serious handicaps for these drugs (see section on "Anxiolytic Drugs" for details). Thus, even the triazolobenzodiazepines appear to have little future as monotherapy for the treatment of major depression. However, in clinical practice, benzodiazepines and tricyclics are often used in combination. Benzodiazepines appear to contribute to an overall antidepressant response in the first two weeks of treatment because of the rapid effects on anxiety and sleep disorders. Beyond the first few weeks of treatment, combination therapy does not seem to be superior to monotherapy with tricyclics (Birkenhäger et al., 1995).

Future Directions

The development of new antidepressant drugs has been motivated by the lack of efficacy of the "first generation" of antidepressant drugs (e.g., the tricyclics in approximately 30% of depressed patients), their slow onset of therapeutic action, the frequency and severity of adverse effects, and toxicity in overdose. In terms of side-effect profile and toxicity in overdose, significant progress has been made with the introduction of "second generation" antidepressants (e.g., the SSRIs and SNRIs). Despite the fact that the SSRIs do not exhibit superior efficacy or a more rapid onset of clinical action than the tricyclics, they have been an outstanding success—the "gold standard" against which any new drug entering the market place must compete. It remains to be seen whether venlafaxine and milnacipran (SNRIs), and mirtazapine, nefazodone, and reboxetine (NARI) will be as successful as the SSRIs. However, none of these compounds is more efficacious or rapidly acting than the tricyclics. Thus, the search goes on for a "third generation" of antidepressants, which retain the gains already made in minimizing side-effects and toxicity in overdose but which tackle the "two" key unmet needs of current antidepressant therapy, efficacy in treatment resistant patients and rapid onset of action.

It appears paradoxical that the SSRIs (e.g. fluoxetine) rapidly inhibit the reuptake of 5-HT whereas treatment for three to eight weeks is required for antidepressant efficacy. It has been postulated that the therapeutic action of the SSRIs to increase 5-HT function within the synapse is initially blunted by an attenuation of 5-HT neuronal firing, resulting from somatodendritically released 5-HT activating somatodendritic 5-HT_{1A} autoreceptors and/or a reduction in 5-HT release in terminal areas resulting from increased 5-HT in the synapse activating prejunctional 5-HT_{1B} autoreceptors. Thus, it has been suggested that the clinical effects of the SSRIs only become apparent when these autoreceptors are desensitized (reviewed by Blier and de Montigny, 1994). As a consequence, two approaches that are being taken in the search for antidepressants that are more rapidly acting and efficacious in treatment-resistant patients are the combination of 5-HT_{1A} antagonism or 5-HT_{1B} antagonism and 5-HT reuptake inhibition.

Clinical trials have demonstrated that pindolol produces a dramatic and rapid improvement in the antidepressant efficacy of the SSRIs (Artigas et al., 1994; Blier and Bergeron, 1995; Pérez et al., 1997; Tome et al., 1997). Pindolol exhibits affinity for β_1- and β_2-adrenoceptors and 5-HT_{1A} receptors (Middlemiss et al., 1977; Hoyer et al., 1985; Rainbow et al., 1985). Thus, it has been postulated that pindolol augments the antidepressant effects of the SSRIs by antagonizing the inhibitory effects of 5-HT at somatodendritic 5-HT_{1A} autoreceptors (Artigas et al., 1994; 1996). This hypothesis is supported by the finding that pindolol decreases the firing activity of 5-HT–containing neurones produced by an SSRI (Blier et al., 1994; Romero et al., 1996). In addition, 5-HT_{1A} antagonists have been shown to potentiate the increase in extracellular 5-HT concentrations produced by an SSRI (Hjorth, 1993; Gartside et al., 1995; Romero et al., 1996). However, 5-HT_{1A} receptors are located both pre- and postsynaptically (Pazos and Palacios, 1985; Sotelo et al., 1990). Furthermore, postsynaptic 5-HT_{1A} receptors have been implicated in the antidepressant actions of SSRIs (Blier and De Montigny, 1994; De Vry, 1996). Thus, antagonist actions at postsynaptically located 5-HT_{1A} receptors would be expected to be counterproductive and to eliminate any beneficial effects that blockade of presynaptically located 5HT_{1A} receptors in the raphé has on 5-HT neurotransmission. Thus, the ideal drug would combine 5-HT_{1A} autoreceptor antagonism with 5-HT reuptake inhibition; it would also either have no effects or be an agonist at postsynaptically located 5-HT_{1A} receptors. Interestingly, electrophysiological studies have shown that pindolol does not alter the responsiveness of postsynaptic neurones to microiontophoretic applications of 5-HT or the 5-HT_{1A} agonist 8-hydroxy-2-(di-n-propylamino)tetralin (Blier et al., 1994; Romero et al., 1996).

On the basis of this information, several companies have filed patents claiming combinations of "specified" SSRIs and SNRIs and "specified" 5-HT_{1A} antagonists. These include Eli Lilly (EP-687472-1995; EP-714663-1996; EP-759299-1997), Astra (WO-722941-1996), American Home Products (GB 2303303-1997), and E. Merck (DE 19615232-1997). However, due to the considerable amount of prior art in this area, it remains to be seen if these patent applications will be successful.

Now also appearing are patents that claim series of compounds combining 5-HT reuptake inhibition with 5-HT_{1A} receptor affinity in the same molecule. Companies with patents in this area include Eli Lilly, Bristol-Myers Squibb, American Home Products, Knoll Pharmaceuticals, and E. Merck. At present, there is only one compound with this pharmacological profile in development, EMD 68843 (E. Merck), which is currently in phase II trials. However, EMD 68843 combines 5-HT reuptake inhibition with 5-HT_{1A} agonism (Bartoszyk et al., 1997). It has been postulated that rapid desensitization of 5-HT_{1A} autoreceptors could be achieved by administration of a 5-HT_{1A} full agonist. Furthermore, 5-HT_{1A} full agonists would directly activate postsynaptically located 5-HT_{1A} receptors; thereby counteracting the initial decrease in function due to activation of presynaptic receptors (De Vry, 1996). Thus, the combination of 5-HT reuptake inhibition with 5-HT_{1A} agonism may also have the potential to provide an antidepressant drug that is rapidly acting and/or efficacious in treatment-resistant depression. However, initial preclinical studies with EMD 68843 indicate that it is an agonist at presynaptic but is without effect at postsynaptic 5-HT_{1A} receptors (Bartoszyk et al., 1997). Results from the phase II trials are eagerly awaited.

The second strategy for circumventing the autoinhibitory actions of the SSRIs is to block the actions of 5-HT at terminal 5-HT_{1B} autoreceptors. This hypothesis is supported by the observation that the effect of the SSRIs on 5-HT efflux is potentiated by the $5\text{-HT}_{1B/1D}$ antagonist GR 127935 (Rollema et al., 1996; Davidson and Stamford, 1997; Gobert et al., 1997). Pfizer has filed a patent claiming combinations of the SSRI sertraline with 5-HT_{1B} agonists or antagonists (EP 0701 819 A2-1995). Currently, there are no compounds in clinical development with this pharmacological profile.

Another approach that is being pursued is blockade of both types of autoreceptors, namely, 5-HT_{1A} to increase 5-HT neuronal firing and 5-HT_{1B} to increase terminal 5-HT release. Microdialysis studies have shown that the combination of the 5-HT_{1A} antagonist WAY 100635 and the 5-$HT_{1B/1D}$ antagonist GR 127935 leads to a dramatic increase in extraneuronal 5-HT in the frontal cortex of guinea-pigs (Roberts et al., 1997). SmithKline Beecham have filed a patent claiming combinations of "specified" 5-HT_{1A} antagonists and 5-HT_{1B} antagonists (WO 9531988-A).

Several companies, including SmithKline Beecham, Eli Lilly, Merrell Dow, Pierre Fabre, Synthélabo, Duphar, and Pfizer have filed patents in this area. However, a number of these patents relate to agonists rather than antagonists. There are no compounds in clinical development.

A slightly different approach aimed at rapid onset of action and improved efficacy is the combination of 5-HT_{1A} agonism and α_2-antagonism. As already postulated, 5-HT_{1A} agonists enhance 5-HT function by stimulating postsynaptically located 5-HT_{1A} receptors, a process that more than compensates for the decrease in 5-HT neuronal firing caused by activation of 5-HT_{1A} autoreceptors in the raphé (De Vry, 1996). 5-HT_{1A} agonists also increase central noradrenaline function by decreasing the firing of 5-HT neurones in the raphé resulting in increased noradrenaline firing in the locus coeruleus (Broderick and Piercey, 1991; Haddjeri, 1997). α_2-Antagonists increase noradrenaline function by antagonizing α_2-autoreceptors on noradrenaline cell bodies and nerve terminals thereby preventing noradrenaline from inhibiting its own release (Baumann and Waldmeier, 1978; L'Heureux et al., 1986; Dennis et al., 1987). CP 93393 (Sinepitron, Pfizer) is a 5-HT_{1A} agonist/α_2-antagonist, which is currently in phase III trials for depression and anxiety. No other compounds with this pharmacological profile appear to be in clinical development.

In conclusion, therefore, the search for a "third generation" of antidepressants continues to focus largely on enhanced 5-HT and, to a much lesser extent, on noradrenaline neurotransmission. These mechanisms represent the approaches most likely to achieve a rapidly acting and more efficacious antidepressants in the short-term. However, other novel approaches are under investigation. The most advanced appears to be MK 869 (Merck & Co.), a substance P antagonist, in phase II trials for depression. Others, such as corticotrophin releasing factor antagonists and inositol monophosphate inhibitors, are at the preclinical stage. However, with the growing interest in gene transcription, the search for a "fourth generation" of antidepressants is likely to have already begun.

ANXIOLYTIC DRUGS

Clinical Features of Anxiety

In common with several categorizations in psychiatry, anxiety is an "umbrella" term for various related mental disorders. There is, in addition, the difficulty from a diagnostic perspective in determining the point at which anxiety ceases to be a response in the repertoire of normal human behavior and it becomes a psychiatric condition. According to the Diagnostic and Statistical Manual of Mental Disorders DSM-III (1980), which was later modified to DSM-III-R (1987) and more recently updated to DSM-IV (1994), anxiety can be subdivided into three major classes:

1. Panic disorder, anxiety is accompanied by the occurrence of panic attacks (often brief, irrational fear responses to situations and/or external stimuli)
2. Generalized anxiety disorder (GAD), panic attacks are not a feature of the clinical syndrome
3. Phobic anxiety, which includes social phobia, general phobia, and/or specific phobias

In addition to these three broad categories, anxiety is an integral factor in about 70% of depressive episodes. There is, however, also the view that the distinctions between anxiety and panic disorder or between anxiety and depression are somewhat artificial (Tyrer, 1985; 1992; Gelder, 1989), and these disorders merely reflect different aspects of a continuum of mood dysfunction, which Tyrer (1985; 1992) and others have described as "general neurotic syndrome".

The core psychological symptoms of anxiety are overwhelming feelings of apprehension, worry, and fear often accompanied by perceptions of helplessness, an inability to cope, social embarrassment, and humiliation.

Anxiety is usually associated with physical signs, which may be the reason for a patient to consult his or her practitioner in the first instance. Physical symptoms include cardiovascular changes (tachycardia, increased blood pressure, palpitations), musculoskeletal problems (tension inducing pain; e.g., headache, backache), gastrointestinal disturbance (dry mouth, diarrhea), respiratory problems (breathlessness, panting, dizziness),and sweating.

It is evident from this brief summary that anxiety is a multifaceted mental disorder that encompasses a spectrum of both physical and mental symptoms. Not surprisingly, there is no unitary pharmacological treatment for all types of anxiety. As such, the various classes of drugs described in this review will also be discussed in terms of their suitability for treating the subclasses of anxiety, viz panic disorder, GAD, and phobic anxiety.

Table 4. Pharmacological Properties of the Ideal Anxiolytic Drug

1.	Efficacy in the three major subdivisions of anxiety, *viz.* panic disorder, GAD, and phobic anxiety
2.	Rapid onset of anxiolytic actions in the clinic
3.	Low side-effect potential, particularly psychomotor impairment, sedation, disinhibition, and amnesia
4.	Safety in overdose
5.	Efficacy without induction of physical or psychological dependence
6.	Lack of abuse potential

Historical Perspectives

In the 1950's, the pharmacological treatment of anxiety did not draw clear lines between this mental disorder and depression. As discussed in the sections of this chapter describing the clinical aspects of depression and anxiety, there is very considerable overlap and co-existence between these two psychiatric conditions, and it is therefore not surprising to discover that antidepressant drugs have played and continue to play a significant role in the treatment of anxiety. It was the pioneering work of two groups, Kline and Kuhn in 1958, which respectively introduced the irreversible MAOI iproniazid and the tricyclic imipramine into the treatment of depression and anxiety. Although early testing revealed the MAOIs' efficacy in the treatment of anxiety (King, 1962; Sargant, 1962) and imipramine's ability to reduce panic (Klein and Fink, 1962), the advent of the benzodiazepines caused the potential value of these drugs in the treatment of anxiety largely to be ignored until the 1980s. It was the work of Sternbach in 1957 that revolutionized the treatment of anxiety with the discovery of chlordiazepoxide as the first benzodiazepine anxiolytic (Sternbach and Reeder, 1961a). Chlordiazepoxide was introduced in 1960 under the trade name of Librium (Roche) and it was followed shortly after by diazepam or Valium (Roche) (Sternbach and Reeder, 1961b). The efficacy of the benzodiazepines combined with their perceived lack of toxicity and lack of dependence potential ensured that these drugs would rapidly replace the barbiturates both as anxiolytics and tranquillizers. The number of benzodiazepine anxiolytics and hypnotics increased rapidly (e.g., see Figure 11) and these drugs enjoyed supremacy in the anxiolytic field for approximately two decades, even though they were already known to produce psychomotor impairment, sedation, muscular relaxation, disinhibition, and amnesia. It was only in the 1980s, when serious questions were raised concerning their overusage and their ability to induce dependence, that the quest for alternative anxiolytic drug therapies commenced in earnest. In addition, substance abuse has now become an issue for the benzodiazepines, especially temazepam (Wyeth).

The pharmacological properties of the ideal anxiolytic are given in Table 4, and it is against these criteria that progress in anxiolytic research and development will be assessed.

Developments in Anxiolytic Drugs

The Benzodiazepines

Judged against the criteria of the ideal anxiolytic (Table 4), the benzodiazepines are generally accepted to be efficacious in the treatment of GAD (Rickels et al., 1988; Fineberg and Drummond, 1995). Their efficacy, with the possible exception of alprazolam (Upjohn), in panic disorder is less well accepted (Sheehan, 1987; Nutt, 1990; Jackson and Nutt, 1996), and in simple phobias, they are regarded as being of little therapeutic value (Zitrin et al., 1983; Nutt, 1990). The benzodiazepines have a fairly rapid onset of clinical effect in contrast to the newer anxiolytics (e.g., $5HT_{1A}$ partial agonists) and this may be due in large part to the sedative action of the former. However, this property is also responsible for the benzodiazepines' side-effects of psychomotor impairment, muscle relaxation, and possibly amnesia. These drugs are relatively nontoxic; overdosing with benzodiazepines is a relatively frequent occurrence, but it rarely leads to serious problems unless the benzodiazepines have been taken in a cocktail with other drugs or alcohol. One of the major attractions of the benzodiazepines compared to the barbiturate sedatives was that the former were perceived not to induce physical or psychological dependence. Furthermore, it was believed that these drugs had to be given for prolonged periods at high dose for withdrawal symptoms to occur on cessation of drug treatment (Allquander, 1978; Woods et al., 1987). This is no longer the perception, and it has been suggested by Shader and Greenblatt (1993) that withdrawal effects occur in 5% to 35% of patients who have received benzodiazepines for one month or longer. It is the last revelation and the subsequent litigation that has changed the opinions of both prescribers and the public over the suitability of the benzodiazepines as drugs for the longterm therapy of anxiety.

The pharmacological mode of action of the existing benzodiazepine anxiolytics is to enhance the function of the inhibitory neurotransmitter γ-aminobutyric acid (GABA), by acting at a modulatory site on the $GABA_A$ receptor to increase chloride ion flux through the ion channel–receptor complex (Haefely, 1983; 1985; 1992; Stephenson et al., 1992). This receptor complex is unique in that it exists in a continuum of activation states such that there are benzodiazepine ligands (Figure 12) that act as full agonists (anxiolytic, anticonvulsant), through antagonists (possibly anxiolytic), to full inverse agonists (anxiogenic, proconvulsant).

This unique receptor has provided the opportunity of developing drugs with pharmacological profiles markedly different from those of the existing benzodiazepine anxiolytics (full agonists), and considerable effort has been focused on the partial agonist benzodiazepine ligands. The concept was that these drugs would

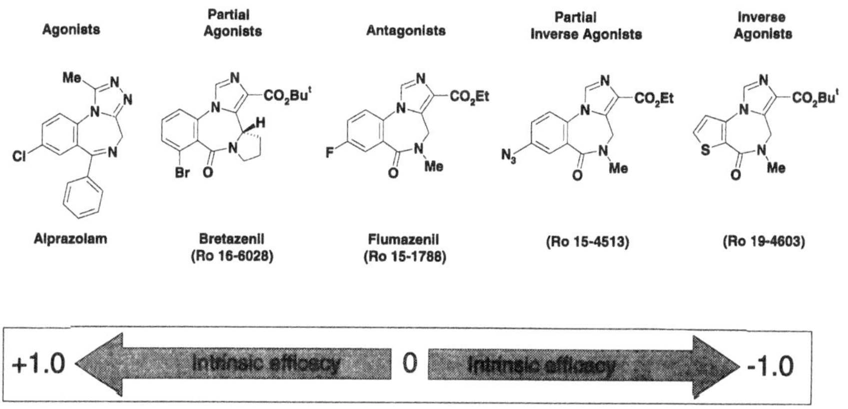

Figure 12. Intrinsic efficacy of various benzodiazepine ligands

provide the anxiolytic effects of the benzodiazepine full agonists without causing psychomotor impairment, sedation, or muscle relaxation. Moreover, as partial agonists, they would not induce tachyphylaxis leading to tolerance and dependence. Abecarnil (Schering AG) and bretazenil (Roche) are benzodiazepine partial agonists that have both shown promising profiles based on preclinical tests in animals. Abecarnil is two to 10 times more potent in rodent anxiety models than diazepam (Stephens et al., 1990), but unlike the latter, it does not evoke signs of dependence on withdrawal (Steppuhn et al., 1993). Similarly, bretazenil has been reported to produce anxiolytic actions in rodents without inducing motor impairment or sedation (Facklam et al., 1992). However, the optimistic predictions based on preclinical data have not come to fruition in the clinical trials. Several double-blind clinical trials have now been performed to determine the efficacy of abecarnil in the treatment of GAD. In the first of them, abecarnil showed some evidence of efficacy, but the results were far from compelling (Ballenger et al., 1991) and a similar pattern has emerged from three further studies. In a group of geriatric patients with long-standing symptoms of anxiety, low-dose abecarnil was found to be rapidly efficacious in alleviating anxiety; however, high-dose abecarnil was ineffective (Small and Bystritsky, 1997). In a six-week trial comparing abecarnil against the 5-HT_{1A} partial agonist buspirone and placebo, low and high doses of abecarnil were significantly more efficacious than placebo during the first three weeks of the study but not at week six. Although this may have been due to high rates of response in the placebo group, buspirone was nevertheless effective at this time-point (Pollack et al., 1997). The latter result is consistent with the view that 5-HT_{1A} partial agonists are not rapidly acting anxiolytics, and there is a lag of several weeks between initiation of treatment and clinical improvement. It also suggests that while abecarnil may be more rapidly acting, it is also less efficacious than buspirone. In a comparison of abecarnil with the benzodiazepine full agonist alprazolam in patients suffering from GAD, subjects on both active treatments showed significant improvement in this four-week trial (Lydiard et al., 1997). However, the benzodiazepine partial agonist abecarnil performed less well than the full agonist alprazolam, with patients who received the latter drug showing faster onset and, on some scales, greater improvement.

In addition to evidence that the benzodiazepine partial agonists may be less efficacious than full agonists (and possibly 5-HT_{1A} partial agonists), the greatest disappointment has come with the discovery that the partial agonist strategy has not eliminated the problems of psychomotor side-effects, dependence, and withdrawal. Thus, in volunteers, abecarnil was found to produce sedation, psychomotor impairment, dizziness, and ataxia (Duka et al., 1993), and this profile has been confirmed in patient studies (Ballenger et al., 1991; Lydiard et al., 1997; Pollack et al., 1997; Small and Bystritsky, 1997). Furthermore, symptoms of withdrawal have been unequivocally observed on discontinuation of treatment (Pollack et al., 1997; Small and Bystritsky, 1997) and during the period of downward tapering of dosing (Lydiard et al., 1997). It may be a

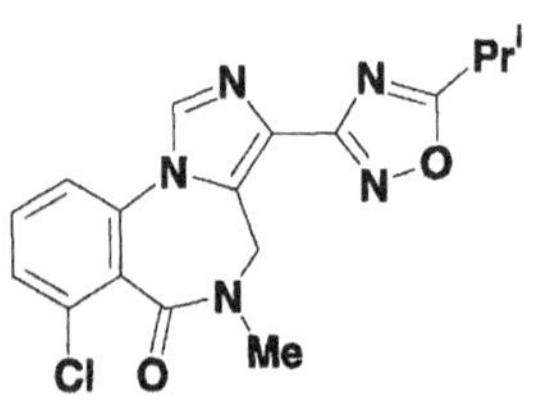

L-663581

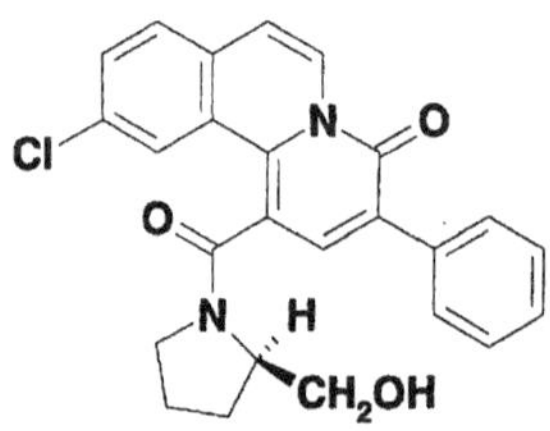

Ro 198022

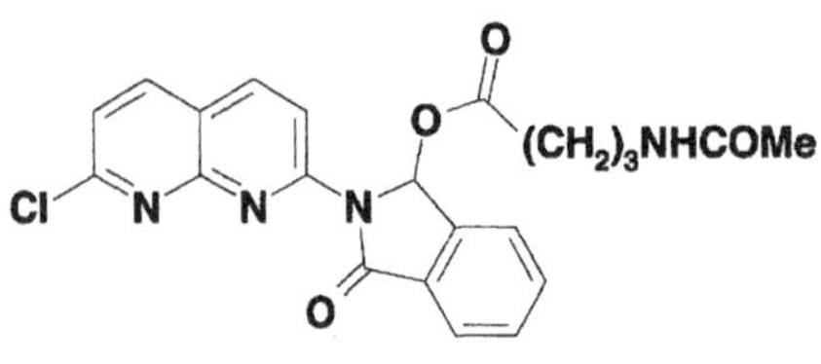

RP 60503

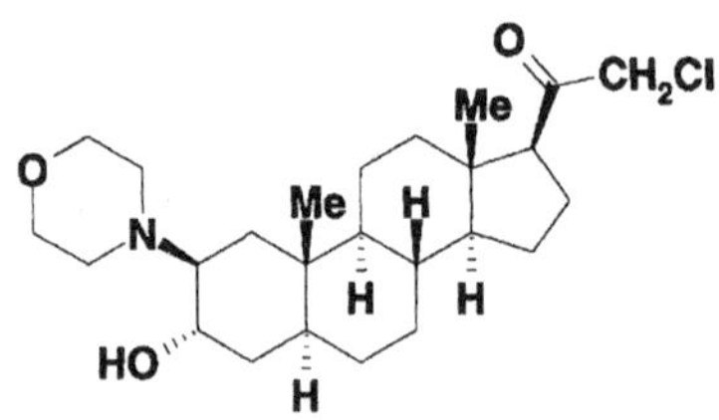

ORG 20599

Figure 13. Several benzodiazepine partial agonists in preclinical development

reflection of this fact that bretazenil has been discontinued in development. Several other benzodiazepine partial agonists are in preclinical development (e.g., L 663,581 [Merck], Ro 198022 [Roche], RP 60503 [Rhone-Poulenc Rorer]; Figure 13); it will be interesting to learn how well tolerated in volunteers and patients these drugs are and how efficacious they are in clinical trials.

One interesting new approach in the benzodiazepine field has been to evaluate clonazepam (Roche), which has been in use since 1973 as a treatment for epilepsy (particularly absence seizures) as a potential antipanic drug. Roche obtained U.S. approval for the use of clonazepam in panic disorder in 1994. Other benzodiazepines (e.g., diazepam [Roche]), which are effective in the control of *status epilepticus*, are not used prophylactically to treat epilepsy because of the problem of tolerance to their anticonvulsant effects. Clonazepam appears different in that its anticonvulsant actions are maintained during prolonged administration. It is unclear whether this atypical profile derives from partial agonism at the benzodiazepine binding site or, less likely, is a reflection of its unusual pharmacokinetics (Cmax = 2–3 h; $t^{1/2}$ = 20–50 h). In clinical trials, Pols and colleagues (1991) reported after an open-label four-week trial of 10 patients with panic disorder or agoraphobia with panic attacks that clonazepam reduced the frequency of panic attacks by approximately 60% and it also significantly improved patient self-rating of agoraphobia. In a placebo-controlled double-blind clinical trial in 29 patients with panic disorder, positive effects of clonazepam were observed with significant reductions in the frequency, duration, and severity of panic attacks (Beauclair et al., 1994). And, in a second six-week, placebo-controlled, double-blind trial with 72 patients suffering from panic disorder, clonazepam showed efficacy equivalent to that observed with the active comparator alprazolam. Consistent with findings obtained with other benzodiazepines, clonazepam and alprazolam had a rapid onset of action showing efficacy after one week of treatment (Tesar et al., 1991). In a longer six-month, double-blind trial comparing clonazepam with imipramine in 12 patients with panic disorder, efficacy was again observed early on in treatment with clonazepam (and also imipramine) and this effect was maintained throughout treatment. In common with the benzodiazepine full agonists and partial agonists were clonazepam's side-effects, which predominantly consist of impairment of psychomotor function (e.g., sedation, drowsiness, memory and concentration problems, ataxia; Tesar et al., 1991; Beauclair et al., 1994). From this perspective, clonazepam does not appear to have improved on the accepted shortcomings of the benzodiazepine anxiolytics. Similarly with benzodiazepines, there are the serious issues of dependence and withdrawl effects and, on this count, there is no evidence to demonstrate that clonazepam has managed to overcome these problems.

The isolation of the $GABA_A$ receptor complex from neuronal tissue was achieved precisely because of its high affinity for benzodiazepines that retained the whole $GABA_A$ receptor complex on affinity columns (Stephenson, 1990; Olsen and Tobin, 1990). The complex was revealed to be a pentameric hetero-oligomer composed of at least three subunits: α, β, and γ. Subsequent research has revealed not only the existence of multiple subforms of α (α_1-α_6), β (β_1-β_4), and γ (γ_1-γ_2), but also δ and ε subunits (Stephenson et al., 1992). The human brain has been shown to express multiple $GABA_A$ receptor subtypes, and this has yielded the possibility that there may be different subtypes that mediate the therapeutic and side-effect properties of the benzodiazepines. Although it is clear that certain benzodiazepines (e.g., zolpidem [Lorex]) show some selectivity for particular $GABA_A$ receptor subtypes, it is not clear whether specific subtypes mediate particular physiological and psychological effects and, if so, which are ideal targets for developing a novel anxiolytic. While this approach is an attractive avenue for developing the ideal benzodiazepine anxiolytic, it will undoubtedly require extensive research to resolve the above issues before drug discovery can be initiated. In light of the unexpectedly poor showing, thusfar, of the benzodiazepine partial agonists, it will be interesting to see whether pharmaceutical companies are prepared to follow this new strategy.

The $GABA_A$ receptor complex has recently been shown to bind neurosteroids (McNeil et al., 1992). ORG 20599 (Organon; Figure 13) is a novel ligand for this modulatory allosteric site (Hill Venning et al., 1994). Although ORG 20599 is being developed for the treatment of sleep disorders and not anxiety, the neurosteroid allosteric site provides nevertheless an interesting alternative pharmacological target for enhancing $GABA_A$-mediated chloride ion flux that is independent of the benzodiazepine receptor.

Judging the new approaches in the benzodiazepine field against the criteria of the ideal anxiolytic is difficult because of the paucity of clinical information. In the case of the benzodiazepine partial agonists, it is probable that, as a result of their reduced pharmacological efficacy, they will not have as broad a spectrum of utility in the treatment of anxiety as the traditional benzodiazepines. For example, they may be of less value in treating panic disorder than the potent benzodiazepine full agonists like alprazolam. The benzodiazepine partial agonists may also have a delayed onset of therapeutic action compared with the benzodiazepine full agonists. What is apparent from the current clinical evidence is that benzodiazepine partial agonists do produce psychomotor impairment and sedation, and their side-effect profile is identical, although less pronounced, than those of the benzodiazepine full agonists. There is also clear evidence to show that the partial agonist strategy has not eliminated the serious problems of benzodiazepine dependence and withdrawal syndromes. Furthermore, depending on the "street" view of their interaction with alcohol, abuse also could be an issue for the benzodiazepine partial agonists. Overall, therefore, the benzodiazepine partial agonists will probably show some minor clinical advantages over the traditional benzodiazepine anxiolytics, but they are unlikely to have shifted the balance sufficiently to make this new generation of benzodiazepines the drugs of choice in the treatment of anxiety.

Monoamine Reuptake Inhibitors

Numerous studies support the view that tricyclic antidepressants (Figure 1) are effective in the treatment of anxiety (Nutt and Glue, 1989; Nutt, 1990). These drugs have a broad spectrum of efficacy in anxiety states that includes even certain types of phobic anxiety (e.g., agoraphobia; Escobar and Landbloom, 1976; Johnston et al., 1988) and social phobia (Allsopp et al., 1984; Benca et al., 1986), but their usefulness is particularly valuable in the treatment of panic disorder (Nutt, 1990). However, the tricyclic antidepressants played a relatively minor role in the treatment of anxiety until the 1980s because of the efficacy and perceived safety of the benzodiazepines. More recently, concerns over the benzodiazepines have shifted the focus back to the monoamine reuptake inhibitor antidepressants, particularly the "second generation" drugs. Analogous to the situation with the treatment of depression, it is the SSRIs, which are now being developed and used for the treatment of several anxiety disorders. For example, paroxetine has been shown to be efficacious in double-blind, placebo-controlled trials for treatment of panic disorder (Dunbar et al., 1993; Oehrburg et al., 1995), and this SSRI has been registered for this indication in both the United Kingdom and the United States. Similar positive findings have also been observed in clinical trials for panic disorder conducted with other SSRIs including fluoxetine (Gorman et al., 1987) and fluvoxamine (Westenberg and Den Boer, 1993). However, the pattern of efficacy produced by the SSRIs is very different from that obtained from the benzodiazepines, because the former include an initial exacerbation of anxiety and an increase in panic attacks before they produce clinical improvement. Currently, sertraline (Pfizer), fluoxetine (Lilly), fluvoxamine (Solvay), citalopram (Lundbeck), and nefazodone (an SSRI + 5-HT_2 antagonist; Bristol-Myers Squibb), are all in late-stage development for the treatment of panic disorder. It is anticipated that these drugs given in combination with behavioral therapy will gradually replace not only the tricyclics but also potent benzodiazepine full agonists (e.g., alprazolam) as the first-line treatment for panic disorder.

The potential effectiveness of the SSRIs to treat social phobia is also being investigated. In open label clinical trials, sertraline and fluoxetine have been reported to be of value in the treatment of social phobia (Van Ameringen et al., 1993; Czepowicz et al., 1995). The efficacy observed in both trials was moderate, but its efficacy was less pronounced in more severely socially phobic patients (Czepowicz et al., 1995) and in those patients where the disorder was of the longest standing (Van Ameringen et al., 1993). In a double-blind, placebo-controlled study of fluvoxamine in 30 patients with social phobia, this SSRI produced significant improvements in anxiety associated with social phobia but not social avoidance (Van Vliet et al., 1994). In this 12-week trial, significant efficacy was only at the endpoint analysis indicating a lack of early onset of efficacy. While this may have been an issue of dosing, the authors also noted that in this anxiety disorder, the onset and extent of efficacy may be lower than that observed with reversible MAOIs (Van Vliet et al., 1994).

Judged against the criteria of the ideal anxiolytic, the monoamine reuptake inhibitors are effective anxiolytics, particularly in the treatment of panic disorder. However, they have a slow onset of action and often exacerbate the symptoms of anxiety before producing clinical improvement. As extensively discussed in the "Antidepressants" section of this chapter, the tricyclics and the SSRIs both have significant side-effect profiles, and the former are unsafe when taken in overdose. On the positive side, however, there is little evidence of dependence, major withdrawal effects, or abuse liability with monoamine reuptake inhibitors. Thus, although the drugs in the class including SSRIs are still a long way from being the ideal anxiolytics, their failure to induce dependence and their lack of abuse potential compared with the benzodiazepines has been, and will continue to be, a major factor in promoting their position in the treatment of anxiety.

Monoamine Oxidase Inhibitors

The irreversible, nonselective MAOIs (e.g., phenelzine) were shown early on to be effective in the treatment of anxiety (King, 1962; Sargant, 1962), and this finding has been confirmed by subsequent clinical investigations (Kelly et al., 1970; Liebowitz et al., 1986; Nutt, 1990). This class of drugs is particularly effective when the anxiety state has a high phobic component (Nutt, 1990). Their onset of therapeutic efficacy in anxiety mirrors the situation in depression with a three- to four-week lag before the appearance of marked signs of clinical improvement. The disadvantaged side-effect profile of the MAOIs, particularly the "cheese reaction" (discussed in detail in the "Antidepressants" section of this chapter), has relegated the irreversible MAOIs to a minor role both in the treatment of anxiety and depression.

Attempts to circumvent the problems associated with the older irreversible MAOIs have focused on specificity for the subforms of this enzyme and its competitive, reversible inhibition. Monoamine oxidase exists in two forms: MAO_A, which preferentially catabolizes noradrenaline and 5-HT, and MAO_B, which preferentially catabolizes phenylethylamine and dopamine (Murphy et al., 1987). There are several reversible MAO_A inhibitors (RIMAs) which have been taken into development, namely, moclobemide (Roche), toloxatone (Synthélabo), brofaromine (Ciba-Geigy; discontinued), and befloxatone (Synthélabo) (Figure 7). Of these, moclobemide and toloxatone have been registered for the treatment of depression. Despite the fact that the irreversible, nonselective MAOIs had previously been shown to be of value in the treatment of certain anxiety states, relatively few studies have been conducted with the RIMAs. Brofaromine and moclobemide have been evaluated for the treatment of panic disorder (with and without agoraphobia) and social phobia; no data on efficacy in anxiety are currently available for toloxatone and befloxatone. In panic disorder, double-blind clinical trials with brofaromine have been conducted versus placebo (Van Vliet et al., 1993) and versus placebo and clomipramine (Bakish et al., 1993). While showing efficacy in terms both of a reduction in panic attacks and anxiety symptoms, the results suggest that brofaromine may be anxiogenic during the early stages of treatment (Van Vliet et al., 1993) and be slower in onset and less efficacious than clomipramine (Bakish et al., 1993). The drop-out rate in the latter study was high for both active treatment groups but higher for brofaromine (63% versus 49%). The major side-effects reported for brofaromine were sleep disturbance, nausea, orthostatic dizziness or lowering of blood pressure, loss of appetite, and dry mouth.

In social phobia, double-blind clinical trails have been conducted with both moclobemide (Versiani et al., 1992) and brofaromine (Fahlén et al., 1992). Here the RIMA moclobemide was found to have a slower onset of therapeutic action than phenelzine. In addition, at week 16, the overall response rate also favored the nonselective irreversible MAOI (73% versus 54% responders). There were also five withdrawals for lack of efficacy with moclobemide but none for phenelzine. The major side-effects produced by moclobemide were similar but much less pronounced than those of phenelzine (i.e., insomnia, somnolence/fatigue, dizziness, headache, and dry mouth). Results with brofaromine have only been published in preliminary form; brofaromine was claimed to significantly reduce both the number of panic attacks and the anxiety scores (Fahlén et al., 1992).

On the basis of the very limited clinical information available, there is little to indicate that the RIMAs will be any closer to the ideal anxiolytic than the older nonselective, irreversible MAOIs. The RIMAs have gained in terms of safety with respect to the "cheese reaction" and the severity of their side-effects, although RIMAs are still not well tolerated by anxious patients. On the other hand, RIMAs have probably lost out to the older MAOIs in terms of therapeutic efficacy and speed of action. Overall, therefore, RIMAs will never be the first choice in the treatment of anxiety and they are likely to be used only after the failure of other treatments such as benzodiazepines and SSRIs.

5-HT$_{1A}$ Ligands

The first new class of drug to enter the field of anxiolytic therapy was the 5-HT$_{1A}$ partial agonists. Buspirone (Bristol-Myers Squibb; Figure 8) was the progenitor (Riblet et al., 1982), and this drug has been extensively registered for the treatment of anxiety. Tandospirone (which is already registered in Japan; Sumitomo and Pfizer), gepirone (Fabre-Kramer), and flesinoxan (Solvay-Duphar) are also in the late stages of clinical development (Figure 8). Ipsapirone (Bayer-Troponwerke; Figure 8), which was at a similar phase of development, now has been discontinued.

The 5-HT$_{1A}$ partial agonists have been reported to show anxiolytic effects in traditional-punished conflict models and also more modern predictive tests involving conditioned and unconditioned stress, novelty, exploration, and social behavior. However, there is considerable disagreement concerning the efficacy of the 5-HT_{1A} partial agonists in these models, with reports that this class of drugs is effective, ineffective, and even anxiogenic (Howard and Pollard, 1990; Pollard and Howard, 1991; Handley et al., 1993; Schreiber and de Vry, 1993).

5-HT_{1A} inhibitory autoreceptors are situated on the soma and dendrites of 5-hydroxytryptaminergic neurons. The pharmacological rationale used to explain the anxiolytic actions of the 5-HT_{1A} partial agonists is that enhancement of 5-hydroxytryptaminergic function is anxiogenic, at least in the shortterm—hence, the increased incidence of panic attacks when patients first receive treatment with the SSRIs. The 5-HT_{1A} partial agonists inhibit 5-hydroxytryptaminergic function in the raphé (Vergé et al., 1985; Basse-Tomusk and Rebec, 1986; Blier et al., 1989), and this in turn attenuates 5-hydroxytryptaminergic function in the terminal fields. The shortcoming of this hypothesis is that the anxiolytic action of 5-HT_{1A} partial agonists in the clinic is not immediate but requires several weeks to occur, which points toward an adaptive change in neuronal function, rather than the inhibition of 5-hydroxytryptaminergic activity, as being responsible for the anxiolytic actions of the 5-HT_{1A} partial agonists. Buspirone, ipsapirone, tandospirone, and gepirone have all been shown to be efficacious in the treatment of GAD (Riblet et al., 1982, 1984; Rickels et al., 1982b, 1997, Kurtz et al., 1990; Murasaki et al., 1993). However, in trials of panic disorder, buspirone was found to have no therapeutic value (Pohl et al., 1989; Sheehan et al., 1990), and flesinoxan (Duphar; Figure 8), which is claimed to be a $5HT_{1A}$ full agonist (Boddeke et al., 1992), was found to exacerbate panic symptoms even when given for long periods (Westenberg et al., 1992; Den Boer and Westenberg, 1995; Van Vliet et al., 1996). To date, the 5-HT_{1A} partial agonists do not appear to have been evaluated for the treatment of phobic anxiety. The onset of anxiolytic effect with the 5-HT_{1A} partial agonists is relatively slow; for example, Pecknold and colleagues (1985) reported that, at weeks one and two, buspirone was significantly less efficacious in alleviating anxiety than diazepam in GAD patients. It has also become apparent that patients who have previously received benzodiazepine anxiolytic therapy respond much less well to 5-HT_{1A} partial agonists than do naïve patients (Schweizer et al., 1986). Although the reasons for this are presently unknown, it nevertheless has a major negative impact on the utility of these drugs in the clinic. The side-effect profile of the 5-HT_{1A} partial agonists is similar to that of the SSRIs, with the former inducing nausea, diarrhea, dizziness, light-headedness, and occasional headache (Feighner and Boyer, 1988). Unlike the benzodiazepines, 5-HT_{1A} partial agonists do not produce psychomotor impairment, muscle relaxation, sedation, or amnesia (Riblet et al., 1982, 1984; Taylor et al., 1985). Furthermore, there is no evidence that 5-HT_{1A} partial agonists have abuse potential, induce drug dependence, or produce relapse anxiety on withdrawal (Lader, 1987; Murphy et al., 1989). Overall, it is evident that the 5-HT_{1A} partial agonists fall very short of the criteria of the ideal anxiolytic. However, for the minority of anxious patients who do benefit from treatment with the 5-HT_{1A} partial agonists, these drugs have some marked advantages over the benzodiazepines such as side-effect profile, lack of dependence, and abuse potential. As such, they will continue to fill an important niche in the anxiolytic market for the foreseeable future.

One recent variant to the 5-HT_{1A} anxiolytic approach has been to combine this action with 5-HT_2 receptor antagonism. Examples of this class of drug are adatanserin (Wyeth), FG 5865 (Pharmacia-Upjohn), and S 213571 (Servier) (Figure 9). Although the contribution of 5-HT_2 antagonism may provide additional anxiolytic efficacy, experience with selective 5-HT_2 antagonists (e.g., ritanserin) has revealed little evidence of anxiolytic potential (Deakin, 1992). These findings do not, however, preclude the possibility of synergism between these mechanisms producing an anxiolytic with greater efficacy and a more rapid onset of action than the 5-HT_{1A} agonists. Adatanserin is described as a 5-HT_{1A} partial agonist/5-HT_2 antagonist (Abou-Gharbia and Moyer, 1990). Despite having greater than 50-fold higher affinity for 5-HT_{1A} than 5-HT_2 receptors, it does nevertheless appear to express both pharmacological actions at similar doses (Abou-Gharbia and Moyer, 1990). In contrast to most findings with 5-HT_{1A} agonists, adatanserin has been shown to increase the punished responding of rats in the Geller conflict model (Abou-Gharbia and Moyer ,1990) and this drug has also been shown to be considerably more efficacious than the 5-HT_{1A} agonist, zalospirone, in a pigeon punished-conflict model (Barrett and James, 1991). Although adatanserin has been discontinued from phase II clinical trials in anxiety, FG 5865 is still being evaluated for clinical development. It will be interesting to discover whether these apparent improvements in the anxiolytic profile of the 5-HT_{1A} agonists are manifest in the clinical situation.

Another recent development in this area has been the discovery of 5-HT_{1A} antagonist ligands (e.g., SDZ 216-525; Sandoz), WAY 100135 (Wyeth), and WAY 100635 (Wyeth) (Figure 14). SDZ 216-525 and WAY 100135 both show anxiolytic activity similar to that of the 5-HT_{1A} partial agonists in the mouse black/white box test (Bill and Fletcher, 1994). The term "silent" antagonist, which has been coined to describe 5-HT_{1A} ligands that are devoid of any agonist efficacy, has been used to describe SDZ 216-525 (Boddeke et al., 1992) and WAY 100135 (Fletcher et al., 1993). However, SDZ 216-525 has subsequently been shown to be a partial agonist of somatodendritic 5-HT_{1A} autoreceptors (Lanfumey et al., 1993) and, in the same study, WAY 100135 was also found to suppress raphé firing. The speculation that WAY 100135 is a partial agonist with very low agonist efficacy is reinforced by the finding that WAY 100135 did not enhance 5-hydroxytryptaminergic firing *in vivo* in guinea-pig brain, whereas a newer, more potent "silent" 5-HT_{1A} antagonist, WAY 100635, did enhance firing rate (Fornal et al., 1994).

To date, no clinical trials in anxiety have been performed with the 5-HT_{1A} antagonists. However, the prediction is that potent 5-HT_{1A} antagonists such as WAY 100635 would enhance 5-hydroxytryptaminergic neuronal firing and be

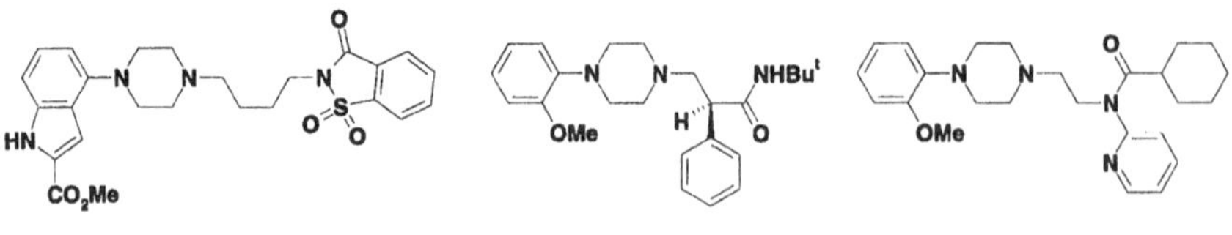

Figure 14. 5-HT_{1A} antagonist ligands

anxiogenic, at least in the shortterm. Whether 5-HT_{1A} antagonists will be anxiolytic when given in the longer term remains to be determined.

Miscellaneous

5-HT_3 antagonists are one of the recent approaches to have been followed in the quest for an alternative to the benzodiazepines in the treatment of anxiety disorders. The 5-HT_3 receptor is analogous to the 5-HT M receptor described by Gaddum and Picarelli (1957). It differs from the other known 5-HT receptor subtypes in being linked directly to calcium channels as opposed to being a member of the G-protein–coupled receptor superfamily (Derkach et al., 1989). 5-HT_3 receptors are located in high density in the area postrema (Kilpatrick et al., 1988), and their inhibition has been shown to have powerful antiemetic actions (Andrews and Hawthorne, 1987; Higgins et al., 1989); two 5-HT_3 antagonists, ondansetron (Glaxo) and granisetron (SmithKline Beecham), are registered for the treatment of emesis. However, 5-HT_3 receptors were also found to be present in the central nervous system (Kilpatrick et al., 1987) and, on the basis of animal experiments, the 5-HT_3 antagonists were predicted to be of potential value in the treatment of anxiety and several psychiatric conditions including anxiety (Costall et al., 1990; Yocca, 1990; Olivier et al., 1992; Blackburn, 1992). Analogous to the 5-HT_{1A} partial agonists, the 5-HT_3 antagonists were found to be inactive in punished-conflict models (Yocca 1990; Olivier et al., 1992), which have traditionally been employed to detect benzodiazepinelike anxiolytics. These compounds were, however, potently active in a number of other models, especially those focused on conditioned and unconditioned stress, novelty, and exploratory activity (Costall et al., 1990; Yocca, 1990; Olivier et al., 1992; Blackburn, 1992). Of particular interest was the finding that ondansetron was not only active in these models but was also effective in treating benzodiazepine-induced withdrawal anxiety (Costall, 1993). This indicated that the 5-HT_3 antagonists would be effective in treating anxious patients who had previously received treatment with benzodiazepines and, importantly, that patients could be switched to these drugs after short-term benzodiazepine therapy. The structures of various 5-HT_3 antagonists are shown in Figure 15.

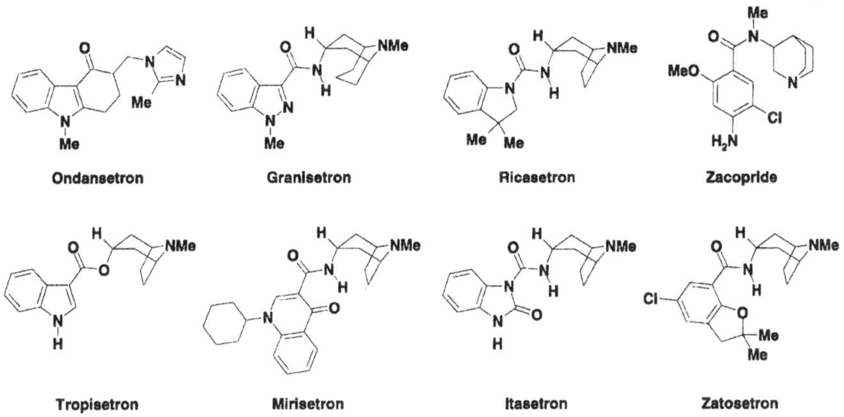

Figure 15. The structures of various 5-HT_3 receptor antagonists

There has been relatively little published on the clinical evaluation of the 5-HT_3 antagonists in anxiety. In volunteers, granisetron, ondansetron, and ricasetron (Smith Kline Beecham) have all been shown to be nonsedative and not to produce psychomotor impairment (Leigh et al., 1991; Link et al., 1993; Van Veggel et al., 1994). Trials in GAD have provided conflicting indications of the value of the 5-HT_3 antagonists as anxiolytics. Schweizer and Rickels (1991) reported a lack of efficacy for ondansetron and zacopride (Delalande; Figure 15) in this patient population. Ansseau and colleagues (1994), on the other hand, claimed that zacopride was anxiolytic in GAD patients when given acutely, and Lecrubier and colleagues (1993) have reported the results of a double-blind, placebo-controlled, three-week trial in GAD patients where tropisetron (Sandoz; Figure 15) showed anxiolytic activity. Encouraging as these findings are, the results in man have proved to be far less impressive than would have been predicted on the basis of animal models. Even assuming that the 5-HT_3 antagonists prove to be anxiolytic in the clinic, the data of Lecrubier and colleagues (1993) indicate that these drugs are not particularly efficacious and they do not have a rapid onset of action and, in a patient population, that is maintained on low dose benzodiazepine therapy. The critical tests for the 5-HT_3 antagonists will be how efficacious they are in patients who have previously received benzodiazepine therapy (cf. the 5-HT_{1A} partial agonists) and in the treatment of benzodiazepine withdrawal. However, additional clinical data have emerged to both confirm and deny the potential value of 5-HT_3 antagonists in the treatment of GAD and related psychiatric disorders. Moreover, the pharmaceutical companies appear to have accepted that these drugs will not have good anxiolytic potential because ondansetron, granisetron, ricasetron, zacopride, mirisetron (Wyeth; Figure 15), itasetron (Pharmacia-Upjohn; Figure 15), and zotosetron (Lilly; Figure 15) have all been discontinued in development for this indication.

Although cholecystokinin (CCK) is an intestinal hormone that regulates pancreatic and bile secretion, it is found in high concentrations in the brain (Vanderhaegen et al., 1981; Rehfeld, 1985). There has been intense interest in its role in anxiety following the discovery that the CCK agonists CCK-4 (tetrapeptide) and pentagastrin cause stress in rodents (Csonka et al., 1988) and panic attacks in man (de Montigny, 1989; Bradwejn et al., 1992; Van Megen et al., 1994). There are two subtypes of CCK receptors—CCK_A and CCK_B—and it is through the latter that the aforementioned peptide agonists exert their behavioral effects. These findings led to speculation that the CCK neuromodulatory system in the CNS may be overactive during anxiety and, furthermore, that attenuating its effects by inhibition of CCK_B receptors may therefore be a useful approach for the treatment of anxiety, and panic attacks in particular. In support of this hypothesis, Bradwejn and colleagues (1994) demonstrated that the CCK_B antagonist CI 988 (Parke-Davis; Figure 16) was moderately effective in alleviating the panic symptoms induced by intravenous injection of CCK-4 into human volunteers, although this was not confirmed in a subsequent study (Van Megen et al., 1997).

Furthermore, in seven panic disorder patients, Cowley and colleagues (1996) found that CI 988 was also ineffective in preventing panic attacks induced by lactate infusions. The CCK_B antagonist approach has suffered further setbacks following the recent reports from three double-blind placebo-controlled trials of these drugs in the treatment of anxiety. L 365, 260 (Merck; Figure 16) was found to be ineffective in reducing the frequency or severity of panic attacks in patients and the associated anxiety (Kramer et al., 1995). CI 988 was evaluated for the treatment of GAD and found to be ineffective (Bammert-Adams et al., 1995), and a six-week trial of CI 988 in panic disorder was terminated early when the interim analysis of efficacy was found to slightly favor placebo (Pande et al., 1996). Consistent with the role of CCK as a gut hormone, the major side-effects of the CCK_B antagonists were diarrhea, dyspepsia, flatulence, and nausea (Kramer et al., 1995;

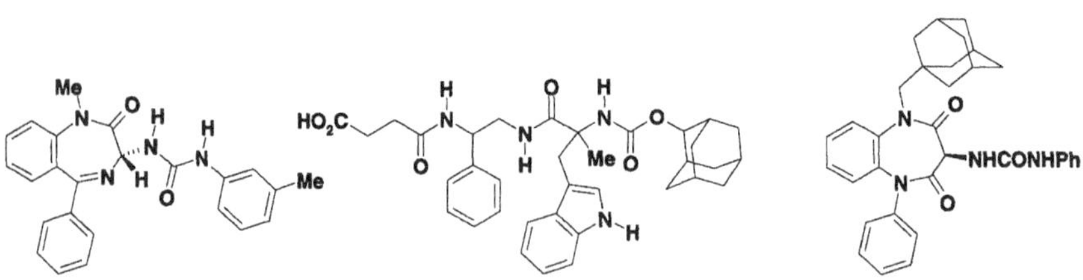

Figure 16. Structure of the CCK_B receptor antagonists, L 365260, CI 988 and GV 150013

Bammert-Adams et al., 1995). No clinical data are available yet on GV 150013 (Glaxo-Wellcome; Figure 16), which is reported to be in phase II clinical trials for anxiety in the USA.

Although the findings with the CCK_B antagonists in the treatment of panic disorder and GAD are insufficient in themselves to demolish the CCK/anxiety hypothesis, they are not encouraging. If subsequent clinical trials confirm these results, the potential role of the CCK_B antagonists in the treatment of anxiety and panic disorder will be at an end.

There are several other new approaches to the drug treatment of anxiety that have not been reviewed here. This is because the drugs involved have not progressed sufficiently far in development to have been extensively tested in the clinic. Examples of these approaches are 5-$HT_{2A/2C}$ antagonists, 5-HT_{2C} antagonists, antagonists of the glycine site on the NMDA receptor, ACE inhibitors, substance P antagonists, and tachykinin antagonists (for reviews, see Deakin, 1992; Blackburn, 1992; Jackson and Nutt, 1996).

Conclusions

When taking an overview of the fields of antidepressant and anxiolytic drug development, it is evident that considerable progress has been made with the greater strides being made in antidepressant research. The SSRIs have been one of the success stories of the 1990s, and the very recent finding that combining a 5-HT_{1A} antagonist with a monoamine reuptake inhibitor has yielded the tantalizing vision of future antidepressant drugs with a genuine early onset of action. In anxiety, a new class of drugs, namely, the 5-HT_{1A} partial agonists, has been introduced with some success, and the monoamine reuptake inhibitors, especially the SSRIs, have found a second existence in the treatment of anxiety. However, other strategies such as the 5-HT_3 antagonists and CCK_B antagonists are looking far from promising. Overall, therefore, when the achievements made by drugs from these new approaches are measured against the criteria of the ideal antidepressant and anxiolytic drugs, it is evident that attaining these goals still remains a major challenge.

ACKNOWLEDGMENTS

We would like to thank Jean Smith, Christine Tyldesley, and Alison Wilson-Osborn for typing this chapter, Dr. Frank Kerrigan for drawing the chemical structures, Professor Roger Buckett for his helpful comments, and Mike Prow for endless proofreading, checking references, and so forth.

REFERENCES

Abou-Gharbia, M. & Moyer, J.A. (1990). Wy-50,324. Drugs of the Future 15, 1093-1094.

Abuzzahab, F.S. (1989). Adinazolam in outpatient depression. In: Psychiatry Today. Accomplishments and Promises (Stefanis, C.N., Soldatos, C.R., & Rabavilas, A.D., Eds.). Excerpta Medica, Int. Congr. Ser. 899, (Abstract, 2538).

Ad Sitsen, J.M.F. & Montgomery, S.A. (1994). The pharmacological treatment of depression and its problems. In: Handbook of Depression and Anxiety: A Biological Approach. (den Boer, J.A. & Ad Sitsen, J.M., Eds.), pp. 349-377. Marcel Dekker Inc., New York.

Ahlenius, S., Larsson, K., & Wijkström, A. (1991). Behavioral and biochemical effects of the $5HT_{1A}$ agonists flesinoxan and 8-OH-DPAT in the rat. Eur. J. Pharmacol. 200, 259-266.

Alby, J.M., Cabane, J., Ferreri, M., & Bougerol, T. (1993). Efficacy and acceptability of tianeptine in major depressive disorder and in dysthymia (DSM-IIIR) with somatic compliants: Double blind study versus fluoxetine. Eur. Neuropsychopharmacol. 3, 333 (abstract).

Allquander, C. (1978). Dependence on sedative and hypnotic drugs. Acta Psychiatrica Scand. 270, 1-120.

Allsopp, L.F., Cooper, G.L., & Poole, P.H. (1984). Clomipramine and diazepam in the treatment of agoraphobia and social phobia in general practice. Curr. Med. Res. Opin. 9, 64-70.

American Psychiatric Association (1980). DSM-III: Diagnostic and Statistical Manual of Mental Disorders, 3rd ed. The American Psychiatric Association, Washington, D.C.

American Psychiatric Association (1987). DSM-III-R: Diagnostic and Statistical Manual of Mental Disorders, 3rd ed. revised. The American Psychiatric Association, Washington, D.C.

American Psychiatric Association (1994). DSM-IV: Diagnostic and Statistical Manual of Mental Disorders, 4th ed. The American Psychiatric Association, Washington, D.C.

Amsterdam, J.D. (1992). Gepirone, a selective serotonin $5\text{-}HT_{1A}$ partial agonist in the treatment of major depression. Prog. Neuropsychopharmacol. Biol. Psychiatry 16, 271-280.

Andrews, P.L.R. & Hawthorn, J. (1987). Evidence for an extra-abdominal site of action for the $5\text{-}HT_3$ receptor antagonist BRL 24924 in the inhibition of radiation-evoked emesis in the ferret. Neuropharmacology 26, 1367-1370.

Ansseau, M., Pitchot, W., & Moreno, A.G. (1994). Placebo-controlled study of the anxiolytic activity of single doses of zacopride among inpatients with generalized anxiety disorder. Neuropsychopharmacology 10, 174-190.

Ansseau, M., Pitchot, W., Moreno, A.G., Wauthy, J., & Papart, P. (1993). Pilot study of flesinoxan, a $5\text{-}HT_{1A}$ agonist, in major depression: Effects on sleep REM latency and body temperature. Hum. Psychopharmacol. 8, 279-283.

Artigas, F., Perez, V., & Alvarez, E. (1994). Pindolol induces a rapid improvement of depressed patients treated with serotonin reuptake inhibitors. Arch. Gen. Psychiatry 51, 248-251.

Artigas, F., Romero, L., De Montigny, C., & Blier P. (1996). Acceleration of the effects of selected antidepressant drugs in major depression by $5\text{-}HT_{1A}$ antagonists. Trends Neurosci. 19, 378-383.

Bakish, D., Bradwejn, J., Nair, N., McClure, J., Remick, R., & Bulger, L. (1992). A comparison of moclobemide, amitriptyline and placebo in depression: A Canadian multicentre study. Psychopharmacology 106, S98-S101.

Bakish, D., Saxena, B.M., Bowen, R., & de Souza, J. (1993). Reversible monoamine oxidase-A inhibitors in panic disorder. Clin. Neuropharmacol. 16 (Suppl. 2), 77-82.

Ballenger, J.C., McDonald, S., Noyes, R., Rickels, K., Sussman, N., Woods, S., Patin, J., & Singer, J. (1991). The first double-blind, placebo-controlled trial of a partial benzodiazepine agonist abecarnil (ZK 112-119) in generalized anxiety disorder. Psychopharmacol. Bull. 27, 171-179.

Bammert-Adams, J., Pyke, R.E., Costa, J., Cutler, N.R., Schweizer, E., Wilcox, C.S., Wisselink, P.G., Greiner, M., Pierce, M.W., & Pande, A.C. (1995). A double-blind, placebo-controlled study of

CCK_B receptor antagonist, CI-998, in patients with generalized anxiety disorder. J. Clin. Psychopharmacol. 15, 428-434.

Barrett, J.E. & Zhang, L. (1991). Anticonflict and discriminative stimulus effects of the $5\text{-}HT_{1A}$ compounds WY-47,846 and WY-48,723 and the mixed $5\text{-}HT_{1A}$ agonist/$5\text{-}HT_2$ antagonist WY-50,324 in pigeons. Drug Dev. Res. 24, 179-188.

Bartoszyk, G.D., Hegenbart, R., & Ziegler, H. (1997). EMD 68843, a serotonin reuptake inhibitor with selective presynaptic $5\text{-}HT_{1A}$ receptor agonistic properties. Eur. J. Pharmacol. 322, 147-153.

Basse-Tomusk, A. & Rebec, G.V. (1986). Isapirone depresses neuronal activity in the dorsal raphé nucleus and the hippocampal formation. Eur. J. Pharmacol. 130, 141-143.

Batra, S. & Björklund, A. (1986). Binding affinities of four tricyclic antidepressive drugs to muscarinic cholinergic receptors in human parotid gland. Psychopharmacology 90, 14.

Baumann, P.A. & Maitre, L. (1977). Blockade of presynaptic α-receptors and of amine uptake in the rat brain by the antidepressant mianserine. Naunyn-Schmiedeberg's Arch. Pharmacol. 300, 31-37.

Baumann, P.A. & Waldmeier, P.C. (1978). The effects of blockade of presynaptic α-receptors by yohimbine and mianserin on the levels of 4-hydroxy-3-methoxyphenethyleneglycol sulphate ($MOPEG\text{-}SO_4$) in different brain regions of the rat. Influence of mianserin on daily rhythm in $MOPEG\text{-}SO_4$ level. Experientia 34, 922.

Baunhackl, U., Biziere, K., Fischbach, R., Geretsegger, C., Hebenstreit, G., Radmayr, E., & Stabl, M. (1989). Efficacy and tolerability of moclobemide compared with imipramine in depressive disorder (DSM-III): An Austrian double-blind, multicentre study. Br. J. Psychiatry 155, 78-83.

Beauclair, L., Fontaine, R., Annable, L., Holobow, N., & Chouinard, G. (1994). Clonazepam in the treatment of panic disorder: A double-blind, placebo-controlled trial investigating the correlation between clonazepam concentrations in plasma and clinical response. J. Clin. Psychopharmacol 14, 111-118.

Benca, R., Matuzas, W., & Al-Sadir, J. (1986). Social phobia, MVP, and response to imipramine. J. Clin. Psychopharmacol. 6, 50-51.

Bertler, A. (1961). Effect of reserpine on the storage of catecholamines in brain and other tissues. Acta Physiol. Scand. 51, 75-83.

Bill, D.J. & Fletcher, A. (1994). Correlation of *in vivo* functional and anxiolytic effects of $5\text{-}HT_{1A}$ receptor ligands in the mouse. Br. J. Pharmacol. 111, 151P.

Birkenhäger, T.K., Moleman, P., & Nolen, W.A. (1995). Benzodiazepines for depression? A review of the literature. Int. Clin. Psychopharmacol. 10, 181-195.

Blackburn, T.P. (1992). 5-HT receptors and anxiolytic drugs. In: Central Serotonin Receptors and Psychotropic Drugs (Marsden, C.A. & Heal, D.J, Eds.), pp. 175-197. Blackwell Scientific Publications, Oxford.

Blackwell, B. (1981). Adverse effects of antidepressant drugs. Part 1. Monoamine oxidase inhibitors and tricyclics. Drugs 21, 201-219.

Blackwell, B., Morley, E., Price, J., & Taylor, D. (1967). Hypertensive interactions between monoamine oxidase inhibitors and foodstuffs. Br. J. Psychiatry 113, 349-365.

Blier, P. & Bergeron, R. (1995). Effectiveness of pindolol with selected antidepressant drugs in the treatment of major depression. J. Clin. Psychopharmacol. 15, 217-222.

Blier, P. & de Montigny, C. (1994). Current advances and trends in the treatment of depression. Trends Pharmacol. Sci. 15, 220-226.

Blier, P., Seletti, B., Bouchard, C., Artigas, F., & de Montigny, C. (1994). Functional evidence for the differential responsiveness of pre- and postsynaptic $5\text{-}HT_{1A}$ receptors in the rat brain. Soc. Neurosci. 20, 1540.

Blier, P., Steinberg, S., Chaput, Y., & de Montigny, C. (1989). Electrophysiological assessment of putative antagonists of 5-HT receptors: A single-cell study in the rat dorsal raphé nucleus. Can. Physiol. Pharmacol. 67, 98-105.

Boddeke, H.W.G.M., Fargin, A., Raymond, J.R., Schoeffter, P., & Hoyer, D. (1992). Agonist/antagonist interactions with cloned human 5-HT_{1A} receptors: Variations in intrinsic activity studied in transfected HeLa cells. Naunyn-Schmiedeberg's Arch. Pharmacol. 345, 257-263.

Bolden-Watson, C. & Richelson, E. (1993). Blockade by newly developed antidepressants of biogenic amine uptake into rat brain synaptosomes. Life Sci. 52, 1023-1029.

Borsini, F., Ceci, A., Bietti, G., & Donetti, A. (1995). BIMT 17, a 5-HT_{1A} receptor agonist/5HT_{2A} receptor antagonist, directly actives postsynaptic 5-HT inhibitory responses in the rat cerebral cortex. Naunyn Schmiedeberg's Arch. Pharmacol. 352, 283-290.

Borsini, F., Cesano, R., Kelly, J., Leonard, B.E., McNamara, M., Richards, J., & Seiden, L. (1997). BIMT 17: A putative antidepressant with a fast onset of action? Psychopharmacology 134, 378-386.

Bradwejn, J., Koszycki, D., Cöuetoux du Tertre, A., Bourin, M., Palmour, R., & Ervin, F. (1992). The cholecystokinin hypothesis of panic and anxiety disorders: A review. J. Psychopharmacol. 6, 345-351.

Bradwejn, J., Paradis, M., Koszychi, D., Reece, P., & Sedman, A. (1994). The effects of CI-988 on CCK-4-panic in healthy volunteers. Can. J. Physiol. Pharmacol. 72, 417.

Bremner, J.D. (1995). A double blind comparison of Org 3770, amitriptyline and placebo in major depression. J. Clin. Psychiatry 56, 519-525.

Briley, M. & Moret, C. (1997). Antidepressant properties of specific serotonin-noradrenaline reuptake inhibitors. In: Antidepressants. New Pharmacological Strategies (Skolnick, P., Ed.), pp. 35-52. Humana Press Inc, Totowa, New Jersey.

Briley, M., Prost, J.F., & Moret, C. (1996). Preclinical pharmacology of milnacipran. Int. Clin. Psychopharmacol. 11, 9-14.

Broderick, P.A. & Piercey, M.F. (1991). 5-HT_{1A} agonists uncouple noradrenergic somatodendritic impulse flow and terminal release. Brain Res. Bull. 27, 693-696.

Brotman, A.W., Falk, W.E., & Gelenberg, A.J. (1987). Pharmacologic treatment of acute depressive subtypes. In: Psychopharmacology: The Third Generation of Progress. (Meltzer, H.Y., Ed.), pp. 1031-1040. Raven Press, New York.

Bunney, W.E. & Davis, J.M. (1965). Norepinephrine in depressive reactions. Arch. Gen. Psychiatry 13, 483-494.

Carlsson, A., Corrodi, H., Fuxe, K., & Hökfelt, T. (1969). Effect of antidepressant drugs on the depletion of intraneuronal brain 5-hydroxytryptamine stores caused by 4-methyl-α-ethyl-m-tyramine. Eur. J. Pharmacol. 5, 357-366.

Carman, J.S., Ahdieh, H., Wyatt-Knowles, E., Warga, E., & Panagides, J. (1991). A controlled study of mianserin in moderately to severely depressed outpatients. Psychopharmacol. Bull. 27, 135-139.

Cassano, G.B., Heinze, G., Lôo, H., Mendlewicz, J., & Paes de Souza, M. (1996). A doubleblind comparison of tianeptine, imipramine and placebo in the treatment of major depressive episodes. Eur. Psychiatry 11, 254-259.

Cassidy, S. & Henry, J. (1987). Fatal toxicity of antidepressant drugs in overdose. Br. Med. J. 295, 1021-1024.

Cerletti, U. & Bini, L. (1938). Un neuvo metodo di shockterapie "l elettroshock." Boll. Acad. Med. Roma 64, 136.

Chaplin, S. (1986). Bone marrow depression due to mianserin, phenylbutazone, oxyphenbutazone and chloramphenicol. Part II. Adverse Drug React. Acute Poisoning Rev. 3, 181-196.

Claghorn, J.L. & Lesem, M.D. (1995). A double-blind placebo-controlled study of Org 3770 in depressed outpatients. J. Affect. Disorder 34, 165-171.

Clerc, G.E., Ruimy, P., & Verdeau-Paillés, J. (1994). A double-blind comparison of venlafaxine and fluoxetine in patients hospitalized for major depression and melancholia. Int. Clin. Psychopharmacol. 9, 139-143.

Cole, J.O. & Brodkin, J.A. (1990). Affective drug side effects. J. Clin. Psychiatry 51, S21-S26.

Cookson, J. (1993). Side effects of antidepressants. Br. J. Psychiatry 163, 20-24.

Cooper, A.J. (1989). Tyramine and irreversible monoamine oxidase inhibitors in clinical practice. Br. J. Psychiatry 155, 38-45.

Coppen, A. (1967). The biochemistry of affective disorders. Br. J. Psychiatry 113, 1237-1264.

Costa e Silva, J.A. & Ruschel, S. (1994). Tianeptine versus placebo study in major depressions and depressive bipolar disorders. Eur. Psychiatry 9, 140s.

Costall, B. (1993). The breadth of action of the 5-HT_3 receptor antagonists. Int. Clin. Psychopharmacol. 8, 3-9.

Costall, B., Naylor, R.J., & Tyers, M.B. (1990). The psychopharmacology of 5-HT_3 receptors. Pharmacol. Ther. 47, 181-202.

Cowley, D.S., Bammert Adams, J., Pyke, R.E., Cook, J. Zaccarias, P., Wingerson, D., & RoyByrn P.P. (1996). Effect of CI-988, a cholecystokinin-B receptor antagonist, on lactate-induced panic. Biol. Psychiatry 40, 550-552.

Crome, P. & Newman, B. (1979). Fatal tricyclic antidepressant poisoning. J. Roy. Soc. Med. 72, 649-653.

Csonka, E., Fekete, M., Nagy, G., Szanto-Fekete, M., Feledgy, G., Penke, B., & Kovaks, K. (1988). Anxiogenic effect of cholecystokinin in rats. In: Peptides. (Penka, B. & Tovok, A., Eds.), pp. 249-252. Walter de Gruyter and Co., New York.

Cusack, B., Nelson, A., & Richelson, E. (1994). Binding of antidepressants to human brain receptors: Focus on new generation compounds. Psychopharmacology 114, 559-565.

Czepowicz, V.D., Johnson, M.R., Lydiard, R.B., Emmanuel, N.P., Ware, M.R., Mintzer, O.B., Walsh, M.D., & Ballenger, J.C. (1995). Sertaline in social phobia. J. Clin. Psychopharmacol. 15, 372-373.

Danish University Antidepressant Group (1986). Citalopram: Clinical effect profile in comparison with clomipramine. A controlled multicentre study. Psychopharmacology 90, 131-138.

Danish University Antidepressant Group (1990). Paroxetine—a selective serotonin reuptake inhibitor showing better tolerance, but weaker antidepressant effect than clomipramine in a controlled multicentre study. J. Affect. Disorders 18, 289-299.

D'Amico, M.F., Roberts, D.L., Robinson, D.S., Schwiderski, U.E., & Copp, J. (1990). Placebo controlled dose-ranging trial designs in phase II development of nefazodone. Psychopharmacol. Bull. 26, 147-150.

Da Prada, M., Kettler, R., Keller, H.H., Burkard, W.P., Muggli-Maniglio, D., & Haefely, W.E. (1989). Neurochemical profile of moclobemide, a short-acting and reversible inhibitor of monoamine oxidase type A. J. Pharmacol. Exp. Ther. 248, 400-414.

Da Prada, M., Zurcher, G., Wuthrich, I., & Haefely, W.E. (1988). On tyramine, food, beverages and the reversible MAO inhibitor moclobemide. J. Neural Transm. 26, S31-S56.

Davidson, C. & Stamford, J.A. (1997). 5-$HT_{1B/D}$ antagonists potentiate paroxetine's effect on 5-HT efflux in the lateral geniculate nucleus; *in vitro* voltametric data. Biochemical Society Transactions 25, 49S.

Davis, J.M. (1975). Tricyclic antidepressants. In: Drug Treatment of Mental Disorders. (Simpson, L.L., Ed.). Raven Press, New York.

Deakin, J.F.W. (1992). Amine mechanisms in anxiety and depression. In: Experimental Approaches to Anxiety and Depression. (Elliott, J.M., Heal, D.J., & Marsden, C.A., Eds.), pp. 233-243. John Wiley & Sons Ltd, Chichester.

De Boer, T., Maura, G., Raiteri, M., De Vos, C.J., Wieringa, J., & Pinder, R.M. (1988). Neurochemical and autonomic pharmacological profiles of the 6-aza-analogue of mianserin, Org 3770 and its enantiomers. Neuropharmacology 27, 399-408.

De Boer, T. & Ruigt, Gé S.F. (1995). The selective α_2-adrenoceptor antagonist mirtazapine (Org 3770) enhances noradrenergic and 5-HT_{1A}-mediated serotonergic neurotransmission. CNS Drugs 4, 29-38.

Dechant, K.L. & Clissold, S.P. (1991). Paroxetine: A review of its pharmacodynamic and pharmacokinetic properties, and therapeutic potential in depressive illness. Drugs 41, 225-253.

De Jongh, R. (1993). Remeron efficacy in major depressive episode: An overview of the principal clinical trials data. 9th World Congress of Psychiatry, pp. 29.

de Montigny, C. (1989). Cholecystokinin tetrapeptide induces panic-like attacks in healthy volunteers. Arch. Gen. Psychiatry 46, 511-517.

Den Boer, J.A. & Westenberg, H.G.M. (1995). Serotonergic compounds in panic disorder, obsessive-compulsive disorder and anxious depression: A concise review. Human Psychopharmacol. 10, S173-S183.

Dennis, T., L'Heureux, R., Carter, C., & Scatton, B. (1987). Presynaptic alpha-2 adrenoceptors play a major role in the effects of idazoxan on cortical noradrenaline release (as measured by in vivo dialysis in the rat). J. Pharmacol. Exp. Ther. 241, 642-649.

Derivan, A., Entsuah, A.R., & Kikta, D. (1995). Venlafaxine: Measuring the onset of antidepressant action. Psychopharmacol. Bull. 31, 439-447.

Derkach, V., Surprenant, A., & North, R.A. (1989). 5-HT_3 receptors are membrane ion channels. Nature 339, 706-709.

De Vry, J. (1996). 5-HT_{1A} receptors in psychopathology and the mechanism of action of clinically effective therapeutic agents. Drug News and Perspectives 9, 270-280.

Dierick, M., Ravizza, L., Realini, R., & Martin, A. (1996). A double blind comparison of venlaxafine and fluoxetine for treatment of major depression in outpatients. Prog. Neuropsychopharmacol. Biol. Psychiatry 20, 57-71.

Dubini, A., Bosc, M., & Polin, V. (1997). Noradrenaline-selective versus serotonin-selective antidepressant therapy: Differential effects on social functioning. J. Psychopharmacol. 11, 517-523.

Duka, T., Schütt, B., Krause, W., Dorow, R., McDonald, S., & Fichte, K. (1993). Human studies on abecarnil a new β-carboline anxiolytic: Safety, tolerability and preliminary pharmacological profile. Br. J. Clin. Pharmacol. 35, 386-394.

Dunbar, G.C., Claghorn, J.L., Kiev, A., & Rickels, K. (1993). A comparison of paroxetine and placebo in depressed out-patients. Acta Psych. Scand. 87, 302-305.

Eison, A.S., Eison, M.S., Torrente, J.R., Wright, R.N., & Yocca, F.D. (1990). Nefazodone: Preclinical pharmacology of a new antidepressant. Psychopharmacol. Bull. 26, 311-315.

Escobar, J.I. & Landbloom, R.P. (1976). Treatment of phobic neurosis with chlorimipramine: A controlled clinical trial. Curr. Ther. Res. 20, 680-685.

Fabre, L.F. (1989). United States experience and perspectives with trazodone. Clin. Neuropharmacol. 12, S11.

Fabre, L.F. (1990). Trazodone dosing regimen: Experience with single daily administration. J. Clin. Psychiatry 51, 23-26.

Facklam, M., Schoch, P., Bonetti, E.P., Jenck, F., Martin, J.R., Moreau, J.-L., & Haefely, W.E. (1992). Relationship between benzodiazepine receptor occupancy and functional effects *in vivo* of four ligands of differing intrinsic efficacies. J. Pharmacol. Exp. Ther. 261, 1113-1121.

Fahlén, T., Humble, M., Koczkas, C., & Nilsson, H.L. (1992). Social phobia and its treatment with brofaromine. Efficacy regarding social phobia symptoms and personality traits. Clin. Neuropharmacol. 15 (Suppl. 1), B64.

Feighner, J.P. & Boyer, W.F. (1988). Serotonin-1A agonists: An overview of anxiolytic activity. Psychopharmacology 96 (Suppl.) (Abstract, WE 05.03.)

Feighner, J.P., Meredith, C.H., Duff, J.E., & Erickson, G. (1982a). A double-blind comparison of lofepramine, imipramine, and placebo in patients with primary depression. Acta Psych. Scand. 66, 100-108.

Feighner, J.P., Meredith, C.H., & Hendrickson, M.A. (1982b). A double-blind comparison of buspirone and diazepam in outpatients with generalized anxiety disorder. J. Clin. Psychiatry 43, 103-107.

Feighner, P., Pambakian, R., Fowler, R.C., Boyer, W.F., & D'Amico, M.F. (1989). A comparison of nefazodone, imipramine, and placebo in patients with moderate to severe depression. Psychopharmacol. Bull. 25, 219-221.

Fineberg, N. & Drummond, L.M. (1995). Anxiety disorders. Drug treatment or behavioural cognitive psychotherapy? CNS Drugs 3, 448-466.

Finley, P.R. (1994). Selective serotonin reuptake inhibitors: Pharmacological profiles and potential therapeutic distinctions. Annals Pharmacotherapy 28, 1359-1369.

Fletcher, A., Cliffe, I.A., & Dourish, C.T. (1993). Silent 5-HT_{1A} receptor antagonists: Utility as research tools and therapeutic agents. Trends Pharmacol. Sci. 14, 441-448.

Fornal, C.A., Metzler, C.W., Veasey, S.C., McCreary, A.C., Dourish, C.T., & Jacobs, B.L. (1994). Single-unit studies in behaving animals provide evidence that WAY-100635, but not (S)-WAY-100135, blocks the action of endogenous serotonin at the 5-HT autoreceptor. Br. J. Pharmacol. 112, 92.

Frazer, A. (1994). Antidepressant drugs. Depression 2, 1-19.

Freis, E.D. (1954). Mental depression in hypertensive patients treated for long periods with large doses of reserpine. N. Engl. J. Med. 251, 1006-1008.

Gaddum, J.H. & Picarelli, Z.P. (1957). Two kinds of tryptamine receptor. Br. J. Pharmacol. 12, 323-328.

Gartside, S.E., Umbers, V., Hajós, M., & Sharp, T. (1995). Interaction between a selective 5HT_{1A} receptor antagonist and an SSRI *in vivo*: Effects on 5-HT cell firing and extracellular 5-HT. Br. J. Pharmacol. 115, 1064-1070.

Gelder, M.G. (1989). Panic disorder: fact or fiction. Psychol. Med. 19, 277-283.

Ginestet, D. (1997). Efficacy of tianeptine in major depressive disorders with or without melancholia. Eur. Neuropsychopharmacol. 7, S341-S345.

Glowinski, J. & Axelrod, J. (1964). Inhibition of uptake of tritiated-noradrenaline in the intact rat brain by imipramine and structurally related compounds. Nature 204, 1318-1319.

Gobert, A., Rivet, J.-M., Cistarelli, L., & Millan, M.J. (1997). Potentiation of the fluoxetine-induced increase in dialysate levels of serotonin (5-HT) in the frontal cortex of freely moving rats by combined blockade of 5-HT_{1A} and 5-HT_{1B} receptors with WAY 100,635 and GR 127,935. J. Neurochem. 68, 1159-1163.

Goldberg, H.L. & Finnerty, R.J. (1979). The comparative efficacy of buspirone and diazepam in the treatment of anxiety. Am. J. Psychiatry 136, 1184-1187.

Gorman, J.M., Liebowitz, M.R., Fyer, A.J., Goetz, D., Campaes, R.B., Fyer, M.J., Orvis, S.O., & Klein, D.F. (1987). An open trial of fluoxetine in the treatment of panic attacks. J. Clin. Psychopharmacol. 7, 329-332.

Grimsley, S.R. & Jann, M.W. (1992). Paroxetine, sertraline, and fluoxetine: New selective serotonin reuptake inhibitors. Clin. Pharmacy 11, 930-957.

Gruter, W. & Poldinger, W. (1982). Maprotiline. Mod. Probl. Pharmacopsychiatry 18, 17-48.

Guelfi, J.D. (1992). Efficacy of tianeptine in comparative trials versus reference antidepressants. An overview. Br. J. Psychiatry 160, 72-75.

Guelfi, J.D., White, C., Hackett, D., Guichoux, J.Y., & Magni, G. (1995). Effectiveness of venlafaxine in patients hospitalized for major depression and melancholia. J. Clin. Psychiatry 56, 450-458.

Guze, S.B. & Robins, E. (1970). Suicide and primary affective disorders. Br. J. Psychiatry 117, 437-438.

Haddjeri, N., de Montigny, C., & Blier, P. (1997). Modulation of the firing activity of noradrenergic neurones in the rat locus coeruleus by the 5-hydroxytryptamine system. Br. J. Pharmacol. 120, 865-875.

Haefely, W. (1983). Tranquillizers. In: Psychopharmacology, Part 1: Preclinical Psychopharmacology. (Grahame-Smith, D.G. & Cowen, P.J., Eds.), pp. 107-151. Excerpta Medica, Amsterdam.

Haefely, W. (1985). Tranquillizers. In: Psychopharmacology 2, Part 1: Preclinical Psychopharmacology. (Grahame-Smith, D.G. & Cowen, P.J., Eds.), pp. 92-182. Excerpta Medica, Amsterdam.

Haefely, W. (1992). The role of GABA in anxiolytic/antidepressant drug action. In: Experimental Approaches to Anxiety and Depression. (Elliott, J.M., Heal, D.J., & Marsden, C.A., Eds.), pp. 151-168. John Wiley & Sons Ltd, Chichester.

Handley, S.L., McBlane, J.W., Critchley, M.A., & Njung'e, K. (1993). Multiple serotonin mechanisms in animal models of anxiety: Environmental, emotional and cognitive factors. Behav. Brain Res. 59, 203-210.

Heal, D., Cheetham, S., Martin, K., Browning, J., Luscombe, G., & Buckett, R. (1992). Comparative pharmacology of dothiepin, its metabolites, and other antidepressant drugs. Drug Dev. Res. 27, 121-135.

Heller, A.H., Beneke, M., Kuemmel, B., Spencer, D., & Kurtz, N.M. (1990). Ipsapirone: Evidence for efficacy in depression. Psychopharmacol. Bull. 26, 58-61.

Henry, J.A. (1989). A fatal toxicity index for antidepressant poisoning. Acta Psych. Scand. 80 (Suppl. 354), 37-45.

Hetzel, W. (1992). Safety of moclobemide taken in overdose for attempted suicide. Psychopharmacology 106, S127-S129.

Higgins, G.A., Kilpatrick, G.J., Bunce, K.T., Jones, B.J., & Tyers, M.B. (1989). 5-HT_3 receptors in the area postrema may mediate the anti-emetic effects of 5-HT_3 antagonist in the ferret. Br. J. Pharmacol. 97, 247-255.

Hill-Venning, C., Callachan, H., Peters, J.A., Lambert, J.J., Gemmell, D.K., & Campbell, A.C. (1994). Modulation of the $GABA_A$ receptor by ORG 20599: A water soluble pregnant steroid. Br. J. Pharmacol. 111, 183.

Hjorth, S. (1993). Serotonin 5-HT_{1A} autoreceptor blockade potentiates the ability of the 5-HT reuptake inhibitor citalopram to increase nerve terminal output of 5-HT *in vivo*: A microdialysis study. J. Neurochem. 60, 776-779.

Holliday, S.M. & Benfield, P. (1995). Venlafaxine. A review of its pharmacology and therapeutic potential in depression. Drugs 49, 280-294.

Holzbauer, M. & Vogt, M. (1956). Depression by reserpine of the noradrenaline concentration in the hypothalamus of the cat. J. Neurochem. 1, 8-11.

Horne, R.L., Ferguson, J.M., & Pope, H.G. (1988). Treatment of bulimia with bupropion: A multicenter controlled trial. J. Clin. Psychiatry 49, 262-266.

Howard, J.L. & Pollard, G.T. (1990). Effects of buspirone in the Geller-Seifter conflict test with incremental shock. Drug Dev. Res. 19, 37-49.

Hoyer, D., (1988a). Functional correlates of serotonin 5-HT_1 recognition sites. J. Receptor Res. 8, 59-81.

Hoyer, D. (1988b). Molecular pharmacology and biology of 5-HT_{1C} receptors. Trends Pharmacol. Sci. 9, 89-94.

Hoyer, D., Engel, G., & Kalkman, H.O. (1985). Molecular pharmacology of 5-HT_1 and 5-HT_2 recognition sites in rat and pig brain membranes; radioligand binding studies with [^{3}H]5-HT, [^{3}H]8-OH-DPAT, (-)[^{125}I]iodocyanopindolol, [^{3}H]mesulergine and [^{3}H]ketanserin. Eur. J. Pharmacol. 188, 13-23.

Hyttel, J. (1982). Citalopram—Pharmacological profile of a specific serotonin uptake inhibitor with antidepressant activity. Prog. Neuropsychopharmacol. Biol. Psychiatry 6, 277-295.

Jackson, H.C. & Nutt, D.J. (1996). Anxiety and panic disorders. In: Neurotherapeutics. Emerging Strategies. (Pullan, L.M. & Patel, J., Eds.), pp. 85-131. Humana Press Inc, Totowa, New Jersey.

Jenkins, S.W., Robinson, D.S., Fabre, L.F. Jr., Andary, J.J., Messina, N.E., & Reich, L.A. (1990). Gepirone in the treatment of major depression. J. Clin. Psychopharmacol. 10, S77-S85.

Johnston, J.P. (1968). Some observations upon a new inhibitor of monoamine oxidase in brain tissues. Biochem. Pharmacol. 17, 1285-1297.

Johnston, J.A., Lineberry, C.G., Asher, J.A., Davidson, J., Khayrallah, M.A., Feighner, J.P., & Stark, P. (1991). A 102-centre prospective study of seizure in association with bupropion. J. Clin. Psychiatry 52, 450-456.

Johnston, D.G., Troyer, I.E., & Whitsett, S.F. (1988). Clomipramine treatment of agoraphobic women. Arch. Gen. Psychiatry 45, 453-459.

Jue, S.G., Dawson, G.W., & Brogden, R.N. (1982). Amoxapine: A review of its pharmacology and efficacy in depressed states. Drugs 24, 1-23.

Kapur, S., Mieczkowski, T., & Mann, J.J. (1992). Antidepressant medications and the relative risk of suicide attempt and suicide. JAMA 268, 3441-3445.

Kasper, S., Fuger, J., & Moller H.J. (1992). Comparative efficacy of antidepressants. Drugs 43, 11-23.

Katz, M.M., Koslow, S.H., Maas, J.W., Frazer, A., Bowden, C., Casper, R., Croughan, J., Kocsis, J., & Redmond, E., Jr. (1987). The timing, specificity and clinical prediction of tricyclic drug effects in depression. Psychol. Med. 17, 297-309.

Kelly, D., Guirguis, W., Frommer, E., Mitchell-Heggs, N., & Sargant, W. (1970). Treatment of phobic states with antidepressants. Br. J. Psychiatry 116, 387-398.

Khan, A., Fabre, L.F., & Rudolph, R. (1991). Venlafaxine in depressed outpatients. Psychopharmacol. Bull. 27, 141-144.

Kilpatrick, G.J., Jones, B.J., & Tyers, M.B. (1987). Identification and distribution of 5-HT_3 receptors in rat brain using radioligand binding. Nature 330, 746-748.

Kilpatrick, G.J., Jones, B.J., & Tyers, M.B. (1988). The distribution of the specific binding of 5-HT_3 receptor ligand [^{3}H]GR65630 in rat brain using quantitative autoradiography. Neuroscience Letts. 94, 156-160.

King, A. (1962). Phenelzine treatment of Roth's Calamity Syndrome. Med. J. Aust. 23, 879-883.

Kinney, J.L. & Evans, R.L. Jr. (1982). Evaluation of amoxapine. Clin. Pharmacy 1, 417-424.

Klein, D.F. & Fink, M. (1962). Psychiatric reaction patterns to imipramine. Am. J. Psychiatry 119, 432-438.

Klerman, G.L., & Cole, J.O. (1965). Clinical pharmacology of imipramine and related antidepressant compounds. Pharmacol. Rev. 17, 101-141.

Kline, A. (1967). Drug treatment of phobic disorders. Am. J. Psychiatry 123, 1447-1450.

Kline, N.S. (1958). Clinical experience with Iproniazid (Marsilid). J. Clin. Exp. Psychopath. 19, 72-79.

Knoll, J. & Magyar, K. (1972). Some puzzling pharmacological effects of monoamine oxidase inhibitors. Adv. Biochem. Psychopharmacol. 5, 393-408.

Kramer, P.D. (1993). Listening to Prozac. A psychiatrist explores antidepressant drugs and the remaking of the self. Viking Penguin, New York.

Kramer, M.S., Cutler, N.R., Ballenger, J.C., Patterson, W.M., Mendels, J., Chenault, A., Shrivastava, R., Matzura-Wolfe, D., Lines, C., & Reines, S. (1995). A placebo controlled trial of L-365,260, a CCK-B antagonist, in panic disorder. Biol. Psychiatry 37, 462-466.

Kuhn, R. (1958). The treatment of depressive states with G22355 (imipramine hydrochloride). Am. J. Psychiatry 115, 459-464.

Kupfer, D.J. & Detre, T.P. (1978). Tricyclic and monoamine oxidase-inhibitor antidepressants; Clinical use. In: Handbook of Psychopharmacology, Volume 14. Affective Disorders: Drug Actions in Animals and Man (Iversen, L.L., Iversen, S.D. & Snyder, S.H., Eds.), pp.199-232. Plenum Press, New York.

Kurtz, N.M. (1990). Monoamine oxidase inhibiting drugs. In: Pharmacotherapy of Depression—Applications for the out-patient practitioner. (Amsterdam, J.D., Ed.), pp. 93-109. Marcel Dekker, New York.

Kurtz, N.M., Keppel-Hesselink, J., Beneke, M., & Heller, A.H. (1990). Preliminary studies of $5HT_{1A}$ agonists. Clin. Neuropharmacol. 13 (Suppl. 2), 95-96.

Lader, M. (1987). Assessing the potential for buspirone dependence or abuse and effects of its withdrawal. Am. J. Medicine 82 (Suppl. 5A), 20-26.

Lancaster, S.G. & Gonzalez, J.P. (1989). Lofepramine. Drugs 37, 123-140.

Lane, R., Baldwin, D., & Preskorn, S. (1995). The SSRIs: Advantages, disadvantages and differences. J. Psychopharmacol. 9, 163-178.

Lanfumey, L., Haj-Dahmane, S., & Hamon, M. (1993). Further assessment of the antagonist properties of the novel and selective 5-HT_{1A} receptor ligands (+)-WAY 100 135 and SDZ 216-525. Eur. J. Pharmacol. 249, 25-35.

Lapin, I.P. & Oxenkrug, G.F. (1969). Intensification of the central serotoninergic processes as a possible determinant of the thymoleptic effect. Lancet i, 132-136.

Laux, G., Becker, T., Classen, W., Lesch, K.-P., Sofic, E., Carl, G., Struck, M., Riederer, P, & Beckmann, H. (1989). Clinical and biochemical findings with new MAO-A-inhibitors in major depressive disorder. In: Psychiatry Today. Accomplishments and Promises (Stefanis, C.N., Soldatos, C.R., & Rabavilas, A.D., Eds.), Excerpta Medica, Int. Congr. Ser. 899 (Abstract, 1386.)

Le Bricquir, Y., Larrey, D., Blanc, P., Pageaux, G.P., & Michel, H. (1994). Tianeptine—an instance of drug-induced hepatotoxicity predicted by prospective experimental studies. J. Hepatology 21, 771-773.

Lecrubier, Y. & Guelfi, J.D. (1990). Efficacy of reversible inhibitors of monoamine oxidase-A in various forms of depression. Acta Psych. Scand. 360, S18-S23.

Lecrubier, Y., Puech, A.J., Azcona, A., Bailey, P.E., & Lataste, X. (1993). A randomized double-blind placebo-controlled study of tropisetron in the treatment of outpatients with generalized anxiety disorder. Psychopharmacology 112, 129-133.

Leigh, T.J., Link, C.G.G., & Fell, G.L. (1991). Effects of granisetron and lorazepam, alone and in combination, on psychometric performance. Br. J. Clin. Pharmacol. 31, 333-336.

L' Heureux, R., Dennis, T., Curet, O., & Scatton, B. (1986). Measurement of endogenous noradrenaline release in the rat cerebral cortex *in vivo* by transcortical dialysis: Effects of drugs affecting noradrenergic transmission. J. Neurochem. 46, 1794-1801.

Leonard, B.E. (1986). Toxicity of antidepressants. Lancet ii, 1105.

Liebowitz, M.R., Fyer, A.J., Gorman, J.M., Campeas, R., & Levin, A. (1986). Phenelzine in social phobia. J. Clin. Psychopharmacol. 6, 93-98.

Liebowitz, M.R., Karoum, F., Quitkin, F.M., Davies, S.O., Schwartz, D., Levitt, M., & Linnoila, M. (1985). Biochemical effects of *L*-deprenyl in atypical depressives. Biol. Psychiatry 20, 558-565.

Liebowitz, M.R., Quitkin, F.M., Stewart, J.W., McGrath, P.J., Harrison, W.M., Markowitz, J.S., Rabkin, J.G., Tricamo, E., Goetz, D., & Klein, D.F. (1988). Antidepressant specificity in atypical depression. Arch. Gen. Psychiatry 45, 129-139.

Lingl, F.A. (1964). Combined drug therapy compared with electric shock in psychiatric depression. Am. J. Psychiatry 120, 808-810.

Link, C.G.G., Leigh, T.J., & Dennison, J.K. (1993). The effects of BRL 46470A, a novel $5HT_3$ receptor antagonist, and lorazepam on psychometric performance and the EEG. Br. J. Clin. Pharmacol. 35, 395-399.

Loomer, H.P., Saunders, J.C., & Kline, N.S. (1957). A clinical and pharmacodynamic evaluation of Iproniazid as a psychic energizer. Psychiatric Res. Rep. Amer. Psychiatric Ass. 8, 129-141.

Lopez-Ibor, J., Guelfi, J.D., Pletan, Y., Tournoux, A. & Prost, J.F. (1996). Milnacipran and selective serotonin reuptake inhibitors in major depression. Int. Clin. Psychopharmacol. 11, 41-46.

Lydiard, R.B., Ballenger, J.C., & Rickels, K. (1997). A double-blind evaluation of the safety and efficacy of abecarnil, alprazolam, and placebo in outpatients with generalized anxiety disorder. J. Clin. Psychiatry 58 (Suppl. 11), 11-18.

Mann, J.J., Aarons, S.F., Wilner, P.J., Keilp, J.G., Sweeney, J.A., Pearlstein, T., Frances, A.J., Kocsis, J.H., & Brown, R.P. (1989). A controlled study of the antidepressant efficacy and side-effects of (-)-deprenyl. Arch. Gen. Psychiatry 46, 45-50.

Mann, J. & Gershon, S. (1980). *L*-Deprenyl, a selective monoamine oxidase type B inhibitor in endogenous depression. Life Sci. 26, 877-882.

Mathur, G.N. (1975). A double-blind comparative clinical trial of maprotiline (Ludiomil) and amitriptyline. J. Int. Med. Res. 3 (Suppl. 2), 71-74.

McGrath, P.J., Stewart, J.W., Nunes, E.V., Ocepek-Welikson, K., Rabkin, J.G., Quitkin, F.M., & Klein, D.F. (1993). A double-blind crossover trial of imipramine and phenelzine for outpatients with treatment-refractory depression. Am. J. Psychiatry 150, 118-123.

McNeil, R.G., Gee, K.W., Bolger, M.B., Lan, N.C., Weiland, S., Belelli, D., Purdy, R.H., & Paul, S.M. (1992). Neuroactive steroids that act at $GABA_A$ receptors. Drug News Persp. 5, 145-152.

Mendels, J., Johnston, R., Mattes, J., & Riesenberg, R. (1993). Efficacy and safety of b.i.d. doses of venlafaxine in a dose-response study. Psychopharmacol. Bull. 29, 169-174.

Mendels, J., Reimherr, F., Roberts, D., Ecker, J., Marcus, R., Schwiderski, U., & Robinson, D. (1992). A double-blind comparison of high dose nefazodone, low dose nefazodone, and placebo in the treatment of depressed outpatients. Eur. Neuropsychopharmacol. 1, 451-452.

Mendis, N., Pare, C.M.B., Sandler, M., Glover, V., & Stern. G. (1981). (-)-Deprenyl in treatment of depression. In: Monoamine Oxidase Inhibition—The State of the Art. (Youdim, M.B.H. & Paykel, E.S., Eds.), pp. 171-176. John Wiley & Sons Ltd, Chichester.

Meredith, T.J. & Vale, J.A. (1985). Poisoning due to psychotropic agents. Adverse Drug React. Acute Poisoning Rev. 4, 83-126.

Middleton, R.S.W. (1975). A comparison between maprotiline (Ludiomil) and imipramine in the treatment of depressive illness in the elderly. J. Int. Med. Res. 3 (Suppl. 2), 7983.

Middlemiss, D.N., Blakeborough, L., & Leather, S.R. (1977). Direct evidence for an interaction of β-adrenergic blockers with the 5-HT receptors. Nature 267, 289-290.

Montgomery, S.A. (1991). Clinical significance of 5-HT reuptake inhibitors. Hum. Psychopharmacol. 6 (Suppl. 3), S3.

Montgomery, S.A. (1993). Venlafaxine : A new dimension in antidepressant pharmacotherapy. J. Clin. Psychiatry 54, 119-126.

Montgomery, S.A. (1997). Reboxetine: Additional benefits to the depressed patients. J. Psychopharmacol. 11, S9-S15.

Montgomery, S.A. & Fineberg, N. (1989). Is there a relationship between serotonin receptor subtypes and selectivity of response in specific psychiatric illness? Br. J. Psychiatry 155 (Suppl. 8), 63-70.

Montgomery, S.A., Prost, J.F., Solles, A., & Briley, M. (1996). Efficacy and tolerability of milnacipran: An overview. Int. Clin. Psychopharmacol. 11, 47-57.

Moret, C., Charveron, M., Finberg, J.P., Couzinier, J.P., & Briley, M. (1985). Biochemical profile of midalcipran (F 2207), 1-phenyl-1-diethyl-aminocarbonyl-2-aminomethyl-cyclopropane (Z) hydrochloride, a potential fourth generation antidepressant drug. Neuropharmacology 24, 1211-1219.

Morris, J.B. & Beck, A.T. (1974). The efficacy of antidepressant drugs: A review of research (1958-1972). Arch. Gen. Psychiatry 30, 667-674.

Mucci, M. (1997). Reboxetine: A review of antidepressant tolerability. J. Psychopharmacol. 11, S33-S37.

Muller, J.C., Pryor, W.W., Gibbons, J.E., & Orgain, E.S. (1955). Depression and anxiety occurring during Rauwolfia therapy. JAMA 159, 836-839.

Murasaki, M., Mori, A., and Kudo, Y. (1993). Efficacy of a new anxiolytic, tandospirone (SM3997) on neurosis—a comparative double-blind study with diazepam. Neuropsychopharmacology 9, S2.

Murdoch, D. & McTavish, D. (1992). Sertraline: A review of its pharmacodynamic and pharmacokinetic properties and therapeutic potential in depression and obsessive-compulsive disorder. Drugs 44, 604-624.

Murphy, D.L., Aulakh, C.S., Garrick, N.A., & Sunderland, T. (1987). Monoamine oxidase inhibitors as antidepressants: Implications for the mechanism of action of antidepressants and the psychobiology of the affective disorders and some related disorders. In: Psychopharmacology: The Third Generation of Progress. (Meltzer H.Y., Ed.), New York, Raven Press.

Murphy, S.M., Owen, R., & Tyrer, P. (1989). Comparative assessment of efficacy and withdrawal symptoms after 6 and 12 weeks treatment with diazepam or buspirone. Br. J. Psychiatry 154, 529-534.

Muth, E.A., Haskins, J.T., Moyer, J.A., Husbands, G.E.M., Nielson, S.T., & Sigg, E.B. (1986). Antidepressant biochemical profile of the novel bicyclic compound Wy-45.030, an ethyl cyclohexanol derivative. Biochem. Pharmacol. 35, 4493-4497.

Myrenfors, P.G., Eriksson, T., Sandsted, C.S., & Sjoberg, G. (1993). Moclobemide overdose. J. Intern. Med. 233, 113-115.

Nolen, W.A. & Haffmans, J. (1989). Treatment resistant depression: Review on the efficacy of various biological treatments, specifically in major depression resistant to cyclic antidepressants. Int. Clin. Psychopharmacol. 4, 217-228.

Nutt, D.J. (1990). The pharmacology of human anxiety. Pharmacol. Ther. 47, 233-266.

Nutt, D.J. & Glue, P. (1989). Clinical pharmacology of anxiolytics and antidepressants: A psychopharmacological perspective. Pharmacol. Ther. 44, 309-334.

Oehrberg, S., Christiansen, P.E., Behnke, K., Borup, A.L., Severin, B., Soegaard, J., Calberg, H., Judge, R., Ohrstrom, J.K., & Manniche, P.M. (1995). Paroxetine in the treatment of panic disorder. A randomized, double-blind, placebo-controlled study. Br. J. Psychiatry 167, 374-379.

Olivier, B., Mos, J., Van der Heyden, J., Van der Poel, G., Tulp, M., Slangen, J., & de Jonge, R. (1992). Preclinical evidence for the anxiolytic activity of 5-HT_3 receptor antagonists: A review. Stress Medicine 8, 117-136.

Olsen, R.W. & Tobin, A.J. (1990). Molecular biology of $GABA_A$ receptors. FASEB J. 4, 1469-1480.

Pande, A.C., Greiner, M., Bammert-Adams, J., & Pierce, M.W. (1996). CI-988, a cholecystokinin-B antagonist, in panic disorder. Biol. Psychiatry 39, 619-620.

Paykel, E.S. & Hale, A.S. (1985). Recent advances in the treatment of depression. In: The Biology of Depression. (Deakin, J.F.W., Ed.), pp. 153-173. Gaskell, London.

Pazos, A. & Palacios, J.M. (1985). Quantitative autoradiographic mapping of serotonin receptors in the rat brain. I. Serotonin-1 receptors. Brain Res. 346, 205-230.

Peck, A.W., Stern, W.C., & Watkinson, C. (1983). Incidence of seizures during treatment with tricyclic antidepressant drugs and bupropion. J. Clin. Psychiatry 44, 197-201.

Pecknold, J.C., Familamiri, P., Chang, H., Wilson, R., Alarcia, J., & McClure, D.J. (1985). Buspirone: anxiolytic? Prog. Neuropsychopharmacol. Biol. Psychiatry 9, 639-642.

Peet, M. (1988). Lofepramine in clinical practice. Int. Clin. Psychopharmacol. 3 (Suppl. 2), 39-48.

Pérez, V., Gilaberte, I., Faries, D., Alvarez, E., & Artigas, F. (1997). Randomized, double blind placebo-controlled trial of pindolol in combination with fluoxetine antidepressant treatment. The Lancet 349, 1594-1597.

Pletscher, A., Shore, P.A., & Brodie, B.B. (1956). Serotonin as a mediator of reserpine action in brain. J. Pharmacol. 116, 84-89.

Pohl, R., Balon, R., Yeragani, V.K., & Gershon, S. (1989). Serotonergic anxiolytics in the treatment of panic disorder: A controlled study with buspirone. Psychopathology 22, 60-67.

Pollack, M.H., Worthington, J.J., Gus Mantro, G., Otto, M.W., & Zucker, B.G. (1997). Abecarnil for the treatment of generalized anxiety disorder: A placebo-controlled comparison of two dosage ranges of abecarnil and buspirone. J. Clin. Psychiatry 58 (Suppl. 11), 19-23.

Pollard, G.T. & Howard, J.L. (1991). Cork gnawing in the rat as a screening method for buspirone-like anxiolytics. Drug Dev. Res. 22, 179-187.

Pols, H., Zandbergen, J., Lousberg, H., de Loof, C. and Griez, E. (1991). Low doses of clonazepam in the treatment of panic disorder. Can. J. Psychiatry 36, 302-303.

Puech, A., Montgomery, S.A., Prost, J.F., Solles, A., & Briley, M. (1997). Milnacipran, a new serotonin and noradrenaline reuptake inhibitor: An overview of its antidepressant activity and clinical tolerability. Int. Clin. Psychopharmacol. 12, 99-108.

Quitkin, F.M., Harrison, W., Stewart, J.W., McGrath, P.J., Tricamo, E., Ocepek-Welikson, K., Rabkin, J.G., Wager, S.G., Nunes, E., & Klein, D.F. (1991). Response to phenelzine and imipramine in placebo nonresponders with atypical depression: A new application of the crossover design. Arch. Gen. Psychiatry 48, 319-323.

Quitkin, F.M., Liebowitz, M.R., Stewart, J.W. McGrath, P.J., Harrison, W., Rabkin, J.G., Markowitz, J., & Davies, S.O. (1984). *L*-Deprenyl in atypical depressives. Arch. Gen. Psychiatry 41, 777-781.

Rainbow, T.C., Parsons, B., & Wolfe, B.B. (1985). Quantitative autoradiography of β_1- and β_2-adrenergic receptors in rat brain. Proc. Natl. Acad. Sci. 81, 1585-1589.

Rausch, J.L., Ruegg, R., & Moeller, F.G. (1990). Gepirone as a 5-HT_{1A} agonist in the treatment of major depression. Psychopharmacol. Bull. 26, 169-171.

Rehfeld, J.F. (1985). Neuronal cholecystokinin: One or multiple transmitters. J. Neurochem. 44, 1-10.

Riblet, L.A., Eison, A.S., Eison, M.S., Taylor, D.P., Temple, D.L., & VanderMaelen, C.P. (1984). Neuropharmacology of buspirone. Psychopathology 17, 69-79.

Riblet, L.L., Taylor, D.P., & Eison, M.S. (1982). Pharmacology and neurochemistry of buspirone. J. Clin. Psychiatry 43, 11-16.

Richelson, E. & Nelson, A. (1984). Antagonism by antidepressants of neurotransmitter receptors of normal human brain *in vitro*. J. Pharmacol. Exp. Ther. 230, 94-102.

Richelson, E. & Pfenning, M. (1984). Blockade by antidepressants and related compounds of biogenic amine uptake into rat brain synaptosomes: Most antidepressants selectively block norepinephrine uptake. Eur. J. Pharmacol. 104, 277-286.

Rickels, K. (1991). Venlafaxine: A new potent antidepressant agent with putative fast onset of action. In: Biological Psychiatry. Volume 1 (Racagni, G., Brunello, N., & Fukuda, T., Eds.) Elsevier Science Publishers, Amsterdam.

Rickels, K. & Schweizer, E. (1990). Clinical overview of serotonin reuptake inhibitors. J. Clin. Psychiatry 51 (12 Suppl. B), 9-12.

Rickels, K., Schweizer, E., Clary, C., Fox, I., & Weise, C. (1994). Nefazodone and imipramine in major depression: A placebo-controlled trial. Br. J. Psychiatry 164, 802-805.

Rickels, K., Schweizer, E., Csanalosi, I., Case, G., & Chung, H. (1988). Long-term treatment of anxiety and risk of withdrawal. Arch. Gen. Psychiatry 45, 444-450.

Rickels, K., Schweizer, E., DeMartinis, N., Mandos, L., & Mercer, C. (1997). Gepirone and diazepam in generalized anxiety disorder: A placebo-controlled trial. J. Clin. Psychopharmacol. 17, 272-277.

Rickels, K., Weise, C.C., Zal, H.M., Csanalosi, I., & Werblowsky, J. (1982a). Lofepramine and imipramine in unipolar depressed out-patients : A placebo controlled study. Acta Psych. Scand. 66, 109-120.

Rickels, K., Weisman, K., Norstard, N., Singer, M., Stolz, D., Brown, A., & Danton, J. (1982b). Buspirone and diazepam in anxiety: A controlled study. J. Clin. Psychiatry 43, 81-86.

Rifflet, H., Vuillemin, E., Rifflet, I., Oberti, F., Pilette, C., & Calés P. (1996). Atteinte hépatique aiguë douloureuse et fébrille associée à la prise de tianeptine. Gastroenterol. Clin. Biol. 20, 606-607.

Riva, M., Brunello, N., Rovescalli, A.C., Galimerti, R., Carfagna, N., Carminati, P., Pozzi, O., Ricciardi, S., Roncucci, R., Rossi, A., & Racagni, G. (1989). Effect of reboxetine, a new antidepressant, on the central noradrenergic system: Behavioural and biochemical studies. J. Drug. Dev. 1, 243-253.

Roberts, C., Price, G.W., & Jones, B.J. (1997). The role of 5-$HT_{1B/1D}$ receptors in the modulation of 5-hydroxytryptamine levels in the frontal cortex of the conscious guinea pig. Eur. J. Pharmacol. 326, 23-30.

Robinson, D.S., Rickels, K., Feighner, J., Fabre, L.F., Gammans, R.E., Shrotriya, R.C., Alms, D.R., Andary, J.J., & Messina, M.E. (1990). Clinical effects of the 5-HT_{1A} partial agonists in depression: A composite analysis of buspirone in the treatment of depression. J. Clin. Psychopharmacol. 10, S67-S76.

Rollema, H., Clarke, T., Sprouse, J.S., & Schule, D.W. (1996). Combined administration of a 5-hydroxytryptamine $(5\text{-HT})_{1D}$ antagonist and a 5-HT reuptake inhibitor synergistically increases 5-HT release in guinea-pig hypothalamus *in vivo*. J. Neurochem. 67, 2204-2207.

Roose, S.P., Glassman, A.H., Giardina, E.G., Johnson, L.L., Walsh, B.T., & Bigger, J.T., Jr. (1987). Cardiovascular effects of imipramine and bupropion in depressed patients with congestive heart failure. J. Clin. Pharmacol. 7, 247-251.

Romero, L., Bel, N., Artigas, F., De Montigny, C., & Blier, P. (1996). Effect of pindolol on the function of pre- and postsynaptic 5-HT_{1A} receptors: *in vivo* microdialysis and electrophysiological studies in the rat brain. Neuropsychopharmacology 15, 349-360.

Roth, J.A. & Gillis, C.N. (1975). Inhibition of rabbit mitochondrial monoamine oxidase by iprindole. Biochem. Pharmacol. 24, 151-152.

Rudolph, R., Entsuah, R., & Derivan, A. (1991). Early clinical response in depression to venlafaxine hydrochloride. Biol. Psychiatry 29, S630.

Rudorfer, M.V. & Potter, W.Z. (1989). Antidepressants. A comparative review of the clinical pharmacology and therapeutic use of the "newer" versus "older" drugs. Drugs 37, 713-738.

Samuelian, J.C., Tatossian, A., & Hackett, D. (1992). A randomized, double-blind, parallel group comparison of venlafaxine and clomipramine in outpatients with major depression. Clin. Neuropharmacol. 15 (Suppl. 1), B324.

Sargant, W. (1962). The treatment of anxiety states and atypical depressions by the monoamine oxidase inhibitor drugs. J. Neuropsychiatry 3 (Suppl. 1), 97-103.

Scher, M., Kreiger, J.N., & Jurgens, S. (1983). Trazodone and priapism. Am. J. Psychiatry 140, 1362-1363.

Schildkraut, J.J. (1965). The catecholamine hypothesis of affective disorders: A review of supporting evidence. Am. J. Psychiatry 122, 509-522.

Schipper, J., Tulp, M.Th.M. Berkelmans, B., Mos, J., Van der Heijden, J.A.M., & Olivier, B. (1991). Preclinical pharmacology of flesinoxan: A potential anxiolytic and antidepressant drug. Human Psychopharmacol. 6, S53-S61.

Schmidt, A.W. & Peroutka, S.J. (1989). Antidepressant interactions with 5-hydroxytryptamine receptor binding sites. Eur. J. Pharmacol. 163, 397-398.

Schreiber, R. & De Vry J. (1993). 5-HT_{1A} receptor ligands in animal models of anxiety, impulsivity, and depression: Multiple mechanisms of action? Prog. Neuropsychopharmacol. Biol. Psychiatry 17, 87-104.

Schweizer, E., Feighner, J., Mandos, L.A., & Rickels, K. (1994). Comparison of venlafaxine and imipramine in the acute treatment of depression in outpatients. J. Clin. Psychiatry 55, 104-108.

Schweizer, E. & Rickels, K. (1991). Serotonergic anxiolytics: A review of their clinical efficacy. In: 5-HT_{1A} agonists, 5-HT_3 antagonists, and benzodiazepines: Their comparative behavioural pharmacology. (Rodgers, R.J. & Cooper S.J., Eds.), pp. 365-376. John Wiley & Sons, Chichester.

Schweizer, E., Rickels, K., & Lucki, I. (1986). Resistance to the anti-anxiety effects of buspirone in patients with a history of benzodiazepine use. N. Engl. J. Med. 314, 719-720.

Schweizer, E., Weise, C., Clary, C., Fox, I., & Rickels, K. (1991). Placebo-controlled trial of venlafaxine for the treatment of major depression. J. Clin. Psychopharmacol. 11, 233-236.

Selikoff, I.J., Robitzek, E.H., & Ornstein, G.G. (1952). Toxicity of hydrazine derivatives of isonicotinic acid in the chemotherapy of human tuberculosis. Quart. Bull. Sea View Hosp. 13, 17-26.

Shader, R.I. & Greenblatt, D.J. (1993). Use of benzodiazepines in anxiety disorders. N. Engl. J. Med. 328, 1398-1405.

Sheehan, D.V. (1987). Benzodiazepines in panic disorder and agoraphobia. J. Affect. Disorder 13, 169-181.

Sheehan, D.V., Raj, A.B., Sheehan, K.H., & Soto, S. (1990). Is buspirone effective for panic disorder? J. Clin. Psychopharmacol. 10, 3-11.

Shein, K. & Smith, S.E. (1978). Structure activity relationships for the anticholinoceptor action of tricyclic antidepressants. Br. J. Pharmacol. 62, 567-571.

Simpson, G.M. & de Leon, J. (1989). Tyramine and new monoamine oxidase inhibitor drugs. Br. J. Psychiatry 155, 32-37.

Skowron, D.M. & Stimmel, G.L. (1992). Antidepressants and the risk of seizures. Pharmacotherapy 12, 18-22.

Small. G.W. & Bystritsky, A., (1997). Double-blind, placebo controlled trial of two doses of abecarnil for geriatric anxiety. J. Clin. Psychiatry 58 (Suppl. 11), 24-29.

Smith, W.T., Glaudin, V., Panagides, J. & Gilvary, E. (1990). Mirtazapine vs. amitriptyline vs. placebo in the treatment of major depressive disorder. Psychopharmacol. Bull. 26, 191-196.

Sotelo, C., Cholley, B., El Mestikawy, S., Gozlan, H., & Hamon, M. (1990). Direct immunohistochemical evidence of the existence of 5-HT_{1A} autoreceptors on serotonergic neurons in the midbrain raphe nuclei. Eur. J. Neurosci. 2, 1144-1154.

Staner, L., Bertolino, A., Cassano, G.B., De Wilde, J., Ferreira. A., Kasas, A., Mendlewicz, J., Mertens, C., Paes de Sousa, M., Sarteschi, P., & Van H. Loo. H.A. (1994). European multicentre study of tianeptine versus imipramine and placebo in the treatment of major depression and depressed bipolar disorder. Eur. Psychiatry 9, 140s.

Stephens, D.N., Schneider, H.H., Kehr, W., Andrews, J.S., Rettig, K.-J., Turski, L., Schmiechen, R., Turner, J.D., Jensen, L.H., Petersen, E.N., Honore, T., & Bondo Hansen, J. (1990). Abecarnil, a metabolically stable, anxioselective β-carboline acting at benzodiazepine receptors. J. Pharmacol. Exp. Ther. 253, 334-343.

Stephenson, F.A. (1990). The $GABA_A$ receptors: Structure and function. In: Current Aspects of the Neurosciences. (Osborne, N.N., Ed.), Vol. 3, pp. 177-193. MacMillan Press, London.

Stephenson, F.A., Duggan, M.J., Pollard, S., & Thompson, C.L. (1992). GABA receptor composition and function in relation to drug action. In: Experimental Approaches to Anxiety and Depression. (Elliott, J.M., Heal, D.J., & Marsden, C.A., Eds.), pp. 169-182. John Wiley & Sons Ltd, Chichester.

Steppuhn, K.G., Schneider, H.H., Turski, L., & Stephens, D.N. (1993). Long-term treatment with abecarnil does not induce diazepam-like dependence in mice. J. Pharmacol. Exp. Ther. 264, 1395-1400.

Sternbach, L.H. & Reeder, E. (1961a). Quinazolines and 1,4-benzodiazepines. II. The rearrangement of 6-chloro-2-chloromethyl-4-phenylquinazoline 3-oxidase into 2-amino derivatives of 7-chloro-5-phenyl-3H-1,4-benzodiazepine 4-oxide. J. Org. Chem. 26, 1111-1118.

Sternbach, L.H. & Reeder, E. (1961b). Quinazolines and 1,4-benzodiazepines. IV. Transformations of 7-chloro-2-methyl-amino-5-phenyl-3H-1,4-benzodiazepine 4-oxide. J. Org. Chem. 26, 4936-4941.

Tang, S.W. & Seeman, P. (1980). Effect of antidepressant drugs on serotonergic and adrenergic receptors. Naunyn-Schmiedeberg's Arch. Pharmacol. 311, 255-261.

Taylor, D.P., Eison, M.S., Riblet, L.A., & VanderMaelen, C.P. (1985). Pharmacology and clinical effects of buspirone. Pharmacol. Biochem. Behav. 23, 687-694.

Tesar, G.E, Rosenbaum, J.F., Pollack, M.K., Otto, M.W., Sachs, G.S., Herman, J.B., Cohen, L.S., & Spier, S.A. (1991). Double-blind, placebo-controlled comparison of clonazepam and alprazolam for panic disorder. J. Clin. Psychiatry 52, 69-76.

Thase, M.E., Frank, E., Mallinger, A.G., Hamer, T., & Kupfer. D.J. (1992). Treatment of imipramine-resistant recurrent depression. III: Efficacy of monoamine oxidase inhibitors. J. Clin. Psychiatry 53, 5-11.

Thornton, J.E. & Stahl, S.M. (1984). Case report of tardive dyskinesia and parkinsonism associated with amoxapine therapy. Am. J. Psychiatry 141, 704-705.

Tollefson, G.D. (1991). Antidepressant treatment and side effect considerations. J. Clin. Psychiatry 52, S4-S13.

Tome, M.B., Isaac, M.T., Harte, R., & Holland, C. (1997). Paroxetine and pindolol: A randomized trial of serotonergic autoreceptor blockade in the reduction of antidepressant latency. Int. Clin. Psychopharmacol. 12, 81-89.

Tyrer, P. (1979). Clinical use of monoamine oxidase inhibitors. In: Psychopharmacology of Affective Disorders. (Paykel, E.S. & Coppen, A., Eds.), pp.159-178. Oxford University Press, Oxford.

Tyrer, P. (1985). Neurosis divisible? Lancet i, 685-688.

Tyrer, P. (1992). Anxiety and depression: A clinical profile. In: Experimental Approaches to Anxiety and Depression. (Elliott, J.M., Heal, D.J., & Marsden, C.A., Eds.), pp. 9-23. John Wiley & Sons Ltd., Chichester.

Van Ameringen, M., Mancini, C., & Streiner, D.L. (1993). Fluoxetine efficacy in social phobia. J. Clin. Psychiatry 54, 27-32.

Vanderhaegen, J.-J., Lofstra, F., Vierendreels, G., Gilles, C., Deschepper, C., & Verbauck, P. (1981). Cholecystokinin in the central nervous system and neurohypophysis. Peptides 2, 81-88.

Van Megen, H.J.G.M., Westenberg, H.G.M., Den Boer, J.A., Haigh, J.R.M., & Traub, M. (1994). Pentagastrin induced panic attacks: Enhanced sensitivity in panic disorder patients. Psychopharmacology 114, 449-455.

Van Moffaert, M., de Wilde, J., Vereecken, A., Dierick, M., Evrard, J.L., Wilmotte, J. & Mendlewicz, J. (1995). Mirtazapine is more effective than trazodone: A double-blind controlled study in hospitalized patients with major depression. Int. Clin. Psychopharmacol. 10, 3-9.

Van Moffaert, M., Vanbrabant, E., von Frenckell, R., Kensler, T., & Roberts, D. (1992). A double-blind comparison of nefazodone and imipramine in the treatment of depressed patients. Eur. Neuropsychopharmacol. 2, 296-297.

Van Veggel, L.M.A., Preston, C. & O'Hanlon, J.F. (1994). Ondansetron and diazepam effects on actual driving and psychomotor performance. Neuropsychopharmacology 10, O83-O104.

Van Vliet, I.M., den Boer, J.A., & Westenberg, H.G.M. (1994). Psychopharmacological treatment of social phobia: A double blind placebo controlled study with fluvoxamine. Psychopharmacology 115, 128-134.

Van Vliet, I.M., Westenberg, H.G.M., & den Boer, J.A. (1993). MAO inhibitors in panic disorder: Clinical effects of treatment with brofaromine. A double blind placebo controlled study. Psychopharmacology 112, 483-489.

Van Vliet, I.M., Westenberg, H.G.M., & den Boer, J.A. (1996). Effects of the 5-HT_{1A} receptor agonist flesinoxan in panic disorder. Psychopharmacology 124, 174-180.

Van Wyck Fleet, J., Manberg, P.J., Miller, L.L., Harto-Truax, N., Sato, T., Fleck, R.J., Stern, W.C., & Cato, A.E. (1983). Overview of clinically significant adverse reactions to bupropion. J. Clin. Psychiatry 44, 191-196.

Vergé, D., Daval, G., Patey, A., Gozlan, H., El Mestikawy, S., & Hamon, M. (1985). Presynaptic 5-HT autoreceptors on serotoninergic cell bodies and/or dendrites, but not terminals, are of the 5-HT_{1A} subtype. Eur. J. Pharmacol. 113, 463-464.

Versiani, M., Nardi, A.E., Figueria, I.L.V., & Stabl, M. (1990). Tolerability of moclobemide, a new reversible inhibitor of monoamine oxidase-A, compared with other antidepressants and placebo. Acta Psych. Scand. 360, S24-S28.

Versiani, M., Nardi, A.E., Mundim, F.D., Alves, A.B., Liebowitz, M.R., & Amrein, R. (1992). Pharmacotherapy of social phobia. Br. J. Psychiatry 161, 353-360.

Versiani, M., Oggero, U., Alterwain, P., Capponi, R., Dajas, F., Heinze-Martine, G., Marquez, C.A., Poleo, M.A., Rivero-Almanzor, L.E., Rossel, L., Schmid-Burgk, W., & Ucha Udabe, R. (1989). A double blind comparative trial of moclobemide v. imipramine and placebo in major depressive episodes. Br. J. Psychiatry 155, 72-77.

Vinar, O., Klein, D.F., Potter, W.Z., & Gause, E.M. (1991). A survey of psychotropic medications not available in the United States. Neuropsychopharmacology 5, 201-217.

Wagner, W., Plekkenpol, B., Gray, T.E., Vlaskamp, H., & Essers, H. (1992). Review of fluvoxamine safety database. Drugs 43, 48-54.

Wander, T.J., Nelson, A., Okazaki, H., & Richelson, E. (1986). Antagonism by antidepressants of serotonin S_1 and S_2 receptors of normal human brain *in vitro*. Eur. J. Pharmacol. 132, 115-121.

Westenberg, H.G.M. & den Boer, J.A. (1993). New findings in the treatment of panic disorder. Pharmacopsychiatry 26, 30-33.

Westenberg, H.G.M., van Vliet, I.M., & Den Boer, J.A. (1992). Flesinoxan, a selective $5\text{-}HT_{1A}$ agonist, in the treatment of panic disorder. Clin. Neuropharmacol. 15 (1B), B60.

Wilde, M.I. & Benfield, P. (1995). Tianeptine. A review of its pharmacodynamic and pharmacokinetic properties, and therapeutic efficacy in depression and coexisting anxiety and depression. Drugs 49, 411-439.

Wong, D.T., Bymaster, F.P., Reid, L.R., & Threlkeld, P.G. (1983). Fluoxetine and two other serotonin uptake inhibitors without affinity for neuronal receptors. Biochem. Pharmacol. 32, 1287-1293.

Woods, J.H., Katz, J.L., & Winger G. (1987). Abuse liability of benzodiazepines. Pharmacol. Res. 39, 251-419.

Workman, E.A. & Short, D.D. (1993). Atypical antidepressants versus imipramine in the treatment of major depression: A meta-analysis. J. Clin. Psychiatry 54, 5-12.

Yocca, F.D. (1990). Novel anxiolytic agents: Actions on specific subtypes of central 5-HT receptors. In: Current and Future Trends in Anticonvulsant, Anxiety, and Stroke Therapy. (Meldrum, B.S. & Williams, M., Eds.), pp. 145-177. Wiley-Liss Inc.

Youdim, M.B.H. & Finberg, J.P.M. (1985). Monoamine oxidase inhibitor antidepressants. In: Psychopharmacology 2. Part I. Preclinical Psychopharmacology. (Grahame-Smith, D.G. & Cowen, P.J., Eds.), pp. 35-70. Excerpta Medica, Amsterdam.

Zeller, E.A., Barsky, J., Fouts, J.R., Kirshheimer, W.F., & Van Orden, L.S. (1952). Influence of isonicotinic acid hydrazide (INH) and isonicotinoyl-2-isopropylhydrazine (IIH) on bacterial and mammalian enzymes. Experientia 8, 349-350.

Zis, P. & Goodwin, F.K. (1979). Novel antidepressants and the biogenic amine hypothesis of depression. Arch. Gen. Psychiatry 36, 1097-1107.

Zitrin, C.M., Klein, D.F., Woerner, M.G., & Ross, D.C. (1983). Treatment of phobias. I. Comparison of imipramine hydrochloride and placebo. Arch. Gen. Psychiatry 40, 125-138.

Zivkov, M. & De Jongh, G.D. (1995). Org 3770 versus amitriptyline: A 6-week randomized double-blind multicentre trial in hospitalized depressed patients. Hum. Psychopharmacology: Clin. Exp. 10, 173-180.

Chapter 24

Lithium Use in Clinical Practice

DAVID L. DUNNER and JOHN F. NEUMAIER

Biological Psychiatry, pages 569-583.

ISBN: 1-55938-819-6

INTRODUCTION

The purpose of this chapter is to discuss the use of lithium in medical practice and particularly in psychiatry. The use of lithium salts for the treatment of bipolar disorder (manic-depressive illness) has resulted in a marked change in thinking about psychiatric conditions in terms of their etiology and has been an instrumental factor in changing the diagnostic scheme used in the United States. Indeed, one might parallel the history of modern psychiatry with the use of lithium in psychiatric practice. This chapter will survey the history of lithium, its medical uses in the 1940s, the development of its use in bipolar disorder, early European studies regarding efficacy in acute mania, and the delay in utilization of lithium in psychiatric practice in the United States. In addition, we will discuss studies demonstrating both acute and maintenance effects of lithium for bipolar disorder, uses of lithium outside of its primary indications for acute treatment of mania and maintenance treatment for recurrent depression in bipolar patients, and some studies of the drug's mechanism of action. General reference guides for this chapter include *Manic Depressive Illness* (Goodwin and Jamison, 1990) and *Moodswing* (Fieve, 1975b).

LITHIUM METABOLISM IN MAN

Lithium is usually administered as a salt, most frequently as lithium carbonate. Lithium is absorbed from the gastrointestinal tract and is dispersed throughout all body fluids. Lithium is excreted by the kidneys and the elimination half life is about one day. Repeated doses of lithium will produce a steady-state situation such that blood lithium concentrations will be stable over time at a stable administered dose. Concentrations of lithium in the serum are higher than intracellular concentrations and concentrations within the brain are lower than concentrations of lithium in peripheral tissues such as erythrocytes (Frazer et al., 1978).

Lithium (Li^+) administration will produce excretion of water and electrolytes, particularly sodium, potassium, magnesium, and calcium. Sodium is necessary for the proper excretion of lithium by the kidney (Baer et al., 1970). Changing sodium metabolism by reducing sodium intake or increasing sodium excretion through the use of diuretics while a patient is being treated with lithium will result in lithium retention. Higher blood lithium levels will result in higher intracellular lithium concentrations and the possibility of lithium toxicity. The usual side effects of lithium administration include nausea, diarrhea, tremor, and increased frequency of urination. Lithium toxicity is a serious medical condition that can be fatal. The person is usually dehydrated, confused, and hyperreflexic and may show choreiform movements.

HISTORY OF LITHIUM USE IN MEDICINE

Lithium is an alkali earth metal first discovered by Arfwedson in 1817. As a free substance, it is a silvery white metal. The lightness of this metal has led to its use commercially in ceramic metallurgy. More recently, lithium had been used in the development of hydrogen bombs. The early medical uses of lithium salts included treatment for gout. Lithium urate is the most soluble of the uric acid salts and, on this basis, attempts were made to treat gouty deposits by administration of lithium. Bromides of lithium were also used in medicine as a sodium replacement. Lithium chloride was substituted for sodium chloride in patients with cardiac disease in order to produce a low sodium diet. In 1949, there were reports of deaths associated with this type of lithium use and lithium substitutes were withdrawn from the U.S. marketplace (Hanlon et al., 1949; Corcoran et al., 1949; Stern, 1949). The year 1949 is of some interest also because it was when lithium use in psychiatry was initiated.

Early Use of Lithium in Psychiatry

The first use of lithium for the treatment of acute mania was reported by Cade (1949) of Australia. Cade, as did many others, hypothesized that mania was related to an excess of some unknown substance and that depression was related to a deficit of some substance. He thought that urea might be the cause of the mental illness. Cade injected urine from patients and controls into guinea pigs and then gave the most soluble salt of uric acid, lithium urate, and urea to guinea pigs. Lithium protected the guinea pigs from the convulsant effects of urea. He also noted that lithium salts (urate, carbonate) caused guinea pigs to become lethargic and less responsive. Cade next gave lithium salts to chronic manic patients and noted improvement in their mania. The same year that lithium was withdrawn as a marketed drug for medicine in the United States because of its toxicity as a sodium substitute, Cade reported lithium to be effective for the treatment of manic patients in uncontrolled trials.

Early Controlled Trials of Lithium in Acute Mania

The early controlled trials of lithium took place in Europe, particularly in Denmark, through the work of Schou and colleagues (1954) who conducted a double-blind trial of lithium in acutely manic patients. Patients were diagnosed according to research criteria and the effects were measured with a mania rating scale. Lithium or placebo was administered to patients over two weeks in an alternating randomized design. The results of this study published in the early 1950s also coincided with the time of the introduction of chlorpromazine as a major psychiatric treatment in the United States. Importantly, in the mid-1960s, Schou's group published research demonstrating that the longterm use of lithium seemed to decrease subsequent future attacks of bipolar disorder in patients (Baastrup and

Schou, 1967; Schou 1968). Thus lithium seemed to have "prophylactic" or maintenance effects.

A second European trial of lithium in acute mania also showed positive effects (Maggs, 1963). As a result of reports from Europe of the effects of lithium in mania, studies were undertaken in the United States both at the New York State Psychiatric Institute and the National Institute of Mental Health (NIMH). In New York, Wharton and Fieve (1966) studied patients who were acutely manic and unable to take phenothiazines because of allergy or treatment resistance. Patients were treated with lithium after placebo washout, and response during treatment was compared to response of prior episodes where lithium had not been used. Positive results with lithium were found.

The NIMH researchers headed by Bunney and co-workers used a longitudinal approach and administered lithium chronically to manic patients with placebo substitution on selected days. Ratings of behavior were performed by nurses. The NIMH studies demonstrated that higher ratings for mania occurred on days when placebo was substituted for lithium (Bunney et al., 1968; Goodwin et al., 1969). Following the early comparisons of lithium to placebo, there were also controlled comparisons of lithium to chlorpromazine or other antipsychotic drugs for the treatment of acute mania. (Table 1)

Treatment of acute mania with a variety of medications can be effective (Dunner and Clayton, 1987). The evidence from double-blind studies supports the use of lithium for the treatment of acute mania. Two of these studies are of special clinical relevance: the study by Prien and colleagues (1972) showed lithium to be superior to chlorpromazine for the less aggressive manic patient whereas chlorpromazine was superior to lithium for the more aggressive patient. Chlorpromazine and other antipsychotic medications are quite sedating. One of the usual symptoms of acute mania is decreased need for sleep. The early sedative effect of antipsychotic medications (and other medications such as benzodiazepines) used in the treatment of acute mania may provide an earlier relief from manic symptoms than lithium, since it takes several days of treatment with lithium to achieve effective blood levels. Another explanation would be that lithium exerts multiple effects, one of which may be modulation of gene expression (Jope, 1999).

The second study of note is that of Taskahashi et al (1975). Most of the studies indicated in Table 1 are two to three week studies, whereas Takahashi et al (1975) treated patients for 5 weeks. Some patients who had not responded by three weeks responded to continued treatment.

SYMPTOMS AND COURSE OF BIPOLAR DISORDER

Manic psychosis (Bipolar I disorder) is a mental disorder affecting 0.5 to 1 percent of the population and is characterized by recurrent episodes of disturbances in psychosocial function. Manic attacks, untreated, last an average of

Table 1. Controlled Studies of Lithium Treatment of Acute Mania

	Year of Publication	*Control*	*N*	*Result*
Schou et al.	1954	Plac	30	Li > Plac
Maggs	1963	Plac	28	Li > Plac
Johnson et al.	1968	CPZ	29	Li > Plac
Goodwin et al.	1969	Plac	12	Li > Plac
Spring et al.	1970	Plac	15	Li > CPZ
Platman	1970	CPZ	23	Li > CPZ
Johnson et al	1971	CPZ	21	Li > CPZ
Stokes et al.	1971	Plac	38	Li > Plac
Prien et al.	1972	CPZ	255	Li > CPZ
Takahashi et al.	1975	CPZ	71	Li > CPZ
Shopsin et al.	1975	CPZ,HAL	30	Li > HAL > CPZ
Bowden et al.	1994	Plac,Val	179	Li = Val> Plac

Notes: Li = Lithium; HAL = Halopendol; CPZ = Chlorpromazine; Plac = Placebo; Val = Divalproex Sodium; N = number of subjects.

six to nine months and are characterized by symptoms such as elated mood, decreased need for sleep, rapid speech, racing thoughts, impulsive behavior, increase in activity, and grandiosity. Delusions of grandeur and/or paranoid delusions are seen. The manic episode is often preceded or followed by low mood, decreased interest in usual activity, weight change, slowed motor activity, fatigue, worthlessness and guilt, difficult concentrating and suicidal thoughts and attempts. The cycles (mania-depression) are usually recurrent but the psychosocial functioning between episodes is normal. Manic depressive cycles tend to reoccur on an average frequency of four episodes in 10 years (Winokur et al 1969;Dunner et al 1979).

The clinical implications of these studies are that treatment of the patient with acute mania should begin with lithium and a sedating agent. Lithium should be administered so that blood levels achieve the therapeutic range (which is about 0.8-1.5 mEq/L for acute manic patients). Patients may not respond rapidly, but most patients will respond to treatment. The predictors of greatest likelihood of response to lithium in acute mania are the absence of dysphoric symptoms, previous history of personal or family response to lithium, and the absence of a mixed state. In less treatment-responsive forms of bipolar disorder such as mixed states or mania complicated by substance abuse, the anticonvulsants carbamazepine or valproic acid may have a higher rate of responsiveness. However, even these patients may respond very well to lithium. A number of small studies with differing methodologies have compared these treatments in bipolar disorder. However, current practice guidelines still leave uncertainly as to the best treatment choice for a given patient.

DIAGNOSTIC DEVELOPMENTS IN THE UNITED STATES

It is of some interest that, until the 1970s, mania was often under-diagnosed in the United States compared to schizophrenia. Indeed this was demonstrated in the "U.S./U.K. project" in which psychiatrists from England and American psychiatrists diagnosed the same patients quite differently (Cooper et al., 1972). The treatment of patients with bipolar disorder could not proceed unless there was adequate means to identify such patients, and in the United States the tendency was to over-diagnose schizophrenia and to apply the newly discovered antipsychotic drug chlorpromazine for the treatment of patients who had psychotic conditions. In England, in contrast, the tendency was to diagnose a variety of disorders including mania and to apply different pharmacotherapies, including lithium for bipolar disorder.

One of the few places in the United States where mania was rather reliably diagnosed was at the Washington University unit of psychiatry headed by Robins, who developed and followed a set of diagnostic criteria. These were based on empirical observations of patients thought to have a disorder, that is, inclusion criteria of symptoms characterizing the disorder, eliciting the course of illness of such patients, determining their family history, and then developing a set of exclusion criteria thus separating the condition from other disorders (Feighner et al., 1972). The early descriptive studies of mania at Washington University were reported by Clayton and colleagues (1965). The criteria laid down on the basis of their descriptive studies form the basis of current diagnosis of bipolar disorder.

It is important to be able to diagnose bipolar disorder so that patients might benefit from treatment with lithium. To this end, it would be important to have a diagnostic schema that narrowed the diagnosis of schizophrenia and broadened that of affective disorders (and particularly mania) in the United States. In fact, this approach resulted in DSM-III, a major advance in psychiatric diagnosis. The changes from DSM-II to DSM-III involve the elaboration of objective criteria for symptoms, exclusion criteria, a multi-axial approach in order to exclude personality disorders from major psychiatric disorders, a triage approach to diagnosis resulting in a narrowing in the ability to diagnose schizophrenia and a broadening of the diagnosis of affective disorders (especially mania). In addition, diagnostic inter-rater reliability was enhanced because of the use of symptoms for the criteria. In essence, DSM-III includes a variation of the Feighner and colleagues (1972) criteria called the "Research Diagnostic Criteria" (Spitzer et al., 1978). Furthermore, the development of structured instruments in order to more systematically assess symptoms changed the style of diagnostic interviewing to be more precise regarding elucidation of symptoms. If what one needed to do was to assess symptoms in patients, then the best way to do this would be through a systematic approach, that is, the development of structured instruments such as the Schedule for Affective Disorders and Schizophrenia (SADS) (Spitzer and Endicott, 1978).

Thus, not only has lithium affected the treatment of affective disorders, it has also had a major impact on the diagnosis of mental conditions.

DSM has been revised twice since and is currently in its fourth edition. Several major changes relevant to bipolar disorder were made in DSM-IV such as the inclusion of Bipolar II disorder. Mixed episode is now formally recognized, requiring the patient to meet criteria for both mania and depression for at least one week on most days. Rapid cycling is a course modifier that can apply to Bipolar I or Bipolar II disorder and carries a greater likelihood of resistance to lithium treatment. Hypomania has specific diagnostic criteria involving four-day or longer periods of symptoms, which are not disabling but are clearly different from euthymia.

Bipolar II disorder is characterized by episodes of major depression and periods of hypomania lasting at least four days. Bipolar II disorder is sometimes encountered when an undiagnosed bipolar patient receives an antidepressant for depression and becomes hypomanic. Prior episodes of hypomania only then are recognized retrospectively, as patients often do not consider these episodes as pathological.

MAINTENANCE EFFECTS OF LITHIUM IN BIPOLAR DISORDERS

Early work demonstrated a novel feature of lithium treatment, namely, that some patients who continued the treatment after resolution of their acute mania noticed a reduction in the severity of their post–manic-depressive episodes and subsequent manias (Schou et al., 1954; Hartigan, 1963; Baastrup and Schou, 1967). Thus, lithium was continued for long periods of time in bipolar patients. What became clear from these open trials was a marked reduction in frequency and severity of both manic and depressive episodes with the continued administration of lithium treatment. Indeed, the novelty of lithium for American psychiatry is probably less related to its use as a treatment for acute mania than as a preventive treatment for bipolar disorder in patients with this condition.

The open studies of lithium maintenance effects were criticized on several grounds among which are the lack of a control group and that patients whose episodes clustered at a certain period of time might be less likely to have episodes afterward (Blackwell and Shepherd, 1968).

For these reasons, a number of longitudinal double-blind studies were undertaken to determine maintenance effects of lithium (Table 2). For these studies, patients were diagnosed as having bipolar disorder and entered into treatment when they were euthymic for a continuation phase involving lithium or placebo.

The blood lithium concentrations required for maintenance treatment have proven problematic. Some studies suggested that maintenance blood levels below 0.4 mEq/L were largely ineffective. Since blood levels above 1.0 mEq/L may be

Table 2. Controlled Lithium Maintenance Trials in Bipolar Disorder

	Year of Publication	*N*	*Duration (months)*
Baastrup et al.	1970	50	5
Melia	1970	15	24
Coppen et al.	1971	65	24
Cundall et al.	1972	24	12
Stallone et al.	1973	52	24
Prien et al.	1973a	205	24

Note: N = number of subjects.

associated with a higher degree of renal problems, the optimal dose for lithium maintenance has not yet been precisely determined, although levels close to 1 mEq/L may be preferable (Gelenberg et al., 1989).

LITHIUM TREATMENT FAILURE: RAPID CYCLERS

As a parallel to these maintenance studies, it was found that some subjects do not do well at all with lithium maintenance. They show early and repeated failures both for hypomanic and depressive episodes. In categorizing such patients, it was found that a history of frequent prior episodes among them was common and they were termed rapid cyclers (Dunner and Fieve, 1974). Thus, a subgroup of bipolar patients was identified. Further investigation of this subgroup showed that not only were they likely early lithium treatment failures but also they were predominantly women in many of whom were abnormal thyroid function tests (Dunner et al., 1977; Bauer and Whybrow, 1990; Bauer and Whybrow, 1993).

EFFECTS AGAINST DEPRESSION

Although the effect of lithium to reduce the severity and frequency of mania was amply demonstrated, it was not clear that lithium had maintenance effects against depression. One way of studying maintenance effects against depression is to study patients with bipolar II. Patients with severe mania (Bipolar I) differ from those with depression and mild hypomania (Bipolar II) in several characteristics (Dunner et al., 1976a; Dunner, 1993). Bipolar II patients have recurrent depression and are unlikely to develop severe mania during a placebo-controlled trial. Placebo-controlled studies of bipolar II patients suggest that lithium maintenance effects against future depressive episodes could be demonstrated (Dunner et al., 1976b; Dunner et al., 1982; Prien et al., 1984). However, these treatment trials took a longer time than studies of maintenance effects against mania for a statistically significant outcome to be demonstrated.

Research studies of acute depression have led to evidence of the beneficial effects of pharmacotherapy with a variety of antidepressants. After decades of such studies, it became apparent that psychotherapy, the standard treatment for acute depression, was not supported by data-based treatment outcome studies. It is noteworthy that the development of manual-based psychotherapies (cognitive and interpersonal) and their assessment by controlled studies in depression occurred after the use of lithium and other pharmacotherapies in depression (Elkin et al., 1989).

When depressed patients fail to respond fully to an adequate antidepressant trial, add-on "augmentation" strategies can offer the most efficient next step. Lithium has been shown in several small double-blind, placebo-controlled studies to improve the response of standard antidepressants when patients initially have not responded satisfactorily (de Montigny et al., 1981; Heninger et al., 1983). When using lithium augmentation therapy in treatment-resistant depression, moderate dosages in the 600-mg to 1200-mg range are usually sufficient to induce such a synergistic response.

CLINICAL USE OF LITHIUM IN THE UNITED STATES

A model for lithium treatment of mood disorder was that pursued by the Lithium Clinic at New York State Psychiatric Institute (Fieve, 1975a). In this clinic, patients were evaluated by nursing personnel who were trained to use depression and mania rating scales. The patients were given white capsules containing lithium or placebo. Neither the nurse raters nor the patients knew who was receiving which substance. Blood samples for measurement of serum lithium were obtained from the patients monthly. For those patients randomly assigned to placebo fictitious serum lithium levels were entered into the clinical record. If the patients behavior was markedly different than usual or if the patient had experienced symptoms to justify concern based on predetermined parameters involving the rating scales, a physician blind to treatment assignment evaluated the patient. Patients were considered treatment failures if additional medication was prescribed for either emergent depression or mania (hypomania) or if hospitalization was required.

This clinic provided treatment with pharmacotherapy. The design of lithium studies was to avoid other treatments including psychotherapy. Trained staff rather than medical personnel were used to follow patients. Blood level monitoring, assessment of severity of mood disorder by use of structured rating scales, and assessment of severity of side-effects by systematic questioning were standard features of the Lithium Clinic. Patients returned to the clinic on a regular basis for a brief follow-up—perhaps 15 minutes in duration.

In order to economize the use of personnel, the Lithium Clinic generally met on a half day or two half days on selected days of the week, say, Tuesday afternoon and Wednesday afternoon. Because lithium was not generally available at that time for the treatment of bipolar disorder, patients came to the Lithium Clinic for treatment and participated in treatment outcome studies as well as other research.

The controlled maintenance lithium trial in New York evaluated patients who had two episodes prior to randomization and entered patients when they were euthymic. The design was to follow these patients prospectively over a period of two years. Patients were randomly given lithium or placebo, and the notion was that, after two years, the treatment assignment code could be broken and maintenance effects could be assessed. Patients who participated in this study generally stayed in treatment if they were maintained on lithium even though mild breakthroughs of mania (hypomania) or depression occurred. Hospitalization was unusual for lithium-treated patients, and these patients generally stayed in treatment but perhaps required additional medication to manage symptoms. However, placebo-treated patients generally became severely manic and dropped out of treatment, often flying to other states, requiring hospitalization for severe manic recurrences (Stallone et al., 1973).

What became clear from this study was how severe mania is. Untreated mania has an episode frequency of approximately four episodes in 10 years, and 80% of these episodes involve severe mania requiring hospitalization (Winokur et al., 1969; Dunner et al., 1979). Thus untreated bipolar disorder has recurrences, many of which are experienced as severe episodes requiring hospitalization.

This study also demonstrated a number of important facts about lithium maintenance. First of all, it is quite effective in reducing frequency and severity of manic and depressive episodes in patients with bipolar disorder. Secondly, prolonged lithium treatment further reduces the chance of subsequent episodes. Important too is that lithium exerts its effects against future manic episodes earlier than its effects against the depressed phase.

Life table curves were developed from the combined data of multiple studies. They showed clear maintenance effects of lithium with a biphasic portion to the life table analysis (Fleiss et al., 1978). During the first six months of treatment, there was a higher failure rate while the subsequent two and a half years of follow-up had a steady but lower failure rate. These data suggested that patients be followed more closely during the initial six months after they achieve euthymia and then followed somewhat less closely but still maintained on lithium. Additionally, patients can have their first failure after two, three, or five or more years of treatment. Thus, lithium is not a 100% effective drug for maintenance treatment, but indeed almost all patients will show a positive effect in that they will have a reduction of severity and/or frequency of subsequent attacks. The effect seems to be more pronounced against mania early in treatment and the effect against depression seems to take longer.

MEDICAL SIDE EFFECTS OF LITHIUM

Lithium has several medical side effects, most of which are gastrointestinal, but renal, cardiac, and central nervous system effects can occur. Side effects early in treatment generally include nausea, diarrhea, and tremor. These can be alleviated by changing the preparation of lithium into a slow-release preparation or changing the time of lithium administration, giving the medication with food, or giving the medication all at bedtime. Side-effects occurring later in treatment include tremor, polyuria, and thirst. The renal effects attributable to lithium include nephrogenic diabetes insipidus. Additionally, lithium has anti-thyroid effects. About 15% of patients develop hypothyroidism from lithium's antagonism of thyroid hormone production and release. Hypothyroid status can be associated with rapid cycling and treatment resistance, and thyroid function should be routinely monitored by assessing serum thyroid stimulating hormone (TSH) levels. Patients who become hypothyroid require exogenous thyroid supplementation, which may improve their general well being and psychiatric status.

Other effects of lithium include cardiac dysfunction (sinus node dysfunction) (Roose et al., 1979). This is an interesting effect and seems to occur in patients who are elderly and presumably might be susceptible to develop sinus node dysfunction anyway. Lithium seems to unmask this, and the use of pacemakers may be indicated. Certainly, if an elderly patient who is treated with lithium falls or has a syncopal episode, the patient should be brought to the attention of their clinician for possible cardiac monitoring.

Severe lithium poisoning is a neurotoxic manifestation characterized by high intracellular lithium levels. Lithium toxicity can be associated with persistent neurotoxicity or death. Although the mechanism of lithium neurotoxicity is unknown, it is likely that lithium enters neurons and replaces other intracellular cations, especially sodium—a mineral also necessary for lithium to be excreted by the kidneys. Thus, in order to avoid neurotoxicity, lithium-treated patients should neither restrict their sodium nor use sodium diuretics. Treatment of lithium toxicity should include discontinuation of lithium, administration of fluids, and administration of potassium. Dialysis may be necessary in extreme cases. In addition there are reports of persistent neurotoxicity resulting from patients who are treated with electroconvulsive therapy while being treated with lithium (Small et al., 1980).

Problems with lithium treatment are common in the elderly. Because of reduced renal clearance, doses should be divided and given throughout the day instead of using single daily doses. The blood level needs to be maintained at a lower level as the therapeutic index in elderly patients is lower as compared to younger patients.

Lithium use in children may be problematic since lithium can replace calcium in bones and potentially cause osteoporosis. It is interesting that manic symptoms in childhood are most frequently a misdiagnosis of neurological conditions. Mania does not generally begin or have an onset prior to puberty.

MECHANISM OF ACTION OF LITHIUM IN BIPOLAR DISORDER

The mechanism of action of lithium is unknown, but the most promising current theories involve lithium's interactions with second messenger systems, especially the phosphotidylinositol (PI) pathway (Avissar and Schreiber, 1989). Lithium inhibits inositol monophosphatase, a key enzyme in the PI pathway, leading to depletion of substrate for the pathway and attenuation of PI signaling. The PI second messenger system is coupled to a variety of neurotransmitter receptors in brain and leads to activation of protein kinase C and release of intracellular calcium stores, thereby transducing the neurotransmitter message. Lithium has a low therapeutic index, and it is possible that lithium toxicity involves either excessive activity through this proposed site of action or through actions on other aspects of cellular physiology, including membrane transport and conductances of other ions such as sodium, calcium, magnesium, and potassium. More recently, evidence has been produced that mice fed Li^+ show an increase in brain IP_3 in association with an increase in the release of glutamate (Hokin). This occurs with lithium in therapeutic levels.

SUMMARY

Lithium salts have had a major impact in the medical treatment of bipolar affective disorder. Longterm treatment of bipolar patients has resulted in considerable reduction in the re-hospitalization rates for this condition. Lithium clinics provided an early model for the psychopharmacologic treatment of individuals with other mental disorders and the medical approach to the treatment of depression.

REFERENCES

Avissar, S. & Schreiber, G. (1989). Muscarinic receptor subclassification and G-proteins: Significance for lithium action in affective disorders and for the treatment of extrapyramidal side effects of neuroleptics. Biol. Psychiatry 26, 113-130.

Baastrup, P.C. & Schou, M. (1967). Lithium as a prophylactic agent: Its effect against recurrent depression and manic-depressive psychosis. Arch. Gen. Psychiatry 16, 162-172.

Baastrup, P.C. , Poulsen, J.C., Schou, M., Thomsen, K,. & Amidsen, A. (1970). Prophylactic lithium: Double-blind discontinuation in manic-depressive and recurrent depressive disorders. Lancet 2, 326-330.

Baer, L., Durell, J., Bunney, W.E., Jr., Levy, B.S., Murphy, D.L., Greenspan, K., & Cardon, P.V., (1970). Sodium balance and distribution in lithium carbonate therapy. Arch. Gen. Psychiatry 22, 40-44.

Bauer, M. & Whybrow, P. (1990). Rapid cycling bipolar affective disorder II: Treatment of refactory rapid cycling with high dose thyroxine, a preliminary study. Arch. Gen. Psychiatry: 47, 435-440.

Bauer, M. & Whybrow, P. (1993). Validity of rapid cycling as a modifier for bipolar disorder in DSM-IV. Depression 1, 11-19.

Blackwell, B. & Shepherd, M. (1968). Prophylactic lithium: Another therapeutic myth? Lancet 1, 968-971.

Bowden, C.L., Brugger, A.M., Swann, A.C., Calabrese, J.R., Janicak, P.G., Petty, F., Dilsaver, S.C., Davis, J.M., Rush, A.J., Small, J.G., Garza-Trevino, E.S., Risch, S.C., Goodnick, P.J., & Morris, D.D. (for the Depakote Mania Study Group)(1994). Efficacy of divalproex vs. lithium and placebo in the treatment of mania. JAMA 271, 918-924.

Bunney, W.E. Jr., Goodwin, F.K., Davis, J.M., & Fawcett, J.A. (1968). A behavioral study of lithium treatment. Amer. J. Psychiatry 125, 499-512.

Cade, J.F.J. (1949). Lithium salts in the treatment of psychotic excitement. Med. J. Aust. 36, 313-322.

Clayton, P.J., Pitts, F.N., Jr., & Winokur, G. (1965). Affective disorder in mania. Compr. Psychiatry 6, 313-322.

Cooper, J.E., Kendell, R.E., Gurland, B.J.., Sharpe, L., Copeland, J.R.M., & Simon, R. (1972). Psychiatric diagnosis in New York and London: A comparative study of mental hospital admissions. Maudsley Monograph No. 20. Oxford University Press, London.

Coppen, A., Noguera, R., Bailey, J., Burns, B.H., Swani, M.S., Hare, E.H., Gardner, R.., & Maggs, R. (1971). Prophylactic lithium in affective disorders. Lancet 2, 275-279.

Corcoran, A.C.., Taylor, R.D., & Page, I. (1949). Lithium poisoning from the use of salt substitutes. JAMA 1139, 685-688.

Cundall, R.L., Brooks, P.W., & Murray, L.G. (1972). A controlled evaluation of lithium prophylaxis in affective disorders. Psychol. Med. 2, 308-311.

de Montigny, C., Grunberg, F., Mayer, A., & Deschenes, J.P. (1981). Lithium induces rapid relief of depression in tricyclic antidepressant drug non-responders. Br. J. Psychiatry 138, 252-256.

Dunner, D.L. (1993). A review of the diagnostic status of "bipolar II" for the DSM-IV work group on mood disorders. Depression 1, 2-10.

Dunner, D.L. & Clayton P.J. (1987). Drug treatment of bipolar disorder. In: Psychopharmacology: The Third Generation of Progress. (Meltzer, H.Y., Ed.), pp. 1077-1083. Raven Press, New York.

Dunner, D.L. & Fieve, R.R. (1974). Clinical factors in lithium carbonate prophylaxis failure. Arch. Gen. Psychiatry 30, 229-233.

Dunner, D.L., Gershon, E.S., & Goodwin, F.K. (1976a). Heritable factors in the severity of affective illness. Biol. Psychiatry, 11, 31-42.

Dunner, D.L., Murphy, D., Stallone, F., & Fieve, R.R. (1979). Episode frequency prior to lithium treatment in bipolar manic-depressive patients. Compr. Psychiatry 20, 511-515.

Dunner, D.L., Patrick, V., & Fieve, R.R. (1977). Rapid cycling manic depressive patients. Compr. Psychiatry: 18, 561-566.

Dunner, D.L., Stallone, F., & Fieve, R.R. (1976b). Lithium carbonate and affective disorders: V. A double-blind study of prophylaxis of depression in bipolar illness. Arch. Gen. Psychiatry 33, 117-130.

Elkin, I., Shea, M., Watkins, J.T., Pilkonis, P.A., Leber, W.R., Docherty, J. P., Fiester, S.J., & Parloff, M.B. (1989). National Institute of Mental Health Treatment of Depression Collaborative Research Program: General effectiveness of treatment. Arch. Gen. Psychiatry 45, 971-982.

Feighner, J.P., Robins, E., Guze., S.B., Woodruff, R.A., Jr., Winokur, G., & Munoz, R. (1972). Diagnostic criteria for use in psychiatric research. Arch. Gen. Psychiatry 26, 57-63.

Fieve, R.R. (1975a). The Lithium Clinic: A new model for the delivery of psychiatric services. Am. J. Psychiatry 132, 1018-1022.

Fieve, R.R. (1975b). Moodswing. William Morrow, New York.

Fleiss, J.L., Prien, R.F., Dunner, D.L., & Fieve, R.R. (1978). Actuarial studies of the course of manic-depressive illness. Compr. Psychiatry 19, 355-362.

Frazer, A., Mendels, J., Brunswick, D., London, J., Pring, M., Ramsey, T. A., & Rybakowski, J. (1978). Erythrocyte concentrations of the lithium ion: Clinical correlates and mechanisms of action. Am. J. Psychiatry 135, 1065-1069.

Gelenberg, A.J. , Kane, J.M. , Keller, M., Lavori, P., Rosenbaum, J.F. , Cole, K., & Lavelle, J. (1989). Comparison of standard and low serum levels of lithium for maintenance treatment of bipolar disorder. N. Engl. J. Med. 321, 1489-1493.

Goodwin, F.K. & Jamison, K.R. (1990). Manic-Depressive Illness. Oxford University Press, New York.

Goodwin, F.K., Murphy, D.L., & Bunney, W.E., Jr. (1969). Lithium carbonate treatment in depression and mania: A longitudinal double-blind study. Arch. Gen. Psychiatry 21, 486-496.

Hanlon, L.W., Romaine, M., Gilroy, F.J., & Dietrick, J.E., (1949). Lithium chloride as substitute for sodium chloride in the diet. JAMA 139, 688, 692.

Hartigan, G.P. (1963). The use of lithium salts in affective disorders. Brit. J. Psychiatry 109, 810-814.

Heninger, G.R., Charney, D.S., & Sternberg, D.E. (1983). Lithium carbonate augmentation of antidepressant treatment: An effective prescription for treatment-refractory depression. Arch. Gen. Psychiatry 40, 1335-1342.

Johnson, G., Gershon, S., & Hekiman, L.J., (1968). Controlled evaluation of lithium and chlorpormazine in the treatment of manic states: An interim report. Compr. Psychiatry 9, 563-573.

Johnson, G., Gershon, S., Burdock, E.I. Floyd, A., & Hekiman, L., (1971). Comparative effects of lithium and chlorpromazine in the treatment of acute manic states. Br. J. Psychiatry 119, 267-276.

Maggs, R. (1963). Treatment of manic illness with lithium carbonate. Br. J. Psychiatry 109, 59-65.

Melia, P.I. (1970). Prophylactic lithium: A double-blind trial in recurrent affective disorders. Br. J. Psychiatry 116, 621-624.

Platman, S.R. (1970). A comparison of lithium carbonate and chlorpromazine in mania. Am. J. Psychiatry 127, 351-353.

Prien, R.F., Caffey, E.M., Jr., & Klett, C.J. (1972). Comparison of lithium carbonate and chlorpromazine in the treatment of mania: Report of the Veterans Administration and National Institute of Mental Health Collaborative Study Group. Arch. Gen. Psychiatry 26, 146-153.

Prien, R.F., Caffey, E.M., Jr., & Klett, C.J. (1973). Prophylactic efficacy of lithium carbonate in manic-depressive illness. Arch. Gen. Psych. 28, 337-341.

Prien, R.F., Kupfer, D.J., Mansky, P.A., Small, J.G., Tuason, V.B., Voss, C.B., & Johnson, W.E. (1984). Drug therapy in the prevention of recurrences in unipolar and bipolar affective disorders: Report of the NIMH Collaborative Study Group comparing lithium carbonate, imipramine, and a lithium carbonate-imipramine combination. Arch. Gen. Psychiatry 41, 1096-1104.

Roose, S.P., Nurnberger, J. I., Dunner, D.L., Blood, D.K., Fieve, R.R. (1979). Cardiac sinus node dysfunction during lithium treatment. Amer. J. Psychiatry 136, 804-807.

Schou, M. (1968). Lithium in psychiatric therapy and prophylaxis. J. Psychiatr. Res. 6, 67-95.

Schou, M., Juel-Nielson, N., Stromgren, E., & Voldby, H. (1954). The treatment of manic psychosis by administration of lithium salts. J. Neurol. Neurosurg. Psych. 17, 250-260.

Shopsin, B., Gershon, S., Thompson, H., & Collins, P. (1975). Psychoactive drugs in mania: A controlled comparison of lithium carbonate, chlorpromazine, and haloperidol. Arch. Gen. Psychiatry 32, 34-42.

Small, J.G., Kellams, J.J., Milstein, V., & Small, I.F. (1980). Complications with electroconvulsant treatment combined with lithium. Biol. Psychiatry 15, 103-112.

Spitzer, R.L., & Endicott, J. (1978). Schedule for Affective Disorders and Schizophrenia. Biometrics Research Evaluation Section, State Psychiatric Institute, New York.

Spitzer, R.L. Endicott, J., & Robins, E. (1978). Research Diagnostic Criteria. Biometrics Research Evaluation Section, State Psychiatric Institute, New York.

Spring, G., Schweld, D., Gray, C. Steinberg, J., & Horwitz, M., (1970). A double-blind comparison of lithium and chlorpromazine in the treatment of manic states. Am. J. Psychiatry 126, 1306-1310.

Stallone, F., Shelley, E., Mendlewicz, J., & Fieve, R.R., (1973). Thus use of lithium in affective disorders. III: A double-blind study of prophylaxis in bipolar illness. Am. J. Psychiatry 130, 1006-1010.

Stern, R.L. (1949). Severe lithium chloride poisoning with complete recovery. JAMA 139, 710-711.

Stokes, P.E., Shamoian, C.A., Stoll, P. M., & Patton, M.J. (1971). Efficacy of lithium as acute tratment of manic-depressive illness. Lancet 1, 1319-1325.

Takahashi, R., Skauma, A., Itoh, K., Itoh, H., & Kurihara, M., (1975). Comparison of efficacy of lithium carbonate and chlorpromazine in mania: Report of collaborative study group on treatment of mania in Japan. Arch. Gen. Psychiatry 32, 1310-1318.

Wharton, R.N. & Fieve, R.R. (1966). The use of lithium in the affective psychoses. Amer. J. Psychiatry 123, 706-712.

Winokur, G., Clayton, P.J., & Reich, T. (1969). Manic-Depressive Illness. CV Mosby, St. Louis, MO.

RECOMMENDED READINGS

Jope, R.S. (1999). Anti-bipolar therapy: Mechanism of action of lithium. Mol. Psychiatry 4, 117-128.

Chapter 25

Antipsychotic Drugs

BRENDA K. COLASANTI

Biological Psychiatry, pages 585-609.

ISBN: 1-55938-819-6

INTRODUCTION AND BACKGROUND

The idea that psychotic disorders represent an abnormality of brain function originated in ancient history. In ancient Greece, the Hippocratic teaching of an organic cause for insanity vied with the commonly held belief of a supernatural origin for psychotic states.

The utilization of a variety of chemical substances for the treatment of psychosis has likewise been practiced since ancient times. In the 17th century, extracts from the pimpernel plant were recommended for the treatment of insanity. In the 19th century, preparations derived from marijuana were utilized for the treatment of mental disorders. It is recognized today that the smoking of marijuana actually exacerbates psychotic conditions.

Major advances in the treatment of psychoses awaited the 20th century. In the early 1930s, extracts from the plant *Rauwolfia serpentina* were used both for the treatment of hypertension and of psychotic symptoms. Reserpine, the active constituent of this plant, remains on the market today for the treatment of both conditions, although it is not used extensively in psychiatry.

The treatment of psychotic conditions was revolutionalized by the introduction of chlorpromazine in the early 1950s. Chlorpromazine was synthesized by Laborit and colleagues (1952) in France in an effort to find a drug that would reduce the autonomic effects of surgical stress. The calming effect of the drug was first noticed in this context. Shortly thereafter, Delay and Deniker (1952) reported that chlorpromazine had a dramatic effect in alleviating the symptoms of schizophrenic patients. Utilization of chlorpromazine subsequently became widespread.

Chemically, chlorpromazine is a phenothiazine derivative. Today, about one dozen phenothiazine derivatives are available for use as antipsychotic drugs. Four additional chemical groups of drugs acting similarly to the phenothiazines—the thioxanthenes, butyrophenones, dibenzoxazepines, and dihydroindolones—are also available. A group of new drugs with properties markedly different from these standard agents has just recently been marketed. Because of the differences, these newer agents are known as atypical antipsychotic drugs. Clozapine, a dibenzodiazepine compound, was the first agent made available in this group in the United States around 1990. An additional dibenzodiazepine, olanzapine, and a benzisoxazole compound, risperidone, have since received FDA approval for marketing.

The antipsychotic drugs are also known as neuroleptics. This term derives from the behavioral effects induced by the drugs. Psychomotor slowing, emotional quieting, and affective indifference are produced. After administration of an antipsychotic drug to the normal individual, the person typically sits quietly and shows little interest in the surrounding environment. External stimuli evoke no emotional response. After administration to a psychotic patient, there is an immediate calming effect that is more pronounced in the agitated patient.

MECHANISM OF ACTION

Although the antipsychotic drugs are quite diverse chemically, they all share one common property. All antipsychotic drugs produce dopamine receptor blockade in the brain. Furthermore, there is a good correlation between *in vitro* antagonism of dopamine receptor sites and clinical antipsychotic potency for the standard antipsychotic drugs (Creese, 1985). The butyrophenone haloperidol is one of the most potent antipsychotics, while chlorpromazine is a low potency agent.

In conjunction with their blockade of dopamine receptors, standard antipsychotic drugs block behavioral arousal induced by dopamine agonists such as apomorphine and amphetamine. In contrast, they increase sensitivity to dopamine agonists after prolonged treatment and prolonged receptor blockade, presumably as an adaptive mechanism. The abundance of dopamine receptors in the forebrain is also increased after chronic treatment. It is the D2 dopamine receptor type for which the antipsychotic drugs have affinity; this affinity parallels their clinical potency. The antipsychotic drugs have variable affinity for the D1 dopamine receptor.

There are three major areas of dopamine-containing cells in the brain: the nigrostriatal tract, the mesolimbic pathway, and the tuberoinfundibular pathway. It is the mesolimbic pathway that is thought to be involved in the production of antipsychotic effects by neuroleptic drugs. This dopamine pathway originates in the midbrain and projects to the limbic system and the forebrain. The limbic system is an important site for the regulation of emotions and behavior.

The nigrostriatal tract originates in the substantia nigra and projects to the caudate nucleus and putamen, which are part of the basal ganglia and extrapyramidal system. Degeneration of this pathway causes Parkinson's disease. All standard antipsychotic drugs cause major motor side-effects to varying degrees by blockade of dopamine receptors within this pathway.

The tuberoinfundibular dopamine pathway has its origin in the arcuate nucleus of the hypothalamus and projects to the median eminence. This pathway is involved in hypothalamic control of endocrine input to the pituitary. Chronic blockade of dopamine receptors in this pathway gives rise to the endocrine side-effects commonly observed after the administration of antipsychotic drugs.

The major sites of action of the antipsychotic drugs within the dopamine pathways of the brain are summarized in Figure 1.

Because all standard antipsychotic drugs block dopamine receptors in the brain, it has been hypothesized that functional hyperactivity of limbic dopamine pathways gives rise to the symptoms of psychosis. Prominent positive symptoms of

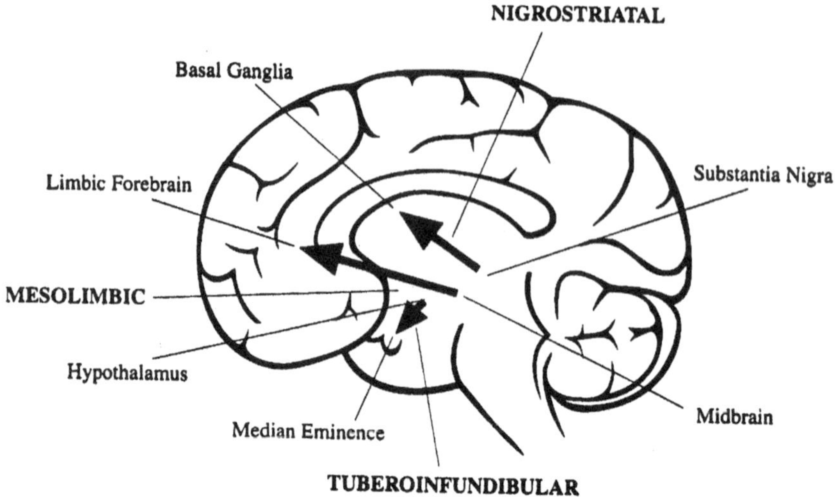

Figure 1. Major sites of action of the antipsychotic drugs within the dopamine pathways of the brain. The antipsychotic effect is presumed to be mediated by the mesolimbic pathway, while drug action within the nigrostriatal and tuberoinfundibular pathways gives rise to extrapyramidal and neuroendocrine side-effects, respectively.

schizophrenia (e.g.,delusions and hallucinations) respond relatively well to neuroleptic drugs. Chronic abuse of high doses of amphetamine likewise produces a toxic psychosis closely resembling schizophrenia.

Genetic studies support the hypothesis that a biochemical abnormality is involved in the etiology of schizophrenia. Experiments in which offspring of both schizophrenic and nonpsychotic individuals were raised by foster parents have demonstrated a higher incidence of schizophrenia in the former group. Studies with identical twins have likewise shown that the incidence of schizophrenia in both monozygotic individuals who are raised in different environments approximates that in those raised together.

The biochemical bases for schizophrenia are discussed in detail in Chapter 12.

The newer atypical antipsychotic drugs possess a pharmacological profile different from that of the standard antipsychotic drugs. One major difference is that these compounds exert only minimal adverse extrapyramidal motor effects. These compounds are likewise largely devoid of some of the prominent endocrine effects of standard antipsychotics, such as elevation of serum prolactin levels.

The pharmacology of clozapine, the first atypical antipsychotic to be utilized clinically, has been examined extensively. Clozapine differs markedly from the standard neuroleptics in its interactions with brain dopamine (Baldessarini and Frankenburg, 1991). In contrast with the standard drugs, clozapine does not block behavioral arousal induced by dopamine agonists such as amphetamine. Chronic administration of clozapine increases neither the abundance nor the sensitivity of dopamine receptors; in fact, clozapine may actually prevent the increased sensitivity produced by chronic administration of the standard drugs. With regard to receptor binding, clozapine has similar, but relatively low, affinity for both the D1 and D2 dopamine receptor types. Clozapine also exerts antagonistic actions at other receptor sites in the brain, including those for serotonin, norepinephrine, acetylcholine, and histamine.

With the development of additional atypical antipsychotic drugs, the mechanism of action of this group of agents has more extensively been examined. These drugs, like clozapine, act as antagonists at a multitude of central receptor sites. It is currently believed, however, that antagonistic action at serotonergic (5HT2) receptors accounts for the better side-effect profile of these drugs. Retention of sufficient dopaminergic (D2) blockade within the limbic system presumably mediates the antipsychotic effect.

CLASSIFICATION AND PHARMACOLOGY

Chlorpromazine was the first chemical compound recognized to possess specific antipsychotic activity. This compound is a phenothiazine derivative. Since the advent of chlorpromazine, a number of phenothiazines have been marketed as antipsychotics. Four additional distinct chemical structures have yielded a limited number of compounds that act similarly to chlorpromazine. Compounds within these five groups are considered to be standard antipsychotic drugs.

Recently, newer drugs with a pharmacological profile different from chlorpromazine have been found to be effective in the treatment of schizophrenia. These agents are known as atypical antipsychotic drugs (Holland et al., 1991). Clozapine was the first agent from this group to be made available in the United States today. One chemically similar compound, olanzapine, and one from a novel drug group, risperidone, have since been marketed. Additional atypical neuroleptics are currently undergoing the preclinical trials necessary for FDA approval. These include quetiapine, iloperidone, zotepine, and ziprasidone.

Phenothiazines

The phenothiazine nucleus is a three-ringed structure (Figure 2). There are three chemical subgroups of phenothiazines based on the side chain present on the nitrogen substituent of the central ring: aliphatic (e.g., chlorpromazine), piperidine (e.g.,thioridazine and mesoridazine), and piperazine (e.g., trifluoperazine and congeners) phenothiazines. The aliphatic and the piperidine phenothiazines are low potency drugs; doses of several hundred milligrams are required for a pharmacological effect. The piperazine phenothiazines, on the other hand, are

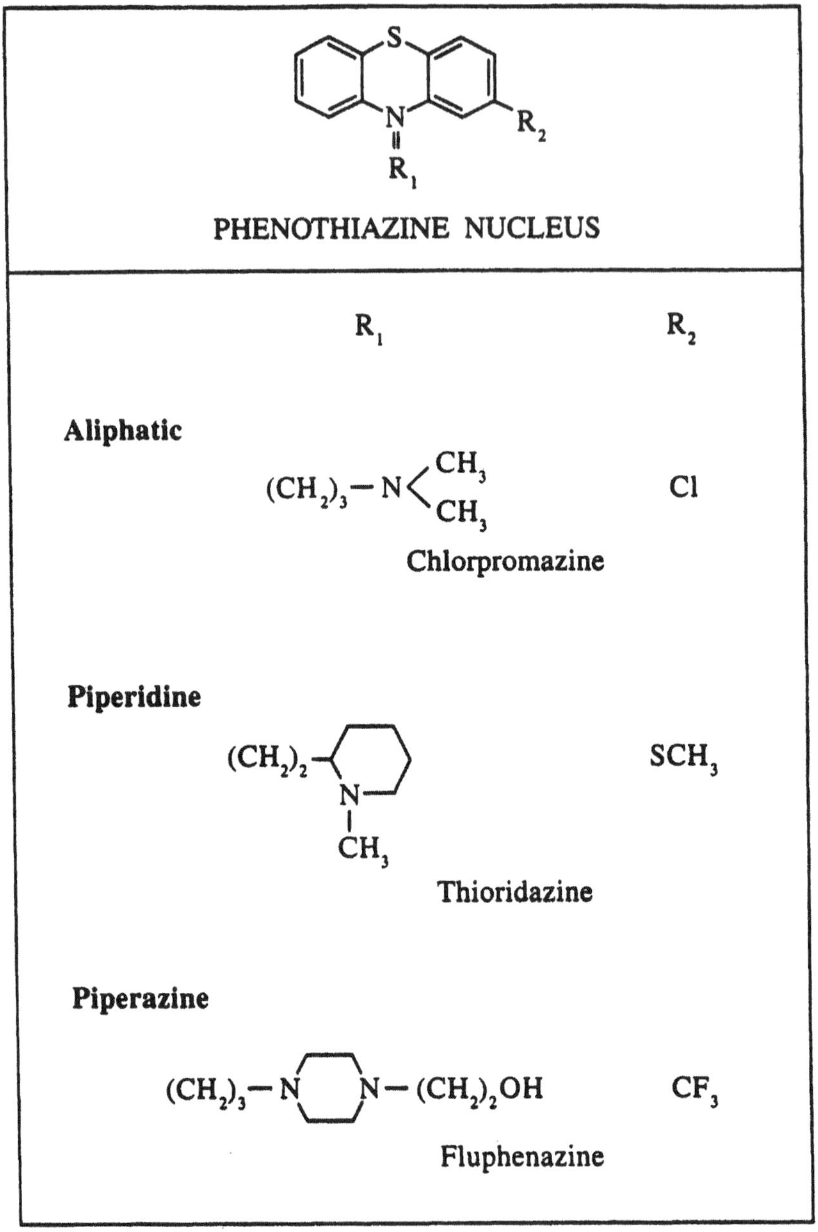

Figure **2.** Chemical structures of three representative phenothiazine derivatives.

high potency drugs requiring a much lower total dose. Compounds with the piperidine moiety produce a much lower incidence of extrapyramidal sde-effects than the other phenothiazines.

The phenothiazine derivatives produce a wide variety of both central and peripheral pharmacological effects. It is for this reason that chlorpromazine was initially given the trade name Largactil.

Central Nervous System

The phenothiazines produce sedation and tranquilization of varying degrees by an action within the central nervous system. Sedation is marked after administration of chlorpromazine and the piperidine phenothiazines. On the other hand, most of the piperazine compounds (e.g., prochlorperazine, fluphenazine, and trifluoperazine) produce only mild sedative effects.

By an action within the central nervous system, the phenothiazines also exert a prominent antiemetic effect. These compounds also interfere with the regulation of body temperature by an action within the hypothalamus. The existing ambient temperature determines the direction of the change in body temperature. Low environmental temperatures result in hypothermia, whereas higher temperatures cause hyperthermia.

The antipsychotic effect of the phenothiazines is presumably due to dopamine receptor blockade within the limbic system and frontal cortex. Dopamine receptor blockade within the basal ganglia produces prominent extrapyramidal side-effects. The incidence of motor side-effects is the greatest with the piperazine phenothiazines and the least with the piperidine phenothiazines.

The neuroendocrine system is also affected as a result of dopamine receptor blockade by the phenothiazines. One prominent effect is an elevation of serum prolactin levels and an increase in lactation. All the phenothiazines likewise cause increased appetite and weight gain.

Chlorpromazine increases the susceptibility of both animals and humans to epileptic seizures. This effect does not occur with the piperazine or the piperidine phenothiazines. In the case of chlorpromazine, incidents of seizure activity have been limited to patients with a poorly controlled seizure disorder or those prone to seizures.

Peripheral Effects

By action both centrally and peripherally, the phenothiazines produce lowering of blood pressure and orthostatic hypotension. Reflex sinus tachycardia also occurs. Vasomotor reflexes controlled by the hypothalamus and the brainstem are depressed centrally. Peripherally, these agents affect the sympathetic nervous system, causing α-adrenergic blockade. The effects of chlorpromazine and the piperidine phenothiazines are more marked than those of the piperazine compounds.

The phenothiazines also affect the parasympathetic branch of the autonomic nervous system, producing anticholinergic activity to varying degrees. The end result is dry mouth, blurring of vision, decreased gastric secretion and mobility, constipation, and diminished sweating and salivation. The piperidine phenothiazines produce the most prominent effects. Chlorpromazine possesses mild anticholinergic activity, while the effects of the piperazine compounds are weaker.

Adverse Reactions

All standard antipsychotic drugs produce a similar array of side-effects of varying severity as a result of their similar pharmacological properties (see below). In addition, the phenothiazines produce several adverse reactions that occur only rarely after the administration of most other groups of antipsychotic drugs.

Jaundice is a potential adverse effect of several of the phenothiazine derivatives and is presumably a hypersensitivity reaction. The incidence of hepatotoxicity has been greatest with chlorpromazine, but liver complications have also been reported after the administration of prochlorperazine and trifluoperazine. Jaundice, which is of the intrahepatic obstructive type, is most likely to occur from the second to fourth week of drug treatment. It is slowly reversible after discontinuation of the drug.

A variety of dermatological reactions have also been associated with phenothiazine treatment—usually with chlorpromazine. Drug eruptions are the most common, occurring in 5% to 10% of patients receiving chlorpromazine. Photosensitivity has also occurred, with about a 3% incidence. Finally, after years of exposure to chlorpromazine or other phenothiazines, hyperpigmentation may occur. Areas of the skin exposed to the sun assume a gray to purplish coloration.

Ophthalmological changes have also been reported after treatment with high doses of the phenothiazines. Pigmentary retinopathy, with pigment deposits observed fundoscopically, may occur one to two months after initiation of treatment with the piperidine phenothiazines—usually with thioridazine. It is reversible after drug discontinuation. After longterm therapy with chlorpromazine, corneal and lenticular opacities due to pigment deposition have occurred. Dermatological pigmentation usually occurs in conjunction with the ocular changes.

Two phenothiazines, chlorpromazine and thioridazine, have been reported to have caused disturbances in the rhythm of the heart. Abnormalities of ventricular repolarization are seen most frequently. These are reflected by a lengthening of the QT interval in the electrocardiogram (EKG). At higher doses, changes in the T wave have also occurred. The piperazine phenothiazines have rarely produced these effects.

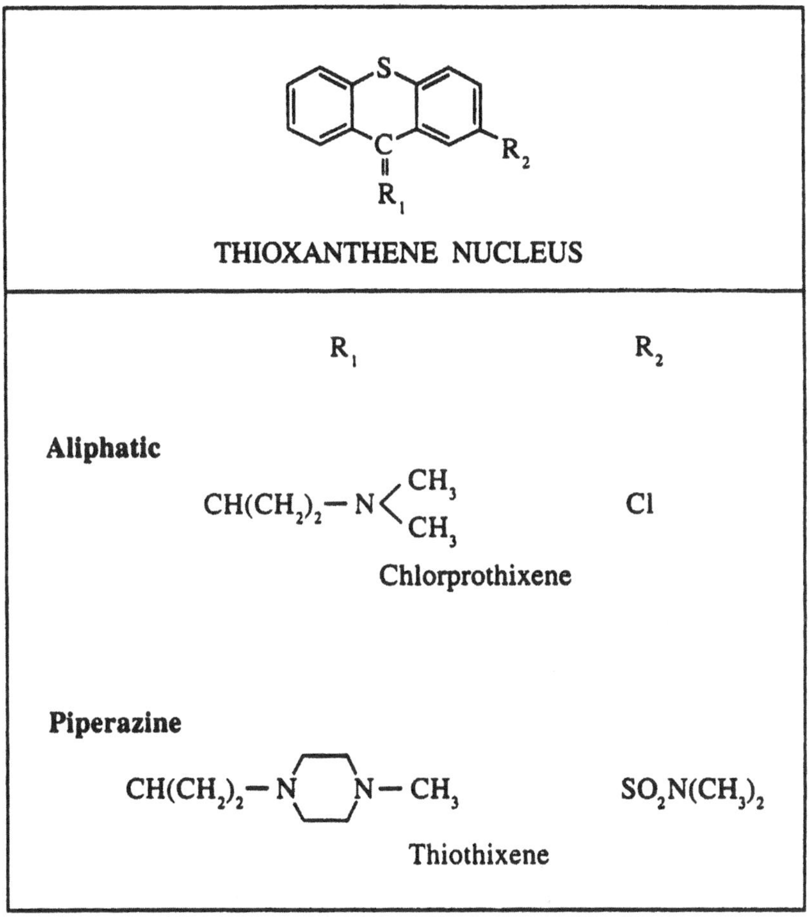

Figure 3. Chemical structure of the two thioxanthene derivatives currently available for clinical use.

Thioxanthenes

Chemically, the thioxanthenes closely resemble the phenothiazine derivatives. The nitrogen atom of the middle ring of the phenothiazine nucleus has been replaced by a carbon atom, which is joined to a side chain by a double bond (Figure 3). Only two thioxanthenes—chlorprothixene and thiothixene—are available for commercial use in the United States.

Chlorprothixene bears the same aliphatic side chain on the middle ring of the thioxanthene nucleus as does chlorpromazine. It is thus not surprising that the pharmacological properties of chlorprothixene closely resemble those of chlorpromazine. Like chlorpromazine, chlorprothixene is a low potency antipsychotic drug.

As with chlorpromazine, sedation is marked after initial doses of chlorprothixene. The occurrence of extrapyramidal side-effects is moderate in incidence. Chlorprothixene also lowers seizure susceptibility, and seizures may occur in prone individuals. Like chlorpromazine, chlorprothixene produces moderate hypotension and mild anticholinergic activity.

The same spectrum of unusual adverse effects observed with chlorpromazine has also occurred after chronic treatment with chlorprothixene, although the incidence has been lower. Thus, jaundice, dermatological changes, ocular changes, and distortions of the T wave of the electrocardiogram have all been reported.

Thiothixene bears a piperazine side chain on the middle ring of the thioxanthene nucleus. The pharmacological properties of thiothixene likewise resemble those of the piperazine phenothiazines. Like the piperazine phenothiazines, thiothixene is a high potency antipsychotic drug.

As with the piperazine phenothiazines, only slight sedation occurs after administration of thiothixene. Similarly, the frequency of occurrence of extrapyramidal side-effects is quite high. Thiothixene may also lower the convulsive threshold to seizures. As with the piperazine phenothiazines, the degree of orthostatic hypotension is mild. Anticholinergic activity, on the other hand, appears to be more marked than with the corresponding phenothiazine compounds.

Clinically documented cholestatic jaundice has not occurred after administration of thiothixene. Dermatological changes, however, have been reported. Lenticular pigmentation has likewise occurred in a small number of patients. Nonspecific changes in the electrocardiogram have also been reported.

Butyrophenones

One structural similarity between the butyrophenones and the phenothiazine derivatives is the possession of a piperidine ring. The nitrogen atom of this ring in the butyrophenones is required for antipsychotic activity. Haloperidol (Figure 4) is the sole butyrophenone available clinically as an antipsychotic in the United States.

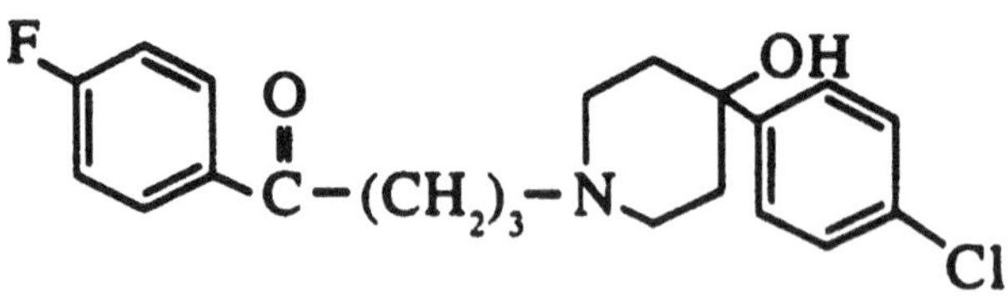

Figure 4. Chemical structure of haloperidol, a butyrophenone derivative.

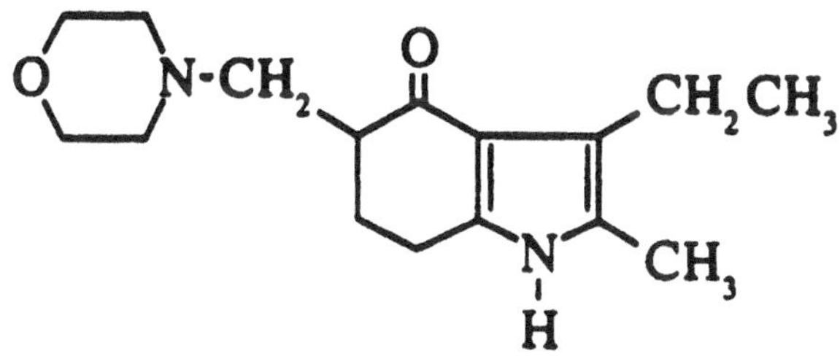

Figure 5. Chemical structure of molindone, a dihydroindolone compound.

The pharmacological properties of haloperidol resemble those of the piperazine phenothiazines. Sedation is not a prominent feature. However, the frequency of occurrence of extrapyramidal sde-effects is quite high. Lowering of the convulsive threshold may also occur with haloperidol. Orthostatic hypotension is less severe than that occurring in response to the phenothiazines. Anticholinergic activity is likewise milder in degree.

Nonspecific adverse reactions, including impaired liver function, dermatological changes, and ocular effects have occurred rarely after haloperidol.

Dihydroindolones

The indole derivative molindone (Figure 5) is the sole dihydroindolone on the market as an antipsychotic drug in the United States. This compound does not structurally resemble either the phenothiazine derivatives or haloperidol. It is a newer agent that has been on the market for about a decade.

Initial doses of molindone produce a moderate degree of sedation. In contrast, the frequency of occurrence of extrapyramidal effects is relatively low. In experimental animals, lowering of the seizure threshold has been less than that observed with older antipsychotic drugs. Nevertheless, there have been several reports of convulsive phenomena in humans prone to seizures.

Anticholinergic activity after molindone administration is similar in degree to that produced by chlorpromazine. Unlike other standard antipsychotics, molindone does not produce orthostatic hypotension. Tachycardia, however, still occurs. The weight gain that occurs after administration of other antipsychotic drugs likewise does not occur after molindone.

Jaundice, lens opacities, and pigmentary retinopathy have not been reported in patients receiving molindone. Although skin rashes have rarely been reported, other dermatological changes have likewise not occurred.

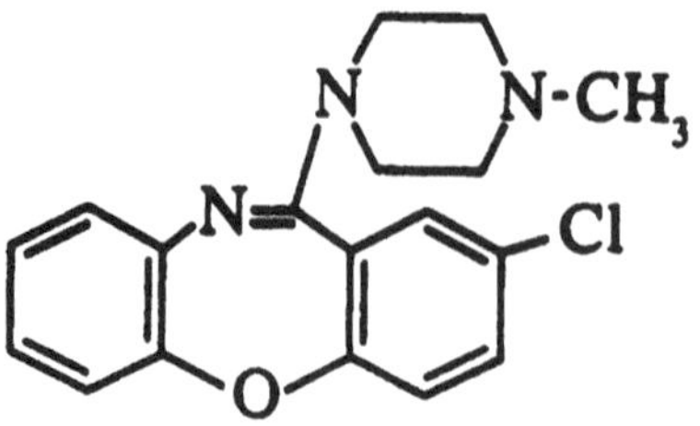

Figure 6. Chemical structure of loxapine, a dibenzoxazepine derivative.

Dibenzoxazepines

Loxapine (Figure 6) is the sole dibenzoxazepine derivative on the market in the United States today. Loxapine contains seven members in its central ring, as do the tricyclic antidepressants. This newer agent has been on the market for about a decade.

Initial doses of loxapine produce only mild sedative effects. The frequency of occurrence of extrapyramidal effects is moderate and comparable to that of chlorpromazine. Lowering of the seizure threshold, on the other hand, appears to be more pronounced with loxapine than with older antipsychotic drugs. Indeed, loxapine has induced seizures in epileptic patients whose conditions were well controlled by anticonvulsants.

Anticholinergic activity after loxapine administration is similar to that after chlorpromazine. Hypotensive activity, however, is somewhat milder in degree, although tachycardia is similar in both degree and frequency. Weight gain appears to occur less frequently than with chlorpromazine.

Neither jaundice nor ocular changes have been reported after loxapine administration. Changes in the electrocardiogram have been reported, although their relation to the administration of loxapine was unclear. Skin rashes have also been rarely reported.

Dibenzodiazepines

Although a number of dibenzodiazepine compounds have been synthesized, only two—clozapine and olanzapine (Figure 7)—are available clinically to date in the United States. Clozapine was introduced into medical therapy for limited use in 1990. Olanzapine received FDA approval for marketing in 1997.

Although clozapine was synthesized and initially examined in the 1970s, extreme toxicity has precluded its widespread clinical use. This drug poses a risk for production of bone marrow suppression and fatal agranulocytosis. Clozapine was made available for limited clinical use in 1990 through a monitoring and

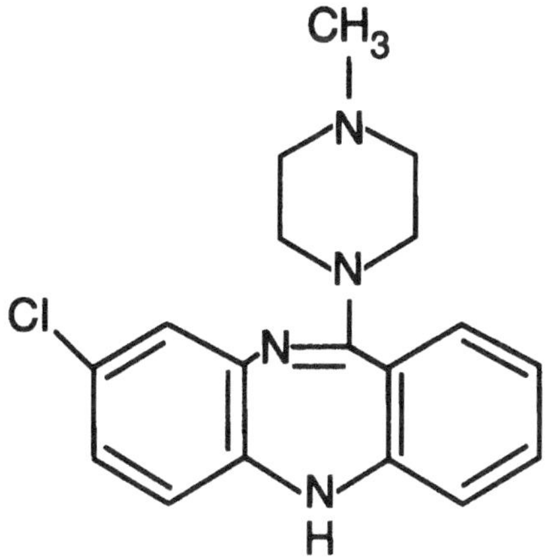

Clozapine

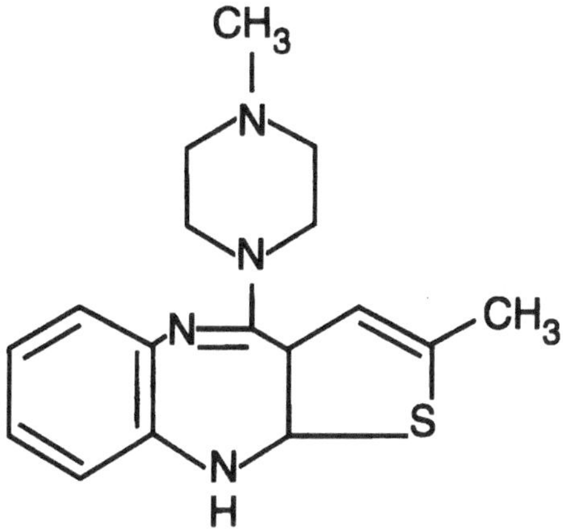

Olanzapine

Figure 7. Chemical structures of clozapine and olanzapine, two dibenzodiazepine atypical antipsychotic drugs.

dispensing system designed to minimize this risk. White cell counts had to be performed on a weekly basis. Furthermore, clozapine was to be prescribed only to severely ill schizophrenic patients who had not responded to treatment with standard antipsychotic drugs.

Unlike the standard neuroleptics available previously, clozapine so far has not caused the appearance of appreciable extrapyramidal side-effects. This drug has likewise not exerted significant neuroendocrine side-effects such as elevation of serum prolactin levels. It is primarily for this reason that clozapine and similar agents under development are considered atypical antipsychotic drugs. These effects are presumably due both to weak central dopamine receptor blockade and relatively strong antagonism of serotonergic 5-HT_2 receptor sites within the relevant neural pathways.

Like standard antipsychotic drugs, clozapine has sedative properties, and indeed sedation is a prominent feature after initial doses. Clozapine likewise exerts a hypotensive effect and possesses anticholinergic activity. Hypotension is moderate in degree, while the anticholinergic activity is strong. Paradoxically, clozapine has frequently caused excessive salivation, usually during sleep. Weight gain has also occurred.

The risk of grand mal seizures is relatively high with clozapine, particularly after higher doses. Sodium valproate or some other anticonvulsant is sometimes prescribed in conjunction with clozapine for this reason.

Olanzapine, a dibenzodiazepine compound just recently marketed, has a much improved side-effect profile in comparison with both clozapine and standard antipsychotic drugs in preclinical testing (Casey, 1996; Richelson, 1996). Like clozapine, this atypical new drug has not produced either significant extrapyramidal or neuroendocrine side-effects thus far. Unlike clozapine, olanzapine has not produced any signs of agranulocytosis or increased seizure frequency.

The degree of sedation associated with olanzapine administration is mild. Mild orthostatic hypotension is also produced. Anticholinergic activity is weak, and the unpleasant drooling observed with clozapine has been eliminated. However, weight gain does still occur with olanzapine. More extensive clinical usage of olanzapine will further define its utility in the treatment of psychotic conditions.

Benzisoxazoles

Risperidone, a novel compound with a benzisoxazole structure (Figure 8), represents the latest chemical grouping to be introduced as antipsychotic therapy. Although risperidone is classified as an atypical drug, its qualification in this regard is dose-related. At the lower end of the therapeutic dose spectrum (i.e., 2 to 6 mg/day), extrapyramidal motor symptoms and serum prolactin levels are no different from values for placebo. At higher therapeutic doses, however, both of these adverse effects increase proportionately.

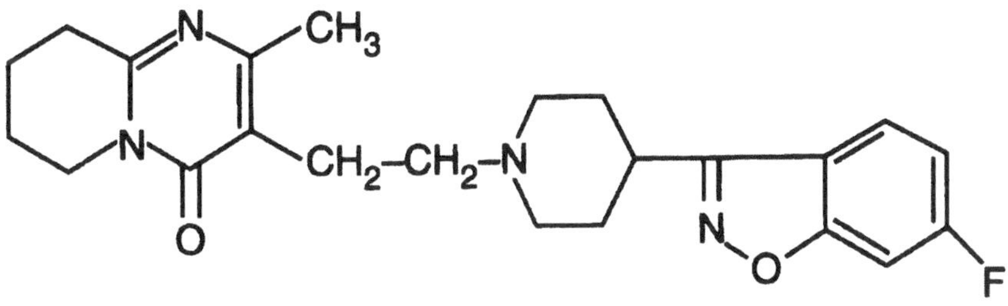

Figure 8. Chemical structure of risperidone, a novel benzisoxazole compound.

Risperidone has the highest affinity for serotonergic 5-HT_2 receptors. It also has good affinity for dopaminergic receptors and produces stronger D_2 blockade than does clozapine.

Sedation after administration of risperidone is mild in intensity. No increase in seizure risk in response to risperidone has been reported. However, significant weight gain has occurred.

Risperidone possesses very little anticholinergic activity. However, the drug may produce mild orthostatic hypotension with associated tachycardia. As noted earlier with several standard antipsychotic drugs, lengthening of the QT interval of the EKG has also been observed, which was not accompanied by clinical symptoms.

COMMON SIDE-EFFECTS

The majority of the side-effects associated with the administration of the antipsychotic drugs are direct extensions of their pharmacological effects. Accordingly, most of the standard antipsychotic drugs produce the same array of side-effects. While the newer atypical antipsychotic drugs conspicuously lack prominent extrapyramidal and neuroendocrine side-effects, they do share other common side-effects with the standard antipsychotic drugs.

Sedation

Sedation is one prominent central nervous system side-effect exerted to varying degrees by the antipsychotic drugs. Table 1 summarizes the intensity of the sedation produced by some of the phenothiazine derivatives. In general, the aliphatic and piperidine phenothiazines are the most sedating, while the piperazine compounds exert a lesser effect. The relative intensities of sedation produced by the remaining antipsychotic drugs are summarized in Table 2.

Table 1. Some Central Side Effects of the Phenothiazines[a]

Compound		Sedation	Extrapyramidal Effects
Aliphatic	Chlorpromazine	+++	++
Piperidine	Mesoridazine	+++	+
	Thioridazine	+++	+
Piperazine	Fluphenazine	+	+++
	Perphenazine	++	+++
	Prochlorperazine	+	+++
	Trifluoperazine	+	+++

Note: [a] Symbols indicate relative degree of effect and range from + for lowest to +++ for highest.

Table 2. Central Side Effects of Some Antipsychotics in Comparison with Chlorpromazine[a]

Compound	Sedation	Extrapyramidal Effects
Chlorpromazine	+++	++
Chlorprothixene	+++	++
Thiothixene	+	++
Haloperidol	+	+++
Molindone	++	+
Loxapine	+	++
Clozapine	+++	0
Olanzapine	+	0
Risperidone	+	0 to ++

Note: [a] Symbols indicate relative degree of effect and range from 0 for no effect to +++ for highest.

Extrapyramidal Effects

The most serious central side-effects produced by the antipsychotic drugs are the various extrapyramidal motor syndromes (Baldessarini, 1984). Tables 1 and 2 summarize the relative propensities of the individual antipsychotic drugs to produce these neurological syndromes. In general, the most potent antipsychotic drugs are the most likely to cause extrapyramidal manifestations.

One neurological syndrome appearing early after initiation of antipsychotic drug therapy (1 to 5 days) is acute dystonia. The symptoms are sudden in onset and include involuntary turning or twisting movements induced by massive sustained muscle contractions. The neck muscles, mouth, and back are usually involved. These symptoms respond dramatically to treatment with antiparkinsonian drugs.

A second syndrome occurring early after initiation of antipsychotic drug treatment (5 to 60 days) is akathisia. With this condition, the patient exhibits motor restlessness and feels compelled to move about continuously. Antiparkinsonian drugs do not always ameliorate this condition, and it may be necessary to reduce the dosage or change to another antipsychotic drug.

A drug-induced Parkinson's syndrome is probably the most common extrapyramidal disorder produced by the antipsychotic drugs. The period of maximal risk for this syndrome is from five to 30 days after initiation of antipsychotic drug therapy. The drug-induced parkinsonism is often indistinguishable from ideopathic syndromes. Symptoms include rigidity and tremor at rest, shuffling gait, bradykinesia, and difficulty in starting and stopping movement. This condition is ameliorated by antiparkinsonian drugs.

A rare but extremely serious neurological disorder induced by the antipsychotic drugs is the neuroleptic malignant syndrome. Unfortunately, cases of its manifestation have been reported after both older and some newer atypical drugs including clozapine and risperidone. This syndrome may appear several weeks after initiation of antipsychotic drug therapy and resembles a severe form of parkinsonism. Severe catatonia and fluctuations in the intensity of tremor are apparent. Autonomic signs are unstable, with labile pulse and blood pressure, and hyperthermia is marked. The mortality rate is greater than 10%. Neuroleptic administration must be stopped immediately upon the appearance of this syndrome. Antiparkinsonian drugs are not effective in alleviating the symptoms, and treatment is primarily supportive.

A final neurological syndrome associated with the usage of antipsychotic drugs is tardive dyskinesia. In contrast with the other syndromes, this condition occurs during or after prolonged treatment (months to years) with antipsychotic drugs and is thought to be due to compensatory excesses in dopamine rather than to dopamine antagonism. Symptoms include stereotyped involuntary movements such as sucking and smacking of the lips, lateral jaw movements, and darting and twisting of the tongue. Purposeless quick movements of the extremities may also be present. The condition is exacerbated by antiparkinsonian drugs. Mild cases of tardive dyskinesia are more readily reversed by drug discontinuation or reduction in dosage than are severe cases, which may take years to improve.

Neuroendocrine Effects

Chronic treatment with the antipsychotic drugs produces a variety of changes in the neuroendocrine system. One prominent effect is depression of the release of prolactin release-inhibiting hormone from the hypothalamus. The end result is increased serum prolactin concentrations. This causes breast engorgement and galactorrhea in both male and female patients.

The antipsychotic drugs also decrease pituitary gonadotropin output. Inhibition of ovulation results, and amenorrhea may occur as well. Release of pituitary growth hormone and neurohypophyseal hormones is also inhibited. However, neuroleptics do not appear to retard growth and development in children.

Chlorpromazine has been shown to reduce corticotropin release. This results in decreased secretion of corticosteroids from the adrenal gland. All the

Table 3. Some Peripheral Side Effects of the Phenothiazines[a]

Compound		Hypotension	Anticholinergic Activity
Aliphatic	Chlorpromazine	++	++
Piperidine	Mesoridazine	++	+++
	Thioridazine	++	+++
Piperazine	Fluphenazine	+	+
	Perphenazine	+	+
	Prochlorperazine	+	+
	Trifluoperazine	+	+

Note: [a] Symbols indicate relative degree of effect and range from + for lowest to +++ for highest.

Table 4. Peripheral Side Effects of Some Antipsychotics in Comparison with Chlorpromazine[a]

Compound	Hypotension	Anticholinergic Activity
Chlorpromazine	++	++
Chlorprothixene	++	++
Thiothixene	++	++
Haloperidol	+	+
Molindone	0	++
Loxapine	+	++
Clozapine	++	+++
Olanzapine	+	+
Risperidone	+	0

Note: [a] Symbols indicate relative degree of effect and range from 0 for no effect to +++ for highest.

phenothiazines cause an increase in appetite and weight gain. Other classes of antipsychotic drugs may cause this effect as well.

Peripheral Effects

All the antipsychotic drugs exert several noteworthy peripheral effects to varying degrees. One important effect is a lowering of blood pressure and orthostatic hypotension. This effect is due in part to alpha adrenergic blockade.

A second important peripheral effect, anticholinergic activity, is also exerted to varying degrees by the antipsychotic drugs. The anticholinergic activity gives rise to symptoms such as dry mouth, blurring of vision, decreased gastric secretion, decreased gastric motility, constipation, and diminished sweating and salivation. Exact symptoms may vary from patient to patient.

Table 3 summarizes the relative degree of production of peripheral side-effects by some of the phenothiazine derivatives. The relative production of these side-effects by the remaining antipsychotic drugs is summarized in Table 4 in comparison to those for chlorpromazine.

CLINICAL USAGE

The antipsychotic drugs are effective in the treatment of a variety of psychotic conditions. Utilization of these agents since the 1950s has vastly decreased the number of institutionalized patients (Figure 9) and has markedly increased their functioning in the community.

A diagnosis of psychosis is entertained whenever the mental functioning and emotional responsiveness of a person are impaired to such an extent that the ordinary demands of life cannot be met. Some psychoses are organic, that is, a toxic, metabolic, or neuropathological cause can usually be found. The causes of other psychotic conditions are unknown, and these are considered functional.

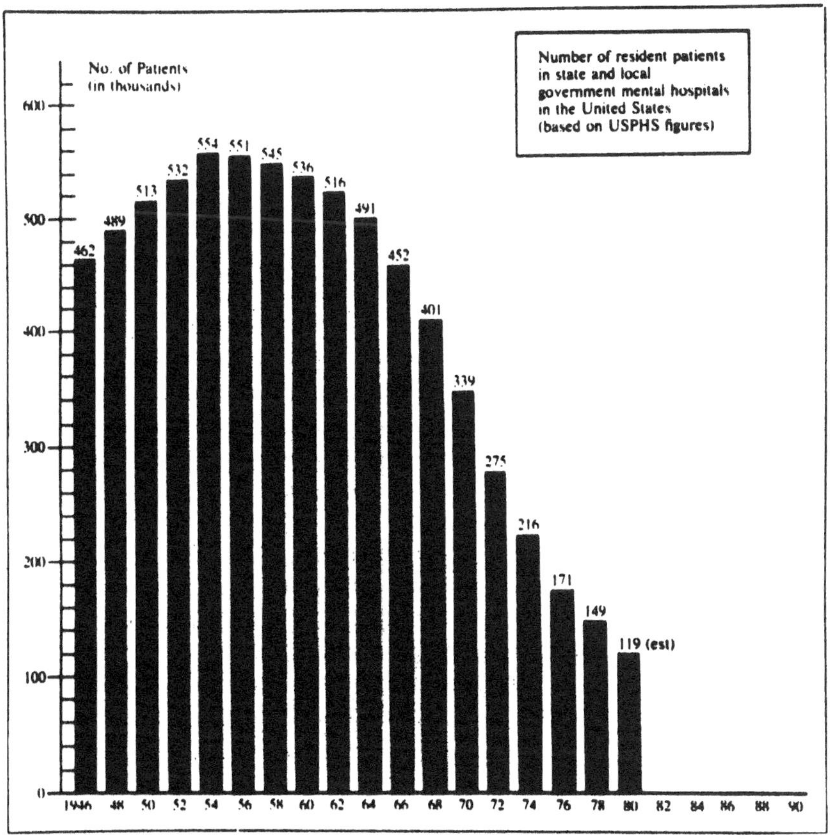

Figure 9. Influence of widespread use of the first antipsychotic drugs on the number of hospitalized mental patients in state and local hospitals in the United States (USPHS statistics). (Provided by Virginia R. Hannon, Sc.D., Principal Research Scientist, The Nathan S. Kline Institute for Psychiatric Research, Orangeburg, New York.)

The most prevalent functional psychoses are the various forms of schizophrenia, a condition characterized by chronically disordered thinking, blunted affect, and emotional withdrawal (see Chapter 13). Affective psychoses in which alterations in mood and emotion are the predominant symptoms are also functional in nature.

The antipsychotic drugs are quite effective in reducing the positive symptoms of psychoses (i.e., those that represent an excess of normal function; Pickar et al., 1990). Thus, hallucinations, agitation, hostility, severe delusions, marked incoherence, illogicality, and bizarre or disorganized behavior are quite amenable to the actions of antipsychotic drugs. On the other hand, negative symptoms (i.e., those that represent a diminution of normal function) have not traditionally responded well to standard neuroleptic therapy. Negative symptoms include poverty of speech, affective flattening, anhedonia, apathy, asociality, and attentional impairment. It is encouraging that some studies have demonstrated increased effectiveness of the newer atypical drugs in treating these negative symptoms, although results in this regard have not been uniform.

Choice of Drug

The vast number of phenothiazine derivatives on the market has contributed to the availability of a large number of antipsychotic drugs for clinical use. It has been recommended that physicians narrow this choice of drugs by becoming thoroughly familiar with the use of one of each of the three chemical subtypes of phenothiazines and one of each of the remaining chemical classes of drugs. In the case of the thioxanthenes, thiothixene is more widely used than chlorprothixene.

In many cases, the previous psychiatric history of the patient may be helpful in determination of drug choice. The rate of readmission in many mental hospitals is over 50%. In these cases, it is usually best to reinstitute therapy with a drug that has worked in the past.

In patients with no previous psychiatric history, the initial drug of choice may be ineffective in alleviating psychotic symptoms. In this case, the drug should be discontinued and a second agent employed. Two agents should not be simultaneously employed because this causes problems in isolating the source of side-effects. Differences in effectiveness of antipsychotic drugs in individual patients are presumed to be due to differences in absorption and in the metabolic handling of the drugs.

The initial choice of an antipsychotic drug in newly diagnosed psychotic patients is based on the previous medical history. Every effort should be made to balance the side-effects of the neuroleptic drugs with the medical condition of the patient.

The sedative side-effect of the antipsychotics offers no advantage in the treatment of agitated psychotic patients. It is better to rely on conventional antianxiety drugs and hypnotics to alleviate excited states or sleeplessness. Less sedating antipsychotics should be used when the medical condition of the patient warrants this. When the more sedative agents are utilized, tolerance to this effect can be expected to develop after several weeks.

The hypotension induced to varying degrees by the antipsychotic drugs is due in part to alpha adrenergic blockade. Patients with a history of cardiovascular disease and the elderly should be treated with agents exerting less pronounced hypotensive effects. The possibility of drug interactions due to hypotension should also be considered.

Other side-effects of the antipsychotics warrant consideration in drug choice. Anticholinergic effects are more likely to be disruptive in the treatment of the elderly. A strong degree of alpha adrenergic blockade may cause inhibition of orgasm and may thus interfere with sexual function. Finally, while galactorrhea may be acceptable to females, it may be more objectionable to males.

The extrapyramidal side-effects produced by the antipsychotic drugs constitute another primary consideration in choice of drug. Parkinsonian symptoms are apt to occur more frequently in the elderly, and particularly elderly women, while acute dystonic reactions are observed more frequently in young males. Controversy still exists regarding the utilization of antiparkinsonian drugs such as benztropine in the control of extrapyramidal side-effects. Some clinicians advocate the simultaneous use of benztropine when large doses of antipsychotics with a high incidence of motor side-effects must be utilized. Similarly, when extrapyramidal effects appear after longterm administration of lower maintenance doses, some clinicians recommend the addition of benztropine to the treatment regimen while others advocate immediate lowering of the antipsychotic dosage.

The atypical antipsychotic drugs are a logical choice for the treatment of newly diagnosed psychotic patients with a propensity for development of extrapyramidal signs. However, in the case of established patients, clinical experience with risperidone has indicated that caution is needed when switching from standard drugs to the newer agents (Marder, 1996). Abrupt discontinuation of the older drugs resulted in both a worsening of the psychosis and the appearance of autonomic effects. It is therefore recommended that the older agent be gradually withdrawn over two to four weeks while slowly increasing the dose of the newer drug.

Clinical experience with risperidone has been disappointing because of the narrow dose range between atypicality and the appearance of extrapyramidal effects. While results showing improvement of the negative symptoms of schizophrenia with all three of the atypical drugs currently available were initially encouraging, problems arising from "awakenings" phenomena (in which patients become more aware of their inner feelings) have recently become recognized (Weiden et al., 1996). More widespread usage of the newer atypical drugs is needed to further define their longterm utility in the management of psychotic disorders.

Table 5. Usual Dosage Ranges for Some Antipsychotic Drugs

Generic Name	*Trade Name*	*Total Daily Dosage (mg)*
Chlorpromazine	Thorazine®	100–800
Mesoridazine	Serentil®	50–400
Thioridazine	Mellaril®	100–600
Fluphenazine	Prolixin®	2–20
Trifluoperazine	Stelazine®	5–30
Thiothixene	Navane®	5–30
Haloperidol	Haldol®	2–20
Loxapine	Loxitane®	20–100
Molindone	Moban®	50–200
Clozapine	Clozaril®	100–900
Olanzapine	Zypreza®	5–20
Risperidone	Risperdal®	2–16

Dosage and Administration

The antipsychotic group of drugs is one of those for which the dosage needs to be determined on an individual basis. These drugs have quite high therapeutic indices, and a wide range of doses may thus be utilized. Recommended dosage ranges for some of the more commonly used antipsychotic drugs are given in Table 5.

In the initial treatment of an acutely ill psychotic patient (Baldessarini, 1984; Bernstein, 1988), the goal is to reduce the attendant symptomatology. Low doses of an antipsychotic agent are initially administered three to four times daily. The dose is gradually increased within the first week or two until there is improvement in psychotic symptoms. It is during this initial phase of treatment that the highest doses of an antipsychotic drug are needed.

After initial symptom control, the patient is then stabilized on the antipsychotic drug. During this phase of the treatment, the dose of the drug is gradually diminished, and the frequency of administration is reduced to twice a day. If symptoms recur, the dose is again increased for several days before further reduction. This stabilization phase usually requires one to three weeks. The goal during this period is to find the lowest effective dose for symptom control.

The final stage of treatment with antipsychotic drugs is the maintenance phase. During this phase, the lowest dose effective in suppressing symptoms is usually administered once daily at bedtime. This is feasible because of the long half-lives of the majority of the antipsychotic drugs.

It is generally recommended that the maintenance phase be continued for six months to one year after an initial psychotic episode. The dosage of the antipsychotic drug is th' . gradually reduced to nothing over a period of time. Abrupt discontinuation of the drug may lead to the appearance of a variety of extrapyramidal symptoms.

The rate of relapse in schizophrenic patients after drug discontinuation is quite high and has ranged from 40% to 70%. In these patients, longer periods of maintenance therapy may be indicated, and careful monitoring upon drug discontinuation is warranted. In some patients, maintenance therapy may have to be continued indefinitely, with utilization of the lowest dose possible to reduce the chance of emergence of extrapyramidal side-effects.

Injectable forms of most of the antipsychotic drugs listed in Table 5 are available. These preparations, which are administered intramuscularly, may be needed in the initial treatment of agitated patients with florid symptoms. Long acting depot preparations of several of these drugs including fluphenazine and haloperidol are also available. These preparations need only to be administered every three to four weeks. They are useful for patients who fail to take their medication regularly and thus require repeated hospitalization.

It has been estimated that approximately 20% of patients with psychotic conditions fail to respond to treatment with the standard antipsychotic drugs. The atypical antipsychotic drugs now offer alternative treatments for these patients. In addition, the older agent reserpine, either alone or in combination with one of the neuroleptics, may be useful in the treatment of some refractory cases as a last resort (Berlant, 1986).

SUMMARY

The antipsychotic drugs, also known as neuroleptics, have allowed functioning in the community of many mental patients who formerly would have been institutionalized. The antipsychotic drugs presumably produce their therapeutic effects by blockade of dopamine receptors in the limbic system of the brain. Unfortunately, dopamine receptor blockade within the extrapyramidal system of the brain gives rise to severe motor side-effects such as parkinsonism. Such side-effects occur with varying frequency in response to standard antipsychotic drugs such as chlorpromazine, thioridazine, fluphenazine, thiothixene, haloperidol, loxapine, and molindone. Newer agents known as atypical antipsychotic drugs, of which clozapine and olanzapine are representative, appear to lack these motor side-effects. All antipsychotic drugs exert in varying degrees other common side-effects, including sedation, hypotension, and anticholinergic activity. In clinical practice, the side-effects of the neuroleptics must be considered carefully in conjunction with the medical condition of the patient before therapy with an individual agent is instituted.

REFERENCES

Baldessarini, R.J. (1984). In: The Psychiatric Therapies (Karasu, B., Ed.), pp. 119-170, American Psychiatric Association, Washington, D.C.

Baldessarini, R.J. & Frankenburg, F.R. (1991). Clozapine: A novel antipsychotic drug. N. Engl. J. Med. 324, 746-754.

Berlant, J.L. (1986). Neuroleptics and reserpine in refractory psychoses. J. Clin. Psychopharmacol. 6, 180-184.

Bernstein, J.G. (1988). In: Handbook of Drug Therapy in Psychiatry, 2nd ed. (Bernstein, J.G., Ed.), pp. 79-122, PSG Publishing Co., Littleton, Massachusetts.

Casey, D.E. (1996). Side effect profiles of new antipsychotic agents. J. Clin. Psychiatry 57 (Suppl. 11), 40-45.

Creese, I. (1985). In: American Psychiatric Association Annual Review (Hales, R.E. & Frances, A.J., Eds.), pp. 17-36, American Psychiatric Press, Washington, D.C.

Delay, J. & Deniker, P. (1952). In: Compte rendu du Congres, Masson et Cie, Paris.

Holland, D., Watanabe, M.D., & Sharma, R. (1991). Atypical antipsychotics. Psychiat. Med. 9, 5-24.

Laborit, H., Huguenard, P., & Alluaume, R. (1952). Un nouveau stabilisateur vegetatif, le 4560 RP. Presse Med. 60, 206-208.

Marder, S.R. (1996). Clinical experience with risperidone. J. Clin. Psychiatry 57 (Suppl. 9), 57-61.

Pickar, D., Litman, R.E., Konicki, P.E., Wolkowitz, O.M., & Breier, A. (1990). In: Schizophrenia: Positive and Negative Symptoms and Syndromes (Andreasen, N.C., Ed.), pp. 124-151, Karger, Basel, Switzerland.

Richelson, E. (1996). Preclinical pharmacology of neuroleptics: Focus on new generation compounds. J. Clin. Psychiatry 57 (Suppl. 11), 4-11.

Weiden, P., Aquila, R., & Standard, J. (1996). Atypical antipsychotic drugs and long-term outcome in schizophrenia. J. Clin. Psychiatry 57 (Suppl. 11), 53-60.

SELECTED READINGS

Baldessarini, R.J. (1990). In: The Pharmacological Basis of Therapeutics (Goodman, L.S. Gilman, A., Rall, T.W., Nies, A.S., & Taylor, P., Eds.), pp. 383-404, Pergamon Press, New York.

Casey, D.E. (1991). Neuroleptic drug-induced extrapyramidal syndromes and tardive dyskinesia. Schizophr. Res. 4, 109-120.

Deutch, A.Y., Moghaddam, B., Innis, R.B., Krystal, J.H., Aghajanian, G.K., Bunney, B.S., & Charney, D.S. (1991). Mechanisms of action of atypical antipsychotic drugs. Implications for novel therapeutic strategies for schizophrenia. Schizophr. Res. 4, 121-156.

Gerlach, J. (1991). New antipsychotics: Classification, efficacy, and adverse effects. Schizophr. Bull. 17, 289-309.

Gerlach, J. & Casey, D.E. (Eds.)(1996). New antipsychotics: Preclinical and clinical research. Psychopharmacology 124, 1-204.

Kane, J.M. (1996). Antipsychotic drug therapy: Current status. J. Clin. Psychiatry 57 (Suppl. 9), 1-76.

Levinson, D.F. (1991). Pharmacologic treatment of schizophrenia. Clin. Ther. 13, 326-352.

Marks, R.C. & Luchinsi, D.J. (1991). Antipsychotic medications and seizures. Psychiatr. Med. 9, 37-52.

Meltzer, H.Y. (1991). The mechanism of action of novel antipsychotic drugs. Schizophr. Bull. 17, 263-287.

Pickar, D. (1996). New strategies for treating psychosis. J. Clin. Psychiatry 57 (Suppl. 11), 1-93.

Ryan, P.M. (1991). Epidemiology, etiology, diagnosis, and treatment of schizophrenia. Am. J. Hosp. Pharm. 48, 1271-1280.

Van Kammen, D.P. (1991). The biochemical basis of relapse and drug response in schizophrenia: Review and hypothesis. Psychol. Med. 21, 881-895.

Chapter 26

Introduction to Electroconvulsive Therapy

ATHANASIOS P. ZIS

Biological Psychiatry, pages 611-621.

ISBN: 1-55938-819-6

INTRODUCTION

The discovery in the 1950s of phenothiazines for the treatment of schizophrenia, and of monoamine oxidase inhibitors and tricyclic antidepressants for the treatment of depression, led to the development of biochemical theories about the pathophysiology of severe psychiatric syndromes and, on the basis of these theories and discoveries, to the continuing production of numerous psychoactive drugs. As a result, empirical somatic treatments such as electroconvulsive therapy (ECT) and insulin-induced coma therapy, which were used widely for the treatment of the mentally disabled before and for a brief period after the second world war, appeared destined for extinction. However, despite the advent of modern psychopharmacology, ECT remains a useful therapeutic modality for the severely mentally ill patient.

ECT originated in the 1930s with the studies and clinical observations of the Hungarian psychiatrist Von Meduna, who used chemical means (initially camphor in oil and later pentylenetetrazol injections) to produce convulsions in schizophrenic patients and reported positive results. The first use of an electrical stimulus as the method of seizure induction is credited to the Italians Cerletti and Bini (Kalinowski, 1986). Over the last sixty years, both the indications and method of administration of ECT have changed and evolved considerably. As practiced today, ECT is a safe, effective, and useful form of treatment. The purpose of this chapter is to serve as an introduction to the practice of ECT with particular emphasis on developments responsible for its continued use and acceptance as a modern medical procedure (Potter and Rudorfer, 1993).

INDICATIONS AND CONTRAINDICATIONS OF ECT

Indications

Although ECT was originally introduced to treat schizophrenia, this is no longer recommended, with the exception of cases with acute onset or those with significant affective symptomatology. Today, ECT is primarily used in the treatment of depression, an application based on the results of several controlled studies, which compared real with simulated or sham ECT, and which have demonstrated that ECT is a particularly effective treatment for this group of patients (Weiner and Coffey, 1988; Fink, 1993). In addition, ECT has been found useful in the treatment of catatonia and mania, while recent observations suggest that it also alleviates (at least temporarily) the motor symptoms of Parkinson's disease (Rasmussen and Abrams, 1991). ECT has also been used successfully in the treatment of neuroleptic malignant syndrome, organic psychosis, and even toxic or metabolic delirium.

In North America, the practice of ECT follows the guidelines of the American Psychiatric Association Task Force on ECT. According to these recommendations, ECT should be considered either after failure to respond to psychotropic drugs or as a first line of treatment if the following conditions are met:

1. A history of poor drug response and/or good ECT response exists for previous episodes of the illness.
2. The risks of other treatments outweigh the risks of ECT.
3. There is a need for rapid, definite response on either medical or psychiatric grounds.
4. Patient preference favors it.

Specific clinical syndromes for which psychiatrists often recommend the use of ECT as a first line of treatment include psychotic depression, depression with severe psychomotor retardation or stupor, depression with severe suicidal ideation, catatonia, and severe manic excitement (Fink, 1993).

Contraindications

The mortality rate attributed to ECT is very low and does not exceed the mortality rate associated with brief general anesthesia for minor surgery (Abrams, 1991). There are no absolute contraindications to ECT, but there are certain medical conditions that are associated with increased morbidity. These include increased intracranial pressure, brain tumor, recent myocardial infarction, recent intracerebral hemorrhage, unstable aneurysm, and other vascular malformations as well as retinal detachment, pheochromocytoma, and conditions considered as high risk from the viewpoint of anesthesia.

THE EFFECTS OF ECT

Central Nervous System

When an electrical stimulus of adequate energy is applied to the skull and depolarizes a sufficiently large number of neurons, a generalized seizure ensues. Behaviorally and electroencephalographically, electrically induced seizures are similar to those occurring spontaneously in patients suffering from grand mal epilepsy. These seizures consist of a tonic and clonic phase. The tonic phase is characterized on the electroencephalogram by high-voltage synchronous polyspike activity of high frequency and lasts between 10 and 20 seconds. During the clonic phase, which lasts another 10 to 20 seconds, the high-voltage, polyspike activity is gradually replaced by slow-wave complexes. This second phase of seizure activity is gradually replaced by low amplitude activity lasting approximately 90 seconds. Low frequency δ and rhythmic I waves follow, while preseizure activity gradually begins to reappear. Low frequency activity (δ and θ waves) interspersed with periods of normal rhythms may persist for hours following a single ECT treatment or even for days following a course of treatments.

In association with these electroencephalographic changes, there are transient increases in cerebral blood flow, cerebral permeability, and glucose consumption. Seizure threshold also increases transiently following the administration of ECT. There is no evidence that seizures induced electrically according to procedures recommended for ECT are associated with permanent structural changes, nor is there any proof that a course of ECT produces brain damage (Coffey, 1993).

Cardiovascular System

The effects of the electrically induced seizure on the central nervous system are accompanied by significant changes in cardiovascular function. There is an initial period of bradycardia or even asystole (lasting 1–2 s), which is followed by a period of tachycardia. This terminates with the end of the seizure activity and is occasionally followed by a period of bradycardia. These changes are attributed to the stimulation of the parasympathetic followed by the stimulation of the sympathetic nervous system. Blood pressure (both systolic and diastolic) rises sharply with the onset of seizure activity and declines rapidly to pretreatment levels following the cessation of clonic seizure activity. These cardiovascular effects of ECT are usually benign. On occasion, in the elderly and particularly, in patients with preexisting cardiovascular problems, these effects are more severe and require preventative or *post hoc* intervention; examples include the administration of atropine or glycopyrrolate for bradycardia, or short-acting β-adrenergic blocking agents (e.g., esmolol) for high blood pressure.

Neuroendocrine System

There is a transient rise in plasma prolactin, corticotropin, endorphin, cortisol, oxytocin, and neurophysin but not growth hormone plasma levels in association with ECT-induced seizures (Kamil and Joffe, 1991). While these effects (e.g., the prolactin release) represent some of the most robust and consistently observed neurochemical sequelae of ECT-induced seizures, their relationship to the mechanism of action of ECT has as yet to be demonstrated.

Cognition

The cognitive effects of ECT consist of postictal confusion and anterograde and retrograde memory impairment.

Postictal Confusion

Following the termination of seizure activity and while the patient is recovering consciousness, there is invariably a short period of confusion characterized by clouding of consciousness and disorientation. Typically, this lasts for 30 to 45 minutes but occasionally extends to one to two hours. Rarely, this period of clouding of consciousness and disorientation is accompanied by severe agitation and restlessness (emergence delirium).

Memory Impairment

In addition to these short-lived cognitive side-effects, there are longer lasting effects on memory function. Specifically, there is anterograde memory impairment (rapid forgetting of new information or events that occur after the treatment) and retrograde memory loss (inability to recall previously learned material or events, which occurred prior to the treatment).

The severity (and duration) of the cognitive effects of ECT increases with the number of treatments and is dependent on electrode placement as well as on the type and intensity of the electrical stimulus used to elicit the seizures (Squire, 1986; Calev et al., 1993). Under optimal treatment conditions most patients exhibit only a mild and transient memory disturbance. The impairment is primarily limited to events that occur during the course of treatment, but it may extend to a few days before and after the course of treatment. With less than optimal treatment conditions, memory testing may reveal memory disturbances extending from a few weeks to a few months before and after a course of ECT. Subjective complaints of memory loss may extend even beyond this period of time.

Table 1. Electrode Placement, Stimulus Parameters, and Cognitive Impairment

A. Increased cognitive impairment is associated with
- Bilateral electrode placement
- Continuous sinewave waveform
- High electrical intensity

B. Decreased cognitive impairment is associated with
- Unilateral electrode placement
- Brief-pulse, square-wave waveform
- Low electrical intensity

OPTIMAL TREATMENT PROCEDURES

The ECT Treatment Team

ECT is administered in a hospital setting with specifically designed and equipped treatment and recovery rooms. The ECT treatment team consists of specially trained professionals including a psychiatrist, an anesthetist, and nurses trained in critical care. The psychiatrist is responsible for evaluating the patient prior to the treatment, for assuring that treatment conditions applied are optimal and according to established procedures, and for determining, in consultation with the patient's attending psychiatrist, the number, frequency, and type of ECT treatments. The anesthetist is responsible for inducing anesthesia and muscle relaxation, managing the airway and ventilating the patient, and for treating potential adverse effects, including the administration of cardiopulmonary resuscitation if necessary. The nursing staff responsibilities include scheduling and escorting or transporting the patient, assisting the treating psychiatrist and anesthetist during the treatment, monitoring the vital signs, and managing the patient in the recovery room as well as being involved in patient and family education.

Anesthesia and Muscle Relaxation

ECT is performed under brief general anesthesia induced by the intravenous administration of a short-acting barbiturate such as thiopental or methohexital. The purpose of anesthesia is to induce a brief period of unconsciousness during which muscle relaxation is induced and the electrical stimulus is applied. Muscle relaxation is produced by the administration of succinylcholine. This drug, which was introduced to the practice of ECT in the early 1950s, produces a state of deep skeletal muscle relaxation by acting as a depolarizing agent at the neuromuscular junction. This state of partial paralysis is necessary to reduce the intensity of the muscle contractions associated with brain seizure activity and therefore to prevent any bone fractures. Because succinylcholine is rapidly metabolized by the plasma enzyme pseudocholinesterase, it has a short duration of action and is ideally suited to a procedure as brief as ECT.

Electrode Placement and Stimulus Parameters

ECT has been used successfully for nearly 60 years, yet its mechanism of action remains poorly understood. Repeated and generalized seizures have traditionally been thought to be the therapeutic feature of ECT, and electricity only a method by which the seizures are produced. This belief is based primarily on observations indicating that chemically induced seizures are also effective therapeutically, while pretreatment with lidocaine, which shortens the duration of electrically induced seizures, also reduces the therapeutic effect of ECT (Sackeim, 1988).

Although these observations constitute considerable evidence that the induction of a generalized seizure is essential to the therapeutic effect of ECT, recent observations challenge the view that the induction of a seizure alone is sufficient. More specifically, evidence has accumulated indicating that electrode placement, stimulus waveform, and stimulus intensity are important variables in the therapeutic and side effects of ECT (Sackeim et al., 1991; 1993).

Stimulus Waveform

In the past, two types of electrical stimuli have been routinely used to induce seizures: continuous sine wave and brief-pulse square wave. Today, however, ECT devices using continuous sine wave stimulation are no longer recommended for use and are considered obsolete. Compared to continuous sine wave, brief-pulse square wave stimulation is a more efficient method for inducing seizures, for it requires considerably less electrical energy and a smaller electrical charge. From a clinical point of view, a course of brief-pulse square wave ECT is associated with significantly fewer and milder cognitive side-effects compared to a course of continuous sine wave ECT (Weiner et al., 1986).

Electrode Placement

According to the method of electrode placement, ECT is characterized as unilateral or bilateral. In unilateral ECT, both electrodes are placed on the same side of the skull, preferably the right, or nondominant side. One electrode is placed by the vertex of the skull and the second electrode is placed frontotemporally, that is, approximately 2 cm above the midpoint of a line connecting the auditory meatus and the external canthus of the eye. For the purpose of bilateral ECT, the two electrodes are placed frontotemporally, one on each side of the skull. A course of unilateral ECT may not be as effective as a course of bilateral ECT as the former may exhibit a slower onset of therapeutic action and/or require a larger number of treatments. On the other hand, bilateral ECT is clearly associated with greater cognitive impairment (Weiner et al., 1986; Sackeim et al., 1993). Therefore, it is recommended that patients first be given a course of unilateral ECT with the option of switching to bilateral electrode placement if clinically necessary. Such cases would include if a more rapid onset of action were desired or the response after a few treatments were judged as inadequate.

Seizure Threshold

Seizure threshold, defined as the minimal stimulus intensity needed to elicit generalized seizure activity of a certain duration, varies considerably among patients. Seizure threshold is determined constitutionally but is also affected by other factors (Sackeim et al., 1987). It increases with age, with the administration

Table 2. Factors that Affect Seizure Threshold

A. Factors associated with high seizure threshold
- Advanced age
- Male sex
- Bilateral electrode placement
- Continuous sinewave stimulus
- A course of ECT
- Anticonvulsants
- Benzodiazepines
- Beta adrenergic blocking agents (?)

B. Factors associated with low seizure threshold
- Younger age
- Female sex
- Unilateral electrode placement
- Brief-pulse, square-wave stimulus
- Methylxanthines
- Phenothiazines (?)
- Heterocyclic antidepressants (?)

of anticonvulsants and benzodiazepines, and with the number of treatments in a course of ECT. Men have a higher seizure threshold than women. Seizure threshold is higher for bilateral compared to unilateral ECT and for continuous sine wave compared to brief-pulse square wave ECT.

The threshold decreases with the administration of methylxanthines including caffeine and possibly in response to certain heterocyclic antidepressants and phenothiazines. Therefore, if a fixed-dose stimulus is applied, many patients will be subjected to either excessive stimulation or too little stimulation. It is hence recommended that an estimate of individual patient seizure threshold is obtained at least at the beginning of a course of ECT and that the stimulus intensity is adjusted for subsequent treatments as necessary (Sackeim et al., 1991).

Dose-Titrated ECT

In addition to the effects of electrode placement and stimulus waveform, there is evidence that stimulus intensity also plays an important role in determining the efficacy and cognitive side-effects of ECT. Higher stimulus intensity is associated with greater efficacy and more rapid therapeutic response. The effects of stimulus intensity on ECT efficacy are particularly pronounced in the case of unilateral electrode placement. Although low intensity (i.e., just above seizure threshold) unilateral ECT causes little cognitive impairment, it is also associated with a poor therapeutic response rate. With unilateral ECT, it is necessary to use a suprathreshold stimulus (at least two to three times above seizure threshold) in order to obtain a rate of clinical improvement comparable to that of bilateral ECT (Sackeim et al., 1993).

Monitoring During ECT

Routine physiological monitoring includes the monitoring of cardiac rhythms by oscilloscopic electrocardiography, the monitoring of peripheral oxygen saturation by pulse oximetry, the monitoring of blood pressure, and monitoring the spread and duration of the seizure.

The careful monitoring of seizure activity during ECT is important for two reasons:

1. Seizures that do not generalize bilaterally or are too short (<20 s) are associated with reduced clinical efficacy.
2. Prolonged seizures (>2–3 min) should be terminated since they do not offer any therapeutic advantage and may be associated with increased cardiovascular and/or neurological toxicity.

Since patients undergoing ECT are in a state of partial paralysis or deep skeletal muscle relaxation due to the administration of succinylcholine, simple clinical observation of motor movement is clearly an insufficient method of monitoring seizure duration. A more accurate estimate of the duration and the extent of bilateral generalization of seizure activity is derived from the pressure-cuff method and/or electroencephalography (EEG)(Fink and Johnson, 1982; Weiner et al., 1991).

For the purpose of electroencephalographic monitoring, one-channel EEG is routinely used. Usually the two recording electrodes are placed bilaterally in the prefrontal area. Alternatively, one of the electrodes is placed prefrontally and the other over the ipsilateral mastoid and contralaterally to the side used to apply the stimulating electrodes. The latter type of placement may be helpful in detecting seizures that are limited to the stimulated hemisphere and that fail to spread to the other side of the brain. The pressure-cuff method consists of placing a blood pressure–cuff over one of the extremities and inflating it above systolic blood pressure prior to the injection of succinylcholine. Seizure activity is then visually monitored by observing the motor activity in the nonparalyzed limb.

The pressure-cuff method is less exact than the EEG and typically underestimates seizure duration by eight to 15 seconds. On the other hand, EEG monitoring requires interpretation, which may detract from attending to the patient and can be affected by artifacts. In fact, the two methods should not be considered mutually exclusive, but rather, complementary to each other.

Frequency and Number of Treatments

In Canada and the United States, ECT treatments are typically given thrice weekly. Increasing the frequency of treatments is associated with an increase in cognitive side-effects whereas, when treatments are given less frequently, the onset of clinical recovery may be delayed. A typical ECT course for the treatment of depression consists of six to 10 treatments. A course of ECT should be discontinued (1) as soon as the symptoms have fully remitted, (2) when indicated because of severe cognitive impairment, and (3)if there is no further clinical response or improvement after 12 treatments.

CONCLUDING REMARKS

This introduction to ECT is neither exhaustive nor complete. It is incomplete as it does not, for example, review the many theories about the mechanism of action of ECT, or discuss issues related to obtaining informed consent and other medicolegal matters. It is not exhaustive since it does not contain a detailed review of the many studies leading to the information presented here essentially in an outline form. This introduction was kept short in order to compete for the medical students' attention with the many other interesting chapters in this volume and series.

ACKNOWLEDGMENT

The editorial effort of Dana Sair is gratefully acknowledged.

REFERENCES

Abrams, R. (1991). Electroconvulsive therapy in the medically compromised patient. In: The Psychiatric Clinics of North America: Electroconvulsive Therapy. (Kellner, C.H., Ed.), Vol. 14, No. 4, pp. 871-885. W.B Saunders, Philadelphia.

Calev, A., Pass, H.L., Shapira, B., Fink, M., Tubi, N., & Lerer, B. (1993). ECT and memory. In: The Clinical Science of Electroconvulsive Therapy. Progress in Psychiatry. (Coffey, C.E., Ed.), No. 38, pp. 125-142. American Psychiatric Press, Washington, D.C.

Coffey, C.E. (1993). Structural brain imaging and ECT. In: The Clinical Science of Electroconvulsive Therapy. Progress in Psychiatry. (Coffey, C.E., Ed.), No. 38, pp. 73-92. American Psychiatric Press, Washington, D.C.

Fink, M. & Johnson, L. (1982). Monitoring the duration of electroconvulsive therapy seizures: "Cuff" and EEG methods compared. Arch. Gen. Psychiatry 39, 1189-1191.

Fink, M. (1993) Who should get ECT? In: The Clinical Science of Electroconvulsive Therapy. Progress in Psychiatry. (Coffey, C.E., Ed.), No. 38, pp. 3-15. American Psychiatric Press, Washington, D.C.

Kalinowski, L.B. (1986). History of convulsive therapy. In: Electroconvulsive Therapy: Clinical and Basic Research Issues. Annals of the New York Academy of Sciences. (Malitz, S., Sackeim, H.A., Eds.), Vol. 462, pp. 1-4. The New York Academy of Sciences, New York.

Kamil, R. & Joffe, R.T. (1991). Neuroendocrine testing in electroconvulsive therapy. In: The Psychiatric Clinics of North America: Electroconvulsive Therapy. (Kellner, C.H., Ed.), Vol. 14, No. 4, pp. 961-970. W.B Saunders, Philadelphia.

Potter, W.Z. & Rudorfer, M.V. (1993). Editorial: Electroconvulsive therapy—A modern medical procedure. N. Engl. J. Med. 328 (12), 882-883.

Rasmussen, K. & Abrams, R. (1991). Treatment of Parkinson's disease with electroconvulsive therapy. In: The Psychiatric Clinics of North America: Electroconvulsive Therapy. (Kellner, C.H., Ed.), Vol. 14, No. 4, pp. 925-933. W.B Saunders, Philadelphia.

Sackeim, H., Decina, P., Prohovnik, I., & Malitz, S. (1987). Seizure threshold in electroconvulsive therapy: Effects of sex, age, electrode placement, and number of treatments. Arch. Gen. Psychiatry 44, 355-360.

Sackeim, H.A. (1988). Mechanisms of action of electroconvulsive therapy. In: Review of Psychiatry. (Frances, A.J. & Hales, R.E., Eds.), Vol. 7, pp. 436-457. American Psychiatric Press, Washington, D.C.

Sackeim, H.A., Devanand, D.P., & Prudic, J. (1991). Stimulus intensity, seizure threshold, and seizure duration: Impact on the efficacy and safety of electroconvulsive therapy. In: The Psychiatric Clinics of North America: Electroconvulsive Therapy. (Kellner, C.H., Ed.), Vol. 14, No. 4, pp. 803-843. W.B Saunders, Philadelphia.

Sackeim, H.A., Prudic, J., Devanand, D.P., Kiersky, J.E., Fitzsimons, L., Moody, B.J., McElhiney, M.C., Coleman, E.A., & Settembrino, J.M. (1993). Effects of stimulus intensity and electrode placement on the efficacy and cognitive effects of electroconvulsive therapy. N. Engl. J. Med. 328(12), 839-846.

Squire, L.R. (1986). Memory functions as affected by electroconvulsive therapy. In: Electroconvulsive Therapy: Clinical and Basic Research Issues. Annals of the New York Academy of Sciences. (Malitz, S. & Sackeim, H.A., Eds.), Vol. 462, pp. 307-314. The New York Academy of Sciences, New York.

Weiner, R.D., Rogers, H.J., Davidson, J.R.T., Squire, L.R. (1986). Effects of stimulus parameters on cognitive side effects. In: Electroconvulsive Therapy: Clinical and Basic Research Issues. Annals of the New York Academy of Sciences. (Malitz, S. & Sackeim, H.A., Eds.), Vol. 462, pp. 315-325. The New York Academy of Sciences, New York.

Weiner, R.D. & Coffey, C.E. (1988). Indications for use of electroconvulsive therapy. In: Review of Psychiatry. (Frances, A.J. & Hales, R.E., Eds.), Vol. 7, pp. 458-480. American Psychiatric Press, Washington, D.C.

Weiner, R.D., Coffey, C.E., & Krystal, A.D. (1991). The monitoring and management of electrically induced seizures. In: The Psychiatric Clinics of North America: Electroconvulsive Therapy. (Kellner, C.H., Ed.), Vol. 14, No. 4, pp. 845-869. W.B Saunders, Philadelphia.

RECOMMENDED READINGS

Abrams, R. (1992). Electroconvulsive Therapy, 2nd ed. Oxford University Press, New York.

Beyer, J.L., Weiner, R.D., & Glenn, M.D. (1998). Electroconvulsive Therapy, A Programmed Text, 2nd ed., American Psychiatric Press, Washington, D.C.

Fink, M. (1979). Convulsive Therapy: Theory and Practice. Raven Press, New York.

Kellner, C., Pritchett, J., Beale, M., & Coffey, C.E. (1997). Handbook of ECT. American Psychiatric Press, Washington, D.C.

Task Force Report of the American Psychiatric Association (1990). The Practice of Electroconvulsive Therapy: Recommendations for Treatment, Training, and Privileging. American Psychiatric Association, Washington, D.C.

INDEX

FSC
www.fsc.org
MIX
Papier | Fördert gute Waldnutzung
FSC® C083411

Kolibri 360 GmbH
Ferdinand-Jühlke-Straße 7
99095 Erfurt, Deutschland
produktsicherheit@kolibri360.de